Welfare of Farmed Fish in Present and Future Production Systems

Hans van de Vis · Anders Kiessling · Gert Flik ·
Simon Mackenzie

Welfare of Farmed Fish in Present and Future Production Systems

Scientific papers and the first paper previously published in
Fish Physiology and Biochemistry,
Volume 38, Issue 1, 2012

Springer

Editors
Dr. Hans van de Vis
IMARES, Wageningen UR,
Institute of Marine Resources and
Ecosystems Studies
Koringaweg 5
4401 NT Yerseke
Netherlands
Hans.vandevis@wur.nl

Prof. Anders Kiessling
Swedish University of Agricultural Science
Department of Wildlife, Fish, and
Environmental Studies
Skogmarksgränd
90183 Umeå
Sweden
Anders.kiessling@slu.se

Prof. Gert Flik
Institute for Water and Wetland Research
Department of Animal Physiology
Heyendaalseweg 135
6525 AJ Nijmegen
Netherlands
g.flik@science.ru.nl

Dr. Simon Mackenzie
Institut de Biotecnologia i de Biomedici
Universitat Autonoma de Barcelona
Bellaterra
08193 Barcelona
Spain
simon.mackenzie@uab.es

ISBN 978-94-007-5382-2

Springer Dordrecht Heidelberg London New York

Library of Congress Control Number: 2012944182

Printed on acid-free paper

Springer is part of Springer Science+Business Media (www.springer.com)

Contents

Preface

This special issue of Fish Physiology and Biochemistry is a result of Cost action 867, entitled "Welfare of Fish in European Aquaculture". During five years - from 2006 until 2011 - a network of over a 100 stakeholders from 26 countries including Canada, New Zealand and Australia, worked together in a meeting platform that was supported by COST, a framework for European Cooperation in the field of Scientific and Technical Research, which facilitates the coordination of nationally funded research on a European level. The stakeholders consisted of representatives from research institutes, NGOs, aquaculture industry and governmental organizations.

By openly discussing and exchanging information, the participating stakeholders worked on the harmonization of the Welfare concept throughout Europe and formulation of tomorrow's production systems, introducing welfare as a management tool. Training schools were an important tool to exchange information: more than 120 students of welfare - addressing theoretical and practical aspects - will take their experiences with them throughout their career as researchers or decision makers in industry, NGOs and governmental institutions. Together with this special issue and the five annual workshops, gathering up to 100 senior and young researchers at each meeting from whole of Europe and countries outside, the training schools are a sustainable contribution of the Cost action to the improve the understanding of the concept of welfare in fish and to exchange and disseminate this knowledge.

The papers in this special issue not only summarize the present status of the field, but also formulate a scientific platform for future production systems, putting the animal in the center of the debate. This issue contains papers with opinionated views as well as research articles with a general interest and broad approach. They are divided into four sections:

- Conceptual framework of welfare in fish
- Factual components of welfare
- Assessment of welfare in fish
- Implementation of welfare in fish

During the time of Cost action 867 a tremendous development in notion, understanding and implementation of welfare as a concept has occurred. At the start of the action, the main focus was on disease and water and feed quality: variables with very direct effects on growth. Now, at the end of the action, further variables have been introduced including fish preferences and biology-based stimulation measures. Among fish farmers there is more interest in preventing disease as opposed to merely treating disease. This Cost action has been only one contributing factor to this shift and to the perception that welfare may set the biological limits for a sustainable production system.

We believe that the results of the efforts combined within this Cost action can help fish farmers and scientists to develop economically viable production systems based on fish welfare.
This special issue gives the reader an overview of the results obtained and can help to guide the way towards welfare of farmed fish in present and future production systems. The first paper gives an overview of the papers that follow, each preceded by a summary for non-experts. These summaries can help you select the results that are best suited to help you in reaching your specific goals.
We wish to thank all contributors for their time and efforts in putting this special issue together. We hope that contributors and readers alike in future years will reap the rewards of the work outlined in this book.

Anders Kiessling, Hans van de Vis, Gert Flik and Simon Mackenzie

Fish Physiol Biochem (2012) 38:1–3
DOI 10.1007/s10695-011-9594-9

Welfare of farmed fish in present and future production systems

Anders Kiessling · Hans van de Vis · Gert Flik · Simon Mackenzie

Received: 2 December 2011 / Accepted: 16 December 2011 / Published online: 14 January 2012

The effort manifested in this special issue, dedicated to the welfare of farmed fish, is the joint result of members of the Cost action 867 "Welfare of Fish in European Aquaculture." This publication is supported by COST. COST is a framework for European Co-operation in the field of Scientific and Technical Research, which facilitates the coordination of nationally funded research on a European level. For the coordination and harmonization of national and EU level research project efforts, a network was established. The network comprised 26 countries, including Canada, New Zealand and Australia, and more than 100 researchers. The action began in 2006 and continued for 5 years, ended with a final meeting in Madrid, Spain, in March 2011. The main aim of the action was to provide a meeting platform supporting (a) discussion and information between active researchers, (b) harmonization throughout Europe of the Welfare concept and (c) formulation of tomorrow's production systems introducing welfare as a management tool.

During the time of the action, a tremendous development in perception, understanding and implementation of welfare as a concept has occurred. At the start of the action, the main focus was on disease as well as water and feed quality, that is, variables with very direct effects on growth. Now, at the end of the action, further variables have been introduced including fish preferences and biological-based stimulation measures. Furthermore, a palpable shift from disease toward prophylactic health has been evident, that is, preventive measures included when farmers discuss welfare. The action as such is in no way responsible for this alone, but has been one contributing factor, contributing to the perception that welfare may set the biological limits for a sustainable production system. In other words, compromising welfare would indicate a production format violating the biological capacity of the animal and thereby demanding nonsustainable

Hans van de Vis—Wageningen Aquaculture is a consortium of IMARES (Institute for Marine Resources & Ecosystem Studies) and AFI (Aquaculture and Fisheries Group, Wageningen University), both part of Wageningen University & Research Centre (WUR).

A. Kiessling (✉)
Departments of Wildlife, Fish, and Environmental Studies, Swedish University of Agricultural Sciences, Umeå, Sweden
e-mail: Anders.kiessling@slu.se

H. van de Vis
IMARES, Wageningen UR, P.O. Box 77, 4400 AB Yerseke, The Netherlands

G. Flik
Department of Animal Physiology, Institute for Water and Wetland Research, Faculty of Science, Radboud University Nijmegen, Heyendaalseweg 135, 6525 AJ Nijmegen, The Netherlands

S. Mackenzie
Institut de Biotecnologia i de Biomedicina, Universitat Autonoma de Barcelona, 08193 Bellaterra, Barcelona, Spain

counter actions such as increased medication, shorter life spans and acceptance of disease.

One could very grossly divide actions to support the welfare of farmed fish into two main approaches. The first concerning actual production systems and how to secure welfare, and the second concerns emerging production systems. Today's intensive aquaculture systems are surprisingly similar often independent of the cultured species and location and could be divided in a several major types. This can be explained by a tradition of technology-driven development in combination with the absence of economic incentives beyond quantity and guaranteed delivery. The question arising at the end of the action is "If biological driven development is the most coherent approach will intensive culture systems, based on the biological needs/abilities of specific species/breeds, be economically viable?" Thus if specific welfare measures (*e.g.* fish preference) are integrated into production practice could they be economically viable potentially via a higher retail price. Many examples from terrestrial farmed animals indicate that systems promoting the natural biological production potential of the animal, if correctly handled, may outperform systems based upon solely technical solutions. The present special issue, being a co-production between the Cost 867 network and Fish Physiology and Biochemistry summarizes not only the present status of the field but also formulates a scientific platform for future production systems, putting the animal in the center of the debate.

This special issue is divided into four sections:

- Conceptual frame work of welfare in fish
- Factual component of welfare
- Assessment of welfare in fish (including some experimental papers)
- Implementation of welfare in fish

The papers are either opinionated reviews or research articles with a general interest and broad approach.

In the first section, the question is asked whether fish are the victim of "speciesism," that is, is there a different ethical framework for fish compared to other animals in the eyes of society. This section also includes articles discussing the fundamentals of welfare and whether controlled laboratory experiments have a direct relevance to the multivariable reality at the farm level.

The second section includes reviews about risk analysis as a tool to quantify welfare and thereby enable comparison of different factors affecting welfare. Also several members of the action have made a very serious attempt to distinguish between veterinarian medicine as an indicator of bad welfare and as a prerequisite for good welfare, that is, how does one distinguish between the two sides of the coin. Another group has approached genomics and welfare, and yet another the issue of nutrition and welfare, resulting in probably the most up-to-date and comprehensive written source of welfare on these subjects available today. Welfare at slaughter and Welfare implications of anesthesia are two other reviews dealing with two very common procedures during both production and research. Cortisol is probably one of the most common welfare indicators especially in applied research. The review addressing cortisol does not only deal with cortisol as indicator for physiological response but also its relationship to psychological condition. Also behavioral indicators and injuries/lesions as objective tools to assess welfare in fish have been addressed as two joint activities of the action and are presented here. Fish live in a protein and fat-dominated energy food chain, in contrast to carbohydrate-driven nutrition in mammals. Nitrogen balance and quality in feed may be an important factor for nutritional welfare in fish and is also reviewed within this special issue.

The more specific research articles include a range of relevant subjects including recovery from surgical procedures, a major interest for any researcher interested in a viable animal when using in situ experiments. Dietary composition and its implications for welfare is yet another question of major interest questioning whether dietary choice and its economics are essential and the implications of a "single diet" in aquaculture, that is, the animals only choice is to eat more or less, as the pre-prepared diet is formulated from an anthropocentric vision of what we think is best for fish. Other subjects included in this section include the importance of the timing of feeding in relation to the natural behavioral rhythm of fish and the effects of live chilling. The issue is then rounded off by a socio-behavioral reflection upon a series of practical activities to implement welfare at farm level together with the farmer.

The organisation COST

COST- the acronym for European Cooperation in Science and Technology- is the oldest and widest

European intergovernmental network for cooperation in research. Established by the Ministerial Conference in November 1971, COST is presently used by the scientific communities of 36 European countries to cooperate in common research projects supported by national funds.

The funds provided by COST - less than 1% of the total value of the projects - support the COST cooperation networks (COST Actions) through which, with EUR 30 million per year, more than 30 000 European scientists are involved in research having a total value which exceeds EUR 2 billion per year. This is the financial worth of the European added value which COST achieves.

A "bottom up approach" (the initiative of launching a COST Action comes from the European scientists themselves), "à la carte participation" (only countries interested in the Action participate), "equality of access" (participation is open also to the scientific communities of countries not belonging to the European Union) and "flexible structure" (easy implementation and light management of the research initiatives) are the main characteristics of COST. As precursor of advanced multidisciplinary research COST has a very important role for the realisation of the European Research Area (ERA) anticipating and complementing the activities of the Framework Programmes, constituting a "bridge" towards the scientific communities of emerging countries, increasing the mobility of researchers across Europe and fostering the establishment of "Networks of Excellence" in many key scientific domains such as: Biomedicine and Molecular Biosciences; Food and Agriculture; Forests, their Products and Services; Materials, Physical and Nanosciences; Chemistry and Molecular Sciences and Technologies; Earth System Science and Environmental Management; Information and Communication Technologies; Transport and Urban Development; Individuals, Societies, Cultures and Health. It covers basic and more applied research and also addresses issues of pre-normative nature or of societal importance.

Web: http://www.cost.eu

Legal notice by the COST Office

Neither the COST Office nor any person acting on its behalf is responsible for the use which might be made of the information contained in this publication. The COST Office is not responsible for the external websites referred to in this publication.

Conceptual Framework of Welfare in Fish

Are Fish the Victims of 'Speciesism'? A Discussion about Fear, Pain and Consciousness – Summary

Stephanie Yue Cottee

A systematic comparison between fish and mammalian animals in the fields of behaviour, physiology and brain anatomy shows that fish may not be so different from their 'higher' counterparts as previously thought. Some fishes produce and release the same stress hormones as mammalian animals, within a comparable biological system, and they show strikingly similar stress responses. In the rainbow trout pain receptors have been identified beneath the skin and in the fins. This information is complemented by the fact that pain-killing drugs administered to fish can dull their pain responses when they are subjected to painful stimuli. The concept of fish not having the ability to feel pain due to their lack of a neocortex is challenged. Research shows that fish have a brain anatomy functionally corresponding to the brain structure of higher vertebrates. The basic building blocks and the patterns of neural connectivity are equivalent. Fish have also successfully demonstrated a variety of avoidance learning, long-term memory, social learning, complex and flexible behaviours as well as problem-solving ability. A growing body of literature supports the argument that fish can indeed experience pain and fear.

Background

Fish welfare is a hotly debated topic. This is mainly due to the issue of whether fish have the capacity for conscious awareness or emotional states. Because of the controversial nature of animal emotions, the subject is often avoided in welfare debates. It is argued that since welfare should be about how animals feel, this issue is unavoidable. There is good reason to believe that the issue of assessing animal emotions is not as insurmountable as some believe. This paper reviews some comparative avoidance behaviour, physiology and brain anatomy.

Scientific Significance

Some fish species have satisfied all the criteria that mammals must fulfil in order to be considered conscious, feeling animals from a physiological, cognitive and behavioural standpoint.

Practical Application

Opinions on fish welfare should not be based on emotional, preconceived beliefs, but rather on logical reasoning. Currently, we have good logical reasons and scientific evidence to start treating fish as animals capable of suffering from pain and distress.

Contact

S. Yue Cottee
Department of Animal and Poultry Science
University of Guelph, Guelph, ON N1G 2W1, Canada
e-mail: syue@uoguelph.ca

Fish Physiol Biochem (2012) 38:5–15
DOI 10.1007/s10695-010-9449-9

Are fish the victims of 'speciesism'? A discussion about fear, pain and animal consciousness

Stephanie Yue Cottee

Received: 13 January 2010 / Accepted: 1 November 2010 / Published online: 18 November 2010

Abstract Fish welfare is currently a hotly debated topic; this is mainly due to the issue of whether or not fish have the capacity for conscious awareness, or subjective states. Because of the contentious nature of animal consciousness, the subject is often avoided in many welfare arguments, but it is argued that since welfare should be about how animals feel, this issue is unavoidable. There is also good reason to believe that the issue of assessing subjective states is not as insurmountable as some believe.

Keywords Fish · Fear · Pain · Consciousness · Animal welfare · Speciesism

Introduction

What is animal welfare?

The scientific community has a difficult time discussing animal welfare matters mainly because there is no exact or universal definition of 'animal welfare' (Duncan 2002). Part of the problem is that there are many approaches to animal welfare science, although the two dominant approaches are the 'biological functioning school' and the 'feelings school'. The biological functioning school suggests that welfare equates to the lack of large physiological stress responses and good physical health. The feelings school proposes that welfare is really about psychological health and how the animal feels (Duncan 2005). Then there are some that take both aspects into consideration. Broom (1998) believes that welfare encompasses "…bad as well as good states and measured by disease, immnosuppression, injury, and abnormal behavior and physiological indicators of comfort, satisfaction, or pleasure and indications of control and social support". Similarly, the OIE considers both schools of thought as they think that "an animal is in a good state of welfare if (as indicated by scientific evidence) it is healthy, comfortable, well nourished, safe, able to express innate behaviour and it is not suffering from unpleasant states such as pain, fear and distress…" (OIE 2010, Article 7.1.1 of the Terrestrial Animal Health Code). Hence, animal welfare scientists struggle to define 'animal welfare'. However, when asked to use a common-sense approach, most people's description of 'animal welfare' is strikingly similar. In surveys of schoolchildren to graduate students to arrays of lay audiences asked to give examples of animals with poor welfare, Duncan (2002) found that responses were consistently alike and comprised of answers such as an animal that is "ill", "in pain", "frightened", "frustrated", "bored". Answers to examples

S. Yue Cottee (✉)
Department of Animal and Poultry Science,
University of Guelph, Guelph, ON N1G 2W1, Canada
e-mail: syue@uoguelph.ca

of poor welfare did not consist of phrases such as "an animal that is showing high levels of cortisol" or "has a significant white blood cell count". When people express concern about animal welfare, it is the conscious experiences of suffering (like pain and fear) that worry them most (Dawkins 1998). If it were not for this psychological component, welfare would not matter. For instance, plant producers worry about the health and growth of their tomatoes in the way a pet owner would worry about the health and growth of his puppy, but it is the belief that dogs are sentient (i.e. have the ability to consciously experience suffering and well-being) that the pet owner may be concerned about his dog's welfare; by contrast, the concept of tomato welfare is nonsense. As Dawkins (1998) put it, "If "welfare" just means "health", then isn't animal welfare simply veterinary science or animal health studies under another name?" Duncan and Dawkins (1983) defined 'suffering' as 'a wide range of unpleasant emotional states' and suggested a broad working description of 'welfare' as that which included the notions of mental and physical health, being in harmony with the environment, being able to adapt without suffering to an artificial environment and somehow taking into account the animal's feelings.

Fish welfare from the feelings perspective

The topic of fish welfare has recently attracted much attention. In fact, the European Food Safety Authority (2010) has recently come out with several documents concerning transport, husbandry and slaughter of an array of farmed fish species. This is because many handling methods, husbandry procedures and management practices in aquaculture can potentially cause pain and fear and therefore reduced welfare. Because welfare should be about how an animal 'feels' rather than how healthy it is or how well it biologically functions, a main point of contention has been the issue of whether or not fish have the capacity to experience mental subjective states, like pain and fear. Thus, two main camps have resulted—those who believe that the feelings-based approach to welfare should not be applied to fish (Arlinghaus et al. 2007) because they lack some neuroanatomical structures, which in humans are associated with conscious subjective states (Rose 2002) and those who believe that fish have the mental capacity as well as the neurological and physiological apparatus that enables them to suffer from negative subjective experiences (Chandroo et al. 2004a, b).

During this debate, it has been noted that fish have a host of cognitive abilities that aid in negotiating different types of aversive situations (Beukema 1970; Gallon 1972; Lima and Dill 1990; Overmier and Hollis 1990; Pyanov 1993; Chivers and Smith 1998). Although some of their cognitive feats are impressive, it is still valid to question whether aversive situations may be accompanied by negative conscious experiences. Basically, do fish show purely 'reflexive' avoidance responses to negative stimuli, or is their avoidance behaviour sometimes motivated by fear? Some would contend that this question cannot be answered due to the fact that it is impossible to directly measure intangible conscious thoughts. Therefore, because of the contentious and somewhat subjective nature of animal consciousness, this topic is often purposely avoided in many welfare arguments (Arlinghaus et al. 2007). However, it is not possible to engage in an in-depth discussion of fish welfare issues without taking into account the feelings-based approach to animal welfare (Duncan 2004), which does attempt to measure and assess (albeit using indirect methods) mental phenomena such as subjective states.

The conundrum of consciousness

The topic of consciousness has had a tumultuous history ranging from Behaviourism's[1] reluctance to accept it as a legitimate scientific subject (see for example James 1904; Watson 1928; Skinner 1975) to the modern resurrection of animal consciousness as an unavoidable issue in the study of animal welfare science (see for example Dawkins 2000; Duncan 2004). It is only in the past 30 years that words like 'consciousness' and 'sentience' have reappeared in the scientific animal literature. Hence, we are only slowly overcoming the taboo of studying conscious thought processes and voluntary behaviour (as

[1] Behaviorism: an approach in psychology (which predominated in the mid 1900s) that believed only in the study of visible and quantifiable aspects of behaviour and rejected subjective experiences such as emotions, motivation or conscious thought processes.

opposed to reflexive or 'instinctive' responses). After centuries of debate about animal consciousness, one would think we should have a better understanding, yet this does not always appear to be the case (Dawkins 2001). Part of the problem is that 'consciousness' is an ambiguous term. It can refer to a physical waking state (i.e. not asleep or comatose), a sensory experience (i.e. being aware of moment to moment events) or the possession of any mental state (i.e. having beliefs, fears, intentions, desires, etc.) (Zeman 2001). Damasio (1999) defined 'extended consciousness' as something that provides the organism with a complex sense of self, ability to think about one's own thoughts, awareness of others and others' minds, ability to ponder about past and the future; it is extended consciousness at its pinnacle that is distinctively human. Whereas when mentioning animal consciousness, we are generally referring to what's known as 'core consciousness', a type of basic sense of self in the immediate moment that is not dependent on reasoning or language. It is this type of simple awareness of one's own state of being that is of interest in teleost fish.

As difficult as this may be, the study of animal welfare science forces scientists to come to grips with this mysterious issue on a daily basis, because subjective experiences such as pain, fear, frustration and pleasure are inextricably linked to welfare. Interestingly, some scientists still seem unable to break free from the ancient Cartesian view of man as intrinsically unique, blessed with a conscious mind, while animals are mindless automatons. Gallup (1985) boldly stated that the only thing defining man as special or even rational is man himself. He also went on to say that our sympathetic tendencies are generally prejudiced depending on whether animals are considered as pests or pets, attractive or ugly, cuddly or slimy or homeothermic versus poikilothermic. The last example of 'cold-bloodedness' is used routinely as a reason why it is often doubted that fish may be capable of subjective experiences as mammals are. However, homeothermy is merely a metabolic distinction, not a psychological one (Gallup 1985); why this obvious fact of thermoregulatory physiology would count as evidence against fish consciousness is quite curious. Upon close examination, all of the above characteristics are unfounded when it comes to animal consciousness. Mendl (2004) shares Gallup's view when he wrote of the human 'speciesist' tendency of attributing human-like conscious abilities to animals more similar to ourselves, than to phylogenetically more distant species even when identical behaviour is displayed. One problem is that most people are content dealing with issues of animal mind based on emotional, preconceived beliefs rather than logical reason. The other problem is the crux of the issue, notably that consciousness is essentially a subjective private experience and as such is not open to observation or empirical measurement. Because one can never access the experiences of anyone else except for their own, one can never be certain that such experiences actually exist in other fellow humans, let alone in non-human animal species. In addition, there are many examples of human behaviour that happens without any apparent conscious awareness. The much cited 'implicit learning' is usually characterized as a passive process, where the subject is exposed to information and subsequently acquires knowledge of that information simply through that exposure. For example, language is something human babies naturally and easily acquire, but as adults, cannot explain how they ever learned to do it. Because of this 'automatic' nature, it has been taken to represent a form of learning devoid of conscious thought. Given the limits in even human consciousness, it is not surprising that there has been little room to welcome the possibility for conscious thought processes in animals. It has been proposed that perhaps animals simply sense and respond to stimuli in an appropriate manner without experiencing the stimuli in a subjective way (Rose 2002); hence, even a type of non-conscious existence could be completely biologically sufficient. Therefore, according to Watsonian principles, why not leave the subject matter of subjective experience out of the scientific realm (Watson 1929). However, only the extreme solipsist would contend that consciousness is some metaphysical phenomenon and that there is simply no way of ever verifying mental states in others. Though, however, challenging, it is necessary to make at least some assumptions about consciousness in other species (Wasserman 1984), because if we close the discussion altogether, we may fail to notice potential animal consciousness all around us (Dawkins 2001).

It has been said that excessive pessimism about the study of animal minds only serves to inhibit scientific progress (Griffin 1995), and the view that subjective experiences are not scientifically accessible is archaic and simply false (Damasio 1999). In fact, in the real world, human beings (and some animals to a certain extent) can and do make reasonable inferences about others' mentality in such a way that it allows us to engage in appropriate social interactions (Griffin 1995). Imagine if we were all convinced that we could not detect intentions, preferences, pain, sadness, anger, etc. in our fellow humans and behaved in a manner consistent with our disbelief of conscious minds, human culture and our orderly, civilized way of living would cease to exist. So, on a practical level, we live out our lives carrying out social interactions under the assumption that others do indeed have separate minds. Yet on the scientific level, we demand full proof of the existence of a phenomenon we already use on a daily basis. Griffin (1998) pointed out the double standard of demanding absolute certainty before accepting the idea of mental experiences in animals (which he termed 'paralytic perfectionism') and that in other scientific disciplines, we customarily make the best of incomplete, ambiguous or intangible data. In the early days of genetics and physics, studies dealt with ideas and predictive models based on what are now known as chromosomes and atomic particles, despite the fact that these elusive components were not even visible to the naked eye; we trusted that these 'things' existed simply, because we could manipulate them and subsequently observe their real effects. In a similar concept, electroencephalograms (EEG) and functional MRI (fMRI) scans can capture correlates of the conscious mind (Klemm 1992; Cavanna and Trimble 2006), even if those correlates are not the mind. Upon being presented a frightening picture or thinking about something sad, the human subject can cause different EEG profiles, giving researchers tools to indirectly probe the neural mechanisms underlying the physical manifestation of emotional experiences. The fact that we must use indirect methods should be no cause for concern and because mental experiences can be only truly accessed by their owners does not make them any less real, nor does it deny its organic and neural origin (Damasio 1999).

Interestingly, it seems that the strongest argument against the case for fish consciousness is based on what we *do not know* (i.e. the argument that we can never know what they perceive due to the private nature of mental subjective states). However, this is not an appropriate argument. Decisions should be based on the evidence of what we presently have, not what we do not have. Science (and society) would never progress, if we refrained from making decisions because of perpetually waiting for infinitely more information. In addition, theories and beliefs are not set in stone. If, in the near future, new experimental evidence firmly contradicts present theories in support of fish consciousness, we are then obliged to re-evaluate our perspectives on the matter. Until then, the growing body of literature supports the argument for subjective experiences in fish.

Pain: a conscious state

Comparative approach to animal minds

Although behavioural research has somewhat relaxed its views about animal consciousness, many scientists still avoid the issue, not because they deny the possibility for subjective experiences, but simply because of the difficulty of the subject (Griffin 1998). Fortunately, there is an ever-growing body of knowledge regarding the biology of diverse types of animals including lower vertebrates and, more recently, invertebrates. Damasio (1999) gives a refreshingly simple and logical means to making reasonable inferences about human private states. His 'three-way linkage' brings together: (1) certain external signs like wakefulness, attention, observable behaviour, (2) corresponding internal signs that we can report on and (3) the internal signs that we, as observers, can verify in ourselves when we are in situations similar to those of the observed individual. This method sums up in part, and sometimes in full, many comparable techniques animal scientists have used in attempts to get a glimpse inside the animal mind. A widely used method of assessing animal experiences is by comparing the similarity of its responses with those of humans and other higher animal species generally accepted as having the capacity for consciousness. This is known as the argument-by-analogy. Though it is important to stress that 'analogous' does not mean 'identical'—for example, all animals have the ability to see, touch, smell, hear and taste, albeit in different manners (Sherwin 2001). Although an animal like a fish might have parts of their physiology vastly differing from their mammalian counterparts, they may still be able to experience mental states, even though those states would likely be qualitatively different. In reality, a quick search of the literature will show that the relevant behaviour and physiology associated with consciousness is strikingly more similar to higher vertebrate animals than previously acknowledged. Until recently, animal welfare studies have been done almost exclusively on mammalian and higher vertebrate animals, so they will be used as the benchmark from which to compare evidence. Beginning with physiology, it appears that teleost fish have the same relevant stress hormones (Donaldson 1981; Mazeaud and Mazeaud 1981) produced and released within a hypothalamic-pituitary-interrenal (HPI) axis (Donaldson 1981), analogous to the mammalian hypothalamic-pituitary-adrenal (HPA) axis. Correspondingly, fish show strikingly similar behavioural stress responses (e.g. freezing/tonic immobility) to other animals ranging from mammals (Davis 1992) to domestic fowl (Gallup 1979). There is, in fact, an exhaustive list of behavioural variables routinely employed to investigate subjective mental states in animals that has likewise been successfully demonstrated in fish. A variety of fish species have shown avoidance learning (Coble et al. 1985), long-term memory (Aronson 1971), conditioned suppression (Geller 1964; Yue et al. 2008), social learning (Brown and Laland 2003), operant responding (Anthouard 1987), nociceptive behaviour (Sneddon et al. 2003), demonstration of complex or flexible behaviour (Topál and Csányi 1999), problem-solving ability (Mizukami et al. 1999), demonstrated preferences (e.g. taste aversion) (Little 1977) and capacity to use 'trade-offs' according to 'consumer demand' theory (Dunlop et al. 2006). It could be argued that if fish perform similarly in these types of tests, they might experience something analogous to mammalian 'emotions' and motivation—an important point in the study of conscious minds. For higher vertebrate and especially mammalian species, we almost always employ argument-by-analogy and accept that if an animal responds with both similar physiological and behavioural mechanisms to negative stimuli, that they are presumably having an analogously negative experience (Sherwin 2001).

Much of the study of animal minds has been done through ethological efforts (Griffin 1976; Duncan 1996; Dawkins 2000; Griffin and Speck 2004), and now more recently, scientists are going beyond the realm of behaviour into the fields of neuroanatomy and neurophysiology. Though researchers often use some form of argument-by-analogy, many now argue that subjective experiences can also be found in animals by methods of homology. Some neurophysiologists maintain that the basic anatomy and physiology of the brain is highly conserved in mammals and even in lower vertebrates (Baars 2005). In humans, conscious perception and various cognitive functions depend on the thalamocortical complex (Seth et al. 2005). Many animals do not posses this thalamocortical system yet have been found to possess functionally equivalent neuroanatomy—which really is the crucial point when comparing conscious states from a neurobiological framework (Edelman et al. 2005). Birds for instance lack a significant cerebral cortex, but developmental, morphological and comparative evidence indicate that part of the avian pallium, known as 'the Wulst', is structurally homologous to the mammalian neocortex (Medina and Reiner 2000). In teleost fish, the lateral and medial pallia are proposed to be homologous to the hippocampus and amygdala, respectively (Broglio et al. 2005; Portavella and Vargas 2005). The amygdala, in particular, is well accepted as a brain region that plays a major role in the production of fear and fear responses (LeDoux 1996). Just as one would expect from comparative neurophysiology, lesions to the teleost lateral pallium produce significant deficits in some forms of learning and memory, but in contrast, lesions to the medial pallium disrupt emotional heart rate conditioning and avoidance learning (Broglio et al. 2005). It has also been found that the teleost forebrain, the telencephalon, receives a pattern of projections from the thalamus incredibly alike to those of land vertebrates, supporting the theory of telencephalic homologies with higher vertebrates (Echteler and Saidel 1981). It is obvious that more distantly related animals to mammals will have fewer homologous traits, but the basic structure of neurons, synapses, neurotransmitters and even patterns of connectivity resembling those of the human cortex remain the same (Baars 2005). Therefore, the neurobiological evidence suggests that there is robust developmental conservation of structure throughout evolution, a conclusion that is easily overlooked if we limit ourselves to looking for gross anatomical and histological comparisons (Edelman et al. 2005).

Indeed, we will see even more similarities between fish and other animals when we compare their usage as subjects in studies of drug development. Vertebrate animals are often employed as models in the development of analgesic and psychoactive drugs. Undeniably, the industry of pharmaceutical drugs and its testing rests on the assumption that these animal subjects do experience conscious experiences (like pain and fear) similar to humans, because it is humans for whom these drugs are ultimately intended (Sherwin 2001). Broiler chickens with leg problems will selectively choose diets containing analgesics, presumably to alleviate their negative mental experience comparable to what we call pain (Danbury et al. 2000). Similarly, studies employing rats have shown that they will self-administer or seek out addictive psychostimulant drugs like amphetamine, morphine, cocaine and heroine (Deminiere et al. 1989; Parker 1992; Klebaur et al. 2001; Kearns et al. 2002; Lenoir and Ahmed 2007). Such animal studies are carried out in hopes of elucidating the nature of human drug addiction. Now because it is believed by some pharmacological researchers that fish, in common with other 'more complex' vertebrates, will seek out pleasure and avoid pain, fish have been used as models to investigate the hedonic effects of addictive drugs (Lett and Grant 1989). Studies of this sort are done in attempt to understand the psychological underpinnings of substance abuse, and more current investigations are also interested in its genetic basis. Darland and Dowling (2001) showed that zebrafish exhibit strong or weak cocaine-induced conditioned place preference behaviour depending on their genetic make-up. Having similar dopaminergic pathways and behavioural responses to more complex animals, zebrafish are "currently the vertebrate of choice" (Darland and Dowling 2001) and can provide a good model for understanding this human affliction (Satora and Morawska 2003).

So, there is a growing body of evidence that fish may have some ability for subjective experiences. Perhaps, the question of whether or not fish can experience fear may become clearer. Although a review of the literature will reveal that there is almost no information regarding this topic, fear, like pain, is a negative mental experience that relies on an animal's capacity for conscious awareness. Also like pain, fear has no direct physical correlate that can be measured. So, by examining the current debate regarding the issue of fish pain, in addition to results from studies of avoidance behaviour, together with the more modern understanding of animal consciousness, we can obtain a reasonable assessment about the topic of subjectively experiencing fear.

Argument for pain as a conscious experience

It should be first mentioned that pain is a negative mental experience (e.g. feeling of distress or agony), whereas nociception is the physical, unconscious response to noxious stimuli that results in a behavioural or physiological change (IASP 1994). In Gregory's (1999) short review on subjective experiences in fish, he mentions the importance of this question, because it can influence our views on how we should manage these animals. Obviously, the question of pain perception in fish is a particularly hot topic, since our current methods of handling, rearing and slaughter could potentially cause them pain. Consequently, the subject of fish pain has, of late, been given substantial attention. Gregory (1999) provided criteria for the assessment of fish pain. First was to establish whether or not fish possess neurotransmitters, neuron types and neuroanatomy known to mediate the pain experience in other animals. Second is to inflict what we consider to be painful stimuli on fish and then assess their responses to the stimuli and then determine whether those responses can be suppressed with analgesic drugs, which can in turn be blocked with analgesic inhibitors. The last method is to teach fish to associate the aversive stimulus with a neutral-conditioned stimulus and examine whether they show avoidance behaviour to the conditioned stimulus alone. Finally, by collating the information from all three approaches, it should be possible to decide whether fish can perceive and experience the negative mental state of pain. With respect to the first criterion, some hypothesize that conscious awareness and therefore mental states such as pain and fear depends specifically and almost solely on the neocortex and that the absence of this brain region makes it impossible for conscious awareness (LaChat 1996; Rose 2002). Rose (2002) argues that decorticate or neocortically damaged humans have no consciousness, yet noxious stimulation applied to the faces of vegetative patients can evoke facial grimaces, flinches and expressions reminiscent of a person in pain. The patient, of course, is not aware of his reflexive responses. This is the argument of the difference between nociception, the sensory response to noxious stimulation, and pain, the negative psychological experience. The fish's behavioural responses to negative stimulation are therefore compared to these types of reflexive withdrawal reactions of decorticate humans. Unfortunately, Rose's (2002) lengthy review of the matter is built completely upon the thesis of the neocortex being the seat of consciousness. Although an interesting proposition, its Achilles heel is that he compares the normal functioning physiology of intact fish to that of a human pathological state. This is simply not a reasonable or logical argument (see Chandroo et al. 2004a, b). Imagine if you will, a fish stating to humans "I have fins. If I am missing my fins, I can no longer swim. Fins are therefore essential for swimming. Humans do not possess fins. Therefore, humans cannot swim". Of course, this is not true; humans, although not possessing fins, can use their arms, legs, hands and feet in order to propel themselves through the water. In addition, although humans do not swim in the exact manner that does a trout, it is an accepted fact that humans can indeed swim. In fact, Sneddon et al. (2003) discovered that rainbow trout possess cutaneous nociceptors capable of detecting noxious stimuli. This electrophysiological study not only identified the physical location of the nociceptors, but also that the nociceptive nerves demonstrate properties similar to those described in the higher-vertebrate pain system. Like in mammals, nerve endings of A-delta and C fibres act as nociceptors in fish. These fibres differ in diameter, myelination and information transmission speed and are associated with different types of pain like pricking or dull, aching pain. Complemented with the behavioural component to the study, it was proposed that the avoidance behaviour induced by painful stimuli is not likely to be merely reflexive, but accompanied with discomfort. With respect to Gregory's second criterion, Chervova et al. (1994) addressed that issue well over a decade ago when they found that trout's strong tail flick avoidance reaction in response to electric shock, needle prick or fin pinching decreased in intensity with increasing dosage of the endogenous mammalian opioid and analgesic, dermorphin. Under a similar set-up with cod, Chervova (1997) found that there was decrease in pain sensitivity (as measured by tail undulation) under the action of even non-opioid substances and suggested the presence of another endogenous analgesia system in fish in addition to the opioid system. In addition, naloxone, an antagonist of opioid receptors, is able to reverse the analgesic effect of morphine in goldfish and carp (Ehrensing et al. 1982; Chervova and Lapshin 2000). These findings are not surprising, given the fact that fish do possess opioid receptors and endogenous opioids in their nervous systems (Sneddon 2004). Sneddon (2003) contributed more supporting evidence when she found that administration of morphine significantly reduced pain-related types of behaviours (induced by acetic acid injection to trout lips) and opercular beat rate. Lastly, with respect to Gregory's (1999) third criterion, there is an abundance of studies demonstrating avoidance learning and is easily found in experimental, zoological, ecological and aquacultural literature (Pfeiffer 1963; Behrend and Bitterman 1964; Atchison et al. 1987; Knudsen et al. 1997; Yue et al. 2004; Dunlop et al. 2006). Since then, there have been plenty more studies investigating noxious stimuli and analgesics on nociceptive responses (Reilly et al. 2008; Nordgreen et al. 2009) all with results showing that various species of fish respond to noxious stimuli as well as analgesia in a manner consistent with pain. Taken collectively, it seems that teleost fish have been able to successfully fulfil all the various criteria and requirements researchers often demand in their quest to assess animal consciousness.

Despite all this evidence, the debate continues. This was seen in the vigorous exchange of views regarding fish welfare that revolved around whether or not fish have feelings. Huntingford et al. (2006) have reviewed a variety of practices, such as commercial fishing, recreational angling and aquaculture, on the welfare of fish and have taken a feelings-based approach in doing this. They concluded that some of these practices will have an adverse effect on fish welfare. Their findings were strongly challenged by Arlinghaus et al. (2007) who were extremely critical of the feelings-based approach and who rejected most of the conclusions of Huntingford et al. (2006). The criticisms of Arlinghaus et al. (2007) were based almost entirely on the fact that there are neuro-anatomical brain differences between fish and the higher vertebrates and in particular that fish lack a neocortex (Rose 2002). However, neuroanatomical similarities or differences play no role in the feelings-based approach.

Since subjective experiences are not available for direct investigation, sentience, or the capacity to feel, is deduced from indirect evidence, mainly behavioural evidence (Duncan 1996, 2004). As aforementioned, in the absence of the capacity to feel, welfare is a non-issue. If fish cannot suffer, then they can be treated like tomatoes. We may try to avoid damaging them, but only because their instrumental value would be reduced.

Avoiding the concept of fear?

Do fish exhibit fear or non-conscious avoidance behaviour? A comprehensive literature review suggests that some finfish demonstrate a combination of both. But first, why should we not simply deal with 'stimuli' and their respective 'responses' and cut out the intervening variable and ambiguous term called 'fear'? Would it not be easier just to acknowledge and measure the stimuli we can manipulate and responses we can measure? Davey (1981) argued that when a previously neutral stimulus becomes a conditioned aversive stimulus by being paired with an aversive event, it is not the stimulus that changes, but *some aspect of the animal* through the mediation of a central state. Anger (1963) agrees that the new conditioned response to the CS be called a 'fear reaction'.

A common idea is that fear is a 'primitive' emotional experience that evolved as a biological adaptation as a simple means of dealing with dangerous situations (Dawkins 2000). In a world full of unpredictable situations, it makes much more sense to use the memory of frightening events to avoid dangerous situations altogether than to repeatedly engage in the same potentially fatal situation in hopes that automatic escape responses will be successful with each encounter. Life would become essentially a game of biological Russian roulette. Avoidance behaviour motivated by fear, a negative emotion, rather than by automatic responses triggered by key stimuli, makes sense. Says Dawkins (1998) "[B]y causing an animal to flee or hide, fear probably evolved and reduced the likelihood of physical injury occurring at all. Just as pain helps to avoid death, fear helps to avoid pain".

Overall, on the basis of probability, it is likely that some teleost species like rainbow trout, for example, are sentient creatures and have at least a rudimentary level of conscious awareness. If one takes a moderately systematic investigation of fear (and pain) in fish, it now seems that the question to be asking is not 'Do fish have conscious awareness', but 'What is the level and nature of their conscious awareness?' Admittedly, what fish are proposed to consciously experience when subjected to fearful or painful stimuli is likely to be qualitatively and perhaps quantitatively very different from human fear. However, this does not matter; the fish's experience may be unpleasant and aversive enough to cause negative mental states. Therefore, their mental experiences (whatever they may be) are important from their perspective. Subsequently, being their stewards, it should also be important from our perspective as human caregivers.

Nonetheless, much of society largely continues to treat fish in a manner that indicates our belief that they do not experience negative mental states and are incapable of suffering. We are curiously unlikely to draw positive conclusions about subjective experiences even when fish fulfil the same criteria satisfied by other animals deemed sentient. In the EFSA's opinion, the concept of welfare should be the same for fish, as it is with mammals and birds. This is because "there is scientific evidence to support the assumption that some fish species have brain structures potentially capable of experiencing pain and fear. The balance of evidence indicates that some fish species have the capacity to experience pain" (EFSA 2009). Also, although we are far from understanding the mechanism of sentience, studies of brain anatomy functionality to behaviour illustrate that there is some evidence of sentience in some species of fish (European Food Safety Authority 2009).

Looking ahead

In a relatively short period of time, the high level of interest in the topic of fish welfare has been matched with an equal amount of scientific research. Nonetheless, many questions remain to be answered, and often, new questions arise in place of answers; the topic of fish welfare is still in its pioneering stages. Scientists have, by no means, solved the origin or the mechanism of consciousness. However, these are not needed to assess whether an animal possesses subjective experiences.

With evidence for the existence of at least a basic capacity for fear and other mental experiences, perhaps it should be concluded that the question of fish awareness (and animal consciousness in general) should remain a philosophical question. For in our real, practical world, it is simply not reasonable to be searching for scientific answers to philosophical questions. In our real world, veterinarians treat animals on a daily basis for a variety of problems—but imagine challenging a veterinarian to practice without ever being allowed to use words, such as 'pain', 'anxiety', 'fear', 'distress'. This would literally be impossible. The time has come when we should be using the knowledge that we *do* know to answer concerns about what the needs of animals really are and how they are to be met (Jennings 1998) and what we do know is that we now have logical reason and scientific evidence to start treating fish[2] as sentient creatures. The conundrum of consciousness is not as hopeless as it may seem. Approaching the subject of mental experiences logically, reasonably and systematically can, to some extent, help demystify it and put it back into a realistic realm alongside other scientific fields of study.

References

Anger D (1963) The role of temporal discriminations in the reinforcement of Sidman avoidance behavior. J Exp Anal Behav 6:477–501

Anthouard M (1987) A study of social transmission in juvenile *Dicentrarchus labrax* (*Pisces, serranidae*), in an operant conditioning situation. Behaviour 103:266–275

Arlinghaus R, Cooke SJ, Schwab A, Cowx IG (2007) Fish welfare: a challenge to the feelings-based approach, with implications for recreational fishing. Fish Fish 8:57–71

Aronson LR (1971) Further studies on orientation and jumping behavior in the Gobiid fish, *Bathygobius soporator*. Ann N Y Acad Sci 18:378–392

Atchison GJ, Henry MG, Sandheinrich MB (1987) Effects of metals in fish behavior: a review. Environ Biol Fishes 18:11–25

[2] The term 'fish' refers to species for which we have sufficient evidence to infer subjective experience using the same objective standards employed to infer consciousness in humans and other phylogenetically advanced vertebrate animals.

Baars BJ (2005) Subjective experience is probably not limited to humans: the evidence from neurobiology and behavior. Conscious Cogn 14:7–21

Behrend ER, Bitterman ME (1964) Avoidance conditioning in the fish: further studies of the CS-US interval. Am J Psychol 77:15–28

Beukema JJ (1970) Acquired hook-avoidance in pike *Esox lucius* L. fished with artificial and natural baits. J Fish Biol 2:155–160

Broglio C, Gómez A, Durán E, Ocaña FM, Jiménez-Moya F, Rodríguez R, Salas C (2005) Hallmarks of a common forebrain vertebrate plan: specialized pallial areas for spatial, temporal and emotional memory in actinopterygian fish. Brain Res Bull 66:277–281

Broom DM (1998) Welfare as a broad scientific concept. J Appl Anim Welf Sci 1(2):149–150

Brown C, Laland KN (2003) Social learning in fishes: a review. Fish Fish 4:280–288

Cavanna AE, Trimble MR (2006) The precuneus: a review of its functional anatomy and behavioural correlates. Brain 129:564–583

Chandroo KP, Duncan IJH, Moccia RD (2004a) Can fish suffer? Perspectives on sentience, pain, fear, stress. Appl Anim Behav Sci 86:225–250

Chandroo KP, Yue S, Moccia RD (2004b) An evaluation of current perspectives on consciousness and pain in fish. Fish Fish 5:1–15

Chervova LS (1997) Pain sensitivity and behavior of fishes. J Ichthyol 37:98–102

Chervova LS, Lapshin DN (2000) Opioid modulation of pain threshold in fish. J Ichthyol 375:590–591

Chervova LS, Lapshin DN, Kamenskii AA (1994) Pain sensitivity of trout and analgesia induced by intranasal administration of dermorphine. Dokl Biol Sci 338:424–425

Chivers DP, Smith RJF (1998) Chemical alarm signaling in aquatic predator-prey systems: a review and prospectus. Ecoscience 5:338–352

Coble DW, Farabee GB, Anderson RO (1985) Comparative learning ability of selected fishes. Can J Fish Aquat Sci 42:791–796

Damasio AR (1999) The feeling of what happens. Harcourt Brace & Company, New York

Danbury TC, Weeks CA, Chambers JP, Waterman-Pearson AE, Kestin SC (2000) Self-selection of the analgesic drug carprofen by lame broiler chickens. Vet Rec 146:307–311

Darland T, Dowling JE (2001) Behavioral screening for cocaine sensitivity in mutagenized zebrafish. Proc Natl Acad Sci 98:11691–11696

Davey G (1981) Animal learning and conditioning. Macmillan Press, London

Davis M (1992) The role of the amygdala in fear and anxiety. Annu Rev Neurosci 15:353–375

Dawkins MS (1998) Evolution and animal welfare. Q Rev Biol 73:305–328

Dawkins MS (2000) Animal minds and animal emotions. Am Zool 40:883–888

Dawkins MS (2001) Who needs consciousness? Anim Welf 10:S19–S29

Deminiere JM, Piazza PV, Le Moal M, Simon H (1989) Experimental approach to individual vulnerability to psychostimulant addiction. Neurosci Biobehav Rev 13:141–147

Donaldson EM (1981) The pituitary-interrenal axis as an indicator of stress in fish. In: Pickering AD (ed) Stress and fish. Academic Press, London, pp 11–47

Duncan IJH (1996) Animal welfare defined in terms of feelings. Acta Agric Scand A Anim Sci 27(Suppl):29–35

Duncan IJH (2002) Poultry welfare: science or subjectivity? Br Poult Sci 43:643–652

Duncan IJH (2004) A concept of welfare based on feelings. In: Benson GJ, Rollin BE (eds) The well-being of farm animals: challenges and solutions. Blackwell, Ames, pp 85–101

Duncan IJH (2005) Science-based assessment of animal welfare: farm animals. Rev Sci Tech Int Epiz 24:483–492

Duncan IJH, Dawkins MS (1983) The problem of assessing 'well-being' and 'suffering' in farm animals. In: Smidt E (ed) In indicators relevant to farm animal welfare. Martinus Nijhoff, The Hague, pp 13–24

Dunlop R, Millsopp S, Laming P (2006) Avoidance learning in goldfish (*Carassius auratus*) and trout (*Oncorhynchus mykiss*) and implications for pain perception. Appl Anim Behav Sci 97:255–271

Echteler SM, Saidel WM (1981) Forebrain connections in the goldfish support telencephalic homologies with land vertebrate. Science 212:638–639

Edelman DB, Baars BJ, Seth AK (2005) Identifying hallmarks of consciousness in non-mammalian species. Conscious Cogn 14:169–187

Ehrensing RH, Michell GF, Kastin A (1982) Similar antagonism of morphine analgesia by MIF-1 and naloxone in *Carassius auratus*. Pharmacol Biochem Behav 17: 757–761

European Food Safety Authority (2009) Scientific opinion of the panel on animal health and welfare on a request from European Commission on general approach to fish welfare and to the concept of sentience in fish. EFSA J 954:3–27

European Food Safety Authority (2010) Fish welfare. http://www.efsa.europa.eu/en/ahawtopics/topic/fishwelfare.htm. Accessed 28 Oct 2010

Gallon RL (1972) Effects of pretraining with fear and escape conditioning on shuttlebox avoidance acquisition by goldfish. Psychol Rep 31:919–924

Gallup GG Jr (1979) Tonic immobility as a measure of fear in domestic fowl. Anim Behav 27:316–317

Gallup GG Jr (1985) Do minds exist in species other than our own? Neurosci Biobehav Rev 9:631–641

Geller I (1964) Conditioned suppression in goldfish as a function of shock-reinforcement schedule. J Exp Anal Behav 7:345–349

Gregory NG (1999) Do fish feel pain? Aust N Z Counc Care Anim Res Teach News 12:1–12

Griffin DR (1976) The question of animal awareness: evolutionary continuity of mental experience. The Rockefeller University Press, New York

Griffin DR (1995) Windows on animal minds. Conscious Cogn 4:194–204

Griffin DR (1998) From cognition to consciousness. Anim Cogn 1:3–16

Griffin DR, Speck GB (2004) New evidence of animal consciousness. Anim Cogn 7:5–18

Huntingford FA, Adams C, Braithwaite VA, Kadri S, Pottinger TG, Sandøe P, Turnbull JF (2006) Current issues in fish welfare. J Fish Biol 68:332–372

International Association for the Study of Pain (IASP) (1994) IASP pain terminology. http://www.iasp-pain.org/AM/Template.cfm?Section=Home&template=/CM/HTMLDisplay.cfm&ContentID=3088#Pain. Accessed 28 Dec 2009

James W (1904) Does 'consciousness' exist? J Philos Psychol Sci Meth 1:477–491

Jennings RC (1998) A philosophical consideration of awareness. Appl Anim Behav Sci 57:201–211

Kearns DN, Weiss SJ, Panlilio LV (2002) Conditioned suppression of behavior maintained by cocaine self-administration. Drug Alcohol Depend 65:253–261

Klebaur JE, Bevins RA, Segar TM, Bardo MT (2001) Individual differences in behavioral responses to novelty and amphetamine self-administration in male and female rats. Behav Pharmacol 12:267–275

Klemm WR (1992) Are there EEG correlates of mental states in animals. Neuropsychobiology 26:151–165

Knudsen FR, Schreck CB, Knapp SM, Enger PS, Sand O (1997) Infrasound produces flight and avoidance responses in Pacific juvenile salmonids. J Fish Biol 51:824–829

LaChat MR (1996) An argument in defense of fishing. Fisheries 21:20–21

LeDoux JE (1996) The emotional brain: the mysterious underpinnings of emotional lift. Simon & Schuster, New York

Lenoir M, Ahmed SH (2007) Heroin-induced reinstatement is specific to compulsive heroin use and dissociable from heroin reward and sensitization. Neuropsychopharmacology 32:616–624

Lett BT, Grant VL (1989) The hedonic effects of amphetamine and pentobarbital in goldfish. Pharmacol Biochem Behav 32:355–356

Lima SL, Dill LM (1990) Behavioral decisions made under the risk of predation: a review and prospectus. Can J Zool 68:619–640

Little E (1977) Conditioned aversion to amino acid flavors in the catfish, *Ictalurus punctatus*. Physiol Behav 19:743–747

Mazeaud MM, Mazeaud F (1981) Adrenergic responses to stress in fish. In: Pickering AD (ed) Stress and fish. Academic Press, London, pp 49–75

Medina L, Reiner A (2000) Do birds possess homologues of mammalian primary visual, somatosensory and motor cortices? Trends Neurosci 23:1–12

Mendl M (2004) Cleverness and consciousness in animals: pronghorns and fiddler crabs. Aust N Z Counc Care Anim Res Teach News 17:4–6

Mizukami E, Gunji YP, Migita M (1999) Learning process by goldfish and its use of a local site as a map. Biosystems 54:91–104

Nordgreen J, Garner JP, Janczak AM, Ranheim B, Muir WM, Horsberg TE (2009) Thermonociception in fish: effects of two different doses of morphine on thermal threshold and post-test behaviour in goldfish (*Carassius auratus*). Appl Anim Behav Sci 119:101–107

OIE (World Organisation of Animal Health) (2010) Terrestrial animal health code. Chapter 7.1. Introduction to the recommendations for animal welfare. Article 7.1.1.

http://www.oie.int/eng/normes/mcode/en_chapitre_1.7.1.htm. Accessed 25 Oct 2010

Overmier JB, Hollis KL (1990) Fish in the think tank: learning, memory, and integrated behavior. In: Kesner RP, Olson DS (eds) Neurobiology of comparative cognition. Lawrence Erlbaum, Hillsdale, pp 205–236

Parker LA (1992) Place conditioning in a three-or four-choice apparatus: role of stimulus novelty in drug-induced place conditioning. Behav Neurosci 106:294–306

Pfeiffer W (1963) The fright reaction in North American fish. Can J Zool 41:69–77

Portavella M, Vargas JP (2005) Emotional and spatial learning in goldfish is dependent on differential telencephalic pallial systems. Eur J Neurosci 21:2800–2806

Pyanov AI (1993) Fish learning in response to trawl fishing. ICES Mar Sci Symp 196:12–16

Reilly SC, Quinn JP, Cossins AR, Sneddon LU (2008) Behavioural analysis of a nociceptive event in fish: comparisons between three species demonstrate specific responses. Appl Anim Behav Sci 114:248–259

Rose JD (2002) The neurobehavioral nature of fishes and the question of awareness and pain. Rev Fish Sci 10:1–38

Satora L, Morawska J (2003) Vaccine, methadone and zebrafish. Przegl Lek 60:311–313

Seth A, Baars BJ, Edelman DB (2005) Criteria for consciousness in humans and other mammals. Conscious Cogn 14:119–139

Sherwin CM (2001) Can invertebrates suffer? Or, how robust is argument-by-analogy? Anim Welf 10:S103–S118

Skinner BF (1975) The steep and thorny path to a science of behaviour. In: Harre R (ed) Problems of scientific revolution. Oxford University Press, Oxford, pp 58–71

Sneddon LU (2003) The evidence for pain in fish: the use of morphine as an analgesic. Appl Anim Behav Sci 83:153–162

Sneddon LU (2004) Evolution of nociception in vertebrates: comparative analysis of lower vertebrates. Brain Res Rev 46:123–130

Sneddon LU, Braithwaite VA, Gentle MJ (2003) Do fishes have nociceptors? Evidence for the evolution of a vertebrate sensory systems. Proc R Soc Lond B 270:1115–1121

Topál J, Csányi V (1999) Interactive learning in the paradise fish (*Macropodus opercularis*): an ethological interpretation of the second-order condition paradigm. Anim Cogn 2:197–206

Wasserman EA (1984) Animal intelligence: understanding the minds of animals through their behavioral "ambassadors". In: Roitblat HL, Bever TG, Terrace HS (eds) Animal cognition. Lawrence Erlbaum Associates, Hillsdale, pp 45–60

Watson JB (1928) Behaviorism. Routledge and Kegan Paul, London

Watson JB (1929) Psychology from the standpoint of a behaviorist. J.B. Lippincott, Philadelphia

Yue S, Moccia RD, Duncan IJH (2004) Investigating fear in domestic rainbow trout, *Oncorhynchus mykiss*, using an avoidance learning task. Appl Anim Behav Sci 87:343–354

Yue S, Duncan IJH, Moccia RD (2008) Investigating fear in rainbow trout (*Oncorhynchus mykiss*) using the conditioned—suppression paradigm. J Appl Anim Welf Sci 11:14–27

Zeman A (2001) Consciousness. Brain 124:1263–1289

Factual Components of Welfare

Behavioural Indicators of Welfare in Farmed Fish – Summary

Catarina Martins, Leonor Galhardo, Chris Noble, Børge Damsgård, Maria Spedicato, Walter Zupa, Marilyn Beauchaud, Ewa Kulczykowska, Jean-Charles Massabuau, Toby Carter, Sònia Rey Planellas and Tore Kristiansen

In this review we identify common behavioural changes used to assess welfare across a range of key farmed fish species. We also discuss what kinds of knowledge we need in the future to foster welfare-friendly aquaculture. Research over the past 10 years suggests that changes in foraging behaviour, ventilatory activity, aggression, individual and group swimming behaviour, stereotypic and abnormal behaviour are linked with acute and chronic stressors in aquaculture. These changes can, therefore, be used as indicators of welfare. Behaviours that are indicative of poor or good welfare seem to be species-specific. It is, therefore, essential to understand the species-specific biology of fish prior to drawing any conclusions in relation to welfare. Nevertheless, an increase of exploratory behaviour, feed anticipatory activity and reward-related operant behaviour are suggested to be indicators of positive emotions and welfare in fish. A major challenge when interpreting behavioural changes is created by the pronounced individual differences in coping strategies exhibited in most farmed fish species. This means that what is tolerated by some individuals may be detrimental to others. The assessment of welfare in a few individuals may not represent the average welfare of a group and vice versa. Therefore, the use of both individual and group-based welfare indicators may provide the best strategy to assess fish welfare.

Background

Behavioural changes are amongst the most common indicators of welfare in farmed animals, including fish. This is not surprising as behaviour represents a reaction to the environment as fish perceive it and, therefore may represent an early warning of environmental changes. In addition, some behavioural changes are fast and relatively easy to observe and therefore good candidates for use "on farm". But how can behavioural changes be used to assess welfare?

Scientific Significance

This review identifies the major research question to be answered in the future if welfare-friendly systems are to be developed: How can positive mental states be measured in fish and enhanced under farming conditions (e.g. through environmental enrichment)? What are the psychological mediators of stress in fish? Is it possible to develop on-farm tools that modulate the way fish interpret their environment (e.g. by making stressors predictable) and therefore alleviate the consequences of unavoidable stressors? What are the behavioural needs of farmed fish? And how can rearing systems be adapted to species-specific behavioural needs?

Practical Application

This review provides a tool for fish farmers to understand how behavioural changes can be interpreted and used to assess welfare in target species. It also highlights the need to accept fish as sentient creatures capable of affective states. It argues how such acceptance may contribute to changing the perspective from "avoiding poor welfare" into "promoting good welfare" in farmed fish.

Contact

C. I. M. Martins
Centro de Ciências do Mar do Algarve (CCMAR),
Universidade do Algarve, Campus de Gambelas,
8005-139 Faro, Portugal
e-mail: cimartins@ualg.pt

Fish Physiol Biochem (2012) 38:17–41
DOI 10.1007/s10695-011-9518-8

Behavioural indicators of welfare in farmed fish

Catarina I. M. Martins · Leonor Galhardo · Chris Noble · Børge Damsgård · Maria T. Spedicato · Walter Zupa · Marilyn Beauchaud · Ewa Kulczykowska · Jean-Charles Massabuau · Toby Carter · Sònia Rey Planellas · Tore Kristiansen

Received: 21 October 2010 / Accepted: 20 May 2011 / Published online: 28 July 2011

Abstract Behaviour represents a reaction to the environment as fish perceive it and is therefore a key element of fish welfare. This review summarises the main findings on how behavioural changes have been used to assess welfare in farmed fish, using both functional and feeling-based approaches. Changes in foraging behaviour, ventilatory activity, aggression, individual and group swimming behaviour, stereotypic and abnormal behaviour have been linked with acute and chronic stressors in aquaculture and can therefore be regarded as likely indicators of poor welfare. On the contrary, measurements of exploratory behaviour, feed anticipatory activity and reward-related operant behaviour are beginning to be considered as indicators of positive emotions and welfare in

C. I. M. Martins (✉)
Centro de Ciências do Mar do Algarve (CCMAR), Universidade do Algarve, Campus de Gambelas, 8005-139 Faro, Portugal
e-mail: cimartins@ualg.pt

C. I. M. Martins
Aquaculture and Fisheries Group, Wageningen University, P.O. Box 338, 6700 AH Wageningen, The Netherlands

L. Galhardo
Instituto Superior de Psicologia Aplicada, Unidade de Investigação em Eco-Etologia, Rua Jardim do Tabaco 34, 1149-041 Lisbon, Portugal

C. Noble · B. Damsgård
Nofima Marin, P.O. Box 6122, 9291 Tromsö, Norway

M. T. Spedicato · W. Zupa
COISPA Tecnologia & Ricerca, via dei Trulli, 18-20, 70126 Bari, Italy

M. Beauchaud
Laboratoire Ecologie et Neuro-Ethologie Sensorielles (EA3988), Université Jean Monnet, 23 rue Dr Paul Michelon, 42023 Saint-Etienne Cedex 02, France

E. Kulczykowska
Institute of Oceanology, Polish Academy of Sciences, 81-712 Sopot, Poland

J.-C. Massabuau
Station Marine, Université Bordeaux 1, CNRS, UMR 5805 EPOC, Place du Dr Peyneau, 33120 Arcachon, France

T. Carter
Anglia Ruskin University, East Road, Cambridge CB1 1PT, UK

S. R. Planellas
Universitat Autònoma de Barcelona, 08193 Cerdanyola del Vallés, Barcelona, Spain

T. Kristiansen
Institute of Marine Research, P.O. Box 1870, 5817 Bergen, Norway

fish. Despite the lack of scientific agreement about the existence of sentience in fish, the possibility that they are capable of both positive and negative emotions may contribute to the development of new strategies (e.g. environmental enrichment) to promote good welfare. Numerous studies that use behavioural indicators of welfare show that behavioural changes can be interpreted as either good or poor welfare depending on the fish species. It is therefore essential to understand the species-specific biology before drawing any conclusions in relation to welfare. In addition, different individuals within the same species may exhibit divergent coping strategies towards stressors, and what is tolerated by some individuals may be detrimental to others. Therefore, the assessment of welfare in a few individuals may not represent the average welfare of a group and vice versa. This underlines the need to develop on-farm, operational behavioural welfare indicators that can be easily used to assess not only the individual welfare but also the welfare of the whole group (e.g. spatial distribution). With the ongoing development of video technology and image processing, the on-farm surveillance of behaviour may in the near future represent a low-cost, noninvasive tool to assess the welfare of farmed fish.

Keywords Fish behaviour · Well-being · Aquaculture · Operational welfare indicators · Mental states · Stress · Appraisal

Introduction

During the last decade, the issue of farmed fish welfare has raised increasing public and scientific concern. This has contributed to the development of policies related to fish welfare (e.g. Council Directive 98/58/EC and the recommendation on the welfare of farmed fish by the council of Europe in 2005). Stakeholders, welfare NGO's and scientists have long recognised the importance of welfare of farmed and captive animals; nevertheless, despite the increasing interest on fish welfare, there is no clear consensus on how fish welfare should be defined or assessed (Huntingford et al. 2006; Ashley 2007).

Most animal welfare definitions can be loosely categorised into 'function-based' or 'feeling-based' definitions. 'Function-based' definitions have the fundamental assumption that welfare is correlated with biological functioning including physiological stress responses (Duncan 2005), while in the 'Feeling-based' definitions, welfare more or less equals the current emotional state of the animal (Duncan and Dawkins 1983) and in the longer term, it represents the balance between positive and negative subjective experiences (Spruijt et al. 2001). The primary basis for the concept of 'animal welfare' is the belief that animals are sentient beings with an ability to experience good or bad feelings or emotional states (Dawkins 1990).

If the fish farmer wants to improve or document fish welfare, they need methods to assess animal welfare that can be feasible on a farm. A welfare assessment system should describe the welfare of farmed fish and allow the farmer to assess the development over time and to respond appropriately. Welfare indicators that are relevant for inclusion in an operational welfare assessment system should be science-based, should measure welfare over extended time periods, should be measurable on a commercial farm within a realistic framework and should be relevant as a decision support system for the farmer. To fulfil these requirements, the welfare indicators must provide information on potential welfare problems and the causes of impaired welfare (Rousing et al. 2001). Behavioural welfare indicators have the advantage of being fast and easy to observe and therefore good candidates for use 'on-farm'. Examples of behaviour that are commonly used as an indicator of welfare are changes in food-anticipatory behaviour, feed intake, swimming activity and ventilation rate (Huntingford et al. 2006). Such qualitative assessment can be converted into quantitative measures via video recordings and software for data collection, analysis and presentation, while image analyses facilitate continuous surveillance of behaviour. In this way, behaviour can be used as a noninvasive and as an early sign of potential welfare problems.

Disadvantages of behavioural welfare indicators can be that they are often variable over time and difficult to quantify, the observers need the sufficient skills to observe the animals 'body language', and there can be large individual differences in behaviour within animal groups. It is also worth noting that some behavioural responses can be considered normal coping activities and therefore contributing to welfare, while other responses may be considered abnormal or maladaptive, therefore hampering welfare. The differences between normal (i.e. part of an adaptive response) and abnormal behaviour are frequently unclear.

To make a robust scientific welfare assessment, other indicators are also needed, and using an integrated approach by combining behaviour, neurophysiology, pathological and molecular indicators may result in an improved welfare assessment (Huntingford et al. 2006).

Over the last few years, advanced biochemical and genomic technology has equipped researchers with robust tools for studying animals' coping systems, and the number of new welfare-related indicators has increased rapidly. Such scientific indicators vary from the measurement of cortisol excreted into water (reviewed in Scott and Ellis 2007 and Scott et al. 2008) into more complex approaches such as the up- and down-regulation of targeted genes (e.g. Terova et al. 2005).

The aim of this review is to show that behaviour can be used to assess the welfare of farmed fish and to provide both aquaculturists and the scientific community with an overview of experimental results that show how certain behaviours can be applied as species-specific welfare indicators. We review behavioural changes whose link with welfare takes a more functional-based approach (foraging behaviour, ventilatory activity, aggression, individual and group swimming behaviour and stereotypic and abnormal behaviour) and those whose link takes a more feeling-based approach (exploratory behaviour, food-anticipatory activity, preferences and reward-related operant behaviour). Whenever the information is available, the link between behavioural indicators and other commonly used welfare indicators will be provided to allow a more comprehensive understanding of welfare assessment.

The diversity of behavioural responses and coping styles

Before addressing how specific behaviours can be used as welfare indicators, it should be noted that behavioural responses under stressful conditions are dependent not only on the type of stressor but also on the coping style that characterises an individual. Individual variation in responsiveness may explain why some conditions that are well tolerated by some individuals may be detrimental to others.

Koolhaas et al. (1999) defined coping style as a correlated set of physiological and behavioural traits consistently linked over time and across situations, which define the ability of the organism to cope with stress. There are two major types of coping styles: proactive (active coping or bold or 'fight-flight') and reactive (passive coping or shy or 'conservation-withdrawal'). Behaviourally, a proactive coping style is characterised by a high level of active avoidance, aggression, exploration, risk-taking and an active attempt to counteract the stressful stimulus as opposed to reactive coping style. Physiologically, the proactive coping style is characterised by low production of corticosteroid hormones in response to stress production, but high sympathetic reactivity, while the opposite is true for the reactive coping style (Koolhaas et al. 1999; Pottinger and Carrick 2001a; Øverli et al. 2007).

The presence of coping styles is now well recognised in fish (reviewed in Øverli et al. 2007), and its implication for aquaculture can be wide-ranging. One of the best examples of such implications comes from studies using selected lines of rainbow trout, *Oncorhynchus mykiss*, differing in stress responsiveness. Low cortisol responders (typical proactive individuals) were shown to exhibit higher growth rates as compared with high cortisol responders (Pottinger 2006; Trenzado et al. 2006). These authors attributed the growth disparity between the two lines to differences in competitiveness/aggressiveness. Low cortisol responders tend to win fights for social dominance (Pottinger and Carrick 2001b), which may provide them with a competitive advantage to access feed when fixed feeding regimes are used. In addition, low cortisol responders have lower feed waste, higher growth rates, higher feed efficiency and lower size variability as compared to high cortisol responders (Øverli et al. 2006a).

These selected lines of rainbow trout were also shown to differ in susceptibility to crowding stress. High-stress responders were shown to be more negatively affected by crowding than the low-stress responders (Trenzado et al. 2006). In addition, divergent coping styles have been shown to exhibit different susceptibility to diseases and inflammatory challenges (Mackenzie et al. 2009).

In general, individuals' coping styles will be related to the stress response of the fish, both the physiological response and the time lag before, e.g. recovering normal feeding (Damsgård and Dill 1998) and swimming behaviour after a stress event Øverli et al. 2006b). If a disturbance causes a fish to swim away or become inactive, such 'lost opportunities' will affect other behaviours, e.g. the time to feed. Small Atlantic salmon, *Salmo salar*, reduce their feeding activities for 2 h after a brief exposure to a trout model (Metcalfe et al. 1987). Also, Dill (1983) showed that the potential risk of predation for a juvenile coho salmon affected the distance the fish was willing to swim from the holding position to take a food item, and the fish seems able to trade-off the relative risk towards the relative value of the food. An individual boldness trait may thus affect where to feed, when to feed, how much time to allocate to feeding and the foraging tactics of the animal. In addition, other authors have also made the link between coping styles and traits relevant to aquaculture such as feed efficiency (e.g. Martins et al. 2006a, 2008).

Coping styles may also affect how fast fish might adapt to new feeding systems and explore new fish farm environments. A study on Atlantic cod, *Gadus morhua*, escaping from net cages revealed consistent individual differences between the fish escaping compared with nonescaping fish, with bold individuals being more willing to swim through small holes in the net, and thus escape more frequently than shy fish (Hansen et al. 2008). These responses are generally linked with the learning ability of farmed fish. In fact, Moreira et al. (2004) showed that lines of rainbow trout selected for high and low responsiveness to stress (HR and LR-lines) also differed in memory retention in addition to cortisol production.

An individual's responsiveness to stress is influenced by its coping style or behavioural syndrome, which should be considered when interpreting the variation in behavioural responses. The extent to which an individual is bold or shy should not be used as a welfare indicator, but one may infer a welfare problem when the behaviour defined under the bold/shy continuum changes; for example, when a bold individual starts exhibiting typical shy behaviours such as decreased exploratory behaviour, one may suspect that it is sick or chronically stressed. Still, one should interpret such changes with care, as coping styles are considerably plastic and may change with context (e.g. Ruiz-Gomez et al. 2008).

Behavioural welfare indicators

Foraging behaviour

In the context of this review, foraging is defined as the search for and exploitation of food resources (Danchin et al. 2008). Foraging behaviour has been recorded in fish in terms of daily feed intake or feeding rate, the latency or time to start of feeding, total feeding time and the activation of self-feeders. Under ideal circumstances, an animal can reach reproductive age or (in case of aquaculture) harvesting age, without having to engage in any aggressive or territorial behaviour, but cannot do so without feeding (Dugatkin 2004). In fact, reduced appetite has been considered one of the main causes for growth reduction after stress exposure (Wendelaar Bonga 1997).

Teleost fish exhibit a wide variety of foraging strategies, which can be affected by the farming systems (Turnbull and Kadri 2007), gender (Øverli et al. 2006b), the genetic strain/family group (Kolstad et al. 2004; Kittilsen et al. 2009) and other factors. Variations in foraging behaviour include where fish feed (bottom feeders, surface feeders and feeding from the water column), when they feed (e.g. nocturnal vs. diurnal as well as scheduled vs. random), how they feed (active predators, scavengers or more passive feeders) and what they feed on (animal or vegetable matter or both). This variation should be taken into consideration when interpreting foraging behaviour.

One of the most relevant aquaculture procedures that have been linked with changes in foraging behaviour and welfare is the feeding method. The techniques for delivering food should be appropriate for the particular species (Alanärä and Brännäs 1996; Alanärä et al. 1998; Andrew et al. 2002; Noble et al. 2007b). For instance, bottom-feeding flatfish have been shown to have improved welfare (measured as feed intake and feeding motivation) when fed sinking pellets compared with floating pellets (Kristiansen and Fernö 2007). In addition, Sánchez et al. (2009)

using sea bream, *Sparus aurata*, as a model showed that altering the feeding time (scheduled vs. random) affects behaviour and physiology. These authors highlight the importance of using scheduled feeding to allow fish to prepare themselves (both behavioural and physiologically) for the next, predictable, meal. Fish exhibited an increase in locomotory activity before feeding, i.e. food-anticipatory activity (see also section "Group swimming behaviour") when a scheduled feeding regime was used as opposed to fish receiving random feeding (Sánchez et al. 2009). However, imposed fixed regimes can increase the incidence of fin damage and injury as compared to demand feeding (Noble et al. 2007b, 2008), so a farmer should bear this in mind when making decisions on utilising a fixed ration or demand feeding strategy. In addition, the use of self-feeders has been shown to improve feed conversion ratio and growth in rainbow trout (Alanärä and Brännäs 1996) and can also be beneficial for their welfare (Suzuki et al. 2008).

Aquaculture stressors can reduce both feeding motivation and feed intake. The speed with which fish resume feeding after a stressor has been used as a foraging-related welfare indicator, for example in rainbow trout (Øverli et al. 2006a). Feeding motivation, measured as the latency to start feeding, has also been used as an indicator of welfare in Nile tilapia *Oreochromis niloticus*, cultured in recirculating aquaculture systems (Martins et al. 2009). More recently, Rubio et al. (2010) used self-feeders in sea bass, *Dicentrarchus labrax*, to understand the effects on refeeding dynamics after fish are exposed to fasting periods (which may occur before harvest). These authors showed that fasting periods can increase both feeding rate and daily feeding times.

In addition to changes in feeding behaviour, changes in feed intake have been repeatedly used as an indicator of reduced welfare. A variety of stressors have been shown to reduce feed intake in different fish species. Examples include changes in water quality (e.g. in Atlantic salmon, Espmark and Baeverfjord 2009; Nile tilapia, Martins et al. 2009; sea bass, Santos et al. 2010), stocking density (reviewed in Ellis et al. 2002 for rainbow trout), anaesthetisation and vaccination (e.g. in Atlantic salmon, Sørum and Damsgård 2004), establishment of dominance hierarchies (e.g. in Arctic char *Salvelinus alpinus*, Alanärä et al. 1998) as well as changes in the social structure of the groups as a consequence of grading or sorting procedures (e.g. in African catfish *Clarias gariepinus*, Martins et al. 2005, 2006b). In addition, cleaning protocols (such as emptying and brushing of the tanks) have been shown to impair self-feeding in sea bass (Rubio et al. 2010).

In brief, any observed changes in foraging-related behaviours, unless coupled with a change in management practices, is highly likely to be an indicator of changes in either the motivational state, health and/or welfare status of fish. This is supported by studies showing similar central signalling systems (corticotrophin-releasing factor, CRF/urocortin activation) involved in both stress-induced anorexia and negative emotions, at least in higher vertebrates (Heinrichs and Koob 2004). Whether this is also the case in fish still needs further study. Changes in foraging behaviour and behaviours associated with foraging can therefore be one of the most important behavioural operational welfare indicators of farmed fish species and are probably the indicator most commonly used by fish farmers. However, one cannot exclude the possibility that reduced feeding is also an adaptive response to a mild stressor or in some cases even a response to positive experiences.

Ventilatory activity

Ventilatory activity is the flow of water ventilated over the gills per unit time. It is performed by bucco-opercular movements. Its aim is the renewal of water in the branchial cavity, and a healthy fish will regulate it precisely. Ventilation ensures the supply of oxygen at the exact rate required by cellular oxygen (O_2) metabolism, neither too much nor too little. Specifically, the partial pressure of O_2 in the blood leaving the gills must be regulated in a very low and narrow range, primarily to avoid oxidative stress at the cellular level. Certainly, a fundamental and central idea is that O_2 deficiency in animal tissues is the result of a mismatch between O_2 needs and O_2 supply. The balance must be finely tuned, and ventilation must be taken as the first firewall protecting the fish tissues against oxidative stress (Massabuau 2001, 2003). Accumulated scientific evidence supports the concept that oxidative damage to tissues and cellular components is either a primary or secondary cause of many human diseases. Considerably less is known about how oxidative stress can affect animal health and well-being, but it is clearly a significant underlying factor related to dysfunctional host immune and inflammatory responses that can increase the susceptibility of any farmed livestock to a variety of health disorders (Sordillo and Aitken 2009). This must be definitively taken into account in fish farming (see for example Ritola et al. 2002; Kucukbay et al. 2006).

Counting the number of opercular movements (ventilatory frequency), and estimating their amplitude, provides an index of ventilatory activity. It is worth noting that besides laboratory tools, there are efforts to develop new field methods to record ventilatory activity by fish telemetry. SmartTags have been developed to measure ventilatory frequency and amplitude of free-swimming fish as an indicator of fish welfare (Damsgård 2008).

Well-balanced ventilatory activity is fundamental to maintaining homoeostasis in terms of O_2 status, blood and tissue acid–base balance (pH) at precise set points. In an acclimated and unstressed fish, the respiratory frequency is adjusted close to a minimum possible value. Therefore, increased ventilatory activity is commonly used as a sign of stress and poor welfare in fish. Importantly, it only takes a few seconds for a fish to change its ventilatory activity in response to water oxygenation or a chemical change. This is mediated by either chemoreceptors in the buccal cavity (taste buds, Ishida et al. 1996) or O_2 chemoreceptors in the gill vascular bed (Burleson and Milsom 2003). How fast it returns to reference levels is fundamentally variable if the health status of the fish is altered. Following a brief water O_2 change, it simply takes seconds (Eclancher and Dejours 1975).

Ventilatory activity can increase as a consequence of several aquaculture procedures or stressors. The key factors affecting ventilation activity include the levels of dissolved oxygen and carbon dioxide in the water with ventilatory activity being inversely related to water oxygenation. Frequency of ventilation increases up to a certain critical low oxygenation level after which it again decreases and the fish suffocate after a while. In contrast, the frequency of ventilation decreases when O_2 is in excess (Randall 1982; Xu et al. 2006). It is known that a slight excess of dissolved carbon dioxide levels leads to variable ventilatory responses according to the water chemistry and the fish physiological status; a large excess can be narcotic resulting in decreased ventilatory activity (Osborne and Muntz 1906).

Several other biotic and abiotic factors have been linked with increased ventilator activity. These include artificial lighting, loud or aversive sound, arousing odours, unsuitable temperatures or substrates, restricted movement, reduced retreat space, forced proximity to humans, reduced feeding opportunities, manipulation of fish groups, transportation (e.g. Barreto and Volpato 2006; Morgan and Tromborg 2007), handling (Brydges et al. 2009), air exposure (White et al. 2008), the presence of predators (Barreto et al. 2003), presence of chemicals (Neely 1979; Israeli-Weinstein and Kimmel 1998; Pane et al. 2004; Scott and Sloman 2004), diseases (Byrne et al. 1991), nanotubes (Smith et al. 2007), cyanobacteria (Ernst et al. 2007) and poor water quality in general (ammonia (NH_4)/nitrate (NO_3) excess, etc.).

In addition, any environmental change stimulating ventilatory activity can increase the vulnerability of fish to toxic chemicals simply by increasing the gills exposure to these: ventilation increases the surface of ventilated gill lamellae, and it decreases the thickness of the unstirred boundary layers (Pierron et al. 2007; Couillard et al. 2008). The enhancement is stronger when the contaminant concentration is low because a very low concentration can then become significant in terms of exposure rate. A high concentration is always a problem because there is so much contaminant in the water that ventilatory activity is no longer a limiting factor.

In brief, during routine inspection, sustained hyperventilatory activity does not tell the observer much about the origin and/or intensity of a stressor (see Barreto and Volpato 2004) but, combined with other observations (such as location in the tanks or reactivity), it can be a sensitive indicator of fish welfare. Furthermore, correlations between increased ventilatory activity and other welfare indicators, such as blood lactate, glucose and haematocrit, are well established (e.g. White et al. 2008). Nevertheless, one should interpret an increase in ventilatory activity with caution as an increase may be linked with positive experiences and therefore not necessarily with poor welfare.

Importantly, considering the profound consequences ventilatory activity has on the homoeostasis of a fish's internal environment such as control of cellular oxygen status, oxidative stress and internal pH regulation, high ventilatory activity can be both a passive consequence of disturbances and a driver promoting various physiological dysfunctions and contaminations. The basic premise that 'ventilatory activity should be low to moderate' should help (1) to minimise the effects and consequences of disease and injury and (2) improve fish welfare.

Aggression

Fish in a natural environment can live alone or in groups (Brännäs et al. 2001). Group living can take the form of short-term aggregations or be more long term and highly structured (Pitcher and Parrish 1993; Grant 1997; Hoare and Krause 2003). Group living confers advantages in foraging, resource defence and predator detection (Pitcher et al. 1982; Ryer and Olla 1991, 1992; Pitcher and Parrish 1993), and shoaling is one of the most common social fish behaviours (Pitcher and Parrish 1993; Parrish et al. 2002; Hoare and Krause 2003). However, when resources or the number or territories are limited, group living becomes a disadvantage as competition for the resource increases (Pitcher and Parrish 1993). Limiting resources may lead to the emergence of competitive behaviours, which may take the form of dominance rank-based hierarchies (Jobling 1983; Jobling and Koskela 1996). This kind of hierarchy is generally established after agonistic encounters between two individuals, and the rank within the hierarchy depends on their ability to fight (Huntingford and Turner 1987). In this review, agonistic behaviours refer to 'all the behavioural units of threat that express the dominance between at least two social partners as well as the submission in response to it'. Agonistic behaviours include attacks, bites or bite attempts, threatening displays and behaviours related to submission such as flight or immobility.

Dominance hierarchies and the associated agonistic behaviours are dependent on several biotic and abiotic factors such as an individual's 'boldness' (see above), size or sex (Ejike and Schreck 1980), the spatial and temporal distribution of food (Bégout-Anras and Lagardère 2004), underfeeding (Symons 1968), feeding frequency (Noble et al. 2007a) and feed delivery rate (Hakoyama and Iguchi 1997), with the latter factors being particularly relevant for aquaculture. The effects of stocking density upon aggression are still unclear, as some studies suggest that high density can reinforce aggressiveness (e.g. Keeley 2000), whereas others suggest that high density reduces aggression (e.g. Almazán-Rueda 2004; van de Nieuwegiessen et al. 2008, 2009).

A subordinate position within a group or social hierarchy may be a stressor (Schreck 1981). A subordinate individual can be subject to social stress resulting from attacks and repeated threats of attack from more dominant individuals for access to resources (food, sexual partners and territory). Social stress leads to marked behavioural and physiological changes in subordinates, who often show a general behavioural inhibition of food intake, aggressiveness, locomotory activity, changes in skin colorations and higher levels of plasma cortisol (Denight and Ward 1982; O'Connor et al. 1999; Winberg and Nilsson 1993a, b; Øverli et al. 1998). It is, however, notable that cortisol elevation can also occur in dominant fish within minutes of the end of aggressive encounters (Øverli et al. 1999).

Not only is the social state of an individual important for welfare but also the social stability of a group. Fox et al. (1997), for example, showed that cortisol levels are dependent on both individual dominance and social stability in the African cichlid fish *Haplochromis burtoni*.

Proxies of aggression can be used as an operational welfare indicator by counting the number of skin lesions. Almazán-Rueda et al. (2004) and Cañon Jones et al. (2010) showed a significant positive correlation between the number of scars and/or wounds and the number of aggressive acts measured by direct observation. Almazán-Rueda et al. 2004 also validated the use of skin lesion frequency to assess the welfare of African catfish in two studies (Almazán-Rueda et al. 2004; 2005) and showed that feeding methods, photoperiod and light intensity affect skin lesions frequency as well as other welfare indicators (such as swimming activity, growth, plasma cortisol and free fatty acids). Similarly, Martins et al. (2006b) used skin lesions as an indicator of aggression to assess the welfare of African catfish subjected to a simulated grading. Considering the time-consuming characteristics of direct or video-recording observations, counting skin lesions may be used as an easy and fast indicator of welfare, when it can be combined with other handling procedures. Although skin lesions serve as an indicator of poor welfare, aggression per se is not necessarily an indicator of poor welfare, unless coupled with lesions and fin damage. A consequence of this is that adjusting the aquaculture system to reduce aggression does not simply mean that one affects the aggressors' behaviour, but depending on the species, it may allow a certain level of aggression to occur while providing the target individuals with means to protect themselves.

The measurement of the nonapeptides vasotocin (AVT) and isotocin (IT) has also been suggested as an indirect way to measure aggression. AVT and IT are synthesised in hypothalamic neurones, from where they are transported to the neurohypophysis for storage and subsequent release. In fish, arginine vasotocin and isotocin have several functions, including being related to stressful conditions (Kulczykowska et al. 2009; Kleszczyńska et al. 2006; Mancera et al. 2008). Changes in hypothalamic, pituitary and plasma AVT/IT concentrations are found in many fish species subjected to different types of stress, e.g. confinement, disturbance, high density, food deprivation or rapid osmotic challenge, which are accompanied by aggressive acts. The AVT and IT response depends on both stress stimulus type and the level of aggression. In many teleost species, AVT has been shown to affect complex social behaviours such as courtship and territoriality (Semsar et al. 2001; Salek et al. 2002; Kleszczyńska et al. 2008), and the release of both AVT and IT is associated with different levels of sociality (Thompson and Watson 2004). Neuromodulatory actions of AVT and IT probably influence basic emotional regulatory mechanisms related to approach/withdrawal behaviours and aggression in social shoaling species. In Mozambique tilapia *Oreochromis mossambicus*, neurohormonal changes are different in dominant and subordinate individuals in social groups (Almeida et al. 2009). These authors suggest that the social stress activates the AVT system in the forebrain of subordinate males in this species, yet the pituitary and olfactory bulbs in both subordinate and dominant males are the areas associated with the highest levels of both neuropeptides. The outcome of social interactions influences the expression of subsequent social behaviour by changing the internal state of the individual affecting growth rates, breeding status and potentially affecting the welfare status (social stress) of the individuals. The changes in absolute levels of AVT and IT or the balance between them, associated with aggression, may reflect variations in internal state of the individuals. Taking into account a link between level of aggression and AVT/IT response, AVT and IT are proposed as specific neurohormonal measures of level of aggression.

Individual swimming behaviour

Swimming performance was defined by Beamish (1978) as an indicator of the progress of fish manoeuvring in water and is therefore an indicator of the ability of the fish to feed, escape predation and maintain position in a current (Beaumont et al. 1996). Some fish are strictly dependent on their swimming capacity in order to undertake migrations and achieve reproductive success (Plaut 2001; Wolter and Arlinghaus 2004).

Different swimming styles and activities have been classified by Beamish (1978) who distinguished (1) sustained swimming, where a fish is able to maintain a certain speed for long periods (over 200 min), (2) prolonged swimming that is characterised by a shorter duration (between 20 s and 200 min) and ends with fatigue and (3) burst swimming that can be maintained only for very short periods (under 20 s). Burst swimming can also incorporate (a) elevated fast-start, high-energy swimming where fish rapidly accelerate from a resting position or a baseline swimming speed and (b) swimming manoeuvres that involve increased complexity and turning angles (e.g. Domenici and Blake 1997). Every swimming activity can be related to a particular swimming behaviour. Indeed, sustained swimming is characterised by low-frequency tail beats and a slow velocity (ca. one-third of maximum speed), while during prolonged swimming, fish can almost reach their maximum velocity (Lembo et al. 2007). Several different tests have historically been used to quantitatively evaluate the swimming activity of fish (Beamish 1978). The ones most widely applied to assess the physiological condition of fish are critical swimming speed, U_{crit} (Wolter and Arlinghaus 2004), recovery tests (Farrell et al. 1998; Jain et al. 1998; Wagner et al. 2004) and muscular activity measured via electromyogram (EMG) (Cooke et al. 2000; McFarlane et al. 2004; Chandroo et al. 2005; Lembo et al. 2008) or remotely measured by physiological telemetry sensors. EMGs give a quantitative estimate of both swimming speed and metabolic costs of activity during swimming (mostly due to oxygen consumption). Thus, EMGs are a sensitive estimator of the metabolic scope for activity (SFA) (Fry 1971) and therefore, uniquely among these indicators, can be used as an operational, on-farm, welfare indicator by quantitatively measuring swimming activity of fish within cages or tanks.

Changes in swimming behaviour could reflect how a fish is sensing and responding to its environment. However, whether such changes can be used as an indicator of poor or good welfare are species and context dependent (Herbert and Steffensen 2005).

A number of common aquacultural water quality parameters can have an effect upon swimming behaviour. For example, reduced dissolved oxygen levels (hypoxia) can reduce the swimming speeds and activity of a number of species including white sturgeon *Acipenser transmontanus* (Crocker and Cech 1997) and Atlantic cod (Schurmann and Steffensen 1994; Herbert and Steffensen 2005). This reduced activity may enable fish to survive prolonged and widespread exposure to hypoxic conditions (Crocker and Cech 1997). In contrast, Tang and Boisclair (1995) have reported that the brook charr *Salvelinus fontinalis* increase swimming speeds in response to hypoxic conditions, and this is suggested to be an escape response from an area of potentially localised hypoxia. So, any deviations from normal, be it reduced or elevated, swimming speeds can be used as a potential indicator of hypoxia, although other water quality factors can have an effect upon swimming speeds (see below).

Hyperoxic conditions can also affect the swimming behaviour of fish. Atlantic salmon exhibit reduced swimming speeds (measured by reduced tail beat frequency) when held in superoxygenated water for prolonged periods (Espmark and Baeverfjord 2009). Other water quality parameters, in addition to oxygen levels, have been shown to affect swimming behaviour. These include low water pH levels (Day and Butler 1996), increased ammonia levels (Shingles et al. 2001), carbon dioxide exposure (Conte 2004) and trace element exposure (Hopkins et al. 2003).

A number of feed management factors can have an impact upon swimming behaviour, but their effect upon welfare can be rather unclear as elevated swimming speeds can again be indicative of both reduced and improved fish welfare, depending upon the fish species and rearing conditions. For example, elevated swimming speeds during feeding can indicate underfeeding in a number of aquacultural species such as gilthead sea bream (Andrew et al. 2004), turbot, *Scophthalmus maximus* (Huse and Skiftesvik 1985) and Atlantic cod (Björnsson 1993). However, increased swimming speeds during feeding may also indicate an increased motivation to feed and may simply be part of an individual's foraging strategy. For example, Atlantic halibut, *Hippoglossus hippoglossus*, which may be classified as a sit and wait forager, responds to feed by increasing their swimming activity (Kristiansen and Fernö 2007). This potential dichotomy of how a single behaviour can be interpreted as either poor or good welfare is a good example of how farmers and researchers should develop species-specific welfare indicators based upon the feeding ecology of the species under investigation.

Feeding regime can also affect swimming speeds and manoeuvre complexity irrespective of whether fish are fed to satiation or not. For example, cage-held Atlantic salmon parr and smolts exhibited increased swimming speeds and performed sharper turns during feeding when they received a uniform, temporally predictable ration to perceived daily satiation. However, fish fed to daily satiation using demand feeding technology did not significantly change their swimming speeds and performed moderate turns during feeding (Noble et al. 2007b, c). Fish can also exhibit increasingly complex swimming manoeuvres during feeding in response to underfeeding (Andrew et al. 2004). However, the same study also found that manoeuvre complexity increases with a high feed delivery rate irrespective of ration size, and the authors suggested that this behaviour allows the fish to remain within the feed delivery area (Andrew et al. 2004). Therefore, the same behaviour within the same species can be indicative of both poor welfare (underfeeding) or may simply be a foraging strategy to maximise an individual's feed intake. Further, a high feed delivery rate may reduce an individual's swimming speeds during feeding, as fish switch from swimming to processing and handling prey as feed encounter rate increases (e.g. Priyadarshana et al. 2006). In addition, increases in long-term EMG profiles (obtained via an EMG transmitter calibrated using a hard-wired EMG during U_{crit} tests) were also observed as consequences of the administration of three diet formulations (organic vs. conventional) (Carbonara 2009) and in experiments evaluating the consequences of starvation (McFarlane et al. 2004).

Behaviour modifications in response to pathogens, parasites and pollutants have been well documented by the industry and in the scientific literature. These include increased activity, such as flashing on the water surface, as well as decreased activity and lethargy (Warren 1982). Altered host behaviour as a consequence of infection may be an important criterion for identifying infected fish that are experiencing reduced welfare. Disease is detrimental to swimming performance in sockeye salmon, *Oncorhynchus nerka* (Tierney and Farrell 2004), and bull trout, *Salvelinus confluentus* (Jones and Moffitt 2004). An increased parasite load can also reduce swimming performance in Atlantic salmon (Wagner et al. 2003). Mycobacterium-infected delta smelt, *Hypomesus transpacificus*, also showed a 20% reduction in swimming performance during U_{crit} tests (Swanson et al. 2002). However, a wide range of parasites only impact significantly upon the host swimming behaviour patterns when they reach a certain developmental stage or infection intensity threshold; thus, changes in individual swimming behaviour can be correlated with severe infections and could potentially indicate poor welfare (Barber 2007). In addition, injuries such as lordosis (Basaran et al. 2007) can also be detrimental to swimming performance. Therefore, reduced swimming speeds or occasions where fish cease schooling and become immobile or inactive can also be an indicator of fish being sick, a disease outbreak, increased parasite load or an indicator of injury in farmed fish.

Fish can also increase their swimming speeds in response to increased stocking density, e.g. rainbow trout (Cooke et al. 2000) and Atlantic halibut (Kristiansen et al. 2004). Similarly, EMG profiles showed that sea bass reared at higher stocking density used on average 25% more energy than at lower stocking density (Scolamacchia 2009). Other common aquacultural procedures, such as handling, chasing, air exposure and changes in light conditions, can also lead to elevated swimming speeds in Atlantic salmon and sea bass (FASTFISH 2009). Further, EMG monitoring was demonstrated to be an effective indicator of fish welfare for investigating energetic responses and stress levels related to transportation in rainbow trout (Chandroo et al. 2005). In fact, a significant posttransportation swimming activity increment was recorded. Furthermore, after a recovery period of 48 h, the oxygen consumption level continued to be significantly elevated, and the swimming performance (measured as critical speed and endurance) of the transported fish was still impaired. Poli et al. (2005) also reported increased swimming speeds during both crowding procedures and slaughter using CO_2 narcosis. Continuous lighting can increase swimming activity in African catfish (Almazán-Rueda 2004), while the type of rearing systems can have an impact upon swimming speeds. Recent research has shown that Atlantic salmon reared in submerged cages for both short- and long-term submergence periods increase their swimming speeds, which may be a strategy allowing them to cope with reduced buoyancy (Dempster et al. 2009; Korsøen et al. 2009).

In summary, farmers and researchers should have a thorough knowledge of fish swimming activity under normal rearing conditions, noting any short-term changes in swimming, such as reduced or elevated swimming speed or increased manoeuvre complexity, as an acute indicator of potentially detrimental welfare conditions. Fish farmers can monitor swimming activity using, for example, telemetry techniques associated with remote sensors (in cages or tanks) and cameras. Possible strategies to maintain normal swimming activity could be related to appropriate water qualities and stocking densities for each species, appropriate feeding regimes and satisfactory disease prevention.

Group swimming behaviour

Cultured fish are reared in large numbers at high densities, and it is impossible to assess the behaviour or welfare of all individuals. Assessments of group behaviour can be used as an operational on-farm welfare indicator and are what most fish farmers use daily to evaluate the hunger, stress level and health status of fish. Group swimming behaviour is defined as the spatial distribution and swimming activity of groups of fish held within an aquaculture production unit and covers shoal structure, polarisation, the horizontal and vertical distribution of the group and their swimming speed and direction. A shoal is defined as a group of fish that remain together voluntary, while synchronised and polarised fish shoals are termed schools (Pitcher and Parrish 1993).

The motivational state of an individual or the tendency to perform certain behaviours will vary with its internal affective and physiological states, which again is a result of external and internal sensory and neuroendocrine signals and the subjective appraisal of these. The motivational state or welfare state of the fish will affect the 'traffic rules' used by the fish, which also should give observable changes in the group swimming behaviour and emergent properties of the whole group, which can be used as indicators of stress level or welfare. Model studies have shown that only one fish with different 'traffic rules' can affect the behaviour of the whole group (Romey 1996), indicating that this can be a sensitive indicator if we are able to correctly interpret the behaviour.

When held at high density, individual fish have to respond to the behaviour of others, and the group can adopt polarised schooling just by swimming in a way that minimises the risk of collisions (Føre et al. 2009). Polarity, group velocity and interindividual spacing are characteristics of fish shoals that strongly affect individual members (Pitcher and Parrish 1993, Parrish et al. 2002). These characteristics are group-level 'emergent properties': collective outcomes of behavioural interactions among members, not under direct control of any single member. Simulation models show that many of the different emergent structures can be explained by forces that act on each individual and are modulated in response to other individuals' behaviour. Examples of forces include the following: biomechanical drag, aggregative attraction and repulsion and arrayal velocity matching (Parrish et al. 2002).

Farmed species are chosen because of their ability to thrive at high densities. In fish tanks and cages, there will usually be areas that are more favourable than the others and restricting these volumes will result in a trade-off between packing density (number of fish per cubic metre, Juell and Fosseidengen 2004) and environmental conditions. Especially in sea cages, there can be large environmental gradients, with highly variable temperature, salinity, oxygen, current speed and light conditions (Juell 1995, Juell and Fosseidengen 2004, Johansson et al. 2006, 2007, Oppedal et al. 2007). Fish are attracted to the most optimal areas and avoid the unfavourable ones as a result of both attractions to a part of the rearing volume with preferred environmental conditions and avoidance/repulsion from areas with poor environmental conditions. Such conditions cause a trade-off between packing density and the benefits of being in the relatively best areas, and at high stocking densities, the less competitive fish will be forced to stay in suboptimal areas (Juell 1995, Johansson et al. 2006, Oppedal et al. 2007). The behaviour of the fish can also have an effect on environmental conditions, and the high local concentrations of fish can cause reduced water flow and increased oxygen consumption leading to areas with hypoxia. Frequent profiling measurements of these conditions are necessary to relate the behaviour to environmental conditions.

In most fish species, exposure to negative stimuli leads to rapid escape movements towards the bottom of the tanks or cages, and concentration of the group near the bottom can indicate relatively recent exposure to acute stressors (Stien et al. 2007, FASTFISH 2009; Bratland et al. 2010).

In situations of acute stress, the fish react initially with reflexes that transmit through the shoal by fish copying or matching the behaviours of their nearest neighbours (Føre et al. 2009). In salmon cages, it is possible to observe strange emergent structures like fast swimming in 'tornados' or 'hour glass behaviour' of streams of fish moving fast from one location to another, which is also seen in wild fish schools stressed by predators (Pitcher and Parrish 1993). This is probably caused by copying behaviour, confusion and cognitive demands on the individuals, which then surrender their control to that of the school. Emergent structures can also spread to unstressed fish since this behaviour itself signals danger and can cause stress. However, similar (but not equal) behaviour can also be a positive indicator of high-feeding motivation (in sea bass), so these indicators differ according to species, life stage and rearing method.

A common indicator of good health and coping ability is the anticipatory response of the fish to feeding and sounds or signals associated with feeding. In aquaculture cages and tanks, this can be shown as a high concentration of fish near the feeding area and increased schooling activity (Juell et al. 1994, Chen and Purser 2001). A good feed anticipatory response and feed intake can be signs of high feeding motivation and unstressed fish (Fernö et al. 1995, 2006; Folkedal 2010). Experiments with salmon subjected to classical conditioning showed that the anticipatory response after giving a conditioned signal was affected by acute stress, and the time for recovery to normal baseline behaviours was longer than the period of elevated cortisol levels (Stien et al. 2007, Bratland et al. 2010). However, increased swimming activity prior to and during a meal can also be a sign of underfeeding or a poorly designed feeding regime (e.g. Andrew et al. 2004; Noble et al. 2007b, c). Therefore, care must be taken to avoid underfeeding as the ensuing feeding frenzy could be interpreted as a positive indicator of welfare.

Stereotypic and abnormal behaviour

Mason (1991) described stereotypic behaviour as a behavioural pattern that is repetitive, invariant and with no obvious goal or function. This definition has been revised following several studies that suggest stereotypic behaviour has an adaptive goal as it allows, for example, an increase in opioid activity that provides individuals with a reward experience (e.g. Cronin et al. 1986).

Observations of stereotypic behaviour and other abnormal behaviours have been subjected to intense study in zoo animals (Mason and Rushen 2006) and terrestrial farm animals (Von Borell 1995), and it is repeatedly used as an indicator of poor welfare. The link between the appearance of stereotypic behaviour and poor welfare seems obvious when it results in physical injury or arises during suboptimal housing and rearing conditions. However, this link is not so clear cut as these repetitive behaviours may generate positive (reward) feelings as part of a coping strategy and therefore be compatible with good welfare (Dawkins 1998). Despite the debate on stereotypic behaviour and feelings, it is generally accepted that the development of stereotypes arises as a response to frustration, discomfort or a conflict of motivation (Würbel and Stauffacher 1997).

There is very little evidence of stereotypic behaviour in farmed fish. Almazán-Rueda (2004) distinguished two stereotypic patterns in African catfish, which swim continuously and compulsively in fixed circular or triangular patterns of 10–50 s or 60–240 s. Both Almazán-Rueda (2004) and van de Nieuwegiessen et al. (2008) demonstrated that African catfish stereotypic behaviour can be used as an indicator of reduced welfare or chronic stress (related to stocking density). Kristiansen et al. (2004) also reported stereotypic behaviour in Atlantic halibut, described as loops of vertical swimming behaviour. These authors showed an increase in stereotypic behaviour when halibut was kept at high stocking densities. Atlantic halibut were shown to develop stereotypic behaviour in shallow tanks and also when fed floating pellets instead of sinking pellets (Kristiansen et al. 2004, Kristiansen and Fernö 2007), suggesting that the rearing system and feeding method may act as chronic stressor in flatfish.

Other behaviours described as abnormal or a kind of stereotypic behaviour have been identified in Atlantic salmon (e.g. types of circular shoaling in cages Kristiansen et al. 2004) and Mozambique tilapia (e.g. vacuum pit digging when housed in the absence of substrate, Galhardo et al. 2008).

Despite the scarcity of literature, the appearance of stereotypes has been linked with aquaculture procedures and stressors and therefore can be an indicator of poor welfare in farmed fish.

Mental states and possible associated welfare indicators

The data reviewed thus far show that the majority of studies on fish welfare have been focused on identifying behavioural indicators of poor welfare (summarised in Table 1). Very little has been done both on identifying indicators of good welfare and on developing strategies to improve the welfare of farmed fish. The possibility that fish are sentient creatures and capable of affective states or feelings will contribute not only to strategies for avoiding poor welfare but also for promoting good welfare.

Feelings are evolved emergent properties arising from the central nervous system, which interfere with memory and learning processes, providing higher flexibility and better responses to dangers, thus increasing fitness (Broom 1998; Dawkins 1998). In other words, the negative and positive valence of stimuli interpreted in a conscious domain may play an important role in the mechanisms of getting what animals need and avoiding what they do not need. The emergence of consciousness in fish has been a matter of intense scientific debate (e.g. Rose 2002; Chandroo et al. 2004a, b). Some authors, e.g. Rose (2002, 2007) and Cabanac et al. (2009), argue that

Table 1 Examples of behavioural changes in responses to aquaculture procedures or stressors that could be used as species-specific operational welfare indicators in farmed fish

Welfare level	Behaviour	Aquaculture procedures/ stressors	Species	Operational welfare indicator	References
Good	Exploratory activity	Stocking density	Rainbow trout *Oncorhynchus mykiss*	Low stocking densities: ↑ use of self-feeders	Boujard et al. (2002)
	Food-anticipatory activity	Feeding method	Gilthead Sea Bream *Sparus aurata*	Scheduled feeding (vs. random): ↑ food anticipatory activity	Sánchez et al. (2009)
Poor	Aggression	Feeding method	African catfish *Clarias gariepinus*	Hand feeding (vs. self-feeding): ↑ % of fish bitten	Almazán-Rueda et al. (2004)
			Atlantic salmon *Salmo salar*	Fixed ration (vs. demand feeding): ↑ aggressive acts	Noble et al. (2007b)
			Atlantic salmon *Salmo salar*	Underfeeding (vs. satiation): ↑ aggressive acts	Cañon Jones et al. (2010)
		Photoperiod and light intensity	African catfish *Clarias gariepinus*	Long periods of light and high light intensity: ↑ skin lesions	Almazán-Rueda et al. (2004, 2005)
		Stocking density	African catfish *Clarias gariepinus*	Low stocking densities: ↑ skin lesions	Almazán-Rueda et al. (2004) and van de Nieuwegiessen et al. (2008, 2009)
			Rainbow trout *Oncorhynchus mykiss*	High stocking densities: ↑ aggressive behaviour	Ellis et al. (2002)
			White sea bream *Diplodus sargus*	High stocking densities: ↑ social interactions	Karakatsouli et al. (2007)
		Grading	African catfish *Clarias gariepinus*	Homogenous groups of small fish: ↑ % of fish bitten	Martins et al. (2006b)
	Foraging behaviour	Feeding method	Atlantic halibut *Hippoglossus hippoglossus*	Floating pellets (vs. sinking): ↓ number of pellets eaten	Kristiansen and Fernö (2007)
		Water exchange rate in RAS	Nile tilapia *Oreochromis niloticus*	High water exchange rates: ↓ feeding latency in small size classes	Martins et al. (2009)
		Grading	African catfish *Clarias gariepinus*	Homogenous groups of small fish: ↓ total feeding time and feeding rate	Martins et al. (2005)
		Fasting periods	Sea bass *Dicentrarchus labrax*	↑ Feeding rate and daily feeding times after feeding is resumed	Rubio et al. (2010)
		Cleaning protocols	Sea bass *Dicentrarchus labrax*	↓ Self-feeding	Rubio et al. (2010)
	Swimming activity	Hypoxia	Atlantic cod *Gadus morhua*	↓ Swimming speed	Schurmann and Steffensen (1994) and Herbert and Steffensen (2005)
			White sturgeon *Acipenser transmontanus*	↓ Swimming speed	Crocker and Cech (1997)
			Brook charr *Salvelinus fontinalis*	↑ Swimming speed	Tang and Boisclair (1995)
			Sockeye salmon *Oncorhynchus nerka*	↓ Recovery test	Farrell et al. (1998)
			Nile tilapia *Oreochromis niloticus*	↑ and ↓ Swimming activity of schools	Xu et al. (2006)
		Hyperoxia	Atlantic salmon *Salmo salar*	↓ Swimming speed	Espmark and Baeverfjord (2009)

Table 1 continued

Welfare level	Behaviour	Aquaculture procedures/stressors	Species	Operational welfare indicator	References
		Pollutants	Sockeye salmon *O. nerka*	Dehydroabietic acid : ↓ recovery test	Jain et al. (1998)
			Brown trout *Salmo trutta*	Cu: ↓ recovery test	Beaumont et al. (1996)
			Juvenile lake chubsuckers *Erimyzon sucetta*	Ash: ↓ U_{crit}	Hopkins et al. (2003)
		Infections	Delta smelt *Hypomesus transpacificus*	↓ U_{crit}	Swanson et al. (2002)
		Parasites	Atlantic salmon *Salmo salar*	↓ U_{crit}	Wagner et al. (2003)
		Temperature	Largemouth bass *Micropterus salmoides*	Low temperature: ↓ U_{crit}	Hasler et al. (2009)
		Transportation	Rainbow trout *O. mykiss*	↑ EMG	Chandroo et al. (2005)
		Underfeeding	Atlantic cod *Gadus morhua*	↑ Swimming speed	Björnsson (1993)
			Rainbow trout *O. mykiss*	↑ EMG	McFarlane et al. (2004)
			Gilthead sea bream *Sparus aurata*	↑ Swimming speed; ↑ school manoeuver complexity	Andrew et al. (2004)
			Turbot *Scophthalmus maximus*	↑ Swimming speed	Huse and Skiftesvik (1985)
		Short- and long-term cage submergence	Atlantic salmon *Salmo salar*	↑ Swimming speed	Dempster et al. (2009) and Korsøen et al. (2009)
		Photoperiod and light intensity	African catfish *Clarias gariepinus*	Long periods of light and high light intensity: ↑ swimming activity	Almazán-Rueda et al. (2004, 2005)
		Photoperiod	African catfish *Clarias gariepinus*	Long periods of light: ↑ swimming activity	Almazán-Rueda et al. (2005)
		Stocking density	African catfish *Clarias gariepinus*	High stocking density: ↑ swimming activity	van de Nieuwegiessen et al. (2008, 2009)
			European sea bass *Dicentrarchus labrax*	High stocking density: ↓ swimming speed	Santos et al. (2010)
			European sea bass *Dicentrarchus labrax*	High stocking density: ↑ EMG	Carbonara et al. (2008)
			Atlantic halibut *Hippoglossus hippoglossus*	High stocking density: ↑ swimming activity	Kristiansen et al. (2004)
		Environmental gradients	Atlantic salmon *Salmo salar*	Changes in Space use and patchiness	Juell and Fosseidengen (2004), Johansson et al. (2006), and Oppedal et al. (2007, 2011)
		Feeding method	African catfish *Clarias gariepinus*	Hand feeding (vs. self-feeding): ↑ swimming activity	Almazán-Rueda et al. (2004)
			Atlantic salmon *Salmo salar*	Scheduled (vs. on demand): ↑ swimming speed and turning angles	Noble et al. (2007b, c)

Table 1 continued

Welfare level	Behaviour	Aquaculture procedures/stressors	Species	Operational welfare indicator	References
	Ventilatory activity	Stocking density	African catfish *Clarias gariepinus*	High stocking density: $\uparrow$ air breathing	van de Nieuwegiessen et al. (2008, 2009)
		Hypoxia	Nile tilapia *Oreochromis niloticus*	$\uparrow$ Ventilator frequency	Xu et al. (2006)
		Confinement	Nile tilapia *Oreochromis niloticus*	$\uparrow$ Ventilator frequency	Barreto and Volpato (2004)
		Handling	Rainbow trout *Oncorhynchus mykiss*	$\uparrow$ Ventilator frequency	Brydges et al. (2009)
	Stereotypic behaviour and abnormal behaviours	Stocking density	African catfish *Clarias gariepinus*	High stocking density: $\uparrow$ escape attempts	Van de Nieuwegiessen et al. (2008)
			Atlantic halibut *Hippoglossus hippoglossus*	High stocking density: $\uparrow$ loops of vertical swimming; $\uparrow$ surface swimming	Kristiansen et al. (2004)
		Feed characteristics	Atlantic halibut *Hippoglossus hippoglossus*	Floating pellets (vs. sinking): $\uparrow$ loops of vertical swimming	Kristiansen and Fernö (2007)

this is not possible because their behaviour is simple and reflexive and they lack a neocortex. Yet, a growing body of evidence related to cognitive (Braithwaite 2006), neuroanatomic (Portavella and Vargas 2005; Rodriguez et al. 2006) and emotional (Sneddon 2007; Yue et al. 2008) aspects of fish behaviour provides strong support for sentience (the ability to feel) in fish. Good welfare should therefore be related to behaviours associated with attraction and positive anticipation and reward behaviour, while bad welfare should be associated with frustration, aversion, fear, pain and sickness behaviours.

The measurement of exploratory behaviour, food-related anticipatory activity (reward anticipation), preferences and reward-related operant behaviour has been suggested as indicators of positive fish emotions and welfare (Galhardo 2010; Galhardo et al. 2011a). Exploratory behaviour has been suggested to be self-rewarding, increasing the predictability of, and control over, the environment (Schulze Westerath et al. 2009). These characteristics have led exploratory behaviour to be proposed as a candidate for assessing positive welfare in terrestrial farm animals (Schulze Westerath et al. 2009). In farmed fish, very little has been done on assessing positive welfare. Indirect suggestions come from the effect of stressors on the use of self-feeders. Boujard et al. (2002), for example, showed that increased stocking densities of rainbow trout reduced the number of operant tasks (use of self-feeders) executed to obtain a fixed food reward. Other examples, although not in an aquaculture context, come from exposing rainbow trout to noxious stimuli, which has been shown to affect an individual's exploratory behaviour (Sneddon et al. 2003). In addition, Galhardo (2010) showed a reduction in neophobia when Mozambique tilapia are kept under stable social conditions, further supporting the use of exploration behaviour as a tool to assess (positive) welfare.

In certain species and under certain conditions, food-anticipatory activity can also be an indicator of good welfare (Kristiansen and Fernö 2007), as it may indicate a behavioural priority for foraging instead of being hidden or trying to escape. Increased swimming activity around feeding areas prior to feeding time is often observed in fish farms. This food-anticipatory activity may allow fish to predict the exposure to appetitive stimuli and have some awareness of control, which have been suggested to be important appraisal modulators of the fish responses to their environment (Galhardo et al. 2011b).

Preference tests that allow animals to display their needs (Dawkins 2004) are used to assess fish welfare. Galhardo et al. (2009) showed that the lack of substrate is likely to decrease the welfare of breeding males in Mozambique tilapia. This approach considers that good welfare may not be equitable in all individuals. Indeed, recent studies suggest that stress response variability does not only depend on the individual's situation but also its cognitive evaluation of the situation, i.e. the way the stressor is appraised (Galhardo et al. 2009). Such knowledge could contribute to practical applications in animal production, of enhancing positive experiences and long-term positive emotional states, and including positive welfare indicators in on-farm monitoring systems (Boissy et al. 2007).

Reward-related operant behaviours are less used in fish welfare studies. However, Galhardo et al. (2011a) has shown how Mozambique tilapia can push a door in order to access different resources, where the degree of motivation to obtain food or social contact was clearly higher than for access to a suitable substrate.

Promoting positive welfare includes, for example, providing opportunities for exploration, which requires complex environments. Environmental enrichment aims to increase the environmental complexity in its physical, temporal and/or social dimensions. Physical complexity includes structural, visual, auditory, olfactory and gustatory stimuli, whereas temporal complexity refers to the degree to which physical stimuli changes (Carlstead and Shepherdson 2000). These levels of complexity are often missing in an aquaculture context. Indeed, when compared with the environment occupied by their wild conspecifics, intensively farmed fish are kept under highly standardised conditions, which often mean a complete lack of enrichment. Farmed fish are cultured for long periods of time in the same tanks of the same colour(s), the same shape and exposed to the same, potentially harmful, background noises. They are fed with the same diets, which have the same taste, same colours and often given at the same time of the day. In addition, appropriate social complexity is of paramount importance for some species, and often, the group size and composition are altered via grading or sorting procedures. If welfare-friendly housing systems are to be developed for farmed fish, we first need to understand the consequences of environmental enrichment on fish behaviour, physiology and performance. This information is scarce in farmed fish. Galhardo et al. (2008) showed that the use of tank substrate affects the reproductive behaviour and physiology, including behavioural needs (see below), in Mozambique tilapia males. Höglund et al. (2005) showed that providing hiding materials affects the fear response to alarm cues from conspecifics in crucian carp, while tank design and colour have also been shown to improve farmed fish welfare (Ross et al. 1995; Karakatsouli et al. 2007). The use of feeding systems or regimes that deliver a responsive ration in relation to appetite levels can also improve welfare by reducing the chances that fish are underfed (Suzuki et al. 2008), by reducing competition (Andrew et al. 2002; Noble et al. 2007b) and by reducing the frequency of injury and damage (Noble et al. 2008; Suzuki et al. 2008). Juell and Fosseidengen (2004) suggested the use of underwater lamps to avoid crowding and improve the welfare of caged Atlantic salmon. The efficacy of these and other strategies for routine use in aquaculture needs further verification, particularly the impact of environmental enrichment on maintenance and water quality.

The assessment of frustrated behavioural needs is related to indicators of poor welfare (Ashley 2007). Galhardo et al. (2009) showed that Mozambique tilapia perform vacuum pit building in the absence of appropriate substrate, which, under normal circumstances, males use to build their spawning pits. Behavioural integrity, in which animals should be allowed to perform their natural repertoire of behaviour to guarantee their welfare, has also been assessed as an indicator of poor welfare (Würbel 2009). Despite the recognition that an aquaculture environment differs significantly from natural habitats, with a questionable biological relevance of some of the fish behavioural repertoire under natural conditions, natural fish behaviour has generally been used as an indicator of fish welfare. Foraging and swimming behaviours are examples of natural behaviours that have been taken into consideration in aquaculture procedures. As mentioned above, flatfish are bottom feeders and show improved welfare when fed sinking pellets as compared to floating pellets (Kristiansen and Fernö 2007). The swimming behaviour of certain species, such as flatfish, has led to novel tank design, for example raceways to maximise the surface area, and allow flatfish to rest on the bottom, indeed, when denied access to a resting area flatfish develops stereotypic behaviour, while schooling salmonids, sea bream and sea bass need to swim constantly and are often cultured in circular tanks.

There is still room to introduce new techniques or strategies in order to improve the welfare of farmed fish. One of the most obvious approaches is to adapt the rearing systems not only to the physiological but also to the species-specific behavioural needs. At present, the production of the most relevant European farmed fish (i.e. Atlantic salmon, sea bream and sea bass) utilises technology that allows fairly good growth and survival rates, while the welfare level, including the mental dimension, of these species remains unclear. This is partly because there is still a lack of knowledge on the basic principles of behavioural, physiological and cognitive responses to stress under captivity. The basic question 'what is it like to be a fish under aquaculture conditions' remains to be answered. Thus far, the scientific community has made considerable efforts to understand *what* causes stress and *how* a stress response can be characterised in farmed fish. However, the future calls for answers to questions such as *when* is a potential stressor a real stressor, *for how long* a stressor remains a stressor, *why* is a stressor interpreted as such and *why* different individuals react differently to the same stressor?

An emergent area of fish research is the understanding of psychological mediators of stress. For example, predictability has been suggested as a tool to manipulate the way animals interpret stimuli or stressors (Wiepkema 1987; Greiveldinger et al. 2009; Veissier et al. 2009). This suggestion is supported by studies in rats that showed that the predictability of negative events (such as electric shocks) leads to a reduction in gastric ulcers as compared to rats that cannot predict the shock event (Guile 1987). Predictability of negative stimuli such as noise (Prior 2002) or introduction into an unfamiliar cage (Muir and Pfister 1986) has also been shown to result in reduced stress responses and better learning abilities in rats (Muir and Pfister 1986; Orsini et al. 2002; Prior 2002). Carlstead (1986) showed a positive effect of predictability of feeding on growth and aggressive behaviour in pigs, while the predictability of a sudden event was shown to lower subsequent emotional responses in sheep (Greiveldinger et al. 2007). These results encourage the use of predictability to improve farm animal welfare. Very little is known about the effects of predictability on fish. Galhardo et al. (2011b) showed that Mozambique tilapia subjected to predictable confinement exhibit lower cortisol levels than fish that could not predict such stimuli. The relieving effects of predictability have been suggested to rely on a sense of both safety during periods of nonsignalisation and control by allowing self-preparation for the future events (Orsini et al. 2002; Bassett and Buchanan-Smith 2007). However, the use of highly predictable environments may lead to a hypostimulation of the HPI axis as a consequence, resulting in a situation that may be considered analogous to boredom (Wemelsfelder 1993).

The incorporation of concepts such as stability through change, i.e. allostasis (Korte et al. 2007), coping strategies (Øverli et al. 2007), appraisal and cognition (Galhardo et al. 2009) in farmed fish welfare will certainly allow a more comprehensive picture of how farmed fish welfare should be assessed, utilising both positive and negative indicators.

Conclusions

Behaviour represents a reaction to the environment as fish perceive it and therefore is a key element of fish welfare. In a developing and increasingly competitive aquaculture industry, it is of paramount importance that both farmers and researchers use species-specific behavioural signals for an early assessment of poor fish welfare. Furthermore, when considering the potential dichotomy of how behaviour can be interpreted as either good or poor (e.g. individual swimming activity) depending on the fish species, it is essential to develop an a priori understanding of the species-specific biology before drawing any conclusions in relation to welfare. Whether the same behavioural welfare indicator can be consistently used, and means the same throughout the life cycle of a fish, also needs to be addressed by future research.

It is also worth noting that fish differ in their coping strategies and perception of external events. Therefore, while one individual may interpret a situation as being highly stressful, another may interpret it as mildly stressful or even not at all stressful. Such variation in individual responses may make an overall assessment of welfare difficult. However, with the ongoing development of underwater video technology and computer image processing, the possibility of sampling high number of individual fish at a low cost will fall increasingly within reach. Another alternative is to develop group-based welfare indicators, such as spatial distribution.

The variability in coping strategies between individuals also emphasises that not all individuals within a group will have a good welfare status. In fact, group-based indicators may hide poor welfare at an individual level. Providing the culture environment with some sort of modifications (e.g. providing escape routes, hiding places, etc.) may help synchronise the culture environment with the inherent individual variability in stress responses present in farmed fish.

The interpretation of the meaning of behavioural changes in terms of welfare will also demand the willingness to accept that these changes are influenced by other factors besides the stressor itself, including mental states. Accepting that fish can be subjected to psychological stress will contribute to the development of strategies (e.g. environmental enrichment or allowing fish to predict or anticipate changes in the environment) that contribute to good welfare practices under aquaculture conditions.

If the interpretation of welfare in commonly reared farmed fish has yet to be resolved, the welfare of new fish species for aquaculture is even more incipient. There are at present a number of new fish species being considered for fish farming, such as the bluefin tuna *Thunnus thynnus* and the sparid common dentex *Dentex dentex*. While much attention is being placed on larval/broodstock and grow-out technologies, little has been done on the needs of these species that may have an impact on their welfare. Using the same technologies and rearing systems already in use for other fish species may not be enough to ensure the welfare of these new fish for aquaculture.

Acknowledgments This publication benefits from participation of authors in COST action 867 'Welfare of fish in European aquaculture'. The authors would like to acknowledge Øyvind Øverli for commenting on the link between coping styles and welfare indicators. Pino Lembo and Pierluigi Carbonara are gratefully acknowledged for their comments on swimming activity and welfare. C. I. M. Martins acknowledge financial support by Fundação para a Ciência e Tecnologia, Portugal, through the grant SFRH/BPD/42015/2007.

References

Alanärä A, Brännäs E (1996) Dominance in demand-feeding behaviour in Arctic charr and rainbow trout: the effect of stocking density. J Fish Biol 48:242–254

Alanärä A, Winberg S, Brännäs E, Kiessling A, Höglund E, Elofsson U (1998) Feeding behaviour, brain serotonergic activity levels, and energy reserves of Arctic char (*Salvelinus alpinus*) within a dominance hierarchy. Can J Zool 76:212–220

Almazán-Rueda P (2004) Towards assessment of welfare in African catfish *Clarias gariepinus*: the first step. PhD thesis, Wageningen University, 160 pp

Almazán-Rueda P, Schrama JW, Verreth JAJ (2004) Behavioural responses under different feeding methods and light regimes of the African catfish (*Clarias gariepinus*) juveniles. Aquaculture 231:347–359

Almazán-Rueda P, van Helmond ATM, Verreth JAJ, Schrama JW (2005) Photoperiod affects growth, behaviour and stress variables in Clarias gariepinus. J Fish Biol 67: 1029–1039

Almeida OG, Kulczykowska E, Oliveira RF (2009) Changes in brain Arginine-vasotocin and Isotocin levels in dominants and subordinates males of cichlid fish. Society for neuroscience annual meeting, Chicago. Poster session 466.6

Andrew JE, Noble C, Kadri S, Jewell H, Huntingford FA (2002) The effect of demand feeding on swimming speed and feeding responses in Atlantic salmon *Salmo salar* L., gilthead sea bream *Sparus aurata* L. and European sea bass *Dicentrarchus labrax* L. in sea cages. Aquacult Res 33:501–507

Andrew JE, Holm J, Kadri S, Huntingford FA (2004) The effect of competition on the feeding efficiency and feed handling behaviour in gilthead sea bream (*Sparus aurata* L.) held in tanks. Aquaculture 232:317–331

Ashley PJ (2007) Fish welfare: current issues in aquaculture. Appl Anim Behav Sci 104:199–235

Barber I (2007) Parasites, behaviour and welfare in fish. Appl Anim Behav Sci 104:251–264

Barreto RE, Volpato GL (2004) Caution for using ventilatory frequency as an indicator of stress in fish. Behav Process 66:43–51

Barreto RE, Volpato GL (2006) Ventilatory frequency of Nile tilapia subjected to different stressors. J Exp Anim Sci 43:189–196

Barreto RE, Luchiari AC, Marcondes AL (2003) Ventilatory frequency indicates visual recognition of an allopatric predator in naïve Nile tilapia. Behav Process 60:235–239

Basaran F, Ozbilgin H, Ozbilgin YD (2007) Effect of lordosis on the swimming performance of juvenile sea bass (*Dicentrarchus labrax* L.). Aquacult Res 38:870–876

Bassett L, Buchanan-Smith HM (2007) Effects of predictability on the welfare of captive animals. Appl Anim Behav Sci 102:223–245

Beamish FWH (1978) Swimming capacity. In: Hoar WS, Randall DJ (eds) Fish physiology. Academic Press, New York, pp 101–187

Beaumont MW, Butler PJ, Taylor EW, Mair JL, Mohammed MS (1996) Reduction in swimming performance of brown trout following copper exposure in soft acidic water. In: MacKinlay D, Nelson J (eds) International congress on the biology of fish, San Francisco, pp 127-132

Bégout-Anras ML, Lagardère JP (2004) Domestication et comportement chez les poisons téléostéens. INRA Prod Anim 17:211–215

Björnsson B (1993) Swimming speed and swimming metabolism of Atlantic cod (*Gadus morhua*) in relation to available food: a laboratory study. Can J Fish Aquat Sci 50:2542–2551

Boissy A, Manteuffel G, Jensen MB, Moe RO, Spruijt B, Keeling LJ, Winckler C, Forkman B, Dimitrov I, Langbein J, Bakken M, Veissier I, Aubert A (2007) Assessment of positive emotions in animals to improve their welfare. Physiol Behav 92:375–397

Boujard T, Labbé L, Aupérin B (2002) Feeding behaviour, energy expenditure and growth of rainbow trout in relation to stocking density and food accessibility. Aquacult Res 33:1233–1242

Braithwaite VA (2006) Cognitive ability in fish. Fish Physiol 24:1–37

Brännäs E, Alanärä A, Magnhagen C (2001) The social behaviour of fish. In: Keeling LJ, Gonyou HW (eds) Social behaviour of farm animals. CABI Publishing, Wallingford, pp 275–303

Bratland S, Stien LH, Braithwaite V, Juell J-E, Folkedal O, Nilsson J, Oppedal F, Fosseidengen JE, Kristiansen TS (2010) From fear to anticipation. Using aversive light stimuli to train reward conditioning in Atlantic salmon (*Salmo salar* L.). Aquacult Int 18:991–1001

Broom DM (1998) Welfare, stress, and the evolution of feelings. In: Moller AP, Milinski M, Slater PJB (eds) Advances in the study of behavior. Academic Press, San Diego, pp 371–403

Brydges NM, Boulcott P, Ellis T, Braithwaite VA (2009) Quantifying stress responses induced by different handling methods in three species of fish. Appl Anim Behav Sci 116:295–301

Burleson ML, Milsom WK (2003) Comparative aspects of O_2 chemoreception: anatomy, physiology, and environmental adaptations. In: Lahiri S, Semenza GL, Prabhakar NR (eds) Oxygen sensing: responses and adaptation to hypoxia. Marcel Dekker, New York, pp 685–707

Byrne P, Ferguson HW, Lumsden JS, Ostland VE (1991) Blood-chemistry of bacterial gill disease in brook trout Salvelinus fontinalis. Dis Aquat Org 10:1–6

Cabanac M, Cabanac AJ, Parent A (2009) The emergence of consciousness in phylogeny. Behav Brain Res 198: 267–272

Cañon Jones HA, Hansen L, Noble C, Damsgård B, Broom DM, Pearce GP (2010) Social network analysis of behavioural interactions influencing fin damage development in Atlantic salmon (*Salmo salar*) during feed-restriction. Appl Anim Behav Sci 127:139–151

Carbonara P (2009) Un approccio metodologico integrato fra telemetria fisiologica e biomarker per la valutazione del benessere animale nella spigola (*Dicentrarchus labrax* L.). PhD dissertation, University of Siena, 165 pp

Carbonara P, Scolamacchia M, Spedicato MT, McKinley RS, Lembo G (2008) Muscle activity as a potential indicator of welfare in farmed European sea bass: results from an electromyographic (EMG) telemetry study. IFOAM conference on organic aquaculture, Cattolica, Italy, ICEA, pp 20–24

Carlstead K (1986) Predictability of feeding: its effect on agonistic behaviour and growth in grower pigs. Appl Anim Behav Sci 16:25–38

Carlstead K, Shepherdson D (2000) Alleviating stress in zoo animals. In: Moberg GP, Mench JA (eds) The biology of animal stress, basic principles and implications for animal welfare. University of Chicago Press, Chicago

Chandroo KP, Duncan IJH, Moccia RD (2004a) Can fish suffer?: perspectives on sentience, pain, fear and stress. Appl Anim Behav Sci 86:225–250

Chandroo KP, Yue S, Moccia RD (2004b) An evaluation of current perspectives on consciousness and pain in fishes. Fish Fish 5:281–295

Chandroo KP, Cooke SJ, Mckinley RS, Moccia RD (2005) Use of electromyogram telemetry to assess the behavioural and energetic responses of rainbow trout, *Oncorhyncus mykiss* (Walbaum) to transportation stress. Aquacult Res 36:1226–1238

Chen W-M, Purser GJ (2001) The effect of feeding regime on growth, locomotor activity pattern and the development of food anticipatory activity in greenback flounder. J Fish Biol 58:177–187

Conte FS (2004) Stress and the welfare of cultured fish. Appl Anim Behav Sci 86:205–223

Cooke SJ, Chandroo KP, Beddow TA, Moccia RD, McKinley RS (2000) Swimming activity and energetic expenditure of captive rainbow trout *Oncorhynchus mykiss* (Walbaum), estimated by electromyogram telemetry. Aquacult Res 31:495–506

Couillard CM, Courtenay SC, MacDonald RW (2008) Chemical-environment interactions affecting the risk of impacts on aquatic organisms: a review with a Canadian perspective—interactions affecting vulnerability. Environ Rev 16:19–44

Crocker CE, Cech JJ (1997) Effects of environmental hypoxia on oxygen consumption rate and swimming activity in juvenile white sturgeon, *Acipenser transmontanus*, in relation to temperature and life intervals. Environ Biol Fish 50:383–389

Cronin GM, Wiepkema PR, van Ree JM (1986) Andorphins implicated in stereotypies of tethered sows. Experientia 42:198–199

Damsgård B (2008) Husbandry techniques and fish quality. In: Lie Ø (ed) Improved farmed fish and quality. Woodhead Publisher, Cambridge, pp 494–516

Damsgård B, Dill LM (1998) Risk-taking behavior in weight-compensating coho salmon, Oncorhynchus kisutch. Behav Ecol 9:26–32

Danchin E, Giraldeau L, Cézilly F (2008) Behavioural ecology. Oxford University Press, Oxford

Dawkins MS (1990) From an animal's point of view: motivation, fitness, and animal welfare. Behav Brain Sci 13:1–61

Dawkins MS (1998) Evolution and animal welfare. Q Rev Biol 73:305–328

Dawkins MS (2004) Using behaviour to assess animal welfare. Anim Welfare 13:S3–S7

Day N, Butler PJ (1996) Environmental acidity and white muscle recruitment during swimming in the brown trout (*Salmo trutta*). J Exp Biol 199:1947–1959

Dempster T, Korsøen Ø, Oppedal F, Folkedal O, Juell JE (2009) Submergence of Atlantic salmon (*Salmo salar*) in sea-cages: a potential short-term solution to poor surface conditions. Aquaculture 288:254–263

Denight ML, Ward JA (1982) Relationship of chin spot size to dominance in the black-chinned mouthbrooding cichlid fish (*Sarotherodon melanotheron*). Anim Behav 30:1099–1104

Désiré L, Boissy A, Veissier I (2002) Emotions in farm animals: a new approach to animal welfare in applied ethology. Behav Process 60:165–180

Dill LM (1983) Adaptive flexibility in the foraging behavior of fishes (coho salmon *Oncorhynchus kisutch*). Can J Fish Aquat Sci 40:398–408

Domenici PD, Blake RW (1997) The kinematics and performance of fish fast-start swimming. J Exp Biol 200: 1165–1178

Dugatkin LA (2004) Principles of animal behavior, 1st edn. W.W. Norton Publishing, New York

Duncan IJH (2005) Science-based assessment of animal welfare: farm animals. Rev Sci Tech Off Int Epiz 24:483–492

Duncan IJH, Dawkins MS (1983) The problem of assessing 'well-being' and 'suffering' in farm animals. In: Smidt D (ed) Indicators relevant to farm animal welfare. Martinus Nijhoff, The Hague, pp 13–24

Eclancher B, Dejours P (1975) Contrôle de la respiration chez les poisons téléostéens: existence de chémorécepteurs physiologiquement analogues aux chémorécepteurs des vertébrés supérieurs. C R Acad Sci 280D:451–453

Ejike C, Schreck CB (1980) Stress and social hierarchy rank in coho salmon. Trans Am Fish Soc 109:423–426

Ellis T, North B, Scott AP, Bromage NR, Porter M, Gadd D (2002) The relationships between stocking density and welfare in farmed rainbow trout. J Fish Biol 61:493–531

Ernst B, Hoeger SJ, O'Brien E, Dietrich DR (2007) Physiological stress and pathology in European whitefish (*Coregonus lavaretus*) induced by subchronic exposure to environmentally relevant densities of *Planktothrix rubescens*. Aquat Toxicol 82:15–26

Espmark AM, Baeverfjord G (2009) Effects of hyperoxia on behavioural and physiological variables in farmed Atlantic salmon (*Salmo salar*) parr. Aquacult Int 17:341–353

Farrell AP, Gamper K, Birthwell IK (1998) Prolonged swimming, recovery and repeat swimming performance of mature sockeye salmon *Onchorhyncus nerka* exposed to moderate hypoxia and pentachlorophenol. J Exp Biol 201:2183–2193

FASTFISH (2009) On farm assessment of stress level in fish. Final report. Unpublished report. http://fastfish.imr.no/__data/page/7764/FASTFISH_Publishable_Final_Activity_Report_draf_1.pdf

Fernö A, Huse I, Juell J-E, Bjordal Å (1995) Vertical distribution of Atlantic salmon (Salmo salar L.) in net pens: trade-off between surface light avoidance and food attraction. Aquaculture 132:285–296

Fernö A, Huse G, Jakobsen PJ, Kristiansen TS (2006) The role of fish learning skills in fisheries and aquaculture. In: Brown C, Laland K, Krause J (eds) Fish cognition and behaviour. Blackwell Publishing, Oxford, pp 119–138

Folkedal O (2010) Anticipatory behaviour as indicator for stress and welfare in farmed Atlantic salmon at different life stages. PhD thesis, University of Bergen, Norway

Føre M, Dempster T, Alfredsen JA, Johansen V, Johansson D (2009) Modelling of Atlantic salmon (*Salmo salar* L.) behaviour in sea-cages: a Lagrangian approach. Aquaculture 288:196–204

Fox HE, White SA, Kao MHF, Fernald RD (1997) Stress and dominance in a social fish. J Neurosci 17:6463–6469

Fry FEJ (1971) The effect of environmental factors on the physiology of fish. In: Hoar WS, Randall DJ (eds) Fish physiology, vol VI. New York Academy Press, New York, pp 1–99

Galhardo L (2010) Teleost welfare: behavioural, cognitive and physiological aspects in *Oreochromis mossambicus*. PhD thesis, University of Porto, 215 pp

Galhardo L, Correia J, Oliveira RF (2008) The effect of substrate availability on behavioural and physiological indicators of welfare in the African cichlid (*Oreochromis mossambicus*). Anim Welfare 17:239–254

Galhardo L, Almeida O, Oliveira RF (2009) Preference for the presence of substrate in male cichlid fish: effects of social dominance and context. Appl Anim Behav Sci 120:224–230

Galhardo L, Almeida O, Oliveira RF (2011a) Measuring motivation in a cichlid fish: an adaptation of the push-door paradigm. Appl Anim Behav Sci 130:60–70

Galhardo L, Almeida O, Oliveira RF (2011b) The role of predictability in the stress response of a cichlid fish. Physiol Behav 102:367–372

Grant JWA (1997) Territoriality. In: Godin J-GJ (ed) Behavioural ecology of teleost fishes. Oxford University Press, Oxford, pp 81–103

Greiveldinger L, Veissier I, Boissy A (2007) Emotional experience in sheep: predictability of a sudden event lowers subsequent emotional response. Physiol Behav 92:675–683

Greiveldinger L, Veissier I, Boissy A (2009) Behavioural and physiological responses of lambs to controllable vs. uncontrollable aversive events. Psychoneuroendocrinology 34:805–814

Guile MN (1987) Differential gastric ulceration in rats receiving shocks on either fixed-time or variable-time schedules. Behav Neurosci 101:139–140

Hakoyama H, Iguchi K (1997) Why is competition more intense if food is supplied more slowly? Behav Ecol Sociobiol 40:159–168

Hansen LA, Dale T, Damsgård B, Uglem I, Aas K, Bjørn PA (2008) Escape-related behaviour of Atlantic cod, *Gadus morhua* L., in a simulated farm situation. Aquacult Res 40:26–34

Hasler CT, Suski CD, Hanson KC, Cooke SJ, Philipp DP, Tufts BL (2009) Effect of water temperature on laboratory swimming performance and natural activity levels of adult largemouth bass. Can J Zool 87:589–596

Heinrichs SC, Koob GF (2004) Corticotropin-releasing factor in brain: a role in activation, arousal, and affect regulation. J Pharmacol Exp Ther 311:427–440

Herbert NA, Steffensen JF (2005) The response of Atlantic cod, *Gadus morhua*, to progressive hypoxia: fish swimming speed and physiological stress. Mar Biol 147: 1403–1412

Hoare DJ, Krause J (2003) Social organisation, shoal structure and information transfer. Fish Fish 4:269–279

Höglund E, Weltzien FA, Schjolden J, Winberg S, Ursin H, Doving KB (2005) Avoidance behavior and brain monoamines in fish. Brain Res 1032:104–110

Hopkins WA, Snodgrass JW, Staub BP, Jackson BP, Congdon JD (2003) Altered swimming performance of a benthic fish (*Erimyzon sucetta*) exposed to contaminated sediments. Arch Environ Contam Toxicol 44:383–389

Huntingford FA, Turner AK (1987) Animal conflict. Chapman & Hall, London, pp 64–71

Huntingford F, Adams C, Braithwait VA, Kadri S, Pottinger TG, Sandøe P, Turnbull JF (2006) Current issues in fish welfare. J Fish Biol 68:332–372

Huse I, Skiftesvik AB (1985) Qualitative and quantitative behaviour studies in starving and feeding turbot (*Scophthalmus maximus* L.) larvae. ICES 14:108–115

Ishida Y, Yoshikawa H, Kobayashi H (1996) Electrophysiological responses of three chemosensory systems in the carp to pesticides. Physiol Behav 60:633–638

Israeli-Weinstein D, Kimmel E (1998) Behavioral response of carp (*Cyprinus carpio*) to ammonia stress. Aquaculture 165:81–93

Jain KE, Birtwell IK, Farrell AP (1998) Repeat swimming performance of mature sockeye salmon following a brief recovery period: a proposed measure of fish health and water quality. Can J Zool 76:1488–1496

Jobling M (1983) Effect of feeding frequency on food intake and growth of Arctic charr, *Salvelinus alpinus* (L). J Fish Biol 23:177–185

Jobling M, Koskela J (1996) Interindividual variations in feeding and growth in rainbow trout during restricted feeding and in a subsequent period of compensatory growth. J Fish Biol 49:658–667

Johansson D, Ruohonen K, Kiessling A, Oppedal F, Stiansen J-E, Kelly M, Juell J-E (2006) Effect of environmental factors on swimming depth preferences of Atlantic salmon (*Salmo salar* L.) and temporal and spatial variations in oxygen levels in sea cages at a fjord site. Aquaculture 254:594–605

Johansson D, Juell J-E, Oppedal F, Stiansen J-E, Ruohonen K (2007) The influence of the pycnocline and cage resistance on current flow, oxygen flux and swimming behaviour of Atlantic salmon (*Salmo salar* L.) in production cages. Aquaculture 265:271–287

Jones DT, Moffitt CM (2004) Swimming endurance of bull trout, lake trout, Arctic char, and rainbow trout following challenge with *Renibacterium salmoninarum*. J Aquat Anim Health 16:10–22

Juell J-E (1995) The behaviour of Atlantic salmon in relation to efficient cage rearing. Rev Fish Biol Fish 5:320–335

Juell J-E, Fosseidengen JE (2004) Use of artificial light to control swimming depth and fish density of Atlantic salmon (*Salmo salar*) in production cages. Aquaculture 233:269–282

Juell J-E, Fernö A, Furevik D, Huse I (1994) Influence of hunger level and food availability on the spatial distribution of Atlantic salmon, *Salmo salar* L., in sea cages. Aquacult Fish Manage 25:439–451

Karakatsouli N, Papoutsoglou SE, Manolessos G (2007) Combined effects of rearing density and tank colour on the growth and welfare of juvenile White Sea bream *Diplodus sargus* L. in a recirculating water system. Aquacult Res 38:1152–1160

Keeley ER (2000) An experimental analysis of territory size in juvenile steelhead trout. Anim Behav 59:477–490

Kittilsen S, Ellis T, Schjolden J, Braastad B, Øverli Ø (2009) Determining stress responsiveness in family groups of Atlantic salmon (*Salmo salar*) using non-invasive measures. Aquaculture 298:146–152

Kleszczyńska A, Vargas-Chacoff L, Gozdowska M, Kalamarz H, Martínez-Rodríguez G, Mancera JM, Kulczykowska E (2006) Arginine vasotocin, isotocin and melatonin responses following acclimation of gilthead sea bream (*Sparus aurata*) to different environmental salinities. Comp Biochem Physiol A 145:268–273

Kleszczyńska A, Skorkowski EF, Kulczykowska E (2008) Stocking density influences brain arginine vasotocin (AVT) and isotocin (IT) concentrations in the three-spined stickleback. In: Resource management. Natural, human and material resources for the sustainable development in aquaculture. European aquaculture society special publication no. 37, September 2008. Inland Fisheries Institute, Olsztyn, pp 321–322

Kolstad K, Grisdale-Helland B, Gjerde B (2004) Family differences in feed efficiency in Atlantic salmon (*Salmo salar*). Aquaculture 241:169–177

Koolhaas JM, Korte SM, De Boer SF, Van der Vegt BJ, Van Reenen CG, Hopster H, De Jong IC, Ruis MAW, Blokhuis HJ (1999) Coping styles in animals: current status in behavior and stress-physiology. Neurosci Biobehav Rev 23:925–935

Korsøen Ø, Dempster T, Fjelldal PG, Folkedal O, Kristiansen T, Oppedal F (2009) Long-term submergence of Atlantic salmon (*Salmo salar* L.) during winter affects behaviour, growth and condition. Aquaculture 296:373–381

Korte SM, Olivier B, Koolhaas JM (2007) A new animal welfare concept based on allostasis. Physiol Behav 92:422–428

Kristiansen TS, Fernö A (2007) Individual behaviour and growth of halibut (*Hippoglossus hippoglossus* L.) fed sinking and floating feed: evidence of different coping styles. Appl Anim Behav Sci 104:236–250

Kristiansen TS, Fernö A, Holm JC, Privitera L, Bakke S, Fosseidengen JE (2004) Swimming behaviour as an indicator of low growth rate and impaired welfare in

Atlantic halibut (*Hippoglossus hippoglossus* L.) reared at three stocking densities. Aquaculture 230:137–151

Kucukbay Z, Yazlak H, Sahin N, Tuzcun M, Cakmak MN, Gurdogan F, Juturu V, Sahin K (2006) Zinc picolinate supplementation decreases oxidative stress in rainbow trout (*Oncorhynchus mykiss*). Aquaculture 257:465–469

Kulczykowska E, Gozdowska M, Kalamarz H, Kleszczyńska A, Nietrzeba M, Martinez-Rodriguez G, Mancera JM (2009) Hypothalamic arginine vasotocin and isotocin are involved in stress response in fish. Comp Biochem Physiol A 154:S26

Lembo G, Carbonara P, Scolamacchia M, Spedicato MT, McKinley RS (2007) Use of muscle activity indices as a relative measure of well-being in cultured sea bass *Dicentrarchus labrax* (Linnaeus, 1758). Hydrobiologia 582:271–280

Lembo G, Carbonara P, Scolamacchia M, Spedicato MT, Bjørnsen JE, Holand B, McKinley RS (2008) Introduction of a new physiological sensor: an acoustic EMG transmitter. Fish Manage Ecol 15:333–338

MacKenzie S, Ribas L, Pilarczyk M, Capdevila DM, Kadri S, Huntingford FA (2009) Screening for coping style increases the power of gene expression studies. PLoS One 4:e5314

Mancera JM, Vargas-Chacoff L, García-López A, Kleszczyńska A, Kalamarz H, Martínez-Rodríguez G, Kulczykowska E (2008) High density and food-deprivation affect arginine vasotocin, isotocin and melatonin in gilthead sea bream (*Sparus auratus*). Comp Biochem Physiol A 149:92–97

Martins CIM, Aanyu M, Schrama JW, Verreth JAJ (2005) Size distribution in African catfish (*Clarias gariepinus*) affects feeding behaviour but not growth. Aquaculture 250:300–307

Martins CIM, Schrama JW, Verreth JAJ (2006a) The relationship between feed efficiency and stress response in African catfish *Clarias gariepinus*. Aquaculture 256: 588–595

Martins CIM, Schrama JW, Verreth JAJ (2006b) The effect of grading on the welfare of African catfish. Appl Anim Behav Sci 97:323–334

Martins CIM, Hillen B, Schrama JW, Verreth JAJ (2008) A brief note on the relationship between residual feed intake and aggression behaviour in juveniles of African catfish *Clarias gariepinus*. Appl Anim Behav Sci 111:408–413

Martins CIM, Ochola DM, Ende S, Eding E, Verreth JAJ (2009) Is growth retardation present in Nile tilapia *Oreochromis niloticus* cultured in low water exchange recirculating aquaculture systems? Aquaculture 298:43–50

Mason GJ (1991) Stereotypies: a critical review. Anim Behav 41:1015–1037

Mason G, Rushen J (2006) Stereotypies in captive animals: fundamentals and implications for welfare, 2nd edn. CAB International, Wallingford

Massabuau JC (2001) From a low blood- to low tissue-oxygenation strategy, an evolutionary theory. Respir Physiol 128:249–262

Massabuau JC (2003) Primitive, and protective, our cellular oxygenation status? Mech Ageing Dev 124:857–863

McFarlane WJ, Cubitt KF, Williams H, Rowsell D, Moccia R, Gosine R, McKinley RS (2004) Can feeding status and stress level be assessed by analyzing pattern of muscle activity in free swimming rainbow trout (*Oncorhynchus mykiss* Walbaum)? Aquaculture 239:467–484

Metcalfe NB, Huntingford FA, Thorpe JE (1987) The influence of predation risk on the feeding motivation and foraging strategy of juvenile Atlantic salmon. Anim Behav 35:901–911

Moreira PSA, Pulman KGT, Pottinger TG (2004) Extinction of a conditioned response in rainbow trout selected for high or low responsiveness to stress. Hormones Behav 46:450–457

Morgan KN, Tromborg CT (2007) Sources of stress in captivity. Appl Anim Behav Sci 102:262–302

Muir JL, Pfister HP (1986) Corticosterone and prolactin responses to predictable and unpredictable novely stress in rats. Physiol Behav 37:285–288

Neely WB (1979) Estimating rate constants for the uptake and clearance of chemicals by fish. Environ Sci Technol 13:1506–1510

Noble C, Mizusawa K, Suzuki K, Tabata M (2007a) The effect of differing self-feeding regimes on the growth, behaviour and fin damage of rainbow trout held in groups. Aquaculture 264:214–222

Noble C, Kadri S, Mitchell DF, Huntingford FA (2007b) Influence of feeding regime on intraspecific competition, fin damage and growth in 1 + Atlantic salmon parr (*Salmo salar* L.) held in freshwater production cages. Aquacult Res 38:1137–1143

Noble C, Kadri S, Mitchell DF, Huntingford FA (2007c) The effect of feed regime on the growth and behaviour of 1 + Atlantic salmon post-smolts (*Salmo salar* L.) in semi-commercial sea cages. Aquacult Res 38:1686–1691

Noble C, Kadri S, Mitchell DF, Huntingford FA (2008) Growth, production and fin damage in cage-held 0 + Atlantic salmon pre-smolts (*Salmo salar* L.) fed either (a) on-demand, or (b) to a fixed satiation-restriction regime: data from a commercial farm. Aquaculture 275:163–168

O'Connor KI, Metcalfe NB, Taylor AC (1999) Does darkening signal submission in territorial contests between juvenile Atlantic salmon, *Salmo salar*? Anim Behav 58:1269–1276

Oppedal F, Juell J-E, Johansson D (2007) Thermo-and photoregulatory swimming behaviour of caged Atlantic salmon: implications for photoperiod management and fish welfare. Aquaculture 265:70–81

Oppedal F, Dempster T, Stien LH (2011) Environmental drivers of Atlantic salmon behaviour in sea-cages: a review. Aquaculture 311:1–18

Orsini C, Ventura R, Lucchese F, Puglisi-Allegra S, Cabib S (2002) Predictable stress promotes place preference and low mesoaccumbens dopamine response. Physiol Behav 75:135–141

Osborne WA, Muntz E (1906) The action of carbon di-oxide on the respiration of the gold fish. Biochem J 1:377–382

Øverli Ø, Winberg S, Damsgård B, Jobling M (1998) Food intake and spontaneous swimming activity in Arctic charr (*Salvelinus alpinus* L.): role of brain serotonergic activity and social interaction. Can J Zool 76:1366–1370

Øverli Ø, Olsen RE, Løvik F, Ringø E (1999) Dominance hierarchies in Arctic charr, *Salvelinus alpinus* L.: differential cortisol profiles of dominant and subordinate individuals after handling stress. Aquacult Res 30:259–264

Øverli Ø, Sørensen C, Kiessling A, Pottinger TG, Gjøen HM (2006a) Selection for improved stress tolerance in rainbow trout (*Oncorhynchus mykiss*) leads to reduced feed waste. Aquaculture 261:776–781
Øverli Ø, Sørensen C, Nilsson G (2006b) Behavioral indicators of stress-coping style in rainbow trout: do males and females react differently to novelty? Physiol Behav 87:506–512
Øverli Ø, Sørensen C, Pulman KGT, Pottinger TG, Korzan W, Summers CH, Nilsson GE (2007) Evolutionary background for stress-coping styles: relationships between physiological, behavioral, and cognitive traits in non-mammalian vertebrates. Neurosci Biobehav Rev 31: 396–412
Pane EF, Haque A, Wood CA (2004) Mechanistic analysis of acute, Ni-induced respiratory toxicity in the rainbow trout (*Oncorhynchus mykiss*): an exclusively branchial phenomenon. Aquat Toxicol 69:11–24
Parrish JK, Viscido SV, Grunbaum D (2002) Self-organized fish schools: an examination of emergent properties. Biol Bull 202:296–305
Pierron F, Baudrimont M, Gonzalez P, Bourdineaud JP, Elie P, Massabuau JC (2007) Common pattern of gene expression in response to hypoxia or cadmium in the gills of the European glass eel (*Anguilla anguilla*). Environ Sci Technol 41:3005–3011
Pitcher TJ, Parrish JK (1993) Functions of shoaling behaviour in teleosts. In: Pitcher TJ (ed) Behaviour of teleost fish. Chapman & Hall, London, pp 363–439
Pitcher TJ, Magurran AE, Winfield IJ (1982) Fish in larger shoals find food faster. Behav Ecol Sociobiol 10:149–151
Plaut I (2001) Critical swimming performance: its ecological relevance. Comp Biochem Physiol 131A:41–50
Poli BM, Parisi G, Scappini F, Zampacavallo G (2005) Fish welfare and quality as affected by pre-slaughter and slaughter management. Aquacult Int 13:29–49
Portavella M, Vargas JP (2005) Emotional and spatial learning in goldfish is dependent on different telencephalic pallial systems. Eur J Neurosci 21:2800–2806
Pottinger TG (2006) Context dependent differences in growth of two rainbow trout (*Oncorhynchus mykiss*) lines selected for divergent stress responsiveness. Aquaculture 256:140–147
Pottinger TG, Carrick TR (2001a) ACTH does not mediate divergent stress responsiveness in rainbow trout. Comp Biochem Physiol A 129:399–404
Pottinger TG, Carrick TR (2001b) Stress responsiveness affects dominant-subordinate relationships in rainbow trout. Horm Behav 40:419–427
Prior H (2002) Effects of predictable and unpredictable intermittent noise on spatial learning in rats. Behav Brain Res 133:117–124
Priyadarshana T, Asaeda T, Manatunge J (2006) Hunger-induced foraging behavior of two cyprinid fish: *Pseudorasbora parva* and *Rasbora daniconius*. Hydrobiology 568:341–352
Randall DJ (1982) The control of respiration and circulation in fish during exercise and hypoxia. J Exp Biol 100:275–288
Ritola O, Livimgstone DR, Peters LD, Lindstrom-Seppa P (2002) Antioxidant processes are affected in juvenile rainbow trout (*Oncorhynchus mykiss*) exposed to ozone and oxygen surpersaturated water. Aquaculture 210:1–19
Rodriguez F, Broglio C, Duran E, Gomez A, Salas C (2006) Neural mechanisms of learning in teleost fish. In: Brown C, Laland K, Krause J (eds) Fish cognition and behaviour. Blackwell Publishing, Oxford, pp 243–277
Romey WL (1996) Individual differences make a difference in the trajectories of simulated schools. Ecol Model 92: 65–77
Rose JD (2002) The neurobehavioral nature of fishes and the question of awareness and pain. Rev Fish Sci 10:1–38
Rose JD (2007) Anthropomorphism and 'mental welfare' of fishes. Dis Aquat Organ 75:139–154
Ross RM, Watten BJ, Krise WF, DiLauro MN (1995) Influence of tank design and hydraulic loading on the behavior, growth, and metabolism of rainbow trout (*Oncorhynchus mykiss*). Aquacult Eng 14:29–47
Rousing T, Bonde M, Sørensen JT (2001) Aggregating welfare indicators into an operational welfare assessment system: a bottom up approach. Acta Agric Scand Anim Sci Suppl 30:53–58
Rubio VC, Sánchez E, Cerdá-Reverter JM (2010) Compensatory feeding in the sea bass after fasting and physical stress. Aquaculture 298:332–337
Ruiz-Gomez ML, Kittilsen S, Höglund E, Huntingford FA, Sørensen C, Pottinger TG, Bakken M, Winberg S, Korzan WJ, Øverli Ø (2008) Behavioral plasticity in rainbow trout (*Oncorhynchus mykiss*) with divergent coping styles: when doves become hawks. Hormones Behav 54:534–538
Ryer CH, Olla BL (1991) Information-transfer and the facilitation and inhibition of feeding in a schooling fish. Environ Biol Fish 30:317–323
Ryer CH, Olla BL (1992) Social mechanisms facilitating exploitation of spatially variable ephemeral food patches in a pelagic marine fish. Anim Behav 44:69–74
Salek SJ, Sullivan CV, Godwin J (2002) Arginine vasotocin effects on courtship behavior in male white perch (*Morone americana*). Behav Brain Res 133:177–183
Sánchez JA, López-Olmeda JF, Blanco-Vives B, Sánchez-Vázquez FJ (2009) Effects of feeding schedule on locomotor activity rhythms and stress response in sea bream. Physiol Behav 98:125–129
Santos GA, Schrama JW, Mamauag REP, Rombout JHWM, Verreth JAJ (2010) Chronic stress impairs performance, energy metabolism and welfare indicators in European seabass (Dicentrarchus labrax): the combined effects of fish crowding and water quality deterioration. Aquaculture 299:73–80
Schreck CB (1981) Stress and compensation in teleostean fishes: response to social and physical factors. In: Pickering AD (ed) Stress and fish. Academic Press, New York, pp 294–321
Schulze Westerath H, Laister S, Winckler C, Knierim U (2009) Exploration as an indicator of good welfare in beef bulls: an attempt to develop a test for on-farm assessment. Appl Anim Behav Sci 116:126–133
Schurmann H, Steffensen JF (1994) Spontaneous swimming activity of Atlantic cod *Gadus morhua* exposed to graded hypoxia at three temperatures. J Exp Biol 197:129–142
Scolamacchia M (2009) Eco-physiological indicators and evaluation of stress factors in European sea bass *Dicentrarchus labrax* (Linnaeus, 1758). PhD thesis, University of Bari, Bari, 162 pp

Scott AP, Ellis T (2007) Measurement of fish steroids in water—a review. Gen Comp Endocr 153:392–400

Scott GR, Sloman KA (2004) The effects of environmental pollutants on complex fish behaviour: integrating behavioural and physiological indicators of toxicity. Aquat Toxicol 68:369–392

Scott AP, Hirschenhauser K, Bender N, Oliveira R, Earley RL, Sebire M, Ellis T, Pavlidis M, Hubbard PC, Huertas M, Canario A (2008) Non-invasive measurement of steroids in fish-holding water: important considerations when applying the procedure to behaviour studies. Behaviour 145(suppl SI):1307–1328

Semsar K, Kandel FL, Godwin J (2001) Manipulation of the AVT system shift social status and related courtship and aggressive behavior in the bluehead wrasse. Horm Behav 40:21–31

Shingles A, McKenzie DJ, Taylor EW, Moretti A, Butler PJ, Ceradini S (2001) Effects of sublethal ammonia exposure on swimming performance in rainbow trout (*Oncorhynchus mykiss*). J Exp Biol 204:2691–2698

Smith CJ, Shaw SBJ, Handy RD (2007) Toxicity of single walled carbon nanotubes to rainbow trout, (*Oncorhynchus mykiss*): respiratory toxicity, organ pathologies, and other physiological effects. Aquat Toxicol 82:94–109

Sneddon L (2007) Assessing pain perception in fish from physiology to behaviour. Comp Biochem Physiol A 146:S78

Sneddon LU, Braithwaite VA, Gentle MJ (2003) Novel object test: examining nociception and fear in the rainbow trout. J Pain 4:431–440

Sordillo LM, Aitken SL (2009) Impact of oxidative stress on the health and immune function of dairy cattle. Vet Immunol Immunopathol 128:104–109

Sørum U, Damsgård B (2004) Effects of anaesthetisation and vaccination on feed intake and growth in Atlantic salmon (*Salmo salar* L.). Aquaculture 232:333–341

Spruijt BM, Bos R, Pijlman FTA (2001) A Concept of welfare based on reward evaluating mechanisms in the brain: anticipatory behaviour as an indicator for the state of reward systems. Appl Anim Behav Sci 72:145–171

Stien LH, Bratland S, Austevoll I, Oppedal F, Kristiansen TS (2007) A video analysis procedure for assessing vertical fish distribution in aquaculture tanks. Aquacult Eng 37: 115–124

Suzuki K, Mizusawa K, Noble C, Tabata M (2008) The growth, feed conversion ratio and fin damage of rainbow trout *Oncorhynchus mykiss* under self-feeding and hand-feeding regimes. Fish Sci 74:941–943

Swanson C, Baxa DV, Young PS, Cech JJ, Hedrick RP (2002) Reduced swimming performance in delta smelt infected with *Mycobacterium* spp. J Fish Biol 61: 1012–1020

Symons PEK (1968) Increase in aggression and the strength of the social hierarchy among juvenile Atlantic salmon deprived of food. J Fish Res Board Can 25:2387–2401

Tang M, Boisclair D (1995) Relationship between respiration rate of juvenile brook trout (*Salvelinus fontinalis*), water temperature, and swimming characteristics. Can J Fish Aquat Sci 52:2138–2145

Terova G, Gornati R, Rimoldi S, Bernardini G, Saroglia M (2005) Quantification of a glucocorticoid receptor in sea bass (*Dicentrarchus labrax* L.) reared at high stocking density. Gene 357:144–151

Thompson RR, Watson JC (2004) Peptide effects on social behavior: effects of vasotocin and isotocin on social approach behavior in male goldfish (*Carassius auratus*). Bahav Neurosci 118:620–626

Tierney KB, Farrell AP (2004) The relationships between fish health, metabolic rate, swimming performance and recovery in return-run sockeye salmon, *Oncorhynchus nerka* (Walbaum). J Fish Dis 27:663–671

Trenzado CE, Morales AE, de la Higuera M (2006) Physiological effects of crowding in rainbow trout, *Oncorhynchus mykiss*, selected for low and high stress responsiveness. Aquaculture 258:583–593

Turnbull JF, Kadri S (2007) Safeguarding the many guises of farmed fish welfare. Dis Aquat Org 75:173–182

van de Nieuwegiessen PG, Boerlage AS, Verreth JAJ, Schrama JW (2008) Assessing the effects of a chronic stressor, stocking density, on welfare indicators of juvenile African catfish, *Clarias gariepinus* Burchell. Appl Anim Behav Sci 115:233–243

van de Nieuwegiessen PG, Olwo J, Khong S, Verreth JAJ, Schrama JW (2009) Effects of age and stocking density on the welfare of African catfish, *Clarias gariepinus* Burchell. Aquaculture 288:69–75

Veissier I, Boissy A, Désiré L, Greiveldinger L (2009) Animals' emotions: studies in sheep using appraisal theories. Anim Welfare 18:347–354

von Borell E (1995) Neuroendocrine integration of stress and significance of stress for the performance of farm animals. Appl Anim Behav Sci 44:219–227

Wagner GN, McKinley RS, Bjørn PA, Finstad B (2003) Physiological impact of sea lice on swimming performance of Atlantic salmon. J Fish Biol 62:1000–1009

Wagner GN, Balfry SK, Higgs DA, Lall SP, Farrell AP (2004) Dietary fatty acids composition affects the repeat swimming performance of Atlantic Salmon in seawater. Comp Biochem Physiol 137A:567–576

Warren JW (1982) Diseases of hatchery fish. US fish & wildlife service, region 3, twin cities. Minnesota stock no. 024-010-00-622-6, US Government Printing Office, Washington, DC, 91 pp

Wemelsfelder F (1993) The concept of boredom and its relationship to stereotyped behavior. In: Lawrence AB, Rushen J (eds) Stereotypic behavior: fundamentals and applications to animal welfare. CAB International, Wallingford

Wendelaar Bonga SE (1997) The stress response in fish. Physiol Rev 77:591–625

White AJ, Schreer JF, Cooke SJ (2008) Behavorial and physiological responses of the congeneric largemouth (*Micropterus salmoides*) and smallmouth bass (*M. dolomieu*) to various exercise and air exposure durations. Fish Res 89:9–16

Wiepkema PR (1987) Behavioural aspects of stress. In: Wiepkema PR, Van Adrichem PWM (eds) Biology of stress in farm animals: an integrative approach. Martinus Nijhoff Publishers, Dordrecht, pp 113–133

Winberg S, Nilsson GE (1993a) Roles of brain monoamine neurotransmitters in agonistic behaviour and stress reactions, with particular reference to fish. Comp Biochem Physiol 106:597–614

Winberg S, Nilsson GE (1993b) Time course of changes in brain serotonergic activity and brain tryptophan levels in dominant and subordinate juvenile Arctic charr. J Exp Biol 179:181–195

Wolter C, Arlinghaus R (2004) Burst and critical swimming speed of fish and their ecological relevance in waterways. In: Leibniz-Institut für Gewässerökologie und Binnenfischerei (IGB) (ed) Annual report 2003, pp 77–93

Würbel H (2009) The state of ethological approaches to the assessment of animal suffering and welfare. Appl Anim Behav Sci 118:105–107

Würbel H, Stauffacher M (1997) Age and weight at weaning affect corticosterone level and development of stereotypies in ICR-mice. Anim Behav 53:891–900

Xu J, Liu Y, Cui S, Miao X (2006) Behavioral responses of tilapia (*Oreochromis niloticus*) to acute fluctuations in dissolved oxygen levels as monitored by computer vision. Aquacult Eng 35:207–217

Yue S, Duncan IJH, Moccia RD (2008) Investigating fear in rainbow trout (*Oncorhynchus mykiss*) using the conditioned-suppression paradigm. J Appl Anim Welfare Sci 11:14–27

Fish Welfare and Genomics – Summary

Patrick Prunet, Øyvind Øverli, Jessica Douxfils, Giovanni Bernardini, Patrick Kestemont and Daniel Baron

In the first part of our chapter we present data on the expression of a few candidate genes. Candidate genes, which are part of the cells of a living organism, are generally selected for their known relevance for responses to stressors that may lead to poor welfare. A gene is a part of the cell that can exert an effect, by its expression, on for instance a physiological process in a living organism. Information about candidate genes has been used to develop molecular biomarkers of stress. These markers give only a partial view of fish welfare, as there is only limited knowledge about cognitive or emotional responses.

The second part consists of an overview on analysis of a large set of gene expression in relation to the effects of various stressors (for example changes in salinity, and lack of oxygen in the water).

In the third part we address studies on the level of a large set of proteins in relation to responses to stressors. Studies focused on biomarkers and proteins allow us to investigate the complexity of the molecular mechanisms which underlie these biological responses.

In conclusion, several new developments are presented, as such technological progress in research on proteins, that will help to complete our understanding of the biological processes underlying fish welfare.

Background

There is considerable public and scientific debate concerning welfare of fish in aquaculture. In the scientific community there is a general consensus to consider fish welfare as an integration of physiological, behavioral and cognitive/emotional responses to challenging situations. All these responses constitute adaptation to these changing situations. The development of new technologies which allow us to investigate a large part of the expression products of the genome (a discipline called genomics) provides new approaches to discover new aspects of these adaptive responses. The genome is the genetic material of an organism of which the genes are functional units. We illustrate how genomic approaches may give rise to new biomarkers for fish welfare and lead us to a better understanding of the complexity of these responses.

Scientific Significance

The main output of genomic studies will be to increase our understanding of the complexity of the biological processes underlying fish responses to challenging situations.

Practical Application

An important societal output of genomic studies will be the development of new sensitive biomarkers which will allow us to detect aquaculture situations that have a significant negative impact on fish welfare.

Contact

P. Prunet
UR1037 SCRIBE, IFR140, INRA, Campus de Beaulieu,
35042 Rennes Cedex, France
e-mail: patrick.prunet@rennes.inra.fr

Fish Physiol Biochem (2012) 38:43–60
DOI 10.1007/s10695-011-9522-z

Fish welfare and genomics

P. Prunet · Ø. Øverli · J. Douxfils · G. Bernardini · P. Kestemont · D. Baron

Received: 21 October 2010 / Accepted: 31 May 2011 / Published online: 14 June 2011

Abstract There is a considerable public and scientific debate concerning welfare of fish in aquaculture. In this review, we will consider fish welfare as an integration of physiological, behavioral, and cognitive/emotional responses, all of which are essentially adaptative responses to stressful situations. An overview of fish welfare in this context suggests that understanding will rely on knowledge of all components of allostatic responses to stress and environmental perturbations. The development of genomic technologies provides new approaches to this task, exemplified by how genome-wide analysis of genetic structures and corresponding expression patterns can lead to the discovery of new aspects of adaptative responses. We will illustrate how the genomic approach may give rise to new biomarkers for fish welfare and also increase our understanding of the interaction between physiological, behavioral, and emotional responses. In a first part, we present data on expression of candidate genes selected a priori. This is a common avenue to develop molecular biomarkers capable of diagnosing a stress condition at its earliest onset, in order to allow quick corrective intervention in an aquaculture setting. However, most of these studies address isolated physiological functions and stress responses that may not be truly indicative of animal welfare, and there is only rudimentary understanding of genes related to possible cognitive and emotional responses in fish. We also present an overview on transcriptomic analysis related to the effect of aquaculture stressors, environmental changes (temperature, salinity, hypoxia), or concerning specific behavioral patterns. These studies illustrate the potential of genomic approaches to characterize the complexity of the molecular mechanisms which underlies not only physiological but also behavioral responses in relation to fish welfare. Thirdly, we address proteomic studies on biological responses to stressors such as salinity change and hypoxia. We will also consider proteomic studies developed in mammals in relation to anxiety and depressive status which may lead to new potential candidates in fish. Finally, in the conclusion,

P. Prunet (✉)
UR1037 SCRIBE, IFR140, INRA, Campus de Beaulieu, 35042 Rennes Cedex, France
e-mail: patrick.prunet@rennes.inra.fr

Ø. Øverli
Department of Animal and Aquacultural Sciences, Norwegian University of Life Sciences, 1432 Ås, Norway

J. Douxfils · P. Kestemont
Research Unit in Environmental and Evolutionary Biology (URBE), The University of Namur (FUNDP), Rue de Bruxelles 61, 5000 Namur, Belgium

G. Bernardini
Department of Biotechnology and Molecular Science, University of Insubria, Varese, Italy

D. Baron
INSERM U533, Institut du Thorax, Faculté de Médecine, Université de Nantes, Nantes, France

we will suggest new developments to facilitate an integrated view of fish welfare. This includes use of laser microdissection in the transcriptomic/proteomic studies, development of meta-analysis methods for extracting information from genomic data sets, and implementation of technological advances for high-throughput proteomic studies. Development of these new approaches should be as productive for our understanding of the biological processes underlying fish welfare as it has been for the progress of pathophysiological research.

Keywords Transcriptome · Gene expression · Proteome · Stress responses · Stressors · Allostasis

Introduction: the relationship between stress and animal welfare in aquaculture

Encompassing both an ethical and a factual aspect, animal welfare is a complex subject with prevailing uncertainty in many fundamental areas, including definition and which organisms the concept should apply to (Sandøe and Simonsen 1992; FSBI 2002; Dawkins 2006; Duncan 2006; Fraser 2009). Regarding teleost fishes in particular, there is considerable public and scientific debate whether fishes possess cognitive and emotional capacities that merit welfare concern (Rose 2002, 2007; Braithwaite and Huntingford 2004; Chandroo et al. 2004; Huntingford et al. 2006; Ashley 2007; Arlinghaus et al. 2007; Huntingford and Kadri 2009). Among the few other terms in biology and biomedicine that have an equally conflicting history of debate and definition is "stress" (see, e.g., Selye 1950; Lazarus and Launier 1978; Engel 1985; Levine and Ursin 1991; Moberg 2000), a concept is very much related to animal welfare. We will therefore begin this section with a brief clarification of what meaning could be ascribed to the terms "stress" and "animal welfare."

Fish welfare definition

As outlined in the review from The Fisheries Society of the British Isles (FSBI 2002), one can generally group definitions of animal welfare in three categories: *Feelings-based definitions* refer to subjective mental states, with good welfare corresponding to states where the animal is both free from negative experiences such as pain or fear, and able to experience positive affective states such as reward (e.g., Duncan 2006). This kind of definition elicits debate, due to the supposition that conscious and subjective components of emotional state are present in animals. *Function-based definitions* involve the animal's ability to adapt to its environment and function well biologically, for instance by showing good health and low stress levels. A much-cited definition put forward by (Broom 1986; Broom and Johnson 1993) falls into this category; here, the welfare of an animal is defined as "*its state as regards its attempts to cope with its environment.*" This definition rests on the assumption that an animal whose physiological systems are functioning poorly is suffering. Finally, *nature-based definitions* rest on the idea that good welfare depends on the animals ability to express its natural behavior, and suffering arises from being prevented from expressing the full repertoire of behavior that is shown in the wild (see, e.g., Kiley-Worthington 1989; Rollin 1993).

The "five freedoms" (Brambell Committee 1965), i.e., freedom from hunger and thirst, discomfort, pain, injury, disease, fear and distress, as well as the freedom to express normal behavior, is a commonly cited framework to assess welfare issues which apply to both state- and nature-based concepts of animal welfare. It is generally considered that this framework is also relevant to teleost fishes, but there are some exceptions to consider (Huntingford 2008). For instance, animals living under seasonal conditions may have evolved to exist without food for extended periods. In fishes, reports do not unequivocally document any neurobiological effects associated with starvation alone in northern species (Winberg et al. 1992; Jobling et al. 1999).

Regardless of definition, the concept of animal welfare is based on the belief that animals have subjective emotions and have the ability to feel pleasure and suffer (Dawkins 2006). This view is in contrast to a positivistic and behavioristic scientific view, where science deals only with the material world, and not with emotion. Undoubtedly, our ability to "measure" welfare, beyond objective states of disease and physical injury, is limited. However, even emergent phenomena such as emotion and cognitive processes are rooted in the material world, specifically in measurable physiological processes creating the communication between networked nerve cells. Current animal welfare research, thus, suffers from a schism between reductionistic studies and anthropogenic and ethical views on animal suffering (Fraser 2009). We will suggest that physiological, behavioral, and cognitive/emotional responses should be seen as adaptive responses to changing situations and environments and develop this argument as a framework for how to include the concept of "animal welfare" as a functional element in biological science.

Fish stress and allostasis

Similarly elusive, but yet very much related to animal welfare, is the concept of stress. Moberg (2000) related a definition that has become somewhat of a consensus, when stating that "stress" is "*the biological response elicited when an individual perceives a threat to its homeostasis.*" This definition encompasses the recent notion that psychological processes associated with how an individual assesses a given situation might be equally important factors as the actual physical challenge in determining the severity of the stress response (reviewed by Von Holst 1998). Korte et al. (2007), however, recently advised on a potential pitfall in the relationship between homeostasis and animal welfare: The concept of homeostasis implies that the purpose of physiological regulation is to maintain physiological parameters at pre-defined settings. However, successful acclimation to changing environments involves significant resetting of neural, endocrine and immune mechanisms. This process is referred to as allostasis, or "*stability through change*" (Sterling and Eyer 1988; McEwen and Wingfield 2003, 2010). Allostasis thus describes regulatory mechanisms mediating change through prediction of activity required to meet new demand. In this concept, impaired welfare arise only when "allostatic overload" arises from chronic, unpredictable, or uncontrollable conditions which do not merit successful allostatic adjustment (McEwen and Stellar 1993; Korte et al. 2007). This realization might clarify some of the confounding and unproductive semantic debate on whether and when stress is relevant for animal welfare.

The concept of allostasis is developed in parallel with stress research taking on an integrative and evolutionary approach, i.e., physiological and behavioral stress reactions are increasingly viewed as adaptive responses that are crucial for survival in a continuously changing environment (e.g., Korte et al. 2005; Romero et al. 2009). Typical proactive responses such as aggression and active avoidance, for instance, are likely to increase fitness when stressors are mild, predictable and of short duration (Wingfield 2003). Under chronic, severe, or unpredictable stress, the organism is better off by reducing risk taking and conserving energy by passive behavioral responses. In line with this evolutionary line of thinking, the functionalist approach to emotions (Frijda 1986) holds that emotions have evolved for a particular function, such as inducing appropriate behavioral responses to potentially dangerous stimuli. It follows that cognitive changes and emotional distress are likely an essential component of passive coping. Therefore, individual variation in the threshold for when a challenge becomes inhibiting rather than stimulatory, i.e., coping style (Koolhaas et al. 1999), is likely correlated to the individual's subjective experience of welfare in a given situation. In this view, welfare is the animal overall subjective experience, a real phenomenon emerging from the brain's motivational and cognitive systems, developed through evolution as an integrated part of the survival mechanisms in animals with advanced central neural systems. This stance circumvents any notion that emotions or affective states are inherently outside the realm than can be subject to the scientific scrutiny; on the contrary, such phenomena are evolved adaptations that play important roles in causing behavior and therefore warrant investigation as biological phenomena.

Coping style and individual variation

Understanding individual differences in coping ability and coping style has indeed become a paramount task in psychiatry and stress research (see, e.g., Susman et al. 1999; Bartolomucci et al. 2005; Korte et al. 2005; Koolhaas 2008). Apart from the now well-established connection between coping and individual health, individual variation in responsiveness is of particular importance for animal welfare (see, e.g., Huntingford and Adams 2005), since conditions that are well tolerated by some individuals may be detrimental to others. Even in fishes, cognitive abilities seem to differ between divergent coping styles, with proactive individuals retaining learned responses to external cues longer than reactive individuals (Moreira et al. 2004; Ruiz-Gomez et al. 2011). This observation suggests differences in learning and memory consolidation, an observation that is in line with an increasing amount of evidences from mammalian models showing that differences in stress-coping characteristics are associated with divergence in cognitive function (see, e.g., Tomie et al. 2000; Aguilar et al. 2004).

The proximate and ultimate mechanisms involved in maintaining individual variation in stress-coping style are, however, poorly understood in teleost fishes. Both genetic and environmental factors contribute to extensive inter-individual variation in how stress affects individuals (e.g., Øverli et al. 2007; Prunet et al. 2008). A direct link between coping style and variation in gene expression as a response to adaptive physiological changes was recently described in a teleost fish species, the common carp *Cyprinus carpio* (MacKenzie et al. 2009). This study suggests that combining behavioral screening for coping style with gene expression studies provides a powerful approach to exploring the link between gene expression and adaptive change. Reciprocally, failure to take such variation into account severely limits the interpretative value of gene expression and other physiological data (MacKenzie et al. 2009). It should, however, be kept in mind that some trait correlations are flexible and can be dissociated by experience and modulation of the environment (Ruiz-Gomez et al. 2008).

In view of the above, providing tools to reliably identify individuals with contrasting responses to allostatic load, and providing a causal framework for this variation, may considerably underpin efforts to improve animal welfare in aquaculture. Presently, our understanding of welfare in fish species is limited, and further significant progress on these issues will rely on our ability to (1) complete our knowledge on the various disciplinary aspects of the adaptive responses to each specific stressor (2) to aggregate all this information to provide an integrated biological description of the adjustments made by the animal and (3) to consider individual characteristics and responses and not only group analysis.

Genomic and fish welfare

The recent development of genomic technologies provides new approaches to study gene expression analysis and, more importantly, lead us to significantly modify our approach to biological questions. The possibility to simultaneously study the expression of many genes through RNA analysis (transriptome analysis) (Baron et al. 2007) or through protein analysis (proteome analysis) (Becker and Bern 2010) allows genome-wide surveys. These genomic approaches make no assumption about which genes or other expression products are worthy to study and also favor development of system-wide analysis of biological processes, all issues that are important points when dealing with welfare questions. Association of this wide information with specific phenotypes characterized at the individual level is the basic approach developed in such studies which should lead to discovery of new aspects of the adaptative response. In this context, genomic approaches appear to bring original and important perspectives not only for the discovery of new aspects of the adaptative responses to stressors, but also for providing an integrated view of animal welfare. The present review of the main publications related to gene expression and fish welfare will start with analysis of candidate genes and extend to large-scale gene expression studies of transcriptome and proteome. Along this review, we will illustrate how these approaches may give rise to new biomarkers for fish welfare and also increase our understanding of these complex interactions between physiological, behavioral, and emotional responses.

Expression of candidate genes and fish welfare

A first obvious step in the analysis of fish welfare in the genomic context is the study of candidate gene expression. Gene expression regulation (i.e., the modulation of any stage of the process by which cells transform the information coded in their DNA to functions) is a complex mechanism by which cells perform several processes and respond to different physiological, pathological, and environmental stimuli. Therefore, also, a stressful stimulus or a deviation from a welfare status can elicit a modification in the expression of a given set of genes. Such genes can be utilized as biomarkers for the study and for the evaluation of the welfare status of an organism. In this context, in the last years, several fish genes whose expression is modified by stress factors or by perturbations of the allostatic optimum have been studied mainly with the aim to have molecular biomarkers capable of diagnosing a stress condition at its earliest onset to allow an immediate corrective intervention in an aquaculture facility (Gornati et al. 2005a, b).

The search for these candidate genes can be approached looking for them among those genes whose expression is already known to be modified by stress conditions. In this case, the orthologs of the genes of interest are cloned in each fish species and sequenced in order to obtain specific primers for PCR-based quantification. With this approach, several gene transcripts have been sequenced and tested by semi-quantitative and quantitative PCR. As examples of this approach, we will discuss overcrowding, a stressful condition often encountered in intensive farming and therefore well studied (Santos et al. 2010), and environmental perturbations such as hypoxia and hypercapnia.

In sea bass (*Dicentrarchus labrax*), the expression of genes such as HSP70, CYP4501A, and Metallothioneines is up-regulated following an increase in stoking density (Gornati et al. 2004). The glucocorticoid receptor, whose expression is known to be modified by cortisol (Sathiyaa and Vijayan 2003; Vazzana et al. 2010), is on the other hand down-regulated (Terova et al. 2005). Atlantic cod (*Gadus morhua*) subjected to short-term overcrowding displayed up-regulation of glucose transporter, pro-inflammatory cytokines, and antibacterial genes (Caipang et al. 2008). The increase of glucose transporters occurs in response to the increase of glucose production to meet the stress augmented energy demand as shown in salmonids (Wiseman et al. 2007). Up-regulation of pro-inflammatory cytokines and antibacterial genes are phenomena which set themselves in the frame of a well-known general enhancement of the immune response following short-term stress (Fast et al. 2008).

Aquatic hypoxia is a common event, and fish often have to contend with hypoxic stress in order to survive in environments with low or variable oxygen levels. Abrupt decreases in water oxygen concentrations may occur in intensive fish farming, especially when fish are reared at high densities. Acute and chronic hypoxia dramatically affects mRNA levels of hypoxia inducible factor-1, and this gene has been proposed as a reliable biomarker of exposure (Terova et al. 2008).

A quite subtle threat to fish welfare is hypercapnia (elevated blood CO_2). Hypercapnia often develops under conditions related to intensive aquaculture, and, paradoxically, it occurs frequently in well-oxygenated waters. Rimoldi et al. (2009) have shown that, in sea bass, NHE-1 and c-Fos genes are involved in the response to hypercapnia and that they may be good biomarkers capable of providing useful information about fish welfare.

As an alternative to the candidate gene approach, we can search for genes whose expression was not previously suspected to be modified by a stress condition by means of genomic techniques, such as DNA microarrays or differential display, capable of identifying altered gene expression at the mRNA level.

DNA microarray-based biomarker discovery in fish has been widely used, and this approach allowed for instance to highlight a clear up-regulation of CYP1A enzyme in rainbow trout gill under an oxidative stress condition (Leguen et al. 2010). Using differential display approach, Gornati et al. (2004) selected six genes differentially expressed in sea bass kept at different population densities. One of these genes resulted to be HMG-CoA reductase, a key enzyme of the cholesterol synthesis (Gornati et al. 2005a, b).

With few exceptions such as cortisol receptor expression analyzed in the brain of selected trout line expressing reduced anxiety and enhanced memory retention (Johansen et al. 2011), these genes, and several others activated or inhibited by the most disparate stress conditions, are not utilized to explore a more psychological status taking into consideration individually variable emotional states such as anxiety and discomfort. Brain-derived neurotrophic factor (BDNF), in this respect, could also represent a quite interesting molecule, in fact, it appears to have an essential role in mediating aversive social experiences (Berton et al. 2006), depression, and anxiety (Martinowich et al. 2007). In the zebrafish, although the control of its expression involves multiple tissue and cell-specific promoters (e.g., skeletal muscle) that are differentially regulated (Heinrich 2003), the early onset and cellular sites of its expression suggest that BDNF mainly functions in nervous system and fin development (Hashimoto and Heinrich 1997) acting as a pro-survival primary sensory neuron factor (Diekmann et al. 2009). Very recently, Tognoli et al. (2010), after providing a comprehensive characterization of BDNF gene in sea bass, have studied its transcriptional translational and post-translational regulation after acute stress, and they observed a significant increase in proBDNF levels and reduction in mature BDNF suggesting altered regulation of proBDNF proteolytic processing. They also showed that the proBDNF/totBDNF ratio is a highly predictive novel quantitative biomarker to detect stress in fishes, a process that is likely to be a central mechanism in adaptation to stress conserved throughout evolution of vertebrates. This study brings the first experimental data indicating involvement of genes regulating behavioral activity in the stress response. However, most of the studies on regulation of gene expression are related to physiological functions and to the analysis of stress response. These physiological descriptors may not be sufficiently reliable when studying chronic stress or animal welfare, and more studies on genes related to emotional responses still need to be developed.

Functional genomics studies and fish welfare

While the analysis of candidate genes has provided useful information on particular mechanisms underlying the physiology of stress response and could lead to description of biomarkers for stress responses, genomic screening approaches using microarray technology have been recently applied to the study of the stress responses in fish (see review Prunet et al. 2008). These new approaches provide a system-wide assessment of alterations of gene expression which should generate both overviews of the biological pathways involved in the stress responses (Cossins et al. 2006). When reviewing the literature related to the genomic analysis of the biological responses to stressors, most of the studies were related to physiological responses, and very limited data have been published on behavioral and social responses. This means that most of the functional genomics studies so far published are related to the stress responses and thus are only partly illustrating fish welfare. However, these studies on the effects of biotic and abiotic stressors related to aquaculture can be reviewed, and analysis of genomic responses will be presented under 3 topics: aquaculture stressors, environmental changes, and responses related to behavior. From these studies, we will mainly highlight new biological processes and biological pathways, but it is clear that such information may give rise to new molecular biomarkers of stress (see also paragraph 1).

Genomic responses to aquaculture stressors

Aquaculture practices frequently lead to a range of unnatural stressors which are related to rearing practices. These include overcrowding, handling, and transport. Thus, gene expression analysis of a large set of genes in various tissues of rainbow trout *Oncorhynchus mykiss* led to identification of a complex network of genes involved in responses to handling or confinement (Krasnov et al. 2005). This study identified in the brain genes related to binding and transport of metal ions, chaperone and heat shock proteins, cytoskeleton and microtubules, whereas in the kidney genes related to mitochondria functions, extracellular structures and peptidases were highlighted. Liver transcriptome has been also studied, and significant changes in various functions including inflammation (acute phase proteins), immune response, gluconeogenesis and glycogenolysis, energy metabolism, and protein degradation have been observed (Cairns et al. 2008; Wiseman et al. 2007; Mommoda et al. 2007). Interestingly, some of these functions have been also modified in vitro in rainbow trout hepatocytes treated with cortisol, the stress-indicator hormone (Aluru and Vijayan 2007). A similar study carried out in the liver of gilthead sea bream *Sparus aurata* exposed to confinement led to the characterization of four major temporal gene expression profiles that cover rapid metabolic readjusment followed by tissue repair and remodeling processes and finally re-establishment of cellular homeostasis and regulation of the immune system (Calduch-Giner et al. 2010).

Interestingly, gene expression analysis following confinement or handling stress also indicated involvement of cellular mechanisms not previously associated with stress responses. Thus, a transient change in immune-related genes and a prominent acute phase protein, haptoglobin, was observed in rainbow trout liver exposed to confinement (Wiseman et al. 2007; Mommoda et al. 2007; Cairns et al. 2008). Transcriptome analyses are also suited for the identification of new biomarkers of acute or chronic stress responses. This can be found in the gene expression studies of acute stress responses in rainbow trout and sea bass (Krasnov et al. 2005; Mommoda et al. 2007; Wiseman et al. 2007; Sarropoulou et al. 2005) but appear more clearly in the time-course analysis of confinement stress (acute and chronic) carried out in trout (Cairns et al. 2008).

Despite these transcriptome studies, we are far from having a comprehensive view of the molecular mechanisms that underlie stress responses to a standard stressor such as confinement or handling. Although more genomic analysis should be carried out on a larger variety of tissues, such studies should also be completed by transcriptomic analysis of target cells in primary culture exposed to corticosteroids, the major endocrine factor that is released after exposure to acute stressors. Such in vitro models should help to better distinguish between primary causes and secondary consequences of the stressors.

Genomic responses to environmental stress

Stress in aquaculture may also come from sudden or sufficiently extreme changes of physical characteristics of the aquatic environment. At a cellular level, such environmental stress leads to defensive or immediately adaptative responses followed by a second type of cellular response which restores cellular homeostasis when changes in environmental conditions become permanent (Kültz 2005). Recent genomic studies of environmental stress brought new information on the regulation of the physiological responses to such stressors.

Thus, following exposure to stressful temperature changes, a time-course analysis of gene expression in seven tissues has been carried out in carp and showed a large set of genes that were affected by cold acclimation (Gracey et al. 2004; Gracey 2007). Such studies led to the identification of the major molecular functions that are modified by such stressful situations and highlighted the importance of lipid for a proper function of the central nervous system and the importance of modification of protein folding required for cold acclimation. These studies also highlighted the importance of genes related to thermotolerance and the complexity of the molecular responses to temperature acclimation through several tissues. Another original study was aiming to analyze trancriptome during acclimation to three constant temperatures or to daily fluctuations in killifish (Podrabsky and Somero 2004). Such situations correspond to natural temperatures to which this species is exposed. The transcriptome analysis in liver revealed effects of temperature on several cellular pathways including the high mobility group B1 (HMGB1) which would act as a compensatory modulator of transcription, thus acting as a temperature sensor for global gene expression. Microarray profiling of the gill transcriptome in zebrafish also revealed that the expression of hmgb1 mRNA was strongly up-regulated by cold treatment along with ionoregulation-related genes (Chou et al. 2008). Short-term effects of temperature on cellular pathways have also been studied in gobies (*Gillichthys mirabilis*) exposed to ecologically relevant situations (Buckley et al. 2006). This transcriptome analysis indicated that several different cellular mechanisms are regulated by temperature changes including those related to molecular chaperones. Analysis of correlation between gene expression and protein levels indicated that induction of mRNA production is correlated to increase in protein levels with, however, variation in timing and magnitude of the protein increase.

Hypoxia is an important stress factor to which aquacultured fish are exposed. Several transcriptome studies have been devoted to this stressor in different fish species. Thus, study in muscle and liver of gobies (Gracey et al. 2001) suggested a coherent view of the mechanisms activated following hypoxia exposure: A rapid shut-down of the major energy-requiring processes is followed by a strong induction of genes involved in anaerobic ATP production and gluconeogenesis at the level of liver associated with suppression of cell growth and proliferation and with a down-regulation of genes encoding protein involved in the translation machinery. In medaka *Oryzia latipes* exposed to short-term hypoxia, analysis of gill, liver and brain transcriptomes showed dysregulation of two important cellular functions, ubiquitin–proteasome and phosphatidynositol pathways. Interestingly, such transcriptome analysis of hypoxia carried out in carp by Fraser et al. (2006) allowed the discovery of a myoglobin expression in non-muscle tissues, probably to facilitate oxygen delivery to the tissue when exposed to hypoxia. Such study was a pertinent example to illustrate the perspective offered by such genomic studies, i.e., the discovery of new mechanisms.

Exposure to salinity changes is a frequently studied example of stress situation which led to a wide literature on the osmoregulatory mechanisms developed by euryhaline species to withstand hyper- or hypo-osmotic choc. Numerous studies have identified the major ion transporters involved in freshwater or seawater osmoregulation (see review by Evans et al. 2005), but recent transcriptome analysis brought new information on the molecular mechanisms involved in responses to salinity changes. Thus, an extensive study of the transcriptome in the major osmoregulatory organs (gill, kidney, intestin) and in brain revealed the major cellular processes involved in acclimation of eel *Anguilla anguilla* to seawater (Kalujnaia et al. 2007). This included functions such as cell protection and immune response, energy metabolism, detoxification, transport functions, structural proteins, regulation of gene transcription and translation. Such study both confirms previous known cellular functions associated with response to hyperosmotic environment and suggests new functions which now need to be confirmed. Using both transcriptomic and proteomic approaches Kültz et al. (2007), identified biological functions, cellular pathways, and molecular functions associated with osmotic stress responses. Kinetic analysis of gene expression changes following transfer to hyperosmotic environment indicated that most of these genes showed rapid and transient increase and have all characteristics of immediate-early genes originally described in neuronal tissues: These genes notably control the transcriptional regulation of effector genes necessary for the adaptation to hyperosmotic medium (Fiol et al. 2006; Kültz et al. 2007; Lanahan and Worley 1998). Protein quantification using specific antibodies allowed to confirm the major roles of several candidate genes such as osmotic stress transcription factor 1 (OSTF1) that is likely to mediate osmotic adaptation by means of transcriptional regulation (Fiol and Kültz 2005).

Genomic responses related to behavior

Understanding the molecular mechanisms underlying fish behavior using genomic approach has been engaged during these last years with the objectives to bring new light on various behaviors as diverse as social dominance, mate choice, reproduction, or migration (see review by St-Cyr and Aubin-Horth 2009). The main way to approach these questions is to associate study of changes in gene expressions with various individual behaviors and to relate these changes to variation at other phenotypic levels. Such approach has produced interesting information on dominance associated with maturation in the Daffodil, an African cichlid fish (*Neolamprogus pulcher*) where high expression of AVT in the brain was associated with dominance in females and males (Aubin-Horth et al. 2007). Moreover, this study also suggests that dominant females had masculinized brains at the molecular level and that large molecular changes occurred in the brain of females with a gain of social dominance rank. A similar study was also carried out in another African cichlid fish, the Tilapia burtoni (*Astatotilapia burtoni*) to study gene expression in dominant and subordinate males as well as in brooding females (Renn et al. 2004). Candidate genes as well as specific functional gene ontology (GO) categories associated with dominance or reproductive state have been revealed by such study which also highlighted high degree of individual variation. In rainbow trout, Sneddon et al. (2005) applied a microarray approach to study gene expression profiles in the brain of fish expressing divergent social status. The results show that the expression of as many as 1,165 genes differed between fish of different social status. Given the link between social rank and food intake, feed conversion ratio, and growth rate, it is no surprising that the majority of genes found to be differentially expressed in that study were related to processes like protein turnover and metabolism. However, a number of genes related to stress responses and behavior, which could serve as candidates for future studies, were also identified (Sneddon et al. 2005).

Despite these genomic studies on fish behavior, very limited information has been gained by such approaches on fish welfare. However, Reilly et al. (2008) studied gene expression changed in the forebrain of rainbow trout and carp exposed to nociceptive stimulus. Among the numerous genes that are differentially regulated in the two species in response to this stimulus, several genes are involved in mammalian nociception, but new candidates are also identified. In particular, the Van Gogh-like 2 (*vangl2*) gene, a member of the non-canonical Wnt signaling/planar cell polarity pathway, was identified, and further studies confirm its involvement in nociceptive processing in the adult fish brain (Reilly et al. 2009) where it participates in mechanism that integrates signaling between the facial branchiomotor neuron and its environment, at least during neuron migration (Sittaramane et al. 2009). Interestingly, *vangl2* has also been shown to govern the zebra fish hair-cell planar polarization during development and regeneration (López-Schier and Hudspeth 2006), the orientation of which ultimately determines the sensitivity of hair cells to directional mechanical stressors. Recently, analysis of brain transcript expression in dominant, subdominant and submissive trout showed large differences and experimentally induced transitions from subdominant to dominant position also led to significant changes (Sneddon et al. 2011). Finally, that study also revealed the important role of brain ependymin in the control of aggressive behavior in trout.

Altogether, these first genomic studies illustrated the potential of genomic approach to characterize the complexity of the molecular mechanisms that underlie behavioral response in relation to fish welfare. These genomic approaches are just starting and combination of behavioral and physiological studies should be very helpful for characterizing welfare state in fish exposed to challenging situation.

Proteomic studies of fish welfare

Critical evaluation of current proteomic studies of fish welfare

Although a great deal of knowledge has been gained by transcriptomic studies, they do not provide information about either the subcellular localization of gene products or post-translational protein modifications that may be essential for its function, transport, or activation. Additionally, mRNA abundance and protein levels are not necessarily correlated (Gygi et al. 1999). To this end, the discipline of proteomics has been initiated to often complement functional genomic research. Proteomics, the study of global changes in proteins, can be defined as the qualitative and quantitative comparison of proteomes under different conditions to further unravel biological processes. Proteomics has emerged as a field in which technology is being developed to enable proteins to be analyzed in a high-throughput, automated way similar to how genes are being studied. This technique can thus provide a critical link between the transcriptome and the metabolome, the latter of which is the end result of transcriptional and proteomic activity. Up to now, very few studies have been performed regarding proteomic investigations aiming to evaluate fish welfare. Hereafter, we critically discuss on the results obtained by current proteomic analysis of fish physiology that can be related to welfare.

Salinity exposure

For this stressor analyzed on fish cells in primary culture, proteomic analysis has led to unexpected results for which further confirmation could be needed. Smith et al. (2005) investigated the in vitro effects of hypo-osmotic environment on the resistance of rainbow trout gill epithelium. Primary cultures of gill cells were exposed to simultaneous contact with apical hypotonic and basolateral isotonic solutions. Five proteins were positively modulated in response to hypo-osmotic environment, and only one of them was correctly identified as pre-Apolipoprotein A-I (ApoA-I), which in mammals is the major protein component of high-density lipoprotein (HDL) in plasma notably promoting cholesterol efflux from tissues to the liver for excretion. In fish, although it seems to have crucial role during embryonic development (Babin et al. 1997), the function of this HDL protein in relation to gill resistance is, however, difficult to interpret. In fish, it has been observed that HDL constitutes one of the most abundant plasma proteins. It has been observed that ApoA-I is involved in the regulation of fish complement and displays antimicrobial activity in the micromolar range against gram positive and gram negative bacteria (Villarroel et al. 2007). Evidence also accumulated that ApoA-I is involved in diverse protective roles in mammals including anti-atherosclerotic, anti-oxidant, anti-inflammatory, and antithrombotic activities (Villarroel et al. 2007). In order to test the physiological relevance of such result, Smith et al. (2005) conducted additional experiment during which trout gill cells were treated with a dose of human ApoA-1. It was found that trans-epithelial resistance increased consequently, thus indicating that ApoA-I may really be involved in such process.

Hypoxic conditions

Zhang et al. (2009) reviewed genomic and proteomic studies associated with hypoxic conditions in fish. To date, two proteomic approaches have been applied to study the effects of hypoxia in fish, especially on zebrafish and medaka species. For the zebrafish study, proteomic analysis of skeletal muscle revealed that one protein decreased while five proteins increased in response to hypoxia (Bosworth et al. 2005). However, due to their respective low abundance, none of them were identified with MS (MALDI-TOF). It must be stated here that 2D-PAGE coupled with silver staining methods has some limitations such as low sensitivity, reduced dynamic range, and gel to gel variability. Regarding evaluation of medaka brain proteome, it has been observed that some proteins were up-regulated following exposure to hypoxia: two hemoglobin β-subunits, four carbonic anydrase forms, calbindin, aldolase, sucinate dehydrogenase, and gluthatione-S-transferase (Oehlers et al. 2007). From previous studies, all these proteins have been associated with hypoxia/ischemia in other animals, representing good source of potential hypoxia markers.

In conclusion, few studies have indirectly evaluated fish welfare through assessment of exposure to stressors so that more investigations are required to go further in the mechanisms involved in such processes. Proteomic may be an interesting technique to highlight without a priori potential biomarkers that could be related to animal welfare as some identified proteins were proven in relation with stimuli-specific response pathways. It is also true that proteomic often reveals non-specific and ubiquitous stress proteins. But these may also be informative as response of animals to stress involved both cellular stress response (CSR) and cellular homeostasis response (CHR) (Kültz 2005). CSR is characteristic of all cells and involves proteins that are conserved in all organisms. These proteins share their responsiveness to diverse stresses probably arising from the most striking and common impact of stress: damages to macromolecules, membranes, lipids, proteins, and DNA (Kültz 2005). Some specificity may arise because the type of lesions and damages somewhat depends on the type of stress. In that way, CHR is aimed to restore homeostasis with specific regard to the particular environmental variable that has changed as it is not triggered by macromolecular damage or oxidative burst (such as for CSR) but by stressor-specific sensors. Discrimination between markers of general stress syndrome and more specific markers of fish welfare (that can be subsequently used as operation welfare indicators—OWI) is also needed, but this problem can be highlighted for all techniques used.

Proteomic studies of mammal welfare and possible extrapolation to fish

As afore mentioned, proteomic studies on fish welfare are very scarce, but, based on several comparative studies, it can be assumed that data obtained from mammals may be partly extrapolated to fish or at least may turn research toward similar fish molecules, pathways, or systems. Actually, comparative approaches of animal physiology afford a better understanding of general molecules, functions, and pathways (Singer 2003; Takei et al. 2006). Likewise, Owen et al. (2007) argued that gathering experimental data in fish toxicology may be done more efficiently by considering mammalian data. Therefore, we suggest that similar comparative approaches can be carried out with regard to stress and welfare issue in fish species. Proteomic studies conducted on mammal species may also be an interesting source of information to reveal potential indicators of fish welfare as some modulated and identified proteins were proven in relation with the imposed stimuli. However, information in various mammalian models is scarce, and studies especially focus on rodent models and particularly on the brain proteome response in relation to anxiety and depressive status. In this context, extrapolation of the data from rodents to fish should be carried out with caution particularly when dealing with proteomic studies on specific brain regions as fish do not have the same brain structure and functionality as in mammals (Chandroo et al. 2004; Ashley 2007). Hereafter, the current proteomic studies more or less directly related to animal welfare are summarized, followed by a critical discussion on possible data extrapolation from mammals to fish.

Major depression is a disorder associated with increased morbidity, and it is suggested that such disease may be derived from genetic, biochemical, psychological, and environmental elements. Evidence suggests that adverse events in life such as stress or trauma may predispose individuals to develop depression in later life. In addition, elevation of cortisol has been reported for 43% of patients with major depression. Skynner et al. (2006) used subcutaneous implants of corticosterone pellets to mimic such depression situation in mice and evaluate its potential effects on brain areas with 2D-DIGE technique and subsequent MALDI-TOF analysis. A total of 150 protein spots were altered, and 117 spots were identified, corresponding to 51 different proteins. Proteins were implicated in alteration of cell morphology and death, glycolysis, and gluconeogenesis as well as changes in phenylalanine, glutamate, and nitrogen metabolism.

Constitutive stress response leads to a number of psychiatric disorders, including depression, Alzheimer's disease, post-traumatic stress, and anxiety disorders. Kim and Kim (2007) exposed some rats to a restraint stress and compared their brain proteome response to a control un-stressed group. It appeared that one interesting protein, the hippocampal cholinergic neuro-stimulating peptide precursor protein (HCNP-pp), was down-regulated in stressed rats. This result was also confirmed by HCNP-pp mRNA analysis. However, it is mentioned that further studies are needed to provide more significant information about correlations between HCNP-pp and neurogenerative diseases caused by stress.

Mammalian hippocampus plays critical roles in memory, learning, neuro-plasticity, and emotion. Increasing evidence suggests that hippocampus and particularly decrease in neurogenesis plays a role in major depression disorder (Mu et al. 2007). In animal model, elevated glucocorticoid levels associated with major depression disease negatively regulate neurogenesis, reduce the levels of neurothrophins, and cause excitotoxic damage in the hippocampus. In this way, Mu et al. (2007) examined the alterations of protein expression patterns in the hippocampus of rats depressed by submission to unpredictable chronic stressors. Expression of 27 proteins was altered, and identified proteins ($n = 15$) play a role in neurogenesis, oxidative metabolism, transcription, and signal transduction. One interesting protein, mitogen-activated protein kinase (MAPK), appeared to be down-regulated in depressed rats MAPK plays a pivotal role in the development of the nervous system by mediating neurogenesis and neuronal differentiation. Another study aimed to identify the molecular changes induced in rat hippocampus by repeated exposure to psychosocial stress (social defeat stress against dominant male rat) with proteomic approach (Carboni et al. 2006). A total of 69 spots were differentially expressed, and 21 were unambiguously identified. Proteins were related to chaperoning, signal transduction, synaptic plasticity, cytoskeleton regulation, and energy metabolism. Several proteins have already been associated with stress-related responses while some of them were described for the first time in relation to stress.

Further investigation is needed to elucidate these pathways and identified novel biomarkers in fishes. It should, however, be mentioned that the fish telencephalon contains areas homologous to the mammalian hippocampus (Northcutt 2006; Vargas et al. 2009; Broglio et al. 2010), and studies of neurogenesis and neural plasticity in fish suggest that the underlying molecular mechanism for negative affective states (recognized as clinical depression in humans) is conserved by evolution. For instance, brain cell proliferation was found to be reduced in the telencephali of rainbow trout exposed to chronic social stress (Sørensen et al. 2011). Also similar to mammals, a higher rate of telencephalic cell proliferation was demonstrated in zebrafish kept in enriched aquaria compared to conspecifics kept in a barren environment (von Krogh et al. 2010).

All in all, these aforesaid studies illustrate the interest of proteins characterized in mammals as potential candidates in fish which would deserve detailed analysis. Indeed, given the compelling molecules and systems similarities among the vertebrate group, it can be supposed that data from mammals can be somewhat inferred based on fish results and vice versa. Consequently and with some caution, mammal comparative approaches may be beneficial for highlighting the main mechanisms and potentially new biomarkers that can be exploited toward the assessment of fish welfare.

Perspectives: what future for analyzing genomic data to characterize fish welfare?

In the introduction, we have proposed to associate the definition of fish welfare with adaptive physiological, behavioral, and cognitives/emotional responses to stressful situations and define the concept of welfare as a functional integration of these elements. According to these definitions, keys for understanding fish welfare will rely on our knowledge of all components of allostatic responses to environmental perturbations. As a consequence, analyses of these responses to stress have been difficult to dissect because they require the activation of multiple genes and pathways.

Genomics technologies have emerged during the past decade and are useful in addressing, in an integrated fashion, the complexity of the studied biological mechanisms. Thus, development of such genomic approaches within the field of fish welfare illustrated in the present review by the genome-wide analysis of genes and their expression led to the discovery of new aspects of adaptative responses. This led to the discovery of new candidate genes that are part of the biological responses to stressful situations but, also to highlight specific functional gene ontology (GO) categories corresponding to major molecular processes (Ashburner et al. 2000). Taking into account the growing number of stress-related studies, it can be stated that genomic information will quickly offer a large overview on expression of many genes associated with these adaptative responses within which the role of defined biological pathways or processes and their underlying regulation can be interpreted. Finally, those studies have also yielded a raft of new promising findings that highlight the complex relationship between behavioral and physiological responses when analyzed at the individual level (MacKenzie et al. 2009).

However, this overview also indicates that a majority of the genomic information is targeted toward specific aspects of the physiological or behavioral response to stress and gives only a fragmented picture of fish welfare, i.e., the complexity related to the biological integration of a multitude of separate responses is usually not addressed. Comparative-genomics by pooling results over a growing body of studies analyzing transcriptional response to stress may help to identify common (and specific) response(s) across fish species. Additionally, this field is now opened for another round of significant advances as integrative-genomics research begins to incorporate protein and metabolite profiles and their dynamic changes as well as the effects of DNA polymorphisms. In this context, several new developments should be encouraged to improve our ability to have an integrated view of fish welfare. Thus, the following points should be considered:

(1) One of the first practical issues to be resolved for productive use of transcriptomic or proteomic studies is the use of laser microdissection laser that would allow gene expression analysis at the level of a few cells (Reilly et al. 2004). This dissection technique allows isolation of small population of homogenous cells among a heterogeneous tissue. This might be a significant problem with genomic analysis of gene expression in multi-functional tissues such as brain or gill where it is quite important to have the gene expression signature of only the cells associated with specific functions. This approach has already been producing interesting data in the analysis of fish reproduction and immune response (Haugarvoll et al. 2008; Jørgensen et al. 2009), and a detailed functional genomic analysis of specific area of the brain in relation to specific physiological or behavioral responses to stressors should be of major interest for understanding fish welfare.

(2) A quick look at the list of published functional genomic studies related to fish welfare indicates that, within a near future, it should be possible to have access to numerous sets of transcriptome data characterizing various aspects of physiological, behavioral, and cognitives/emotional responses to stressful situations. This will represent a very interesting opportunity for extracting genomic-relevant information using meta-analysis methods (Moreau et al. 2003) and for improving our understanding of fish welfare. Thanks to the existence of gene homology between species (Baron et al. 2011), such meta-analysis analysis would include not only integration of multiple microarray analysis applied to specific situation in a fish species but also comparisons between different species as demonstrated in mammals by several independent studies (see for instance Chalmel et al. 2007). By using relevant hierarchical clustering methods (Eisen et al. 1998), one should be able to detect within different sets of genomic data clusters of genes showing correlated expression. Carrying out such analysis within several sets of experimental data coming from the same species or from inter-species comparisons (Chalmel et al. 2007) will give rise to robust clusters of genes and GO functional categories and also to new genes that would not be detected in individual experiments. Such an approach has already been used in medical research and for example allowed identification of sets of genes involved in tumor development and metastasis

processes common to different cancers (Segal et al. 2005). In the meta-analysis of genomic data from several primary solid tumors, a new gene expression signature associated to metastasis has been proposed by Ramaswamy et al. (2003). An interesting objective of these meta-analyses is also the identification of bio-markers specific for different biological processes. This would be very interesting for characterizing common responses to different stressful situations, and this approach has been successfully applied to the characterization of a pronostic meta-signature for breast cancer (Shen et al. 2004; Hughes et al. 2000) In addition to specific genes, meta-analysis can also improve our understanding of the functional role of the genes involved in a specific response. Thus, in yeast, using more than 400 genomic analysis of numerous treatments or mutations, Hughes et al. (2000) have been able to identify the biological functions of unknown genes by comparing their expression profiles with cluster of genes involved in known physiological processes. Another benefit of the meta-analysis is the possibility to dissect complex genomic signatures obtained in tissues containing a mixture of different cell types. The characterization of these different cell populations through their respective genomic signature is possible by comparison with gene expression data obtained with isolated cells. This approach that is called "virtual microdissection" has been used by Alizadeh et al. (2000) for the identification of 2 new subtypes of lymphoma within the diffuse large B-cell lymphoma that was previously considered as homogenous. Such kind of analysis can also be achieved by integration of numerous datasets related to the same biological process as demonstrated for the signatures of the immune response by Shaffer et al. (2001). Finally, identification of specific signature and genes co-regulation (Tavazoie et al. 1999) give rise to the characterization of transcriptional regulation network which could be further validated by ChiP-chip studies (see, e.g., Blais et al. 2005; Ren et al. 2000; Bartel 2004).

(3) Such integrative studies may also be performed for proteome data in near future. Indeed, high-throughput proteomics is a rapidly developing field that offers the global profiling of proteins from a biological system. Numerous novel and powerful technologies have been developed around mass spectrometry and proteomics over the past 20 years to globally study complex proteomes (e.g., plasma). These high-throughput technological advances are fueling a revolution in biology, enabling analyses at the scale of entire systems (e.g., whole cells, tissues, or environmental communities) so that as for transcriptomics, large amount of data will be soon available for the scientific community. The same can be noticed for metabolomics (Blow 2008; Dunn et al. 2011; Weckwerth 2010) that is the systematic study of the unique chemical fingerprints that specific cellular processes leave behind. Besides, in view of the rapid progress in these techniques, exploring fish welfare integrating the different approaches (e.g., both transcriptomics and proteomics) may be an added/complementary value to better understand stressor impact on large collections of tissues. Finally, welfare in fish can also be considered as a multifactorial problem. In addition to environmental factors, genetic factors are also to largely contribute to extensive inter-individual variation in how stress affects individuals (Øverli et al. 2005; Prunet et al. 2008). To this end, it would be invaluable to integrate genotyping data to transcriptome data in order to facilitate the discovery of genes affecting stress susceptibility (Jansen and Nap 2001) because genes affected by a same polymorphism locus are likely to share similar biological function. A more integrative-genomics view will combine the power of genetics, transcriptomics, and proteomics in a way that is likely to become instrumental in the further unraveling of metabolic, regulatory and developmental pathways relevant to welfare in fish.

In conclusion, development of these new approaches in the study of fish welfare using genomic tools should be as productive for our understanding of the biological processes underlying the responses to stressful situation as it has been for the progress of pathophysiological research.

References

Aguilar R, Gil L, Fernandez-Teruel A, Tobeña A (2004) Genetically-based behavioral traits influence the effects of shuttle box avoidance overtraining and extinction upon intertrial responding: a study with the Roman rat strains. Behav Process 66:63–72

Alizadeh AA, Eisen MB, Davis RE, Ma C, Lossos IS, Rosenwald A, Boldrick JC, Sabet H, Tran T, Yu X, Powell JI, Yang L, Marti GE, Moore T, Hudson Jr J, Lu L, Lewis DB, Tibshirani R, Sherlock G, Chan WC, Greiner TC, Weisenburger DD, Armitage JO, Warnke R, Levy R, Wilson W, Grever MR, Byrd JC, Botstein D, Brown PO, Staudt LM (2000) Distinct types of diffuse large B-cell lymphoma identified by gene expression profiling. Nature 403(6769):503–511

Aluru N, Vijayan MM (2007) Heaptic transcriptome response to glucocorticoid receptor activation in rainbow trout. Physiol Genom 31:489–491

Arlinghaus R, Cooke SJ, Schwab A, Cowx IG (2007) Fish welfare: a challenge to the feelings-based approach, with implications for recreational fishing. J Fish Biol 8:57–71

Ashburner M, Ball CA, Blake JA, Botstein D, Butler H, Cherry JM, Davis AP, Dolinski K, Dwight SS, Eppig JT, Harris MA, Hill DP, Issel-Tarver L, Kasarskis A, Lewis S, Matese JC, Richardson JE, Ringwald M, Rubin GM, Sherlock G (2000) Gene ontology: tool for the unification of biology. The gene ontology consortium. Nat Genet 25(1):25–29

Ashley PJ (2007) Fish welfare: current issues in aquaculture. Appl Anim Behav Sci 104:199–235

Aubin-Horth N, Desjardins JK, Martei YM, Balshine S, Hofmann HA (2007) Masculinized dominant females in a cooperatively breeding species. Mol Ecol 16:1349–1358

Babin PJ, Thisse C, Durliat M, Andre M, Akimenko MA, Thisse B (1997) Both apolipoprotein E and A-I genes are present in a nonmammalian vertebrate and are highly expressed during embryonic development. Proc Natl Acad Sci USA 94(16):8622–8627

Baron D, Raharijaona M, Houlgatte R (2007) DNA microarrays. IRBM 28(5–6):210–215

Baron D, Bihouee A, Teusan R, Dubois E, Savagner F, Steenman M et al (2011) MADGene: retrieval and processing of gene identifier lists for the analysis of heterogeneous microarray datasets. Bioinformatics 27:725–726

Bartel DP (2004) MicroRNAs: genomics, biogenesis, mechanism, and function. Cell 116:281–297

Bartolomucci A, Palanza P, Sacerdote P, Panerai AE, Sgoifo A, Dantzer R, Parmigiani S (2005) Social factors and individual vulnerability to chronic stress exposure. Neurosci Biobehav Rev 29:67–81

Becker CH, Bern M (2010) Recent developments in quantitative proteomics. Mutat Res (in press)

Berton O, McClung CA, DiLeone RJ, Krishnan V, Renthal W, Russo SJ, Graham D, Tsankova NM, Bolanos CA, Rios M, Monteggia LM, Self DW, Nestler EJ (2006) Essential role of BDNF in the mesolimbic dopamine pathway in social defeat stress. Science 311:864–868

Blais A, Tsikitis M, costa-Alvear D, Sharan R, Kluger Y, Dynlacht BD (2005) An initial blueprint for myogenic differentiation. Genes Dev 19:553–569

Blow N (2008) Metabolomics: biochemistry's new look. Nature 455:697–700

Bosworth CA, Chou C, Cole RB, Rees BB (2005) Protein expression patterns in zebrafish skeletal muscle: initial characterization and the effects of hypoxic exposure. Proteomics 5:1362–1371

Braithwaite VA, Huntingford FA (2004) Fish and welfare: do fish have the capacity for pain perception and suffering? Anim Welf 13:S87–S92

Brambell C (1965) Report of the technical committee to enquire into the welfare of animals kept under intensive livestock husbandry systems. Command Report 2836, Her Majesty's Stationery Office, London

Broglio C, Rodriguez F, Gomez A, Arias JL, Salas C (2010) Selective involvement of the goldfish lateral pallium in spatial memory. Behav Brain Res 210:191–201

Broom D (1986) Indicators of poor welfare. Brit Vet J 142:524–526

Broom D, Johnson K (1993) Stress and animal welfare. Chapman and Hall, London

Buckley BA, Gracey AY, Somero GN (2006) The cellular response to heat stress in the goby Gillichthys mirabilis: a cDNA microarray and protein-level analysis. J Exp Biol 209:2660–2677

Caipang CMA, Brinchmann MF, Berg I, Iversen M, Eliassen R, Kiron V (2008) Changes in selected stress and immune-related genes in Atlantic cod, *Gadus morhua*, following overcrowding. Aquacult Res 39:1533–1540

Cairns MT, Johnson MC, Talbot AT, Pemmasani JK, McNeill RE, Houeix B, Sangrador-Vegas A, Pottinger TG (2008) A cDNA microarray assessment of gene expression in the liver of rainbow trout (Oncorhynchus mykiss) in response to a handling and confinement stressor. Comp Biochem Physiol Part D 3(1):51–66

Calduch-Giner JA, Davey G, Saera-Vila A, Houeix B, Talbot A, Prunet P, Cairns MT, Pérez-Sánchez J (2010) Use of microarray technology to assess the time course of liver stress response after confinement exposure in gilthead sea bream (Sparus aurata L.). BMC Genomics 11:193

Carboni L, Piubelli C, Pozzato C, Astner H, Arban R, Righetti PG, Hamdan M, Domenici E (2006) Proteomic analysis of rat hippocampus after repeated psychosocial stress. Neuroscience 137:1237–1246

Chalmel F, Rolland AD, Niederhauser-Wiederkehr C, Chung SS, Demougin P, Gattiker A, Moore J, Patard JJ, Wolgemuth DJ, Jégou B, Primig M (2007) The conserved transcriptome in human and rodent male gametogenesis. Proc Natl Acad Sci USA 104(20):8346–8351

Chandroo KP, Duncan IJH, Moccia RD (2004) Can fish suffer?: perspectives on sentience, pain, fear and stress. Appl Anim Behav Sci 86:225–250

Chou MY, Hsiao CD, Chen SC, Chen IW, Liu ST, Hwang PP (2008) Effects of hypothermia on gene expression in zebrafish gills: upregulation in differentiation and function of ionocytes as compensatory responses. J Exp Biol 211(19):3077–3084

Cossins A, Fraser J, Hughes J, Gracey A (2006) Post-genomic approaches to understanding the mechanisms of environmentally induced phenotypic plasticity. J Exp Biol 209:2328–2336

Dawkins MS (2006) A user's guide to animal welfare science. Trends Ecol Evol 21:77–82

Diekmann H, Anichtchik O, Fleming A, Futter M, Goldsmith P, Roach A, Rubinsztein DC (2009) Decreased BDNF levels are a major contributor to the embryonic phenotype of Huntingtin knockdown zebrafish. J Neurosci 29(5):1343–1349

Duncan IJH (2006) The changing concept of animal sentience. Appl Anim Behav Sci 100:11–19

Dunn WB, Broadhurst DI, Atherton HJ, Goodacre R, Griffin JL (2011) Systems level studies of mammalian metabolomes: the roles of mass spectrometry and nuclear magnetic resonance spectroscopy. Chem Soc Rev 40:387–426

Eisen MB, Spellman PT, Brown PO, Botstein D (1998) Cluster analysis and display of genome-wide expression patterns. Proc Natl Acad Sci USA 95(25):14863–14868

Engel BT (1985) Stress is a noun! No, a verb! No, an adjective. In: Field TM, McCabe PM, Schneiderman N (eds) Stress and coping. Erlbaum, Hillsdale, NJ, pp 3–12

Evans DH, Piermarini M, Choe KP (2005) The multifunctional fish gill: dominant site of gas exchange, osmoregulation, acid-base regulation and excretion of nitrogenous waste. Physiol Rev 85:97–177

Fast MD, Hosoya S, Johnson SC, Afonso LOB (2008) Cortisol response and immune-related effects of Atlantic salmon (*Salmo salar* Linnaeus) subjected to short- and long-term stress. Fish Shellfish Immunol 24:194–204

Fiol DF, Kültz D (2005) Rapid hyperosmotic coinduction of two tilapia (Oreochromis mossambicus) transcription factors in gill cells. Proc Natl Acad Sci USA 102:927–932

Fiol DF, Chan S, Kültz D (2006) Identification and pathway analysis of immediate hyperosmotic stress responsive molecular mechanisms in tilapia (*Oreochromis mossambicus*) gill. Comp Biochem Physiol Part D 1:344–356

Fraser D (2009) Animal behaviour, animal welfare and the scientific study of affect. Appl Anim Behav Sci 118:108–117

Fraser J, Vieira de Mello L, Ward D, Rees HH, Williams DR, Fang Y, Brass A, Gracey AY, Cossins AR (2006) Hypoxia-inducible myoglobin expression in nonmuscle tissues. Proc Natl Acad Sci 103:2977–2981

Frijda N (1986) The emotions. Cambridge University Press, Cambridge

FSBI (2002) Fish welfare. Briefing paper 2 fisheries society of the British Isles, Granta Information Systems, Cambridge

Gornati R, Papis E, Rimoldi S, Terova G, Saroglia M, Bernardini G (2004) Rearing density influences the expression of stress-related genes in sea bass (*Dicentrarchus labrax*, L). Gene 341:111–118

Gornati R, Cavaliere R, Terova G, Saroglia M, Benardini G (2005a) Molecular biology and fish welfare: a winning combination. Aquacult Int 13:51–55

Gornati R, Papis E, Rimoldi S, Chini V, Terova G, Prati M, Saroglia M, Bernardini G (2005b) Molecular markers for animal biotechnology: sea bass (*Dicentrarchus labrax*, L) HMG-CoA reductase mRNA. Gene 344:299–305

Gracey AY (2007) Interpreting physiological responses to environmental change through gene expression profiling. J Exp Biol 209:1584–1592

Gracey AY, Troll JV, Somero GN (2001) Hypoxia-induced gene expression profiling in the euryoxic fish Gillichthys mirabilis. Proc Natl Acad Sci 98:1993–1998

Gracey AY, Fraser EJ, Li W, Fang Y, Taylor RR, Rogers J, Brass A, Cossins AR (2004) Coping with cold: an integrative, multitissue analysis of the transcriptome of a poikilothermic vertebrate. Proc Natl Acad Sci USA 101:16970–16975

Gygi SP, Rochon Y, Franza BR, Aebersold R (1999) Correlation between protein and mRNA abundance in yeast. Mol Cell Biol 19:1720–1730

Hashimoto M, Heinrich G (1997) Brain-derived neurotrophic factor gene expression in the developing zebrafish. Int J Dev Neurosci 15(8):983–997

Haugarvoll E, Bjerkås I, Nowak BF, Hordvik I, Koppang EO (2008) Identification and characterization of a novel intraepithelial lymphoid tissue in the gills of Atlantic salmon. J Anat 213(2):202–209

Heinrich G (2003) A novel BDNF gene promoter directs expression to skeletal muscle. BMC Neurosci 4:1

Hughes TR, Marton MJ, Jones AR, Roberts CJ, Stoughton R, Armour CD, Bennett HA, Coffey E, Dai H, He YD, Kidd MJ, King AM, Meyer MR, Slade D, Lum PY, Stepaniants SB, Shoemaker DD, Gachotte D, Chakraburtty K, Simon J, Bard M, Friend SH (2000) Functional discovery via a compendium of expression profiles. Cell 102:109–126

Huntingford FA (2008) Animal welfare in aquaculture. In: Culver K, Castle D (eds) Aquaculture innovation and social transformation. Springer, Netherlands, pp 21–33

Huntingford FA, Adams C (2005) Behavioural syndromes in farmed fish: implications for production welfare. Behaviour 142:1207–1221

Huntingford FA, Kadri S (2009) Taking account of fish welfare: lessons from aquaculture. J Fish Biol 75:2862–2867

Huntingford FA, Adams C, Braithwaite VA, Kadri S, Pottinger TG, Sandøe P, Turnbull JF (2006) Current issues in fish welfare. J Fish Biol 68:332–372

Jansen RC, Nap JP (2001) Genetical genomics: the added value from segregation. Trends Genet 17(7):388–391

Jobling M, Koskela J, Winberg S (1999) Feeding and growth of whitefish fed restricted and abundant rations: influences on growth heterogeneity and brain serotonergic activity. J Fish Biol 54:437–449

Johansen IB, Sanvik GK, Nilsson GE, Bakken M, Overli O (2011) Cortisol receptor expression differs in the brains of rainbow trout selected for divergent cortisol responses. Comp Biochem Physiol part D 6:126–132

Jørgensen A, Nielsen JE, Morthorst JE, Bjerregaard P, Leffers H (2009) Laser capture microdissection of gonads from juvenile zebrafish. Reprod Biol Endocrinol 14(7):97

Kalujnaia S, McWilliam IS, Zaguinaiko VA, Feilen AL, Nicholson J, Hazon N, Cutler CP, Cramb G (2007) Transcriptomic approach to the study of osmoregulation in the European eel Anguilla anguilla. Physiol Genomics 31:385–401

Kiley-Worthington M (1989) Ecological, environmental and ethically sound environments for animals: toward symbiosis. J Agr Ethics 2:323–347

Kim HG, Kim KL (2007) Decreased hippocampal cholinergic neuro stimulating peptide precursor protein associated with stress exposure in rat brain by proteomic analysis. J Neurosci Res 85:2898–2908

Koolhaas JM (2008) Coping style and immunity in animals: making sense of individual variation. Brain Behav Immun 22:662–667

Koolhaas JM, Korte SM, De Boer SF, Van Der Vegt BJ, Van Reenen CG, Hopster H, De Jong IC, Ruis MAW, Blokhuis HJ (1999) Coping styles in animals: current status in behavior and stress-physiology. Neuroscie Biobehav Rev 23:925–935

Korte SM, Koolhaas JM, Wingfield JC, McEwen BS (2005) The Darwinian concept of stress: benefits of allostasis and costs of allostatic load and the trade-offs in health and disease. Neurosci Biobehav Rev 29:3–38

Korte SM, Olivier B, Koolhaas JM (2007) A new animal welfare concept based on allostasis. Physiol Behav 92:422–428

Krasnov A, Koskinen H, Pehkonen P, Rexroad III CE, Afanasyev S, Mölsä H (2005) Gene expression in the brain and kidney of rainbow trout in response to handling stress. BMC Genomics 6(3):1–11

Kültz D (2005) Molecular and evolutionary basis of the cellular stress response. Annu Rev Physiol 67:225–257

Kültz D, Fiol D, Valkova N, Gomez-Jimenez S, Chan SY, Lee J (2007) Functional genomics and proteomics of the cellular osmotic stress response in 'non-model' organisms. J Exp Biol 210:1593–1601

Lanahan A, Worley P (1998) Immediate-early genes and synaptic function. Neurobiol Learn Mem 70:37–43

Lazarus R, Launier R (1978) Stress-related transactions between person and environment. In: Pervin LA, Lewis M (eds) Perspectives in interactional psychology. Plenum, New York, pp 287–327

Leguen I, Odjo N, Le Bras Y, Luthringer B, Baron D, Monod G, Prunet P (2010) Effect of seawater transfer on CYP1A gene expression in rainbow trout gills. Comp Biochem Physiol A Mol Integr Physiol 156(2):211–217

Levine S, Ursin H (1991) What is stress? In: Brown MR, Koob GF, Rivier C (eds) Stress: neurobiology and neuroendocrinology. Dekker, New York, pp 3–21

López-Schier H, Hudspeth AJ (2006) A two-step mechanism underlies the planar polarization of regenerating sensory hair cells. Proc Natl Acad Sci USA 103(49):18615–18620

MacKenzie S, Ribas L, Pilarczyk M, Capdevila DM, Kadri S, Huntingford FA (2009) Screening for coping style increases the power of gene expression studies. PLOS One 4:e5314

Martinowich K, Manji H, Lu B (2007) New insights into BDNF function in depression and anxiety. Nat Neurosci 10:1089–1093

McEwen BS, Stellar E (1993) Stress and the individual: mechanisms leading to disease. Arch Intern Med 153:2093–2101

McEwen BS, Wingfield JC (2003) The concept of allostasis in biology and biomedicine. Horm Behav 43:2–15

McEwen BS, Wingfield JC (2010) What is in a name? Integrating homeostasis, allostasis and stress. Horm Behav 57:105–111

Moberg G (2000) Biological responses to stress: implications for animal welfare. In: Moberg G, Mench J (eds) The biology of snimal stress: Basic principles and implications for animal welfare. CABI Publishing, Wallingford, pp 1–22

Mommoda TS, Schwindt AR, Feist GW, Gerwick L, Bayne CJ, Schreck CB (2007) Gene expression in the liver of rainbow trout Oncorhynchus mykiss, during the stress response. Comp Biochem Physiol Part D 2:303–315

Moreau Y, Aerts S, De Moor B, De Strooper B, Dabrowski M (2003) Comparison and meta-analysis of microarray data: from the bench to the computer desk. Trends Genet 19(10):570–577

Moreira PSA, Pulman KGT, Pottinger TG (2004) Extinction of a conditioned response in rainbow trout selected for high or low responsiveness to stress. Horm Behav 46:450–457

Mu J, Xie P, Yang Z-S, Yang D-L, Lv F-J, Luo T-Y, Li Y (2007) Neurogenesis and major depression: implications from proteomic analyses of hippocampal proteins in a rat depression model. Neurosci Lett 416:252–256

Northcutt RG (2006) Connections of the lateral and medial divisions of the goldfish telencephalic pallium. J Comp Neurol 494:903–943

Oehlers LP, Perez AN, Walter RB (2007) Detection of hypoxia-related proteins in medaka (Oryzias latipes) brain tissue by difference gel electrophoresis and de novo sequencing of 4-sulfophenyl isothiocyanate-derivatized peptides by matrix-assisted laser desorption/ionization time-of-flight mass spectrometry. Comp Biochem Physiol C 145:120–133

Øverli Ø, Winberg S, Pottinger TG (2005) Behavioral and neuroendocrine correlates of selection fro stress responsiveness in rainbow trout–a review. Integr Comp Biol 45:463–474

Øverli Ø, Sørensen C, Pulman KGT, Pottinger TG, Korzan WJ, Summers CH, Nilsson GE (2007) Evolutionary background for stress-coping styles: Relationships between physiological, behavioral, and cognitive traits in non-mammalian vertebrates. Neurosci Biobehav Rev 31:396–412

Owen SF, Giltrow E, Huggett DB, Hutchison TH, Saye J, Winter MJ, Sumpter JP (2007) Comparative physiology, pharmacology and toxicology of β-blockers: mammals versus fish. Aquat toxicol 82:145–162

Podrabsky JE, Somero GN (2004) Changes in gene expression associated with acclimatation to constant temperatures and fluctuating daily temperatures in ana annual killifish *Austrofundulus limnaeus*. J Exp Biol 207:2237–2254

Prunet P, Cairns MT, Winberg S, Pottinger TG (2008) Functional genomics of stress responses in fish. Rev Fish Sci 16:157–166

Ramaswamy S, Ross KN, Lander ES, Golub TR (2003) A molecular signature of metastasis in primary solid tumors. Nat Genet 33(1):49–54

Reilly SC, Cossins AR, Quinn JP, Sneddon LU (2004) Discovering genes: the use of microarrays and laser capture microdissection in pain research. Brain Res Rev 46:225–233

Reilly SC, Quinn JP, Cossins AR, Sneddon LU (2008) Novel candidate genes identified in the brain during nociception in common carp (*Cyprinus carpio*) and rainbow trout (*Oncorhynchus mykiss*). Neurosci Lett 437:1350–2138

Reilly SC, Kipar A, Hughes DJ, Quinn JP, Cossins AR, Sneddon LU (2009) Investigation of Van Gogh-like 2 mRNA regulation and localisation in response to nociception in the brain of adult common carp (Cyprinus carpio). Neurosci Lett 465(3):290–294

Ren B, Robert F, Wyrick JJ, Aparicio O, Jennings EG, Simon I et al (2000) Genome-wide location and function of DNA binding proteins. Science 290:2306–2309

Renn SC, Aubin-Horth N, Hofmann HA (2004) Biologically meaningful expression profiling across species using heterologous hybridization to a cDNA microarray. BMC Genomics 5:42

Rimoldi S, Terova G, Brambilla F, Bernardini G, Gornati R, Saroglia M (2009) Molecular characterization and expression analysis of Na+/H+ exchanger (NHE)-1 and c-Fos genes in sea bass (Dicentrarchus labrax, L) exposed to acute and chronic hypercapnia. J Exp Mar Biol 375:32–40

Rollin B (1993) Animal welfare, science and value. J Agr Environ Ethic 6(Suppl 2):44–50

Romero LM, Dickens MJ, Cyr NE (2009) The reactive scope model—a new model integrating homeostasis, allostasis, and stress. Horm Behav 55:375–389

Rose JD (2002) The neurobehavioral nature of fishes and the question of awareness and pain. Rev Fish Sci 10:1–38

Rose JD (2007) Anthropomorphism and 'mental welfare' of fishes. Dis Aquat Org 75:139–154

Ruiz-Gomez MD, Kittilsen S, Höglund E, Huntingford FA, Sørensen C, Pottinger TG, Bakken M, Winberg S, Korzan WJ, Øverli Ø (2008) Behavioral plasticity in rainbow trout (*Oncorhynchus mykiss*) with divergent coping styles: When doves become hawks. Horm Behav 54:534–538

Ruiz-Gomez MD, Huntingford FA, Øverli Ø, Thörnqvist PO (2011) Response to environmental change in rainbow trout selected for divergents coping styles. Physiol Behav 102(3–4):317–322

Sandøe P, Simonsen H (1992) Assessing animal welfare: where does science end and philosophy begin? Anim Welf 1:257–267

Santos GA, Schrama JW, Mamauag REP, Rombout JHWM, Verreth JAJ (2010) Chronic stress impairs performance, energy metabolism and welfare indicators in European seabass (*Dicentrarchus labrax*): The combined effects of fish crowding and water quality deterioration. Aquaculture 299:73–80

Sarropoulou E, Kotoulas G, Power DM, Geisler R (2005) Gene expression profining of gilhead sea bream during early development and detection of stress-related genes by the application of cDNA microarray technology. Physiol Genomics 23:182–191

Sathiyaa R, Vijayan MM (2003) Autoregulation of glucocorticoid receptor by cortisol in rainbow trout hepatocytes. Am J Physiol Cell Physiol 284:C1508–C1515

Segal E, Friedman N, Kaminski N, Regev A, Koller D (2005) From signatures to models: understanding cancer using microarrays. Nat Genet 37(Suppl):S38–S45

Selye H (1950) Stress. Acta, Montreal

Shaffer AL, Rosenwald A, Hurt EM, Giltnane JM, Lam LT, Pickeral OK et al (2001) Signatures of the immune response. Immunity 15:375–385

Shen R, Ghosh D, Chinnayan AM (2004) Prognostic metasignature of breast cancer developed by two-stage mixture modeling of microarray data. BMC Genomics 5(1):94

Singer MA (2003) Do mammals, birds, reptiles and fish have similar nitrogen conserving systems? Comp Biochem Physiol B 134:543–558

Sittaramane V, Sawant A, Wolman MA, Maves L, Halloran MC, Chandrasekhar A (2009) The cell adhesion molecule Tag1, transmembrane protein Stbm/Vangl2, and Lamininalpha1 exhibit genetic interactions during migration of facial branchiomotor neurons in zebrafish. Dev Biol 325(2):363–373

Skynner HA, Amos DP, Murray F, Salim K, Knowles MR, Munoz-Sanjuan I, Camargo LM, Bonnert TP, Guest PC (2006) Proteomic analysis identifies alterations in cellular morphology and cell death pathways in mouse brain after chronic corticosterone treatment. Brain Res 1102:12–26

Smith RW, Wood CM, Cash P, Diao L, Pärt P (2005) Apolipoprotein AI could be a significant determinant of epithelial integrity in rainbow trout gill cell cultures: A study in functional proteomics. Biochim Biophys Acta Proteins Proteomics 1749:81–93

Sneddon LU, Margareto J, Cossins AR (2005) The use of transcriptomics to address questions in behaviour: production of a suppression subtractive hybridisation library form dominance hierarchies of rainbow trout. Physiol Biochem Zool 78(5):695–705

Sneddon LU, Schmidt R, Fang Y, Cossins AR (2011) Molecular correlates of social dominance: a novel role for ependymin in aggression. PLoS One 6(4):e18181

Sørensen C, Nilsson GE, Summers CH, Øverli Ø (2011) Social stress reduces forebrain cell proliferation in rainbow trout (Oncorhynchus mykiss). Behav Brain Res (in press)

St-Cyr S, Aubin-Horth N (2009) Integrative and genomics approaches to uncover the mechanistic bases of fish behavior and its diversity. Comp Biochem Physiol A Mol Integr Physiol 152(1):9–21

Sterling P, Eyer J (1988) Allostasis: a new paradigm to explain arousal pathology. In: Fisher S, Reason J (eds) Handbook of life stress, cognition and health. Wiley, New York, pp 629–649

Susman EJ, Schmeelk KH, Worrall BK, Granger DA, Ponirakis A, Chrousos GP (1999) Corticotropin-releasing hormone and cortisol: longitudinal associations with depression and antisocial behavior in pregnant adolescents. J Am Acad Child Adolesc Psych 38:460–467

Takei Y, Kawakoshi A, Tsukada T, Yuge S, Ogoshi M, Inoue K, Hyodo S, Bannai H, Miyano S (2006) Contribution of comparative fish studies to general endocrinology: structure and function of some osmoregulatory hormones. J Exp Zool 305A:787–798

Tavazoie S, Hughes JD, Campbell MJ, Cho RJ, Church GM (1999) Systematic determination of genetic network architecture. Nat Genet 22:281–285

Terova G, Gornati R, Rimoldi S, Bernardini G, Saroglia M (2005) Quantification of a glucocorticoid receptor in sea bass (*Dicentrarchus labrax*, L.) reared at high stocking density. Gene 357:144–151

Terova G, Rimoldi S, Cora S, Bernardini G, Gornati R, Saroglia M (2008) Acute and chronic hypoxia affects HIF-1 alpha mRNA levels in sea bass (*Dicentrarchus labrax*). Aquaculture 279:150–159

Tognoli C, Rossi F, Di Cola F, Baj G, Tongiorgi E, Terova G, Saroglia M, Bernardini G, Gornati R (2010) Acute stress alters transcript expression pattern and reduces processing of proBDNF to mature BDNF in *Dicentrarchus labrax*. BMC Neurosci 11:4

Tomie A, Aguado AS, Pohorecky LA, Benjamin D (2000) Individual differences in Pavlovian autoshaping of lever pressing in rats predict stress-induced corticosterone release and mesolimbic levels of monoamines. Pharmacol Biochem Behav 65:509–517

Vargas JP, Lopez JC, Portavella M (2009) What are the functions of fish brain pallium? Brain Res Bull 79:436–440

Vazzana M, Vizzini A, Sanfratello MA, Celi M, Salerno G, Parrinello M (2010) Differential expression of two glucocorticoid receptors in seabass (teleost fish) head kidney after exogeneous cortisol inoculation. Comp Biochem Physiol Part A 157:49–54

Villarroel F, Bastías A, Casado A, Amthauer R, Concha MI (2007) Apolipoprotein A-I, an antimicrobial protein in *Oncorhynchus mykiss*: Evaluation of its expression in primary defence barriers and plasma levels in sick and healthy fish. Fish Shellfish Immunol 23:197–209

Von Holst D (1998) The concept of stress and its relevance for animal behavior. Adv Stud Behav 27:1–131

von Krogh K, Sørensen C, Nilsson GE, Øverli Ø (2010) Forebrain cell proliferation, behavior, and physiology of zebrafish, Danio rerio, kept in enriched or barren environments. Physiol Behav 101:32–39

Weckwerth W (2010) Metabolomics: an integral technique in systems biology. Bioanalysis 2:829–836

Winberg S, Nilsson GE, Olsén KH (1992) The effect of stress and starvation on brain serotonin utilization in Arctic charr (Salvelinus alpinus). J Exp Biol 165:229–239

Wingfield JC (2003) Control of behavioural strategies for capricious environments. Anim Behav 66:807–815

Wiseman S, Osachoff H, Bassett E, Malhotra J, Bruno J, VanAggelen G, Mommsen TP, Vijayan MM (2007) Gene expression pattern in the liver during recovery from an acute stressor in rainbow trout. Comp Biochem Physiol D 2:234–244

Zhang Z, Ju Z, Wells MC, Walter RB (2009) Genomic approaches in the identification of hypoxia biomarkers in model fish species. J Exp Mar Biol Ecol 381:S180–S187

Injuries and Deformities in Fish: Their Potential Impacts upon Aquacultural Production and Welfare – Summary

Chris Noble, Hernán Cañon Jones, Børge Damsgård, Matthew Flood, Kjell Midling, Ana Roque, Bjørn-Steinar Sæther and Stephanie Yue Cottee

In our review paper we adopted a risk factor / intervention approach to improve fish welfare by i) identifying the factors that can injure farmed fish and ii) identifying and devising a number of mitigation strategies for reducing the frequency and severity of injuries across a range of farmed species and production systems.
Although the type and severity of numerous injuries can be species-specific, synergies did emerge between risk factors and mitigation strategies across a range of farmed species. For instance, many common aquacultural procedures such as handling and transporting live fish can cause injuries through e.g. abrasion with handling materials or the technologies used to move fish around and between farms. Farmers can implement management practices that minimise handling throughout the production cycle or use technologies that have a low impact upon the health and welfare status of live fish. In addition, the choice of feed type offered to the fish, and the method of feed delivery can affect the performance, behaviour and welfare of fish within the rearing system. For example, increased competition during feeding can lead to numerous injuries and these injuries can be reduced by using feeding technologies that provide fish with a responsive ration in relation to changes in their appetite and their dietary needs.

Background

The incidence of injuries or deformities can be used as robust and rapid indicators of the welfare of farmed fish throughout the production cycle. But fish farmers need to know how to use these indicators to improve their husbandry practices and fish welfare.

Scientific Significance

Our review helps farmers and researchers identify the key procedures that can injure fish, both in terms of their general management strategies and also in their daily husbandry practices. This information can be used to develop short- and long-term strategies for improving the welfare of farmed fish, based upon quantified and robust operational research.

Practical Application

The prevalence and type of injury can be used to identify the aquacultural procedures that are potentially detrimental to fish welfare. Once a farmer has identified potential welfare threats, they can then devise and implement potential short- and long-term management practices to reduce or eliminate these risks.
Our review also compiles recent relevant research on external injuries in fish into one practical, concise paper. Farmers can use the paper to i) identify potential welfare problems on their farm, and ii) devise ways to reduce these problems and improve the welfare of farmed fish.

Contact

C. Noble
Nofima, The Norwegian Institute of Food,
Fisheries and Aquaculture Research, Muninbakken 9-13,
P.O. Box 6122, 9291 Tromsø, Norway
e-mail: chris.noble@nofima.no

Fish Physiol Biochem (2012) 38:61–83
DOI 10.1007/s10695-011-9557-1

Injuries and deformities in fish: their potential impacts upon aquacultural production and welfare

Chris Noble · Hernán A. Cañon Jones · Børge Damsgård · Matthew J. Flood · Kjell Ø. Midling · Ana Roque · Bjørn-Steinar Sæther · Stephanie Yue Cottee

Received: 29 March 2010 / Accepted: 4 September 2011 / Published online: 15 September 2011

Abstract Fish can be the recipients of numerous injuries that are potentially deleterious to aquacultural production performance and welfare. This review will employ a systematic approach that classifies injuries in relation to specific anatomical areas of the fish and will evaluate the effects of injury upon production and welfare. The selected areas include the (1) mouth, (2) eye, (3) epidermis and (4) fins. These areas cover a large number of external anatomical features that can be injured during aquacultural procedures and husbandry practices. In particular, these injuries can be diagnosed on live fish, in a farm environment. For each anatomical feature, this review addresses (a) its structure and function and (b) defines key injuries that can affect the fish from a production and a welfare perspective. Particular attention is then given to (c) defining known and potential aquacultural risk factors before (d) identifying and outlining potential short- and long-term farming practices and mitigation strategies to reduce the incidence and prevalence of these injuries. The review then concludes with an analysis of potential synergies between risk factors the type of injury, in addition to identifying potential synergies in mitigation strategies. The paper covers both aquaculture and capture-based aquaculture.

Keywords Aquaculture · Deformities · Fish welfare · Injuries · Production

C. Noble (✉) · B. Damsgård · K. Ø. Midling · B.-S. Sæther
Nofima, The Norwegian Institute of Food, Fisheries and Aquaculture Research, Muninbakken 9-13, P.O. Box 6122, 9291 Tromsø, Norway
e-mail: chris.noble@nofima.no

H. A. Cañon Jones
Department of Veterinary Medicine, Wildlife Health, Fish and Conservation Medicine Group, University of Cambridge, Cambridge, UK

M. J. Flood
Australian Bureau of Agricultural and Resource Economics and Sciences, Department of Agriculture, Fisheries and Forestry, GPO Box 1563, Canberra, ACT 2601, Australia

A. Roque
Institut de Recerca i Tecnologia Agroalimentàries-Sant Carles de la Rápita, 43540 Sant Carles de la Rápita, Spain

S. Y. Cottee
Department of Animal and Poultry Science, University of Guelph, Guelph, ON N1G 2W1, Canada

Introduction

Injuries and deformities can occur in both wild (Slooff 1982) and farmed fish (Matsuoka 2003) and can occur at any time from the larval (López-Albors et al. 1995) to the adult stage (Korsøen et al. 2009) of the life cycle. In this review, the term injury is defined as 'physical damage', whilst the term deformity is defined as 'an acquired or congenital distortion of an organ or part'

(Collins English Dictionary 1998). Injuries are direct damage to live tissue (Ellis et al. 2008), and deformities are abnormalities that can lead to functional impairment. Both can be detrimental to fish welfare (Huntingford et al. 2006) and production performance. The subject of fish welfare is gaining prominence amongst researchers, aquaculturists, retailers, quality assurance schemes, NGOs and also consumers (e.g. Huntingford et al. 2006). Recent years have seen the aquaculture industry taking an active role in employing strategies that incorporate fish welfare considerations into daily production practices, and a number of national and international legislations and policies provide specific guidelines on preventing and dealing with injuries during common aquacultural production practices (e.g. the Council of Europe recommendations for farmed fish, 2005 http://www.coe.int/t/e/legal_affairs/legal_co-operation/biological_safety,_use_of_animals/Farming/Rec%20fish%20E.asp). A specific EU Council Directive (98/58/EC) also outlines minimum standards for protecting farm animals, including fish http://eur-lex.europa.eu/LexUriServ/LexUriServ.do?uri=CELEX:31998L0058:EN:HTML.

Any good aquaculturist will attempt to protect their fish from potentially injurious situations as part of their general moral and ethical responsibility towards their livestock. In addition, from a purely production perspective, injuries can lead to reduced growth (Miyashita et al. 2000), reduced feeding (Dyková et al. 1998), reduced feeding ability (Kurokawa et al. 2008), increased susceptibility to infection (Turnbull et al. 1996) and increased levels of mortality (Cobcroft and Battaglene 2009). Injuries can also reduce the market value of farmed fish (e.g. Michie 2001; the Norwegian Industry Standard for Fish http://fhl.nsp01cp.nhosp.no/files/Quality_grading_of_farmed_salmon.pdf). Farmers therefore have a vested interest in reducing the occurrence and incidence of injuries in farmed fish. An additional incentive for reducing the prevalence of injuries can be gaining added value, as fish produced to high welfare standards can command a price premium (Olesen et al. 2010).

An important step for improving the welfare of farmed fish is to reduce the occurrence of injuries and deformities. This can be accomplished through a risk factor/welfare intervention approach where one identifies potential aquacultural risk factors in addition to outlining potential short- and long-term farming practices and mitigation strategies. By utilising this approach, a fish culturist can identify procedures or husbandry practices that may be detrimental to fish welfare and production performance, and then devise and implement a number of mitigation strategies to reduce or eliminate these risks. To the authors' knowledge, there is no published review covering this burgeoning field of research, and none have utilised this risk factor/mitigation strategy approach.

This review will classify fish injuries and deformities in relation to specific parts of the anatomy and the selected areas include the (1) mouth, (2) eye, (3) epidermis and (4) fins. These areas cover a large number of external anatomical features that can be injured during aquacultural procedures and husbandry practices. Most importantly, each injury can be diagnosed on live fish and can be easily identified by eye, or by simple veterinary procedures. The review will then conclude with a summary of potential synergies between risk factors and injury type, before identifying potential synergies in operational farming practices and mitigation strategies. The paper will restrict its scope to the aquaculture and capture-based aquaculture of teleostean fish, and citations will not be exhaustive due to space limitations.

Mouth injuries

Mouth structure and function

In fish, the mouth is primarily used to take in water (and in some cases, air) and also to ingest food (Guillaume and Choubert 2001). Its location can be (1) superior, opening dorsally; (2) inferior, opening ventrally; or (3) terminal and located at the front of the head. The mouth consists of an upper and lower jaw, and the oral cavity contains a tongue. The upper jaw bones are classified as the premaxilla at the front, and the maxilla at the sides, and each premaxilla is joined anteriorly by a symphysis, whilst the lower jaw consists of a dentary bone (McEachron and Fechhelm 1998). If the fish possess teeth, they are primarily used for prey capture (Guillaume and Choubert 2001).

Types of mouth damage

Mouth damage and deformities occur in wild fish populations (Slooff 1982) and in aquaculture, but are more commonly seen in cultured fish (Sadler et al.

2001). Their occurrence has been documented throughout the lifecycle, from yolk sac larvae to adults (e.g. Lein et al. 1997; Miyashita et al. 2000).

Mouth deformities include (1) pug-headedness (Matsuoka 2003), (2) lower jaw deformity syndrome LJD (Sadler et al. 2001), (3) double mouth (Swan 1968), (4) cross bite (Barahona-Fernandes 1982) and (5) gaping jaws syndrome (Pittman et al. 1990) along with many documented incidences of deformity not given specific names. Mouth deformities and damage can also include one or more of the following: elongation or lengthening of either the upper or lower jaw (Matsuoka 2003), shortening of the entire snout (Yamauchi et al. 2006), the bending downwards or sideways of the lower jaw (Okamura et al. 2007), total absence of an upper jaw (Sumagaysay et al. 1999), the fusion of jaws in an open mouthed position (Roberts et al. 2001), pinching or twisting of the upper and lower jaws (Fraser and de Nys 2005), and the presence of extra bones or lack of bones (Martinez et al. 2007). Deformities caused by mechanical damage include broken bones (Miyashita et al. 2000), mouth lesions and haemorrhages, tissue erosion around the mouth and snout deformation (Shiau and Suen 1992; Cobcroft and Battaglene 2009). Mechanical erosion and haemorrhaging have been observed on the snout of rainbow trout *Oncorhynchus mykiss* and on the snout and underside of the lower jaw in tank held Atlantic cod *Gadus morhua* (observations of authors). In addition to the immediate detrimental effects of such mechanical damage, its occurrence can also make fish susceptible to infection by pathogens (Barthel et al. 2003). In most cases, mouth deformities occur in conjunction with deformation to the operculum and spine of fish (Sadler et al. 2001).

In Europe, Atlantic salmon is one of the most important aquacultural species by both volume and market value (FEAP statistics; Anon 2008). Two mouth deformities previously highlighted as serious welfare issues for this species are LJD and pug-headedness (or brachygnathia; Sadler et al. 2001; Branson and Turnbull 2008). LJD is characterised by downward curvature of the lower jaw and ankylosis of mandibular articulation which results in the jaw being locked in a permanently open position (Bruno 1990). This deformity can occur in conjunction with other skeletal deformities including short, folded and sometimes softened opercula (Roberts et al. 2001). LJD in Atlantic salmon has been recorded in Scotland (Bruno 1990), Ireland (Quigley 1995), Canada (McGeachy et al. 1996), Norway (Sadler et al. 2001), Australia (Jungawalla 1991) and Chile (Roberts et al. 2001) and affects salmon in both freshwater and seawater (Sadler et al. 2001; Roberts et al. 2001). It affects both males and females (Sadler et al. 2001), and its prevalence is also much higher in triploid than in diploid stocks (Jungawalla 1991).

The second mouth deformity previously highlighted as a welfare issue for Atlantic salmon is pug-headedness and is characterised by a severe underdevelopment of the upper jaw, resulting in the appearance of a much protruded lower jaw (Bæverfjord et al. 1998a). Whilst the number of fish that suffer from this is normally low, there can be serious welfare implications for individual fish with the deformity (Branson and Turnbull 2008).

As stated earlier, mouth deformities can develop throughout the lifecycle, but most literature sources relate to the early life stages. Much of this research specifically focuses on the difficulties of developing suitable husbandry practices for larval and juvenile fish to allow aquaculture production of these species. The production of high value species such as bluefin tuna (Miyashita et al. 2000), striped trumpeter *Latris lineata* (Cobcroft and Battaglene 2009) and Japanese eel *Anguilla japonica* (Okamura et al. 2007) are hindered by the occurrence of mouth damage and deformation during the early life stages.

Effect of mouth injuries on production and welfare

Mouth damage and deformity can affect fish from both a production and a welfare perspective. In terms of production, reduced market prices due to the unappealing appearance of fish with mouth deformities present a major problem for industry (Sadler et al. 2001). Given that mouth deformities often inhibit a fish's ability to ingest food (Branson and Turnbull 2008), growth rates can also be adversely affected and survival rates reduced (Miyashita et al. 2000; Cobcroft and Battaglene 2009).

From a functional perspective, fish with mouth damage and deformations face two main issues resulting from the inability to properly open and close their mouths. The first is an impaired ability to feed such as that seen in fish suffering from LJD and pug-headedness (Branson and Turnbull 2008). In extreme cases, mouth deformities and damage can stop fish from moving their mouths, leading to a total inability to eat, which can in turn leading to starvation and eventual death (Pittman et al. 1990). The second main problem facing these fish is breathing impairment resulting from a reduced ability to use buccal-opercular pumping to properly ventilate their gills. This again has been noted as an issue for fish with LJD and pug-headedness (Lijalad and Powell 2009). It is thought that fish with these mouth deformities may have to swim more than other fish in order to achieve sufficient ram irrigation to ensure adequate water flow over their gills. As a result, high stocking densities that can potentially inhibit swimming may further compromise the welfare of these fish.

It is still under debate as to whether fish can experience pain and suffering, and this is quite a controversial issue (Huntingford et al. 2006). Irrespective of this, previous research has shown that rainbow trout injected with a weak acetic acid solution or bee venom demonstrate behaviour and physiological changes suggesting discomfort (Sneddon 2003). These animals temporarily cease feeding, rest on the substratum and rub their snouts on the walls and base of their tanks. The use of morphine has been shown to reduce these adverse behaviours (Sneddon 2003). Whilst these findings may indicate simply a reflex response to potentially painful stimulus, i.e. nociception (Broom 1998) the possibility it indicates the presence of pain cannot be ruled out. With this in mind, damage to the mouth and jaws of fish should be regarded as a potential concern.

Further, it is likely that an inability to express normal behaviours may help explain the occurrence of mechanical damage to the mouths of cultured fish. In bluefin tuna culture, for example, night-time collisions with tank walls and cage netting are common during the juvenile and early adult stages (Ishibashi et al. 2009). In the wild, these fish would rarely encounter such obstacles as they are generally found in open pelagic waters.

Risk factors for mouth injuries

In the case of LJD in Atlantic salmon, two important risk factors have been identified relating to nutrition and genetics. Investigations into LJD in Chile indicated that its occurrence was associated with feed lacking sufficient phosphorus and vitamin C around the time of introduction to the sea (Roberts et al. 2001). In Australia, LJD has been linked with genetics, specifically ploidy and occurs more frequently in triploid than diploid fish. During the seawater phase, LJD has been shown to affect <30% of commercially produced all-female triploid Atlantic salmon (Jungawalla 1991). In addition, dietary deficiencies in phosphorus and vitamins A, C, D and K have also been suggested as causative factors for triploid fish (King and Lee 1993). In both Chile and Australia, it has been suggested that nutritional deficiencies may arise from the use of feeds designed for slower growing fish. This concept is supported in Chile where problems occur at a time of elevated temperatures with subsequently faster growth (Roberts et al. 2001).

Definitive risk factors for pug-headedness in Atlantic salmon have not been determined, but it has been suggested that the cause may be genetic or epigenetic (Slooff 1982). Two potential risk factors include egg incubation at inappropriate temperatures (Bæverfjord et al. 1998a) and exposure to heavy metals such as zinc (Slooff 1982). In addition to those already listed for LJD in Atlantic salmon, nutritional risk factors demonstrated in other species include diets deficient in niacin (Shiau and Suen 1992) and highly unsaturated fatty acids (HUFA; Kestemont et al. 2007). However, it is not only deficiencies that can cause problems. Excess vitamin A has also been shown to be a risk factor. In hatchery-reared flatfish, supplementation of diets with vitamin A is routinely used to stimulate pigmentation development. The problem is that excess dietary vitamin A can lead to high levels of its metabolite retinoic acid, which has been shown to dramatically increase the occurrence of mouth deformities (Suzuki et al. 2000).

The use of inappropriate rearing temperatures during the early development of fish is also a well-documented risk factor in the development of mouth deformities. During egg incubation and larval development, many fish are particularly susceptible to inappropriate culturing temperatures. Temperatures that are too high, too low or that fluctuate too much can potentially increase the incidence of jaw deformation in these early stages (Lein et al. 1997; Okamura et al. 2007). In addition to this, inappropriate salinities (Okamoto et al. 2009) and light intensities (Bolla and Holmefjord 1988) can also strongly increase the incidence of mouth deformities in these early life stages.

The presence of pollutants presents another risk. It has been reported that exposure to dioxins during the embryonic development of red sea bream *Pagrus major* can result in a shortened snout and deformity of the lower jaw (Yamauchi et al. 2006). Similarly, if the common bream *Abramis brama* is grown in polluted waters, they develop higher levels of mouth deformity, in most cases pug-headedness (Slooff 1982).

There are also many risk factors that specifically relate to the behaviour of fish and how they interact with the rearing environment. Self-inflicted mechanical damage such as that caused by physical contact with tank walls or cage nets is a major cause of mouth damage. In bluefin, tuna darkness is considered a major risk factor given that most collisions occur at night when these structures are not clearly visible (Miyashita et al. 2000), especially given the poor visual function of these fish in dim lighting conditions (Matsumoto et al. 2009). Flashing lights, loud noises and vibrations through the tank walls also cause fish to panic and increase collisions (Miyashita et al. 2000). Another risk factor relating to lighting is the length of dawn and dusk. It has been found that changing too quickly from night-time to daytime light intensities causes visual disorientation, thus increasing the risk of collision (Masuma et al. 2001). In the case of walling behaviour, it has been shown that for larval striped trumpeter, inappropriate tank coloration is a risk factor. The occurrence of walling behaviour and subsequent jaw deformations in this species changed depending on the colour of housing tank used (Cobcroft and Battaglene 2009). Collisions that do not result in instant death often cause injury to the snout and lower jaw (Miyashita et al. 2000).

Netting materials used to make sea cages and trawl nets may also represent a risk of mouth injury in both aquaculture and capture-based aquaculture via mechanical damage. The nets used to contain and capture fish present surfaces against which mechanical damage can occur. In bluefin, tuna fractures to the jaw bones often result from fish catching their mouths on nets (Miyashita et al. 2000). Nets may also cause abrasive damage in other species. This is possibly the case for Atlantic salmon, especially during the time immediately following transfer to sea. If the intensity of lighting in the freshwater hatcheries is markedly lower than that of the sea cages, these fish can exhibit light avoidance behaviours and push into the cage walls and base (Halls 1994). In this case, the cage netting and the increased light intensity may form a combined risk for mouth damage. In both sea cages and tanks, any overly abrasive structures or materials that fish may run into or rub up against also represent a risk for mouth damage. With respect to capture-based aquaculture, previous research has shown that Baltic cod *Gadus morhua* caught in trawl nets sustain damage such as lesions on their snouts and dark net marks behind their heads (Suuronen et al. 1996).

Mitigation strategies to reduce or eliminate risk factors for injury

In order to mitigate against mouth and jaw deformities and their resulting impacts upon welfare, it is important to ensure that where possible, the growing conditions are optimised in relation to the ecology of the fish. Mitigating measures against mouth deformities include ensuring the diets of fish contain appropriate levels of phosphorus, niacin, HUFA, vitamins A, C, D, K (Shiau and Suen 1992; King and Lee 1993; Roberts et al. 2001; Kestemont et al. 2007; Martinez et al. 2007). Rearing temperatures should be optimised to avoid the ranges where each species has proven susceptibility to increased mouth deformities (e.g. Lein et al. 1997). Rearing salinities (Okamoto et al. 2009) and light regimes (Bolla and Holmefjord 1988) that are known to reduce the occurrence of these deformities should be utilised. Fish should also be grown in water free from pollutants such as dioxins and heavy metals (Yamauchi et al. 2006), and where acute exposure to these pollutants occurs, rearing systems should be rapidly flushed with contaminant free water.

In the case of LJD in Atlantic salmon, mitigation measures include the adequate formulation of diets to ensure that nutritional requirements are met, especially when growth rates are higher than normal (Roberts et al. 2001). In the case of triploid fish, it has been suggested that these fish have different dietary utilisation and uptake abilities than diploid fish (Sadler et al. 2001). Mitigation measures may therefore require identifying any such differences and altering diet formulation for triploid fish accordingly.

In order to mitigate against mechanical damage resulting from physical contact with tank walls or cage nets, a number of measures may be appropriate. In bluefin tuna, the use of all-night lighting has been shown to reduce collisions. The removal of excess stimulation such as flashing lights, loud noises and vibrations also helps reduce panic and subsequent collisions (Miyashita et al. 2000). The use of lighting strategies in Atlantic salmon hatcheries that minimise the difference between hatchery intensities and the ambient levels in sea cages may help prevent fish exhibiting light avoidance behaviours when first introduced to cages (Halls 1994). Walling behaviour can be reduced by choosing appropriate tank colours (Cobcroft and Battaglene 2009) or using turbid culturing conditions (Bristow and Summerfelt 1994). As mentioned earlier, appropriate choices of materials used for cages and trawl nets will also help mitigate against mouth damage through abrasion.

Eye injuries

Eye structure and function

The basic eye anatomy of fish is very similar to that of other vertebrates. The globe is maintained in position by three pairs of oculomotor muscles, which are attached to the sclera and are innervated by the third cranial nerve. The sclera is made of laminated fibrous layers reinforced with hyaline cartilage. Eyelids and lacrimal apparatus are not present as they are not needed in the aquatic environment (Branson 1993). The cornea is the anterior transparent window of the eye which permits a sharp image of the external environment to be displayed onto the retina. The cornea consists of the outermost epithelium, stroma and innermost endothelium layers. The corneal epithelium acts as a barrier to the fish's external aqueous environment (Ubels and Edelhauser 1987), and the integrity of the corneal epithelium is essential for normal vision. The cornea contributes to the structural integrity of the globe and protects the inner eye from invasion of unwanted organisms and environmental changes, whilst balancing the intra-and extra-ocular pressure (Collin and Collin 2001). The lens is generally spherical and protrudes through the iris, generating a very wide angle of view. The iris is fixed and has weakly developed sphincter muscles, possibly because fish rarely need to rapidly adapt to changes in light levels (Branson 1993).

Types of eye damage

Probably the most frequently described eye condition found in almost any fish health textbook is the condition of 'pop eye' or exophthalmia. Exophthalmia can be caused by many factors (Bouck 1980), but sometimes it can be caused by gas bubble disease. Characteristic signs of gas bubble trauma are gas bubbles in the blood, lateral line, gills filaments, fins, under the skin and of course, the eyes. Emboli and gas bubbles can only form when the sum of the dissolved gas pressures exceeds the sum of the hydrostatic pressure—simply put super-saturation of gas in the water (Bouck 1980). As a general rule, anything more than 110% air saturation will eventually cause gas bubble disease. If gas bubbles occur between the cornea and the lens of the eye, the eye may swell and protrude or 'pop' out (van Duijn 1973).

Other injuries can be a product of eye snapping. Eye snapping is an odd problem that happens when rays of light enter a tank and reflect off the eyes of fish. The reflected flash attacks from conspecifics and can result in one-sided periocular dermatitis, ulceration of the cornea and even loss of the eye. The angle at which light enters the tank in combination with fish distribution can favour damage to the right or left eye (Speare 2008).

A common form of eye damage is cataracts, which are opaqueness or clouding of the eye lens. Several reported factors can cause cataracts including nutritional deficiencies, toxic agents, parasites, exposure to ultraviolet light, hereditary factors, variation in water temperature and rapid growth (reviewed by Björnsson 2004).

Effect of eye injuries on production and welfare

There is concern that eye wounds can lead to secondary bacterial infections (Greaves and Tuene 2001) and that fish with eye injuries may also be subject to parasite infections. Eye flukes generally decrease the size of the eye lens, which can lead to deleterious effects like cataract formation; the amount of lens area covered by cataracts increases with infection intensity, and in some cases, eye fluke infection can cause blindness (Karvonen and Seppälä 2008). Other changes may also be noticeable with parasite infection, including increased ventilation rate, heart rate and swimming activity (Laitinen et al. 1996).

Regardless of the type or severity, eye damage has detrimental effects upon fish welfare as it can increase behavioural and physiological stress (Thatcher 1979).

At worst, most consider injuries to the eyes as potentially lethal (Ostrand et al. 2006). At best, defective eyes or visual impairment can decrease feeding ability and reduce avoidance behaviour as fish are not able to discern the exact location of their target, whether it be a pellet of food or avoiding a dip-net (Thatcher 1979; Mesa et al. 1994).

Risk factors for eye injuries

Common aquaculture husbandry and management practices such as handling and netting the fish during for example vaccination, grading and bathing procedures can cause eye injuries. For example, if many fish are netted or lifted at once, the weight of the fish bearing down can injure those on the bottom. Lift nets and different pumping technologies can cause injury (Grizzle et al. 1992), and even turbulent shear flow of water directed at fish from certain angles in turbines can cause damage to their eyes (Deng et al. 2005). Grizzle et al. (1992) found higher incidence of injuries such as abrasion and lacerations using a turbine pump and netting accompanied by an increase in injury related enzymatic activity (LDH and AST) in channel catfish. Interestingly, Grizzle and Lovshin (1994) later found that extreme turbine pump speeds (390 rpm vs. 330 rpm) resulted in a higher incidence of injuries in the same species.

Behavioural interactions can lead to eye damage, and eye injuries are a significant concern for halibut producers, given that aggressive behaviours including 'eye snapping' has been known to happen especially amongst juvenile halibut (Ottesen and Strand 1996). In a study of aggressive behaviour in farmed Atlantic halibut, Greaves and Tuene (2001) found that accidental or deliberate aggressive contact between fish primarily occurred during feeding periods, resulting in an estimated 3–5% of young halibut have at least one damaged or absent eye.

UV light exposure is a documented risk factor for cataract formation in a number of farmed species. For example, in a study where juvenile Atlantic cod were reared in recirculating aquaculture facilities, the frequency of cataracts significantly increased after UV water disinfection was utilised (Björnsson 2004). The increase was greatest in the tank closest to the UV light. Interestingly, UV light does not have to shine on fish directly so long as the water is treated with UV. UV-induced cataracts can also be a problem in farmed halibut (Treasurer et al. 2007). Cataracts are also a problem in farmed Atlantic salmon. Histidine deficiency has been suspected of being a contributing factor. It has been shown that increasing dietary histidine and adding 5% NaCl had decreased cataract frequency and severity (Rhodes et al. 2010). Fluctuating water salinity and increases in water temperature can also be factors that contribute to cataract development (Bjerkås and Sveier 2004). Brandt et al. (1986) noted corneal cloudiness and cataract formation following transport in largemouth bass *Micropterus salmoides*. Ubels and Edelhauser (1987) hypothesised that corneal opacity was due to the repeated contact of fish eyes with spines and scales of tank-mates, nets and tank walls during the high-density confinement typical of transport. Besides abrasions caused by collisions with other fish, abrasions can be caused by fish brushing against netting. Fish reared in net pens in marine or lake environments, especially, are contained within netted walls. Sometimes nets harbour hard-shelled organisms and fish become injured after swimming against them, losing scales and increasing their risk of developing bacterial and viral infections (Braithwaite and McEvoy 2005).

Eye flukes of the genus *Diplostomum* possess a stage in their life cycle where they infect the eyes of fish. *Diplostomum* penetrate the lens capsule and access the eye lens; this may cause lens material to leak into the surrounding tissues and cause a subsequent reduction in lens size. These parasites can also establish themselves in the lens, growing and developing for several weeks, during which time, they may damage the physiology of the lens and exploit energy resources that would otherwise be used by the host for normal eye functioning (Karvonen and Seppälä 2008).

Mitigation strategies to reduce or eliminate risk factors for eye injury

As the majority of the injuries described revolve around wounding, damaging or causing abrasions to the eye, it is logical that any husbandry or management practice involving the handling of these animals be carefully considered as to its design; the same applies to equipment. For instance, fish pumps, nets, sorting and grading apparatus come in a variety of materials and designs. Hence, there has been much commercial interest in identifying the methods that reduce the possibilities of damage when moving fish. Moving and transferring fish by pumping them in pipes where they are not removed from water has been considered a less damaging option than netting (Conte 2004) as the pump can move large quantities of fish quickly and without injury. One such pump is a hidrostal pump—a screw-type centrifugal impeller. Helfrich et al. (2004) looked at the effects of live transport on mortality, descaling and injury rate in striped bass *Morone Saxatilis* and rainbow trout using this pump and found the frequency of eye injuries was 2.8% for bass and 0% for trout. Similar studies of large hidrostal pumps on other species of fish, such as splittail *Pogonichthys macrolepidotus* and Chinook salmon *Oncorhynchus tshawytscha,* have yielded comparable results (Helfrich et al. 2001). Helfrich et al. (2004) have therefore suggested that large hidrostal pumps be used to transport live fish at high densities.

Greaves and Tuene (2001) have suggested that injuries can be minimised by reducing aggression around meal times. This can be achieved by dispersing a satiation ration over a wide surface area for example by using appetite-based demand feeding systems. This should diffuse the sudden burst of fish activity and reduce the occurrence of competition, aggression and accidental fish collisions. In addition, manipulating how light or shade hits the top of a tank may reduce or prevent eye-snapping behaviour. With regard to controlling eye fluke infections, it has been suggested to eliminate water snails (the intermediate host in *Diplostomum*'s 3-host life cycle). Molluscicides, filtration and removal of vegetation can help (Southgate 1993). The risk of inducing corneal opacity during live transport can be minimised by regularly monitoring for corneal abrasions in fish operations as abrasions can easily be detected by touching a fish's eye with a sodium fluorescein ophthalmic strip (Ubels and Edelhauser 1987).

Finally, although cataracts can cause serious problems in an array of farmed species, studies have shown that there are different strategies to combat this condition. Nutritional, toxic, parasite, UV light, hereditary and environmental factors have all been associated with cataract formation (reviewed in Björnsson 2004); therefore, it is important to monitor these aspects. Limiting the amount of exposure to direct and indirect sources of UV light has been shown to reduce cataract development (Björnsson 2004), and these authors suggested that systems are designed so that they allow the UV-treated water to first flow though a large enough reservoir tank before flowing into rearing tanks. Another way to help mitigate nutritionally related cataract formation is to balance the diet (Waagbø et al. 2003). Parasite-induced cataracts (Karvonen et al. 2004) may be alleviated by rapidly treating parasite infections. In addition, maintaining an optimal water temperature (e.g. avoiding warm water) and preventing temperature fluctuations can also help to prevent the development and severity of cataracts (Bjerkås et al. 2001).

Epidermal injuries

Epidermal structure and function

The epidermis of fish mainly consists of two types of cells: filamentous cells and mucous production cells. They are arranged according to three distinct but continuous zones (basal, medial and peripheral) of the fish epidermis (Harris and Hunt 1975). The epidermis represents a biological barrier between the fish and the aquatic environment, conferring protection against infectious diseases (Ellis 2001), friction (Daniel 1981) and pollution, and it is actively involved in ion regulation (Hoar and Randall 1969).

Types of epidermal damage

Skins injures and lesions have been defined as a visible loss of epidermis accompanied by changes in skin colour, haemorrhages or ulcers involving dermal, subdermal and/or muscle tissues occurring in any part of the body (Vågsholm and Djupvik 1998). Skin lesion also includes papillomatous lesions characterised as uni- or multi-focal mass of hypertrophied skin, absence of hyperplasia and ulceration (Ottesen et al. 2007).

Effects of epidermal injuries on production and welfare

From a production perspective, a direct decrease in the average price of 2% occurs when injuries are present in 10% or more of the fish (Vågsholm and Djupvik 1998). Also, consumers are currently demanding high-quality seafood products, and gross appearance, including skin, appears to be a quality selection parameter (Huidobro et al. 2000). There is no direct association between skin injuries and fish mortalities, but several bacterial infections such as winter ulcers (Løvoll et al. 2009), piscirickettsiosis (Smith et al. 1999) or infectious salmon anaemia (Totland et al. 1996) can colonise the initial lesion and lead to mortalities. For example, mortalities due to winter ulcers are usually <10%, but can also lead to decreased fillet quality at slaughter, having a major dual impact on the productivity of a farm (Løvoll et al. 2009). Also, infectious salmon anaemia outbreaks have been associated with skin lesions (Nylund et al. 1994) and injuries which occur previous to or concurrently with sea lice infestations of *L. salmonis* and *C. elongatus* have been implicated with high mortalities (Boxshall and Defaye 1993).

Bacterial skin infections and parasitic infestations can be treated with antimicrobials or antiparasitic drugs whose purchase directly increases production costs or can increase production losses through reduced appetite and increased mortalities. This reduced appetite or anorexia can lead to reduced growth (Vågsholm and Djupvik 1998).

Skin injuries and lesions have a direct effect upon nociception, as fish possess free nerve cells close to and throughout the skin surface (Kotrschal et al. 1993). In addition, skin injuries have a direct effect on disease susceptibility and are a fertile environment for the proliferation of bacterial and viral diseases and perpetuation of the skin damage (Nylund et al. 1994).

Risk factors for epidermal injuries

Several aetiological agents for epidermal injuries have been identified, and these can be divided into abiotic and biotic factors (Tørud and Håstein 2008). Causes of skin injuries related to common husbandry practices such as sorting, pumping and netting have been investigated in several farmed fish (see earlier section on eye injuries). The main skin lesions associated with these practices were skin abrasions characterised by a discontinuity of the epidermis and accompanying subcutaneous haemorrhaging. Mechanically induced skin injuries have also been described in the abocular side of flat fish such as Atlantic halibut and southern flounder *Paralichthys lethostigma* under commercial conditions (Ottesen and Strand 1996). In Atlantic halibut, smooth tank surfaces have been associated with the occurrence of papillomas, characterised as masses with multiples nodules and skin erosion consisting mainly of eroded epithelium and hyperplasic areas (Ottesen et al. 2007). Further, skin lesions in Atlantic salmon have been associated with vaccination against *Vibrio anguillarum*, *Vibrio salmonicida* and *Aeromonas salmonicida* sp *salmonicida* using plant oils as vaccine adjuvants instead of mineral oils (Vågsholm and Djupvik 1998). The same authors investigated other risk factors in Atlantic salmon at the slaughter line and found a higher risk of skin injuries in smaller fish, in fish slaughtered in summer or fish that have been held under an extended seawater growth phase.

Other management practices such as preventive or therapeutic formalin baths can cause thinning of the epidermal strata (Sanchez et al. 1998), which leads to mucous cell opening (Buchmann et al. 2004) that can be associated with bacterial infections such as *Flavobacterium psychrophilum* (Madsen and Dalsgaard 1999) that perpetuate the initial skin lesions. High mucosal cell differentiation and openings are found using high doses (200–300 ppm for 1 h) or low doses for longer time periods (50 ppm for 24 h; Buchmann et al. 2004). Ultraviolet radiation has been identified as a risk factor for the development of skin lesions in broodstock rainbow trout in Bolivia (Bullock and Coutts 1985) and during the summer season in Atlantic salmon in Ireland (McArdle and Bullock 1987). Presumably, latitude (associated with season) could also be considered as a risk factor in aquaculture farms located in the southern hemisphere where the protective atmospheric ozone layer is thinning as this has been associated with detrimental effects on other aquatic organisms (Villafane et al. 2001).

Biotic risk factors for skin injuries are varied and can include bacterial, parasitic and fungal fish pathogens, and predators (Vågsholm and Djupvik 1998). *Moritella viscosa* has been described as one of the major bacterial fish pathogens that can cause winter ulcer syndrome, characterised by skin injuries and lesions in Atlantic salmon, Atlantic cod, turbot *Psetta maxima* and Atlantic halibut (Lunder et al. 1995; Björnsdóttir et al. 2004; Løvoll et al. 2009). The clinical signs of the disease appear during the cold season when water temperature drops below 10°C. Associated skin injuries can cover up to one-third of fish body and are characterised by severe ulcerations exposing subcutaneous and muscle tissue accompanied with liquefactive necrosis and haemorrhaging of the surrounding tissues. Parasitic infestation such as sea lice (*Lepeophteirus salmonis*, *Caligus elongatus* and *Caligus rogercresseyi*) can cause severe itching, skin irritation and ulcerations in farmed Atlantic salmon and trout (Boxshall and Defaye 1993). Risk factors include high stocking densities, hydrological factors (tides and water current) and delays to implement biosecurity measures such as cage and farm disinfections (Hardy-Smith 2006). Further, skins injuries and ulcers can be caused by *Saprolegnia* spp. colonisation of the epidermis (Dykstra et al. 1989). Other biological causes of skin lesions in farmed fish include stings from jellyfish blooms leading to ulceration and subsequent secondary bacterial infections (Tørud and Håstein 2008). In addition, in farmed olive flounder, the parasite *Philasterides dicentrarachi* infiltrates the epidermis, producing skin injuries and lesions histologically characterised by the presence of necrosis, red blood cells and a low inflammatory responses with few monocytes, macrophages and thrombocytes (Jin et al. 2009).

Mitigation strategies to reduce or eliminate risk factors for epidermal injury

Potential operational interventions to decrease skin injuries and lesion include mitigation of contributing risk factors and causes. As such, the use of vacuum pumps to transport or sort fish into well boats should be encouraged, as well as the use of intermediate pump velocities whenever turbine pumps are employed. Similarly, tanks with rigged floors should be used in flatfish aquacultural activities. Chemical baths used for the treatment or prevention of diseases should be utilised in low quantities and for the shortest time possible (Buchmann et al. 2004). Whilst prevention of infectious disease is important, farmers should use vaccines based on mineral oil coadjuvant instead of plant oil in order to decrease the risk of skin injuries (Vågsholm and Djupvik 1998). Vaccine development companies should also be encouraged to change their coadjuvant and demonstrate no skin injuries after inoculation. A well-implemented biosecurity plan should be set-up to decrease the spread of fish infectious diseases associated with skin injuries. Finally, although environmental factors such as water temperature, water currents and tides and ultraviolet exposure are difficult or impossible to control in cage aquaculture, these factors should be taken into account when setting up new farm sites or expanding existing farm production facilities.

Fin injuries

Fin structure and function

The teleost fin consists of an epithelial layer or fold of skin supported by fin rays, which form part of a larger skeletal structure (see Videler 1993). The fin rays are controlled by a series of depressor, erector and inclinator muscles (see Winterbottom 1974) and are connected to each other and the endoskeleton by collagenous fibres. The base of the fin rays can be individually rotated (Westneat et al. 2004), and fish can also adjust the curvature of a fin ray (Standen and Lauder 2005), allowing a fish to rotate, spread, undulate, elevate and depress a fin (Videler 1993) in relation to manoeuvring requirements or environmental disturbances such as variable flows or currents.

It is widely accepted that the primary functions of fins are to help a fish to control its posture, and to generate and control propulsion during locomotion (e.g. Standen and Lauder 2005). For example, the caudal fin of teleosts is primarily used to generate forward locomotion, but may also have a role in generating lift (Lauder 2000). The dorsal fin can enhance stability and can also generate thrust in salmonids (Drucker and Lauder 2005), and its elevation or extension may also be used to communicate status during aggressive contests (e.g. Abbott and Dill 1985). The role of the adipose fin in salmonids is somewhat uncertain, but it has a hypothesised role in locomotion (Drucker and Lauder 2005). With regard to the paired fins, the pectoral fins can be utilised in braking manoeuvres, turns and can also aid stability in salmonids (Drucker and Lauder 2003). Both the pectoral and pelvic fins can be used to augment station holding in juvenile salmonids (Arnold et al. 1991), whereas the pelvic fins can be used by males to grasp females during reproductive rituals and also during intra-specific aggressive encounters between males in Atlantic cod (e.g. Brawn 1961).

Types of fin damage

Fin damage has been recorded in wild fish, especially in degraded habitats (see Latremouille 2003 for a review) but is more common under aquaculture conditions (Bosakowski and Wagner 1994). Fin damage can be described in numerous ways, but the basic forms of damage can be divided into three categories: (1) splitting, (2) erosion and (3) thickening (e.g. MacLean et al. 2000). An additional form of fin damage can also be malformed fins (Turnbull et al. 1996). Splitting refers to splits or clefts in epithelial tissue between fin rays. It appears to be a relatively mild form of fin damage that leaves the skeletal fin rays intact and undamaged. Erosion refers to damage that results in loss of both the epithelial fin tissue and also the whole or part of the fin ray (Turnbull et al. 1996). Thickening refers to nodular, opaque thickening along the distal edge of an affected fin. This thickening appears to be the result of a hyperplasia in the fin epithelia and may contain some necrotic cells. Deformed or malformed fins contain twisted or abnormally branched fin rays that can lead to folding or twisting of the entire fin structure (see Turnbull et al. 1996). All forms of fin damage may result in diffuse haemorrhaging throughout the fin tissue or focal haemorrhaging in proximity to the fin base.

Fins have a capacity for healing and regeneration, and the regeneration process has been extensively reviewed by Akimenko et al. (2003). In brief, epithelial healing begins with the rapid migration of epithelial cells to form a wound epidermis. The formation of a blastema then allows the re-growth of fin rays and connective tissue, permitting fin regeneration (Akimenko et al. 2003). Regeneration may not be a straight forward process, as regenerated fins may also contain twisted fin rays, or may exhibit abnormal branching patterns (Turnbull et al. 1996). Previous research has reported that fins may not regenerate if there is complete erosion of fin ray and epithelial tissue (Goss and Stagg 1957). However, a recent study (Shao et al. 2009) reported that four species of fish including zebrafish *Danio rerio* and koi carp *Cyprinus carpio* can regenerate an amputated caudal fin, even when the amputation is at the level of the endoskeleton.

Effect of fin damage on production and welfare

Fin damage can affect fish from both a production and a welfare perspective. Fin damage may have a detrimental effect upon two common production performance parameters, growth and survival and may also affect product quality. It may also impact upon welfare due to being a direct injury to living tissue (Ellis et al. 2008) and may increase susceptibility to infection and potentially reduce swimming ability.

From a production perspective, there is no published, quantifiable data on the effect of different types of fin damage upon product quality. However, some authors have implied from personal experience or unpublished data that fin damage reduces the market value of an affected fish from both table fish processors and live fish buyers for re-stocking purposes (e.g. Hoyle et al. 2007). Further, fin damage can be used as a potential quality indicator (Winfree et al. 1998). These statements are supported by literature from industrial sources and quality assurance schemes. For example, the processing and quality grading scheme of the Alaska Quality Seafood Program http://www.alaskaqualityseafood.com/pdf/h_g_specifications.pdf, incorporates both fin loss and fin erosion into its quality classifications. In addition, the Norwegian Industry Standard for Fish http://fhl.nsp01cp.nhosp.no/files/Quality_grading_of_farmed_salmon.pdf also incorporates the presence/absence of fin damage into their quality scheme. Further, the farmer's standard quality assessment procedure, the Quality Index Method (QIM), incorporates fin haemorrhaging into its assessment criteria (Knowles et al. 2007).

A concise and well-researched review of the potential effect of fin damage upon fish welfare was carried out by Ellis and his co-authors in 2008, and they highlighted the lack of quantified studies that have directly investigated the effects of fin damage upon fish welfare. In the absence of this direct evidence, a number of associative inferences must be drawn from other studies.

To the authors' knowledge, the effect of fin damage upon on-farm growth performance and survival has not been documented. However, the potential effects of fin damage can be inferred from tagging studies investigating the effect of fin clipping upon growth and survival. The effects upon growth and survival are dependent upon the type and number of fins removed. For example, the removal of a single fin does not reduce growth in rainbow trout, but does reduce survival, and the level of mortality depends upon the type of fin removed (Nicola and Cordone 1973). Further, the survival of fish with multiple fin excisions is lower than those which have a single fin removed (Mears and Hatch 1976). In a study by Coble (1967), the growth performance of yellow perch *Perca flavescens* was only affected when the pectoral fin was clipped. The causes and mechanisms responsible for reduced growth and survival have not been elucidated in the above studies.

Fin damage is a direct injury to living tissue (Ellis et al. 2008), and this tissue contains nerve cells both within and between fin rays (Roques et al. 2010). These nerve cells contain C-fibres and A-δ fibres that are involved in pain perception, and Roques et al. (2010) have validated an acute response to a painful stimulus, caudal fin clipping, in Nile tilapia *Oreochromis niloticus*. This response was quantified via increased swimming activity up to 6 h after the damage occurred and also transient increases in both gill Na+/K+-ATPase activity and branchial mucus cell activity 1 h after clipping. Irrespective of the debate on whether fish feel pain or not (see Yue Cottee in press), the fins are nociceptive (Chervova 1996) and are capable of detecting noxious stimuli via the nervous system. Fin damage may also be detrimental to fish health. For example, fin damage may increase a fish's susceptibility to infection via opportunistic pathogens such as *Aeromonas salmonicida* which causes furunculosis (Turnbull et al. 1996). The base of the fin is also a documented entry point for infectious haematopoietic necrosis virus, IHNV (Harmache et al. 2006).

One could surmise that fin damage could reduce fin function and have a detrimental effect upon a fish's capacity to control its posture or reduce its swimming ability (as noted in Barthel et al. 2003) during routine swimming or feeding. However, this would depend upon the severity and type of fin damage (erosion, splitting) and also the frequency and type of fins affected. For example, recent research on the effect of experimentally induced pectoral fin erosion upon the prey-capturing ability of bluegill sunfish found that a reduction in 35% of pectoral fin area did not have an effect upon maximal swimming velocity or braking capacity during prey capture (Higham et al. 2005). The authors suggested bluegill sunfish make behavioural compensations to adjust for reduced pectoral fin size. This same result was also noted in muskellunge *Esox masquinongy* that were subject to clipping of one or both pectoral and pelvic fins or all paired fins (Wagner et al. 2009). However, complete loss of pectoral fins due to experimental amputation can affect fish in other ways for example by reducing the station-holding capacity of Atlantic salmon parr (Arnold et al. 1991).

Risk factors for fin damage

It should first be noted that there are a multitude of factors that may cause or aggravate fin damage amongst farmed fish, and these factors can interact with others to exacerbate this damage. In addition, one risk factor may mask or be related to another for example ration size can affect fin damage (Moutou et al. 1998), but this may be due to competition and increased levels of aggression. For additional summaries of the numerous non-aquacultural factors implicated with fin damage, a reader can refer to two previous reviews (Latremouille 2003; Ellis et al. 2008).

Both the choice of rearing system and the choice of its construction material can affect fin damage. For example, when rainbow trout are reared in re-circulating aquaculture systems versus flow-through tanks, fish in the re-circulation systems had greater pectoral and dorsal fin erosion than their counterparts. The authors suggested this was due to changes in tank flow dynamics (Roque d'Orbcastel et al. 2009). Concrete raceways also lead to greater ventral fin erosion in rainbow trout than raceways that have cobbles incorporated into their substrate (Wagner et al. 1996). Baffles in raceways also lead to greater pectoral erosion (Barnes et al. 1996). Recent research on the effects of submersible cage culture upon fish performance and welfare reported that short-term submersion (ca 3 weeks) has no effect upon fin damage (Dempster et al. 2009), whilst longer-term submersion (>6 weeks) is detrimental to fin condition in Atlantic salmon (Korsøen et al. 2009). The choice of light regime can also influence the incidence of fin damage. When silver catfish *Rhamdia quelen* are held under 24 h darkness, they exhibit no fin damage in comparison with those held under 24 h light or a 10:14 LD cycle (Piaia et al. 1999).

There are numerous husbandry practices that can be detrimental to fin damage such as transferring fish using a turbine pump rather than a vacuum pump (Grizzle et al. 1992) and handling fish with knotted instead of knotless nets (Barthel et al. 2003). Feeding practices and the nutritional content of feed can also influence the incidence of fin injuries. Numerous studies have shown that underfeeding can increase the prevalence of fin damage (e.g. Moutou et al. 1998), whereas others have reported no effect (e.g. Klontz et al. 1991). Feeding frequency can also be a risk factor for fin damage. When rainbow trout receive a daily satiation ration from self-feeders, reducing feeding frequency from 3 meals to 1 meal a day significantly hinders recovery from historical dorsal fin erosion (Noble et al. 2007a). However, when fish receive a fixed satiation ration, an increase in the feeding frequency from 1 meal to 3 meals a day increases pectoral fin damage in rainbow trout (Rasmussen et al. 2007). The authors' attributed this increase to escalated competition around repeated feed events. Numerous studies have also reported that diet formulation, such as the choice of fishmeal versus krill as a protein or lipid source, can increase fin damage (Lellis and Barrows 1997).

The relationship between stocking density and the incidence of fin damage is somewhat unclear and may be related to other factors such as poor water quality or feed access. Some studies have suggested fin damage escalates with increasing stocking density in rainbow trout (North et al. 2006) and sea bass (Person-Le Ruyet and Le Bayon 2009), but others have reported that stocking density has no effect in Atlantic salmon (Hosfeld et al. 2009). The potential detrimental effects of stocking density can also depend upon the species under investigation, as previous work by Siikavuopio and Jobling (1995) has shown that fin damage decreases at increasing stocking densities in Arctic charr *Salvelinus alpinus*. Further work by Rasmussen et al. (2007) found that the effect of stocking density upon fin damage differs depending on which fin is examined. At low densities, the anal fin of rainbow trout was in better condition than the anal fins of fish under higher densities, whereas the dorsal fin was in better condition at high compared to low densities. However, group living does appear to be detrimental to fin damage, as individually held fish can have better fins than those held in groups (Winfree et al. 1998).

Other abiotic factors such as temperature can affect the prevalence of fin damage amongst farmed fish, but the effects again differ with species. Fin damage increases in rainbow trout (Winfree et al. 1998) and sea bass (Person-Le Ruyet and Le Bayon 2009) when temperatures are high, whereas it increases in Atlantic salmon when temperatures are low (Schneider and Nicholson 1980). Wagner et al. (1998) found no clear effects of temperature upon fin damage in Bonneville cutthroat trout, *Oncorhynchus clarki*. Oxygen saturations greater than 100% have also emerged as a risk factor for fin damage in sea bass (Person-Le Ruyet and Le Bayon 2009). Exposure to high light intensity can lead to sunburn and increased damage to the upper caudal and dorsal fins (Bullock 1988). Bosakowski and Wagner (1994) reported increased fin erosion in trout exposed to low alkalinity and high ammonia levels.

Triploid rainbow trout exhibit less fin erosion than diploids (Wagner et al. 2006), and genetic strain can also affect the incidence of fin damage, with albino fish exhibiting less erosion than their pigmented counterparts (Wagner et al. 1996). Sexual maturity can also play a role in the degree of fin damage exhibited by fish. Mature Atlantic salmon male parr (precocious parr) exhibit less fin erosion than immature parr (Mork et al. 1989), but this may be related to behavioural adaptations on the part of mature precocious males to avoid conflict with conspecifics. Intraspecific competition in the form of overt biting has also been identified as a risk factor for dorsal fin damage (Cañon Jones et al. 2010). Physiological stressors may also play a role as cortisol-treated fish exhibited greater fin damage than their counterparts in experimental studies on rainbow trout (Gregory and Wood 1999). Further, parasites such as sea lice, *Caligurus* spp., have also been reported as risk factors for fin damage (Dawson 1998).

Mitigation strategies to reduce or eliminate risk factors for fin damage

In order to mitigate against fin splitting, thickening and erosion, it is important to identify and define straightforward operational strategies that can be carried out either during the setting up of a farm or during its daily operations. The correct choice of tank, net or raceway construction materials can reduce the prevalence and severity of fin erosion. Incorporating cobbles into the substrate of concrete raceways reduces erosion of the ventral fins (Wagner et al. 1996). Introducing some form of environmental enrichment, such as branch structures, overhead cover may also improve fin condition in steel head *Oncorhynchus mykiss* (Berejikian and Tezak 2005). With regard to submersible cage culture, Atlantic salmon cages should only be fully submerged for less than 3 weeks without surface exposure (Dempster et al. 2009). Where possible, water current should be increased in the rearing unit as this reduces fin damage in Arctic charr and Atlantic salmon (Jobling et al.

1993). This has the added benefit of exercising the fish, which can also improve fin condition (Jørgensen and Jobling 1993). In net culture, cages could be located in high flow areas to achieve the same effect, or a farmer could deploy recently developed technologies that induce fish swimming and exercise by utilising their optometric response.

To reduce the amount of fin damage that daily husbandry practices can inflict upon their fish, farmers should use knotless nets when handling their fish (Barthel et al. 2003), and limit the frequency of handling wherever possible. Further, during fish transfers, fish should be pumped rather than netted and fish should be pumped using vacuum pumps rather than turbine pumps (Grizzle et al. 1992). Feeding regimes that incorporate responsive rations via demand feeding technology can reduce fin damage in both cage and tank culture (Noble et al. 2007a, b; Suzuki et al. 2008), and this can in turn improve welfare. These demand feeding systems allow fish to dictate the size, frequency and timing of their daily ration and can be tailored for each rearing system (e.g. Blyth et al. 1993; Alanärä 1992). Formulating diets that match the nutritional requirements of the fish will also reduce the capacity for fin damage within a rearing system, such as replacing fish meal with a krill based diet in rainbow trout (Lellis and Barrows 1997).

Growing the fish within suitable ranges of (1) temperature (e.g. Person-Le Ruyet and Le Bayon 2009), (2) oxygen saturation (Person-Le Ruyet and Le Bayon 2009) and (3) light intensity (Bullock 1988) can decrease the risks for fin damage. This can be achieved through improved oxygen monitoring to maintain levels at an optimum. Raceways can be covered or shaded to reduce light intensity levels or reduce the risk of sunburn if this emerges as a risk factor. Increased water exchange rates will also reduce the risk of fin damage (Good et al. 2009) and will also reduce the risk of a build up of ammonia levels or reduce the risk of low alkalinity (Bosakowski and Wagner 1994).

Triploid rainbow trout also exhibit less fin erosion than diploids (Wagner et al. 2006) and can also regenerate their fins faster than diploids (Alonso et al. 2000). Hybrid fish also exhibit less caudal erosion than their pure strain counterparts (Clayton et al. 1998 in Latre 2003), so this may be an option to some farmers. Finally, research investigating fin damage in both duo- and mono-culture reported a reduced incidence of damage under duo-culture conditions for Atlantic salmon and Arctic charr (Holm 1989; Nortvedt and Holm 1991). However, this is dependent upon the species selected as fin damage can be unaffected (e.g. rainbow trout and masou salmon, Flood et al. 2010) or increase when fish are held in duo-culture (e.g. Baltic salmon and brown trout, Jobling et al. 1998).

Concluding summary: synergies between risk factors and mitigation strategies

As stated at the start of this review, injuries are deleterious to both production and welfare (e.g. Cobcroft and Battaglene 2009). Farmers have an ethical responsibility to reduce the frequency and severity of injuries and by doing so can increase the market value of their fish by direct gains in fillet quality (Vågsholm and Djupvik 1998) and added value for incorporating welfare considerations into their production strategies (Olesen et al. 2010).

The primary aim of this review was to identify risk factors for injuries and a number of potential or existing mitigation strategies. It is one of the first to adopt a mitigation approach, identifying husbandry practices and operational strategies for improving the welfare of farmed fish. This review covers a wide range of existing or emerging aquaculture/capture-based aquaculture species. It also identifies risk factors from numerous types of production and rearing systems and has shown that there are many synergies between risk factors and numerous types of injury (for a summary see Table 1). If a farmer acts upon these, they can actively reduce the frequency and severity of these injuries. For example, as a number of injuries involve abrasion during handling, it is therefore logical that any husbandry or management practice involving the handling of these animals be carefully considered as to its design; the same applies to equipment. For instance, fish pumps, nets, sorting and grading apparatus come in a variety of materials and designs. Giving careful thought to designs less likely to cause injury would be beneficial such as using knotless nets or vacuum pumps instead of turbine pumps when transferring fish. Further, manipulating diet quality and feed management practices can also reduce the prevalence of injuries. For example, providing fish with a balanced diet, containing sufficient mineral and protein requirements can reduce the

Table 1 Showing examples of key risk factors for injuries to a number of different species

Risk factor	Injury	Species	Mitigation strategies
Diet formulation	Lower jaw deformity syndrome (LJD)	Dietary deficiencies in phosphorus and vitamins A, C, D and K in Atlantic salmon (King and Lee 1993; Roberts et al. 2001)	Ensure diets contain appropriate levels of phosphorus and vitamins A, C, D and K
	Mouth lesions, haemorrhaging and erosion of tissue around the mouth	Deficiency of niacin in tilapia (Shiau and Suen 1992)	Ensure diets contain appropriate levels of niacin
	Early life stage jaw deformities	Excess vitamin A in Japanese flounder (Suzuki et al. 2000) and Summer flounder (Martinez et al. 2007) or deficiencies in HUFA in pikeperch (Kestemont et al. 2007)	Ensure diets contain appropriate levels of vitamin A
	Cataracts	Cod, salmon, lake trout, wolf-fish, sea bass and sea bream (see Björnsson 2004)	Balance diet for each specific species
	Fin damage	E.g. steelhead (Lellis and Barrows 1997)	Appropriate selection of protein and lipid sources
Underfeeding	Fin damage	Rainbow trout (Moutou et al. 1998), Atlantic salmon (Ytrestøyl et al. 2005), Atlantic cod (Hatlen et al. 2006)	Feed to satiation
Choice of feed regime	Fin damage	Atlantic salmon (Noble et al. 2007a, b) rainbow trout (Suzuki et al. 2008)	Use demand feeding systems
Parasitic infections	Cataracts	Cod, salmon, lake trout, wolf-fish, sea bass, sea bream (see Björnsson 2004)	Biosecurity measures to prevent pathogen entry
	Epidermal ulcers (winter ulcer)	Prevalence of *Moritella viscosa* on fish farms in Atlantic salmon, cod, turbot and halibut (Lunder et al. 1995; Björnsdóttir et al. 2004; Gudmundsdóttir et al. 2006; Løvoll et al. 2009)	Biosecurity measures to prevent pathogen entry (disinfection, transport restrictions). High salinity (>12–15 ppt). Low water temperature (>10°C) Use of antimicrobials at the correct dose
	Epidermal ulcers	Prevalence of sea lice (*L. salmonis, C. elongatus, C. rogercresseyi*) in Atlantic salmon and rainbow trout (Boxshall and Defaye 1993)	Reduce stocking densities, disinfection of cages and utensils, restrict fish transport and control of natural host Use of antiparasitic drugs, only if infestation is massive
	Epidermal ulcers and necrosis	Prevalence of *Philasterides dicentrarachi* in olive flounder (Jin et al. 2009)	Biosecurity measures to prevent pathogen entry (disinfection, transport restrictions) Use of antimicrobials at the correct dose
	Epidermal ulcers	Prevalence of *Saprolegnia* sp. in Rainbow trout (Dykstra et al. 1989)	Biosecurity measures to prevent pathogen entry (disinfection, transport restrictions)
Choice of rearing system	Fin damage	Re-circulating versus flow-through tanks in rainbow trout (Roque d'Orbcastel et al. 2009)	Farm fish in flow-through tanks or optimise flow dynamics in RAS tanks
		Concrete versus cobbled raceways in rainbow trout (Bosakowski and Wagner 1994; Wagner et al. 1996)	Incorporate cobble substrates into salmonid raceways during construction
		Baffles in raceways in rainbow trout (Barnes et al. 1996)	Omit baffles during raceway construction

Table 1 continued

Risk factor	Injury	Species	Mitigation strategies
		Long-term cage submergence in Atlantic salmon (Korsøen et al. 2009)	Limit cage submergence to periods <6 weeks (Dempster et al. 2009; Korsøen et al. 2009)
	Epidermal injuries	Smooth tank surfaces in Halibut and southern flounder (Ottesen et al. 2007; Ottesen and Strand 1996)	Use of rigged tanks floor
	Broken jaw bones through collisions with tanks walls and catching jaws on cage netting	Inappropriate cage netting material in bluefin tuna (Miyashita et al. 2000). Inappropriate tank coloration in striped trumpeter (Cobcroft and Battaglene 2009)	Use of smaller mesh sizes and rubber or knotless mesh to reduce jaw catching, or use of appropriate tank colours
	Mouth lesions, haemorrhaging and erosion of tissue around the mouth	Rubbing against cage netting, tanks walls or other structures in holding facilities in Atlantic salmon (Halls 1994); rainbow trout (authors observations); Atlantic cod (authors observations; Halls 1994; authors observations)	Ensure appropriate materials are used for cages netting and tanks and make efforts to minimise the presence of abrasive surfaces that may cause damage to the mouths of fish if they rub up against them
Pumping	Fin damage. Epidermal abrasion and laceration	Channel catfish, Ictalurus punctatus (Grizzle et al. 1992)	Use a vacuum pump to transfer large numbers of fish (Grizzle et al. 1992)
	Eye injuries	Striped bass, rainbow trout (Helfrich et al. 2004); splittail and salmon (Helfrich et al. 2001)	Use a vacuum pump to transfer large numbers of fish (Grizzle et al. 1992)
	Epidermal injuries	Rainbow trout an striped bass (Helfrich et al. 2004)	Use of vacuum pumps or turbine pumps at no more than 330 rpm
Handling—net type	Fin damage	Bluegill (Barthel et al. 2003)	Use knot-less nets (Barthel et al. 2003)
	Mouth damage	Baltic cod (Suuronen et al. 1996)	Use of smaller mesh sizes and rubber or knotless mesh to reduce damage
Inappropriate egg incubation temperatures	Pug-headedness (brachygnathia)	Atlantic salmon (Bæverfjord et al. 1998a, b; Branson and Turnbull 2008)	Utilise optimal egg incubation temperatures
Inappropriate rearing temperatures	Early life stage jaw deformities	Atlantic halibut (Lein et al. 1997; Bolla and Holmefjord 1988); Atlantic salmon (Ornsrud et al. 2004); Japanese eel (Okamura et al. 2007)	Utilise appropriate rearing temperatures
	Fin damage	High temperatures in rainbow trout (Winfree et al. 1998) and sea bass (Person-Le Ruyet and Le Bayon 2009); low rearing temperatures in Atlantic salmon (Schneider and Nicholson 1980)	Utilise appropriate rearing temperatures
Inappropriate lighting	Eye injury and snapping (Speare 2008)	Halibut (Greaves and Tuene 2001)	Avoid creating shadows and uneven lighting over tanks.
	Early life stage jaw deformities	Atlantic halibut (Bolla and Holmefjord 1988)	Utilise appropriate light regimes
	Broken jaw bones through collisions with tanks walls and catching jaws on cage netting	Bluefin tuna (Miyashita et al. 2000)	Use of all-night lighting to improve vision. Removal of flashing lights loud noises and vibrations

Table 1 continued

Risk factor	Injury	Species	Mitigation strategies
High UV	Epidermal abrasion	Rainbow trout in Bolivia (Bullock and Coutts 1985)	Avoid using farms located in zones with high ultraviolet incidence
		Atlantic salmon in Ireland during summer (McArdle and Bullock 1987)	
	Cataracts	Cod, salmon, lake trout, wolf-fish, sea bass, sea bream (see Björnsson 2004)	
Gas super-saturation	Exophthalmia	Most fish species, but mainly salmonids (Bouck 1980)	Aeration of water, monitor gas levels in water
High stocking density	Fin damage	Rainbow trout (North et al. 2006); sea bass (Person-Le Ruyet and Le Bayon 2009)	Reduce stocking densities
Low stocking density	Fin damage	Arctic charr (Siikavuopio and Jobling 1995)	Increase stocking densities
Low water current	Fin damage	A. charr and A. salmon (Jobling et al. 1993)	Optimise water current
Direct aggression	Eye injuries	Halibut (Greaves and Tuene 2001); flounder (Carter et al. 1996); sole (Howell 1997)	Reduce feeding aggression and competition by delivering multiple feedings and dispersing feed over wide surface area
	Fin damage	Atlantic salmon (Cañon Jones et al. 2010); rainbow trout (Abbott and Dill 1985)	Reduce feeding aggression and competition by delivering multiple feedings and dispersing feed over wide surface area
Triploidy	Lower Jaw deformity syndrome (LJD)	Atlantic salmon (Jungawalla 1991)	Farm diploids
Vaccination	Epidermal injuries and intra-peritoneal adherences	Atlantic salmon (Vågsholm and Djupvik 1998)	Use of mineral oil–based adjuvant
Therapeutic medicinal treatment	Epidermis thinning and mucous cell opening	Rainbow trout (Buchmann et al. 2004; Madsen and Dalsgaard 1999)	Use of bath with low concentration and for the shortest time
Pollutants	Pug-headedness (brachygnathia)	Common bream (Slooff 1982); Red seabream (Matsuoka 2003)	Use clean water free of pollutants to culture fish. Early detection
	Early life stage jaw deformities	Common bream (Slooff 1982); Red seabream (Yamauchi et al. 2006)	Grow fish in clean waters free of pollutants
	Cataracts	Cod, salmon, lake trout, wolf-fish, sea bass, sea bream (see Björnsson 2004)	Use clean water free of pollutants to culture fish. Monitor water quality
Jellyfish blooms	Epidermal damage	Atlantic salmon (Tørud and Håstein 2008)	Jellyfish traps

Possible operational mitigation strategies for each risk factor are also included

severity of both mouth and fin damage. In addition, providing fish with a responsive ration to satiation can reduce both eye injuries (Greaves and Tuene 2001) and fin damage (Noble et al. 2007a) and provides further evidence that a single mitigation strategy can reduce the prevalence of a number of injuries.

Mitigation strategies can be both acute and long term. For instance, egg incubation temperature is a risk factor for pug-headedness, in addition to gill deformities and spinal injuries (Bæverfjord et al. 1998b; Ornsrud et al. 2004), and this can be remedied by having knowledge of the optimal egg incubating temperatures for a given species. With regard to long-term mitigation strategies, diet formulation, which is a risk factor for lower jaw deformity syndrome (Roberts et al. 2001), mouth lesions, haemorrhaging and erosion of tissue around the mouth (Shiau and Suen 1992), early life stage jaw deformities (Suzuki et al.

2000) and fin damage (Lellis and Barrows 1997), can be mitigated by formulating diets with appropriate levels of vitamins C and K, and utilising optimal sources of protein and lipid.

The literature on mitigating the injuries of farmed fish is scant, but the subject matter is important. Some of the current knowledge about risk factors for injury and their associated mitigation strategies can be applied to existing or emerging aquacultural or capture-based aquacultural species. However, further research must identify species-specific risk factors and potential mitigation schemes. Many more areas need to be addressed than the ones commonly reported in this paper, and controlled quantitative studies in a diverse number of cultured fish species are needed. This requires an integrated approach, utilising inputs from aquaculturists, researchers, veterinarians and aquacultural technology manufacturers.

Acknowledgments This review was carried out as part of EU COST Action 867: Welfare of fish in European aquaculture.

References

Abbott JC, Dill LM (1985) Patterns of aggressive attack in juvenile steelhead trout (*Salmo gairdneri*). Can J Fish Aquat Sci 42:1702–1706

Akimenko M-A, Mari-Beffa M, Becerra J, Géraudie J (2003) Old questions, new tools, and some answers to the mystery of fin regeneration. Dev Dyn 226:190–201

Alanärä A (1992) The effect of time-restricted demand-feeding on feeding activity, growth and feed conversion in rainbow trout (*Oncorhynchus mykiss*). Aquaculture 108:357–368

Alonso M, Tabata YA, Rigolino MG, Tsukamoto RY (2000) Effect of induced triploidy on fin regeneration of juvenile rainbow trout, *Oncorhynchus mykiss*. J Exp Zool 287: 493–502

Anon (2008) Production and price reports of member associations of the FEAP, 2001–2008. Report by the Federation of European Aquaculture Producers. http://www.feap.info/feap/. Accessed 15 March 2010

Arnold GP, Webb PW, Holford BH (1991) The role of the pectoral fins in station-holding of Atlantic salmon (*Salmo salar* L.). J Exp Biol 156:625–629

Bæverfjord G, Lein I, Åsgård T, Rye M, Storset A (1998a) High temperatures during egg incubation may induce malformations in Atlantic salmon (*Salmo salar* L.). Aquaculture and water, EAS special publication, no 26. European Aquaculture Society, Oostend, pp 24–25

Bæverfjord G, Åsgård T, Shearer KD (1998b) Development and detection of phosphorous deficiency in Atlantic salmon, *Salmo salar* L. parr and post-smolts. Aquacult Nutr 4:1–11

Barahona-Fernandes MH (1982) Body deformation in hatchery reared European sea bass *Dicentrarchus labrax* (L.)—types, prevalence and effect on fish survival. J Fish Biol 21:239–249

Barnes ME, Sayler WA, Cordes RJ (1996) Baffle usage in covered raceways. Prog Fish Cult 58:286–288

Barthel BL, Cooke SJ, Suski CD, Philipp DP (2003) Effects of landing net mesh type on injury and mortality in a freshwater recreational fishery. Fish Res 63:275–282

Berejikian BA, Tezak EP (2005) Rearing in enriched hatchery tanks improves dorsal fin quality of juvenile steelhead. N Am J Aquacult 67:289–293

Bjerkås E, Sveier H (2004) The influence of nutritional and environmental factors on osmoregulation and cataracts in Atlantic salmon (*Salmo salar* L). Aquaculture 235:101–122

Bjerkås E, Bjørnestad E, Breck O, Waagbø R (2001) Water temperature regimes affect cataract development in smolting Atlantic salmon, *Salmo salar* L. J Fish Dis 24:281–291

Björnsson B (2004) Can UV-treated seawater cause cataract in juvenile cod (Gadus morhua L.)? Aquaculture 240: 187–199

Björnsdóttir B, Gudmundsdóttir S, Bambir SH, Magnadóttir B, Gudmundsdóttir BK (2004) Experimental infection of turbot, *Scophthalmus maximus* (L.), by *Moritella viscosa*, vaccination effort and vaccine-induced side-effects. J Fish Dis 27:645–655

Blyth PJ, Purser GJ, Russell JF (1993) Detection of feeding rhythms in seacaged Atlantic salmon using new feeder technology. In: Reinertsen H, Dahle LA, Jørgensen L, Tvinnereim K (eds) Fish farming technology. Balkema, Rotterdam, pp 209–216

Bolla S, Holmefjord I (1988) Effect of temperature and light on development of Atlantic halibut larvae. Aquaculture 74: 355–358

Bosakowski T, Wagner EJ (1994) Assessment of fin erosion by comparison of relative fin length in hatchery and wild trout in Utah. Can J Fish Aquat Sci 51:636–641

Bouck GR (1980) Etiology of gas bubble disease. Trans Am Fish Soc 109:703–707

Boxshall GA, Defaye D (1993) Pathogens of wild and farmed fish: sea lice. Taylor & Francis, Routledge

Braithwaite RA, McEvoy LA (2005) Marine biofouling on fish farms and its remediation. Adv Mar Biol 47:215–252

Brandt TM, Jones RM, Koke JR (1986) Corneal cloudiness in transported largemouth bass. Prog Fish Cult 48:199–201

Branson E (1993) Basic anatomy and physiology. In: Brown L (ed) Aquaculture for veterinarians. Fish husbandry and medicine. Pergamon Press, Oxford, pp 1–30

Branson EJ, Turnbull T (2008) Welfare and deformities in fish. In: Branson EJ (ed) Fish welfare. Blackwell, Oxford, pp 202–216

Brawn VM (1961) Reproductive behaviour of the cod (*Gadus callarias* L.). Behaviour 18:177–198

Bristow BT, Summerfelt RC (1994) Performance of larval walleye cultured intensively in clear and turbid water. J World Aquacult Soc 25:454–464

Broom DM (1998) Welfare stress and the evolution of feelings. Stress Behavior 27:371–403

Bruno DW (1990) Jaw deformity associated with farmed Atlantic salmon (*Salmo salar*). Vet Rec 126:402–403

Buchmann K, Bresciani J, Jappe C (2004) Effects of formalin treatment on epithelial structure and mucous cell densities in rainbow trout, *Oncorhynchus mykiss* (Walbaum), skin.

J Fish Dis 27:99–104. doi:10.1111/j.1365-2761.2003.00519.x
Bullock AM (1988) Solar ultraviolet radiation: a potential environmental hazard in the cultivation of farmed finfish. In: Muir JF, Roberts RJ (eds) Recent advances in aquaculture, vol 3. Croom Helm, London, pp 139–226
Bullock AM, Coutts RR (1985) The impact of solar ultraviolet radiation upon the skin of rainbow trout, *Salmo gairdneri* Richardson, farmed at high altitude in Bolivia. J Fish Dis 8:263–272. doi:10.1111/j.1365-2761.1985.tb00942.x
Cañon Jones HA, Hansen LA, Noble C, Damsgård B, Broom DM, Pearce GP (2010). Social network analysis of behavioural interactions influencing fin damage development in *Atlantic salmon* (Salmo salar) during feed-restriction. Appl Anim Behav Sci 127:139–151
Carter CG, Purser GJ, Houlihan DF, Thomas P (1996) The effect of decreased ration on feeding hierarchies in groups of greenback flounder (*Rhombosolea tapirina*: Teleostei). J Mar Biol Assoc UK 76:505–516
Chervova LS (1996) Pain sensitivity and behaviour of fishes. J Ichthyol 37:98–102
Clayton RD, Stevenson TL, Summerfelt RC (1998) Fin erosion in intensively cultured walleyes and hybrid walleyes. Prog Fish Cult 60:114–118
Cobcroft JM, Battaglene SC (2009) Jaw malformation in striped trumpeter *Latris lineata* larvae linked to walling behaviour and tank colour. Aquaculture 289:274–282
Coble DW (1967) Effects of fin-clipping on mortality and growth of yellow perch with a review of similar investigations. J Wildl Manage 31:173–179
Collin SP, Collin HB (2001) The fish cornea: adaptations for different aquatic environments. In: Kapoor BG, Hara TJ (eds) Sensory biology of Jawed fishes—new insights. Science Publishers Inc, Plymouth, pp 57–96
Collins English Dictionary (1998) Millennium edition. HarperCollins Publishers, Glasgow
Conte FS (2004) Stress and the welfare of cultured fish. Appl Anim Behav Sci 86:205–223
Daniel TL (1981) Fish Mucus—in situ measurements of polymer drag reduction. Biol Bull 160:376–382
Dawson LHJ (1998) The physiological effects of salmon lice (*Lepeophtheirus salmonis*) infections on returning post-smolt sea trout (*Salmo trutta* L.) in western Ireland, 1996. ICES J Mar Sci 55:193–200
Dempster T, Korsøen Ø, Folkedal O, Juell JE, Oppedal F (2009) Submergence of Atlantic salmon (*Salmo salar*) in sea-cages: a potential short-term solution to poor surface conditions. Aquaculture 288:254–263
Deng Z, Guensch GR, McKinstry CA, Mueller RP, Dauble DD, Richmond MC (2005) Evaluation of fish-injury mechanisms during exposure to turbulent shear flow. Can J Fish Sci Aquat 62:1513–1522
Drucker EG, Lauder GV (2003) Function of pectoral fins in rainbow trout: behavioral repertoire and hydrodynamic forces. J Exp Biol 206:813–826
Drucker EG, Lauder GV (2005) Locomotor function of the dorsal fin in rainbow trout: kinematic patterns and hydrodynamic forces. J Exp Biol 208:4479–4494
Dyková I, Figueras A, Novoa B, Casal JF (1998) *Paramoeba* sp., an agent of amoebic gill disease of turbot *Scophthalmus maximus*. Dis Aquat Org 33:137–141
Dykstra MJ, Levine JF, Noga EJ, Hawkins JH, Gerdes P, Hargis WJ Jr, Grier HJ, Strake DT (1989) Ulcerative mycosis: a serious menhaden disease of the southeastern coastal fisheries of the United States. J Fish Dis 12:175–178. doi:10.1111/j.1365-2761.1989.tb00289.x
Ellis AE (2001) Innate host defense mechanisms of fish against viruses and bacteria. Dev Comp Imm 25:827–839. doi:10.1016/S0145-305X(01)00038-6
Ellis T, Oidtmann B, St-Hilaire S, Turnbull J, North B, MacIntyre C, Nikolaidis J, Hoyle I, Kestin S, Knowles T (2008) Fin erosion in farmed fish. In: Branson E (ed) Fish welfare. Blackwell, Oxford, pp 121–149
Flood MJ, Noble C, Kagaya R, Damsgård B, Purser GJ, Tabata M (2010) Growing amago and rainbow trout in duoculture with self-feeding systems: implications for production and welfare. Aquaculture 309:137–142
Fraser MR, de Nys R (2005) The morphology and occurrence of jaw and operculum deformities in cultured barramundi (*Lates calcarifer*) larvae. Aquaculture 250:496–503
Good C, Davidson J, Welsh C, Brazil B, Snekvik K, Summerfelt S (2009) The impact of water exchange rate on the health and performance of rainbow trout *Oncorhynchus mykiss* in water recirculation aquaculture systems. Aquaculture 294:80–85
Goss RJ, Stagg MW (1957) The regeneration of fins and fin rays in *Fundulus heteroclitus*. J Exp Zool 136:487–508
Greaves K, Tuene S (2001) The form and context of aggressive behaviour in farmed Atlantic halibut. Aquaculture 193:139–147
Gregory TR, Wood CM (1999) The effects of chronic plasma cortisol elevation on the feeding behavior, growth, competitive ability, and swimming performance of juvenile rainbow trout. Physiol Biochem Zool 72:286–295
Grizzle JM, Lovshin LL (1994) Effect of pump speed on injuries to channel catfish (*Ictalurus punctatus*) during harvest with a turbine pump. Aquac Eng 13:109–114. doi:10.1016/0144-8609(94)90008-6
Grizzle JM, Chen J, Williams JC, Spano JS (1992) Skin injuries and serum enzyme activities of channel catfish (*Ictalurus punctatus*) harvested by fish pumps. Aquaculture 107:333–346. doi:10.1016/0044-8486(92)90081-U
Gudmundsdóttir BK, Björnsdóttir B, Gudmundsdóttir S, Bambir SH (2006) A comparative study of susceptibility and induced pathology of cod, *Gadus morhua* (L.), and halibut, *Hippoglossus hippoglossus* (L.), following experimental infection with *Moritella viscosa*. J Fish Dis 29:481–487
Guillaume J, Choubert G (2001) Digestive physiology and nutrient digestibility in fishes. In: Guillaume J, Kaushik S, Bergot P, Métailler R (eds) Nutrition and feeding of fish and crustaceans. Springer, Chichester, pp 27–58
Halls R (1994) Good water and space to grow. Fish Farmer 17:49–50
Hardy-Smith P (2006) Biosecurity at the farm level-how to create a state of mind. In: Scarfe AD, Lee C-S, O'Bryen PJ (eds) Aquaculture biosecurity. Prevention, control, and eradication of aquatic animal disease, 1st edn. Blackwell, London
Harmache A, LeBerre M, Droineau S, Giovannini M, Brémont M (2006) Bioluminescence imaging of live infected salmonids reveals that the fin bases are the major portal of entry for novirhabdovirus. J Virol 80:3655–3659

Harris JE, Hunt S (1975) The fine structure of the epidermis of two species of salmonid fish, the Atlantic salmon (*Salmo salar* L.) and the brown trout (*Salmo trutta* L.). I. General organization and filament-containing cells. Cell Tiss Res 157:553–565

Hatlen B, Grisdale-Helland B, Helland SJ (2006) Growth variation and fin damage in Atlantic cod (*Gadus morhua* L.) fed at graded levels of feed restriction. Aquaculture 261: 1212–1221

Helfrich LA, Liston CR, Mefford B, Bark R (2001) Survival and injury of splittail and Chinook salmon passed through a large hidrostal pump. N Am J Fish Manage 21:616–623

Helfrich LA, Bark R, Liston CR, Weigmann DL, Mefford B (2004) Live transport of striped bass and rainbow trout using a Hidrostal pump. J World Aquac Soc 35:268–273. doi:10.1111/j.1749-7345.2004.tb01084.x

Higham TE, Malas B, Jayne BC, Lauder GV (2005) Constraints on starting and stopping: behavior compensates for reduced pectoral fin area during braking of the bluegill sunfish *Lepomis macrochirus*. J Exp Biol 208:4735–4746

Hoar WS, Randall DJ (1969) Fish physiology. Academic Press, London

Holm JC (1989) Mono- and duoculture of juvenile Atlantic salmon (*Salmo salar*) and Arctic char (*Salvelinus alpinus*). Can J Fish Aquat Sci 46:697–704

Hosfeld CD, Hammer J, Handeland SO, Fivelstad S, Stefansson SO (2009) Effects of fish density on growth and smoltification in intensive production of Atlantic salmon (*Salmo salar* L.). Aquaculture 294:236–241

Howell BR (1997) A re-appraisal of the potential of the sole, *Solea solea* (L.), for commercial cultivation. Aquaculture 155:355–365

Hoyle I, Oidtmann B, Ellis T, Turnbull J, North B, Nikolaidis J, Knowles TG (2007) A validated macroscopic key to assess fin damage in farmed rainbow trout (*Oncorhynchus mykiss*). Aquaculture 270:142–148

Huidobro A, Pastor A, Tejada M (2000) Quality index method developed for raw gilthead seabream (*Sparus aurata*). J Food Sci 65:1202–1205. doi:10.1111/j.1365-2621.2000.tb10265.x

Huntingford FA, Adams C, Braithwaite VA, Kadri S, Pottinger TG, Sandoe P, Turnbull JF (2006) Current issues in fish welfare. J Fish Biol 68:332–372

Ishibashi Y, Honryo T, Saida K, Hagiwara A, Miyashita S, Sawada Y, Okada T, Kurata M (2009) Artificial lighting prevents high night–time mortality of juvenile Pacific bluefin tuna *Thunnus orientalis* caused by poor scotopic vision. Aquaculture 293:157–163

Jin CN, Harikrishnan R, Moon YC, Kim MC, Kim JS, Balasundaram C, Azad IS, Heo MS (2009) Histopathological changes of Korea cultured olive flounder, *Paralichthys olivaceus* due to scuticociliatosis caused by histophagous scuticociliate, *Philasterides dicentrarachi*. Vet Parasitol 161:292–301. doi:10.1016/j.vetpar.2009.01.033

Jobling M, Baardvik BM, Christiansen JS, Jørgensen EH (1993) The effects of prolonged exercise training on growth performance and production parameters in fish. Aquacult Int 1:95–111

Jobling M, Koskela J, Pirhonen J (1998) Feeding time, feed intake and growth of Baltic salmon, *Salmo salar*, and brown trout, *Salmo trutta*, reared in monoculture and duoculture at constant low temperature. Aquaculture 163: 73–84

Jørgensen EH, Jobling M (1993) The effects of exercise on growth, food utilisation and osmoregulatory capacity of juvenile Atlantic salmon, *Salmo salar*. Aquaculture 116: 233–246

Jungawalla P (1991) Production of non–maturing Atlantic salmon in Tasmania. In: Pepper VA (ed) Proceedings of the Atlantic Canada workshop on methods for the production of non-maturing salmonids. Canadian Dept of Fisheries & Oceans, St Johns, pp 47–71

Karvonen A, Seppälä O (2008) Eye fluke infection and lens size reduction in fish: a quantitative analysis. Dis Aquat Org 80:21–26

Karvonen A, Seppälä O, Valtonen ET (2004) Eye fluke-induced cataract formation in fish: quantitative analysis using an ophthalmological microscope. Parasitology 129: 473–478

Kestemont P, Xu XL, Hamza N, Maboudou J, Toko II (2007) Effect of weaning age and diet on pikeperch larviculture. Aquaculture 264:197–204

King H, Lee P (1993) Progress report: jaw deformity and respiratory physiology of triploids seeking and solving: papers from the SALTAS research and development review seminar. SALTAS. Wayatinah, Tasmania, pp 37–44

Klontz GW, Maskill MG, Kaiser H (1991) Effects of reduced continuous versus intermittent feeding of steelhead. Prog Fish Cult 53:229–235

Knowles TG, Brown SN, Warriss PD, Lines J, Tinarwo A, Bravo A, Carvalho H, Gonçalves A (2007) Effect of electrical stunning at slaughter on the carcass, flesh and eating quality of farmed sea bass (*Dicentrarchus labrax*). Aquac Res 38:1732–1741

Korsøen Ø, Dempster T, Fjelldal PG, Folkedal O, Kristiansen T, Oppedal F (2009) Long-term submergence of Atlantic salmon (*Salmo salar* L.) during winter affects behaviour, growth and condition. Aquaculture 296:373–381

Kotrschal K, Whitear M, Finger TE (1993) Spinal and facial innervation of the skin in the gadid fish *Ciliata mustela* (teleostei). J Comp Neurol 331:407–417. doi:10.1002/cne.903310310

Kurokawa T, Okamoto T, Gen K, Uji S, Murashita K, Unuma T, Nomura K, Matsubara H, Kim SK, Ohta H, Tanaka H (2008) Influence of water temperature on morphological deformities in cultured larvae of Japanese eel *Anguilla japonica* at completion of yolk resorption. J World Aquacult Soc 39:726–735

Laitinen M, Siddall R, Valtonen ET (1996) Bioelectronic monitoring of parasite-induced stress in brown trout and roach. J Fish Biol 48:228–241

Latremouille DN (2003) Fin erosion in aquaculture and natural environments. Res Fish Sci 11:315–335

Lauder GV (2000) Function of the caudal fin during locomotion in fishes: kinematics, flow visualization, and evolutionary patterns. Am Zool 40:101–122

Lein I, Holmefjord I, Rye M (1997) Effects of temperature on yolk sac larvae of Atlantic halibut (*Hippoglossus hippoglossus* L.). Aquaculture 157:123–135

Lellis WA, Barrows FT (1997) The effect of diet on dorsal fin erosion in steelhead trout (*Oncorhynchus mykiss*). Aquaculture 156:233–244

Lijalad M, Powell MD (2009) Effects of lower jaw deformity on swimming performance and recovery from exhaustive exercise in triploid and diploid Atlantic salmon *Salmo salar* L. Aquaculture 290:145–154

López-Albors O, Gil F, Ramirez-Zarzosa G, Latorre R, Garcia-Alcazar A, Abellan E, Blanco A, Vazquez JM, Moreno F (1995) Early muscle injuries in a standard reared stock of sea bass *Dicentrarchus labrax* (L.). Aquaculture 138:69–76

Løvoll M, Wiik-Nielsen CR, Tunsjo HS, Colquhoun D, Lunder T, Sorum H, Grove S (2009) Atlantic salmon bath challenged with *Moritella viscosa*—pathogen invasion and host response. Fish Shell Immunol 26:877–884. doi: 10.1016/j.fsi.2009.03.019

Lunder T, Evensen Ø, Holstad G, Håstein T (1995) 'Winter ulcer' in the Atlantic salmon *Salmo salar*. Pathological and bacteriological investigations and transmission experiments. Dis Aquat Org 23:39–49

MacLean A, Metcalfe NB, Mitchell DF (2000) Alternative competitive strategies in juvenile Atlantic salmon (*Salmo salar*): evidence from fin damage. Aquaculture 184:291–302

Madsen L, Dalsgaard I (1999) Reproducible methods for experimental infection with Flavobacterium psychrophilum in rainbow trout *Oncorhynchus mykiss*. Dis Aquat Org 36:169–176

Martinez GM, Baron MP, Bolker JA (2007) Skeletal and pigmentation defects following retinoic acid exposure in larval summer flounder *Paralichthys dentatus*. J World Aquacult Soc 38:353–366

Masuma S, Kawamura G, Tezuka N, Koiso M, Namba K (2001) Retinomotor responses of juvenile bluefin tuna *Thunnus thynnus*. Fish Sci 67:228–231

Matsumoto T, Ihara H, Ishida Y, Okada T, Kurata M, Sawada Y, Ishibashi Y (2009) Electroretinographic analysis of night vision in juvenile Pacific bluefin tuna (*Thunnus orientalis*). Biol Bull 217:142–150

Matsuoka M (2003) Comparison of meristic variations and bone abnormalities between wild and laboratory-reared red sea bream. Jpn Agric Res Q 37:21–30

McArdle J, Bullock AM (1987) Solar ultraviolet radiation as a causal factor of 'summer syndrome' in cage-reared Atlantic salmon, *Salmo salar* L.: a clinical and histopathological study. J Fish Dis 10:255–264. doi:10.1111/j.1365-2761.1987.tb01070.x

McEachron J, Fechhelm J (1998) Fishes of the Gulf of Mexico, volume 1: myxiniformes to gasterosteiformes. University of Texas Press, Austin

McGeachy SA, O'Flynn FM, Benfey TJ, Friars GW (1996) Seawater performance of triploid Atlantic salmon in New Brunswick aquaculture. Bull Aquacult Assoc Can 96: 24–28

Mears HC, Hatch RW (1976) Overwinter survival of fingerling brook trout with single and multiple fin clips. Trans Am Fish Soc 105:669–674

Mesa MG, Poe TP, Gadomski DM, Petersen JH (1994) Are all prey created equal? A review and synthesis of differential predation on prey in substandard condition. J Fish Biol 45:81–96

Michie I (2001) Causes of downgrading in the salmon farming industry. In: Kestin SC, Warriss PD (eds) Farmed fish quality. Fishing News Books, London, pp 129–136

Miyashita S, Sawada Y, Hattori N, Nakatsukasa H, Okada T, Murata O, Kumai H (2000) Mortality of northern bluefin tuna *Thunnus thynnus* due to trauma caused by collision during growout culture. J World Aquacult Soc 31:632–639

Mork J, Järvi T, Hansen LP (1989) Lower prevalence of fin erosion in mature than in immature Atlantic salmon (*Salmo salar*) parr. Aquaculture 80:223–229

Moutou KA, McCarthy ID, Houlihan DF (1998) The effect of ration level and social rank on the development of fin damage in juvenile rainbow trout. J Fish Biol 52:756–770

Nicola SJ, Cordone AJ (1973) Effect of fin removal on survival and growth of rainbow trout (*Salmo gairdneri*) in a natural environment. Trans Am Fish Soc 102:753–758

Noble C, Mizusawa K, Suzuki K, Tabata M (2007a) The effect of differing self-feeding regimes on the growth, behaviour and fin damage of rainbow trout held in groups. Aquaculture 264:214–222

Noble C, Kadri S, Mitchell DF, Huntingford FA (2007b) Influence of feeding regime on intraspecific competition, fin damage and growth in 1+ Atlantic salmon parr (*Salmo salar* L.) held in freshwater production cages. Aquac Res 38:1137–1143

North BP, Turnbull JF, Ellis T, Porter MJ, Migaud H, Bron J, Bromage NR (2006) The impact of stocking density on the welfare of rainbow trout (*Oncorhynchus mykiss*). Aquaculture 255:466–479

Nortvedt R, Holm JC (1991) Atlantic salmon in duoculture with Arctic charr—decreased aggression enhances growth and stocking density potential. Aquaculture 98:355–361

Nylund A, Hovland T, Hodneland K, Nilsen F, Lovik P (1994) Mechanisms for transmission of infectious salmon anemia (ISA). Dis Aquat Org 19:95–100

Okamoto T, Kurokawa T, Gen K, Murashita K, Nomura K, Kim SK, Matsubara H, Ohta H, Tanaka H (2009) Influence of salinity on morphological deformities in cultured larvae of Japanese eel *Anguilla japonica* at completion of yolk resorption. Aquaculture 293:113–118

Okamura A, Yamada Y, Horie N, Utoh T, Mikawa N, Tanaka S, Tsukamoto K (2007) Effects of water temperature on early development of Japanese eel *Anguilla japonica*. Fish Sci 73:1241–1248

Olesen I, Alfnes F, Røra MB, Kolstad K (2010) Eliciting consumers willingness to pay for organic and welfare-labelled salmon in a non-hypothetical choice experiment. Livestock Sci 27:218–226

Ornsrud R, Gil L, Waagsbø R (2004) Teratogenicity of elevated egg incubation temperature and egg vitamin A status in Atlantis salmon, *Salmo salar* L. J Fish Dis 27:213–223

Ostrand KG, Siepker MJ, Cooke SJ (2006) Capture efficiency of two hook types and associated injury and mortality of juvenile muskellunge angled with live baitfish. N Am J Fish Manage 26:622–627

Ottesen OH, Strand HK (1996) Growth, development, and skin abnormalities of halibut (*Hippoglossus hippoglossus* L.) juveniles kept on different bottom substrates. Aquaculture 146:17–25. doi:10.1016/S0044-8486(96)01359-2

Ottesen OH, Noga EJ, Sandaa W (2007) Effect of substrate on progression and healing of skin erosions and epidermal papillomas of Atlantic halibut, *Hippoglossus hippoglossus* (L.). J Fish Dis 30:43–53. doi:10.1111/j.1365-2761.2007.00780.x

Person-Le Ruyet J, Le Bayon N (2009) Effects of temperature, stocking density and farming conditions on fin damage in European sea bass (*Dicentrarchus labrax*). Aquat Living Resour 22:349–362

Piaia R, Baldisserotto B, Townsend C (1999) Growth and survival of fingerlings of silver catfish exposed to different photoperiods. Aquacult Int 7:201–205

Pittman K, Berg Ø, Opstad I, Skiftesvik AB, Skjolddal L, Strand H (1990) Development of eggs and yolk-sac larvae of halibut (*Hippoglossus hippoglossus* L.). J Appl Ichthyol 6:142–160

Quigley DTG (1995) A lower jaw deformity in juvenile and adult Atlantic salmon (*Salmo salar* L.). Bull Eur Assoc Fish Pathol 15:206–209

Rasmussen RS, Larsen FH, Jensen S (2007) Fin condition and growth among rainbow trout reared at different sizes, densities and feeding frequencies in high temperature re-circulated water. Aquacult Int 15:97–107

Rhodes JD, Breck O, Waagbø R, Bjerkås E, Sanderson S (2010) N-acetylhistidine, a novel osmolyte in the lens of *Atlantic salmon* (Salmo salar L.) Am J Physiol Regul Integr Comp Physiol 299:R1075–R1081

Roberts RJ, Hardy RW, Sugiura SH (2001) Screamer disease in Atlantic salmon *Salmo salar* L. in Chile. J Fish Dis 24: 543–549

Roque d'Orbcastel E, Person-Le Ruyet J, Le Bayon N, Blancheton JP (2009) Comparative growth and welfare in rainbow trout reared in recirculated and flow through rearing systems. Aquac Eng 40:79–86

Roques JAC, Abbink WA, Geurds FB, van de Vis HA, Flik G (2010) Tailfin clipping, a painful procedure: Studies on *Nile tilapia* and common carp. Physiol Behav 101:533–540

Sadler J, Pankhurst PM, King HR (2001) High prevalence of skeletal deformity and reduced gill surface area in triploid Atlantic salmon (*Salmo salar* L.). Aquaculture 198:369–386

Sanchez JG, Speare DJ, Sims DE, Johnson GJ (1998) Morphometric assessment of epidermal and mucous-biofilm changes caused by exposure of trout to chloramine-T or formalin treatment. J Comp Pathol 118:81–87. doi: 10.1016/S0021-9975(98)80032-7

Schneider R, Nicholson BL (1980) Bacteria associated with fin rot disease in hatchery-reared Atlantic salmon (*Salmo salar*). Can J Fish Aquat Sci 37:1505–1513

Shao J, Qian X, Zhang C, Xu Z (2009) Fin regeneration from tail segment with musculature, endoskeleton, and scales. J Exp Zool (Mol Dev Evol) 312B:762–769

Shiau SY, Suen GS (1992) Estimation of the niacin requirements for tilapia fed diets containing glucose or dextrin. J Nutr 122:2030–2036

Siikavuopio SI, Jobling M (1995) The effects of stocking density on survival and growth of wild-caught Arctic char. Nordic J Freshw Res 71:419–423

Slooff W (1982) Skeletal anomalies in fish from polluted surface waters. Aquat Toxicol 2:157–173

Smith PA, Pizarro P, Ojeda P, Contreras J, Oyanedel S, Larenas J (1999) Routes of entry of *Piscirickettsia salmonis* in rainbow trout *Oncorhynchus mykiss*. Dis Aquat Org 37: 165–172

Sneddon LU (2003) The evidence for pain in fish: the use of morphine as an analgesic. Appl Anim Behav Sci 83: 153–162

Southgate PE (1993) Diseases in aquaculture. In: Brown L (ed) Aquaculture for veterinarians. Fish husbandry and medicine. Pergamon Press, Oxford, pp 91–129

Speare DJ (2008) Environmental diseases of intensively reared fish. In: Eiras JC, Segner H, Wahli T, Kapoor BG (eds) Fish diseases, vol 2. Science, Enfield, pp 1225–1253

Standen EM, Lauder GV (2005) Dorsal and anal fin function in bluegill sunfish (*Lepomis macrochirus*): three-dimensional kinematics during propulsion and maneuvering. J Exp Biol 205:2753–2763

Sumagaysay NS, Hilomen-Garcia GV, Garcia LMB (1999) Growth and production of deformed and non-deformed hatchery–bred milkfish (*Chanos chanos*) in brackishwater ponds. Isr J Aquac Bamidgeh 51:106–113

Suuronen P, Lehtonen E, Tschernij V, Larsson PO (1996) Skin injury and mortality of Baltic cod escaping from trawl codends equipped with exit windows. Arch Fish Mar Res 44:165–178

Suzuki T, Srivastava AS, Kurokawa T (2000) Experimental induction of jaw gill and pectoral fin malformations in Japanese flounder *Paralichthys olivaceus* larvae. Aquaculture 185:175–187

Suzuki K, Mizusawa K, Noble C, Tabata M (2008) The growth, feed conversion ratio and fin damage of rainbow trout *Oncorhynchus mykiss* under self-feeding and hand-feeding regimes. Fisheries Sci 74:941–943

Swan MA (1968) Double mouth deformity in a trout (*Salmo trutta*) and its cause. J Zool 156:449–455

Thatcher TO (1979) A morphological defect in shiner perch resulting from chronic exposure to chlorinated sea water. Bull Environm Contam Toxicol 21:433–438

Tørud B, Håstein T (2008) Skin lesions in fish: causes and solutions. Acta Vet Scan 50(Suppl 1):S7. doi:10.1186/1751-0147-50-S1-S7

Totland GK, Hjeltnes BK, Flood PR (1996) Transmission of infectious salmon anaemia (ISA) through natural secretions and excretions from infected smolts of Atlantic salmon *Salmo salar* during their presymptomatic phase. Dis Aquat Org 26:25–31

Treasurer JW, Cox DI, Wall T (2007) Epidemiology of blindness and cataracts in cage reared ongrown Atlantic halibut *Hippoglossus hippoglossus*. Aquaculture 271: 77–84

Turnbull JF, Richards RH, Robertson DA (1996) Gross, histological and scanning electron microscopic appearance of dorsal fin rot in farmed Atlantic salmon, *Salmo salar* L., parr. J Fish Dis 19:415–427

Ubels JL, Edelhauser HF (1987) Effects of corneal epithelial abrasion on corneal transparency, aqueous humor composition, and lens of fish. Prog Fish Cult 49:219–224

Vågsholm I, Djupvik OH (1998) Risk factors for skin lesions in Atlantic salmon, *Salmo salar* L. J Fish Dis 21:449–454. doi:10.1046/j.1365-2761.1998.00123.x

van Duijn C (1973) Diseases of fishes. Iliffe Books, London

Videler JJ (1993) Fish swimming. Fish and fisheries series 10. Chapman & Hall, London

Villafane VE, Helbling EW, Zagarese HE (2001) Solar ultraviolet radiation and its impact on aquatic systems of Patagonia, South America. Ambio 30:112–117

Waagbø R, Hamre K, Bjerkås E, Berge R, Wathne E, Lie Ø, Torstensen B (2003) Cataract formation in Atlantic salmon, *Salmo salar* L., smolt relative to dietary pro- and antioxidants and lipid level. J Fish Dis 26:213–229

Wagner EJ, Routledge MD, Intelmann SS (1996) Fin condition and health profiles of albino rainbow trout reared in concrete raceways with and without a cobble substrate. Prog Fish Cult 58:38–42

Wagner EJ, Arndt R, Routledge MD, Bradwisch Q (1998) Hatchery performance and fin erosion of Bonneville cutthroat trout, *Oncorhynchus clarki*, at two temperatures. J Appl Aquacult 8:1–12

Wagner EJ, Arndt RE, Routledge MD, Latremouille D, Mellenthin RF (2006) Comparison of hatchery performance, agonistic behavior, and poststocking survival between diploid and triploid rainbow trout of three different Utah strains. N Am J Aquacult 68:63–73

Wagner CP, Einfalt LM, Scimone AB, Wahl DH (2009) Effects of fin-clipping on the foraging behavior and growth of age-0 muskellunge. N Am J Fish Manage 29:1644–1652

Westneat M, Thorsen DH, Walker JA, Hale M (2004) Structure, function, and neural control of pectoral fins in fishes. IEEE J Oceanic Eng 29:674–683

Winfree RA, Kindschi GA, Shaw HT (1998) Elevated water temperature, crowding, and food deprivation accelerate fin erosion in juvenile steelhead. Prog Fish Cult 60:192–199

Winterbottom R (1974) A descriptive synonymy of the striated muscles of the Teleostei. Proc Acad Nat Sci Philadelphia 125:225–317

Yamauchi M, Kim EY, Iwata H, Shima Y, Tanabe S (2006) Toxic effects of 2378—tetrachlorodibenzo-*p*-dioxin (TCDD) in developing red seabream (*Pagrus major*) embryo: an association of morphological deformities with AHR1 AHR2 and CYP1A expressions. Aquat Toxicol 80: 166–179

Ytrestøyl T, Struksnæs G, Koppe W, Bjerkeng B (2005) Effects of temperature and feed intake on digestibility and metabolism of astaxanthin in *Atlantic salmon* (Salmo salar). Comp Biochem Physiol 142B:445–455

Yue Cottee S (in press) Are fish the victims of 'speciesism'? A discussion about fear, pain and animal consciousness. Fish Physiol Biochem

Health of Farmed Fish: Its Relation to Fish Welfare and its Utility as Welfare Indicator – Summary

Helmut Segner, Hendrik Sundh, Kurt Buchmann, Jessica Douxfils, Kristina Snuttan Sundell, Cédric Mathieu, Neil Ruane, Fredrik Jutfelt, Hilde Toften and Lloyd Vaughan

Fish with good welfare have a high capability to cope with infectious and non-infectious stressors. This allows them to build sufficient energy reserves to maintain effective barriers against invading pathogens, but at the same time provide the energy needed for acclimation and response to stressor impact. Fish experiencing an increasing load of stressors will eventually wear out their energy reserves and be more prone to acquire a disease through compromised primary barriers (skin, intestine and gills), a situation that characterises bad welfare. Based on existing knowledge in stress physiology, we want to explain in our paper how stressful culture conditions and/or poor welfare can translate into impaired health and how impaired health in turn, can reduce the ability of fish to cope with stressful conditions.

Background

Health parameters are routinely monitored in most aquaculture units in order to control and prevent diseases. They can be used as direct measures of welfare status, since good health is an essential component of good welfare and, vice versa, poor welfare conditions can cause impaired health. The aim of this work is to discuss the relationship between health and welfare status of fish and to examine how alterations of health status translate into impaired welfare.

Scientific Significance

We draw attention to one of the major front lines in the fish response to environmental stressors, the primary barriers provided by skin, intestine and gills. Not only physical injury,but also stressors will negatively affect the biological and immunological functions of these barriers. This in turn leads to more leaky barriers and an increased risk of pathogens invading and infecting the fish. At the same time it leads to changes in the fish internal balance which require, the fish to use energy to counteract these changes. This may indirectly threaten their health status. Primary barrier responses, including ensuing systemic immune responses, are areas urgently in need of further research. This is necessary for a robust scientific backing of health- and function-related operational welfare indicators and the development of new operational welfare indicators.

Practical Application

Well founded and measurable health parameters are arguably the most readily adopted operational welfare indicators for use on fish farms. They have the advantage of being mostly straightforward to measure and, as a consequence, are routinely monitored in most aquaculture units.This permits a comparison between diverse husbandry conditions and, in addition, the same health parameters remain valid and applicable, regardless of whether fish can be regarded as sentient or not.

Contact

H. Segner
Centre for Fish and Wildlife Health, Vetsuisse Faculty,
University of Bern, Bern, Switzerland
e-mail: helmut.segner@vetsuisse.unibe.ch

Fish Physiol Biochem (2012) 38:85–105
DOI 10.1007/s10695-011-9517-9

Health of farmed fish: its relation to fish welfare and its utility as welfare indicator

Helmut Segner · Henrik Sundh · Kurt Buchmann · Jessica Douxfils · Kristina Snuttan Sundell · Cédric Mathieu · Neil Ruane · Fredrik Jutfelt · Hilde Toften · Lloyd Vaughan

Received: 21 October 2010 / Accepted: 20 May 2011 / Published online: 18 June 2011

Abstract This brief review focuses on health and biological function as cornerstones of fish welfare. From the function-based point of view, good welfare is reflected in the ability of the animal to cope with infectious and non-infectious stressors, thereby maintaining homeostasis and good health, whereas stressful husbandry conditions and protracted suffering will lead to the loss of the coping ability and, thus, to impaired health. In the first part of the review, the physiological processes through which stressful husbandry conditions modulate health of farmed fish are examined. If fish are subjected to unfavourable husbandry conditions, the resulting disruption of internal homeostasis necessitates energy-demanding physiological adjustments (allostasis/acclimation). The ensuing energy drain leads to trade-offs with other energy-demanding processes such as the functioning of the primary epithelial barriers (gut, skin, gills) and the immune system. Understanding of the relation between husbandry conditions, allostatic responses and fish health provides the basis for the second theme developed in this review, the potential use of biological function and health parameters as operational welfare indicators (OWIs). Advantages of function- and health-related parameters are that they are relatively straightforward to recognize and to measure and are routinely monitored in most aquaculture units, thereby providing feasible tools to assess fish welfare under practical farming conditions. As the efforts to improve fish welfare and environmental sustainability lead to increasingly diverse solutions, in particular integrated production, it is imperative that we have objective OWIs to compare with other production forms, such as high-density aquaculture. However, to receive the necessary acceptance for legislation, more robust scientific backing of the health- and function-related OWIs is urgently needed.

H. Segner (✉)
Centre for Fish and Wildlife Health, Vetsuisse Faculty, University of Bern, Bern, Switzerland
e-mail: helmut.segner@vetsuisse.unibe.ch

H. Sundh · K. S. Sundell · F. Jutfelt
Fish Endocrinology Laboratory, Department of Zoology/Zoophysiology, Göteborg University, Gothenburg, Sweden

K. Buchmann
Laboratory of Fish Diseases, Department of Veterinary Disease Biology, University of Copenhagen, Copenhagen, Denmark

J. Douxfils · C. Mathieu
Unit of Research in Organismal Biology, University of Namur, Namur, Belgium

N. Ruane
Marine Institute, Oranmore, County Galway, Ireland

H. Toften
Nofima, Tromsø, Norway

L. Vaughan
Institute of Veterinary Pathology, Vetsuisse Faculty, University of Zürich, Zurich, Switzerland

Keywords Fish · Welfare · Health · Disease · Stress · Welfare indicator · Epithelial barrier · Allostasis

Introduction

Welfare of an animal is a complex issue, and a unified definition is hard to find in the literature. Nevertheless, most definitions of animal welfare take into account the three categories: feelings, nature and function. Briefly, the first is concerned with the subjective experience of the animal and implies that the animals are sentient and able to suffer, i.e. experience pain and fear (Chandroo et al. 2004). In this category, good welfare is defined as the absence of negative feelings and the presence of positive feelings (cf. the concept of the five freedoms). The second category requires that an animal is allowed to express its natural behaviour. The third category, based on functional definitions, focuses on the ability of the animal to cope and acclimate to its environment without being forced beyond its physical capacity. In this category, good welfare can be viewed upon as the ability to maintain homeostasis and normal biological functions. Ultimate reflections of good welfare from the function-based point of view are good health and absence of disease, and with respect to aquaculture, good productivity (e.g., Turnbull and Kadri 2007; Volpato et al. 2007). In this review, we focus on health and biological function as cornerstones of fish welfare. Although health and welfare are intimately connected (Moberg 2000; Ashley 2007), we are aware that a health-based approach to fish welfare is in a way reductionistic and does not take into account all components of welfare. While the physical health of an animal is fundamental for good welfare (Ashley 2007; Duncan 2005), the fact that an animal is healthy does not necessarily mean that it has a good welfare status. Thus, welfare is the broader, more overarching concept than the health concept.

In aquaculture, the term 'health' is often interpreted as 'absence of overt disease', and thus, emphasis is given to disease prevention and eradication. In this communication, however, the term 'health' is widened beyond the absence of disease to also cover pathology defined as detrimental arrangements of molecules, cells, tissues and their dysfunction (reviewed by Broom 2007). Health, from this perspective, means the ability of an animal to perform normal physiological functions and to maintain homeostasis, thereby supporting its ability to withstand infectious and non-infectious stressors. As such, good health is essential (but not yet sufficient) for good welfare (Ashley 2007; Duncan 2005). Poor health, i.e. the reduced ability of the animal to perform normal functioning, to acclimate to stressful conditions and to prevent disease, implies a bad welfare status. Importantly, it is often poor welfare itself, which is a precondition to loss of health, with the important exception that healthy animals in an optimal welfare situation may still suffer an acute infection which by definition is regarded as poor health. Stressors (e.g. handling, inappropriate husbandry conditions such as confinement or crowding leading to accentuated social interactions between conspecifics, poor water quality and pathogen occurrences) present in culture systems are potential threats to animal welfare and thereby to animal health (Huntingford et al. 2006), as they impose an allostatic load on the animal, which in the short term will impair its physiological homeostasis and in the long term its condition and health status (Broom and Corke 2002; Roger 2008). The close link between health and welfare is also evident from the 'concept of the five freedoms' that has been developed in order to secure welfare. This concept includes freedom from thirst, hunger and malnutrition, freedom from discomfort by providing adequate environmental quality parameters, freedom from distress by ensuring adequate living conditions and freedom from disease. For example, when an animal has no access to food, its welfare ('freedom of hunger') is impaired, but at the same time, starvation places the fish at risk of developing metabolic problems (Fig. 1, quadrant C) and eventually metabolic diseases and infections, which are health problems (Fig. 1, quadrant D). In this scenario, poor welfare precedes poor health. However, bad health may during certain circumstances also result in bad welfare. In Fig. 1, quadrant A, fish reared in an optimal environment, determined by the biology of the fish, experience a minimal allostatic load, show no signs of health problems, and the welfare is maximized. However, all healthy animals may be struck by an acute infection that reduces the welfare of the fish (Fig. 1,

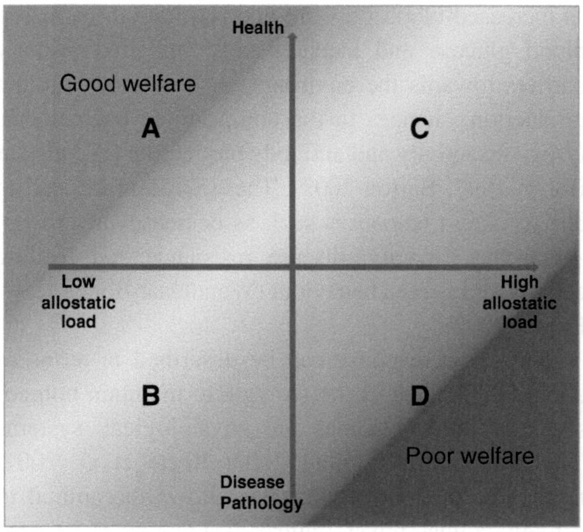

Fig. 1 The graph visualizes the distinctions and links between the concepts of health and welfare from a function-based point of view. When plotting health against allostatic load, four different conditions that fish can experience are discernible (quadrants **a**, **b**, **c** and **d**): In **a** fish are in an optimal physical environment with low level of stressors, while at the same time being healthy. Thus, the welfare is defined as good. In **b** the allostatic load is low, but pathogen-induced disease is present. As a consequence, welfare is reduced for the individual infected fish because of disease symptoms. In **c** stressor intensity is higher, what affects health and the ability of the fish to cope with its environment, but the intensity is not yet strong enough to cause disease—a situation yielding acceptable welfare. However, there is risk of increased disease susceptibility that by definition is reduced welfare. Finally, in **d** there is high allostatic load as well as disease, which lead to the worst welfare status

quadrant B), a scenario where bad health will lead to bad welfare. The inherent link between health and welfare is further emphasized by Dawkins (2006) who suggested that an animal's welfare can be characterized by asking 'does the animal get what it wants' and 'is the animal healthy?' Thus, although close and mutual links exist between welfare and health, the knowledge of how alterations in homeostasis and health status translate into welfare of the fish and vice versa is currently rather limited.

An advantage of assessing health in a broad sense, including both pathological changes and disease, is that it provides a pragmatic approach to the assessment of fish welfare in culture situations. Since health parameters are routinely monitored in most aquaculture units, they can provide information on at least some aspects of the current welfare status of the farmed fish. In addition, the health-related approach allows us to formulate testable hypotheses to develop operational welfare indicators for the practice.

This communication aims to discuss the utility of the health-related approach in fish welfare assessment. Firstly, we will examine the physiological processes through which husbandry conditions modulate health of farmed fish. This mechanistic understanding provides the basis for the second theme we develop in this communication, i.e. the potential of health parameters as operational welfare indicators. Welfare scenarios that ignore the effects on the environment are doomed to non-sustainability, an aspect that if addressed in full would spring the frame of this short review, but which we allude to at the end of Sect. 3. We will not include the question of sentience and mental health. There is an ongoing controversial discussion as to whether fish possess or lack the brain centres believed to be a prerequisite for a sentient animal; this question will be addressed in other chapters of this volume. Thus, our definition of health refers only to physical health.

Processes through which poor husbandry and welfare conditions translate into impaired fish health

Stressful farming conditions may impair welfare of the fish, and this can translate into reduced fish performance and health. The health of farmed fish is of major concern as impaired health or any disease state is neither agreeable regarding welfare considerations nor economically sustainable. Here, we discuss the processes through which poor welfare conditions in the husbandry system can affect the health status of farmed fish and favour diseases. If the fish is subjected to unfavourable husbandry conditions, this impairs diverse basal physiological functions and disrupts internal homeostasis. The adaptive responses needed to counteract this require energy, and the ensuing drain will lead to trade-offs with other energy-demanding functions of the organisms, including defence mechanisms such as immune responses and/or physiological processes such as growth and reproduction (Lupatsch et al. 2010; Anderson et al. 2011).

In order to better understand the relationships between the various processes, the following section addresses the basic necessities for life (homeostasis) and the physiological systems that maintain it (allostasis). This discussion will also consider what the physiological limits are for coping with a sub-optimal environment, i.e. what determines the coping ability of an organism. In this context, the energetic costs of allostasis/acclimation and possible consequences on the key defence system of the fish, the immune system, will be briefly discussed.

Homeostasis, stress and allostasis

All living organisms are adapted to the environment they inhabit through the natural selection of evolution. A prerequisite for life is being able to maintain stability of essential physiological systems (homeostasis) within the ranges enabling survival of the animal, despite constant challenges from the external environment (Stott 1981; McEwen and Wingfield 2003). Stress can be considered as a state of threatened homeostasis that is re-established by a complex suite of neural, behavioural and physiological modifications through the stress response (Barton 2002). The stress responses are perhaps the most fundamental reaction preserved among all vertebrates. The primary stress response is initiated by the CNS perceiving cues from the environment and mediated by the hypothalamic–pituitary–inter-renal (HPI) axis and the autonomic nervous system (ANS), resulting in the release of stress hormones, corticosteroids and catecholamines, from the head kidney (Wendelaar Bonga 1997; Barton 2002). Once released into the blood stream, the stress hormones elicit secondary stress responses in the target tissues expressing the appropriate receptors (Mommsen et al. 1999). In general, the secretion of catecholamines is initiated almost immediately after onset of the stress response and is transiently decreased. For example, rainbow trout subjected to moderate hypoxia (<40% DO) responded with a sharp increase in adrenaline with a peak at 24 h but returned to basal values after 48 h, despite maintenance in an hypoxic environment (Thomas et al. 1991). On the other hand, the release of cortisol starts within minutes of stress onset, and elevated levels can be maintained for a longer period of time compared with adrenaline. In fish subjected to a moderate hypoxia of 50% DO, elevated plasma cortisol levels could be measured for up to 30 days after the initiation of hypoxic treatment (Sundh et al. 2010). Well-documented secondary stress responses of fish include metabolic, cellular, osmoregulatory, haematological and immunological changes, manifested, among others, as increased blood pressure and respiration, increased blood glucose and lactate levels, impaired primary barriers towards the environment, heat shock protein production, changes in ion composition, haematocrit, lysozyme activity and antibody production (Wendelaar Bonga 1997; Barton 2002). These will in turn lead to tertiary stress responses such as decreases in growth, swimming capacity, disease resistance and feeding activity and altered behaviour (Wendelaar Bonga 1997; Barton 2002).

The stress response can be described in terms of allostasis, defined as the struggle to maintain homeostasis through changes in physiological systems (McEwen and Wingfield 2003; Korte et al. 2007; Landys et al. 2006). Allostasis allows the animal to actively adjust physiological systems to meet predictable and unpredictable changes in the environment. Thus, stress is not necessarily detrimental to the individual but is an essential adaptive response to promote the best chance of survival in the face of threatening situations (Iwama et al. 1997; Ashley 2007). However, when the allostatic load turns into allostatic overload, the stress becomes distress and turns detrimental (Moberg 2000).

What then determines the scope to cope of an organism, i.e. the ability to acclimate to conditions outside the 'optimal' range? Basically, what sets the scope are the biological 'needs' shaped by millions of years of natural selection, which are therefore different for each single species. A certain degree of tolerance to the displacement from the optimal range may be present, but eventually the displacement will result in disease and/or death. The ability of a fish to acclimate to a stressor, i.e. being able to maintain homeostasis despite the stressor impact, is further affected by the magnitude and duration of the stressor as well as of the number of additional stressors. During more chronic situations, wear and tear on the allostatic systems may eventually lead to severe disturbance and death by exhaustion.

Coping with threats to homeostasis: the cost of acclimation

Energetically, there are costs associated with allostasis, meaning that if a portion of the fish's energy budget is required to cope with stress, then less energy will be available for other biological functioning components (Iwama et al. 1997; Wendelaar Bonga 1997). This is particularly true when organisms are exposed to stressors from which they cannot escape. The most favourable response to conditions that could constitute an allostatic load on the fish is to avoid it. However, farmed fish are restricted to holding facilities and thus are usually not able to avoid unfavourable conditions. For example, a fish put under allostatic load uses energy reserves in order to acclimate to a stressful environment. Even if the fish has managed to acclimate to a stressor for weeks or months, the energy stores will eventually be depleted. This will indeed affect the ability of fish to meet another stressor or a change in the fish's endogenous development. Thus, it is when energy reserves are depleted, which is highly influenced by severity and duration of the stressor, that the fish is at overload and may no longer be able to acclimatize. When acclimation does not occur, this may give rise to pathology, disease and/or death.

Allostatic load and stress may favour diseases

Empirically, it has been repeatedly demonstrated that standard aquaculture practices can result in increased susceptibility of fish to disease (Mazur and Iwama 1993; Conte 2004; Dror et al. 2006). The questions remain, what are the underlying mechanisms behind these effects and how do allostatic mechanisms contribute? Primary and secondary stress responses are associated with substantial energetic costs, and particularly under conditions of repeated and chronic stress, this will lead to tertiary stress responses such as decreased growth, reduced swimming capacity, impaired disease resistance or lower feeding activity (Wendelaar Bonga 1997; Barton 2002). In this way, stressful husbandry conditions compromise, via the stress response, the health status of the fish and favour disease. In the following sections, the classical stress response and examples of common husbandry conditions will be used as starting points to discuss how husbandry-related stress may affect the primary defence systems of the fish: the epithelial barriers and the immune system.

Consequences of allostatic load/stress on the defence capabilities of the fish: primary barriers

The first lines of defence are the primary barriers, i.e. the mucus and epidermal surfaces in skin, gills and gut, which constitute an interface between the fish and the external world. These barriers regulate, for instance, ion, water and gas exchange with the environment, they are the initial sites of pathogen entry into the fish, and in the case of harmful substances, the barriers are the initial sites of uptake and action of these substances. The nature of the barriers ranges from the physical barrier constituted by the enterocytes and the 'fencing' tight junctions (Sundh et al. 2010), through ABC transporters that protect the fish against uptake of toxic substances into the organism (Sturm and Segner 2005) to the mucosal immune system defending against invading pathogens (Press and Evensen 1999). In this section, we focus on the barrier function of intestine and skin, while for the gills, we refer to recent reviews (e.g., Evans et al. 2005). The following discussion will provide evidence that the functioning of these primary barriers is weakened by poor welfare and stressful husbandry conditions (Fig. 2).

Main characteristics of the intestinal primary barrier The physical intestinal barrier is created by a single layer of epithelial cells, the enterocytes, which are connected to each other on the apical side of the cells through the tight junctions (TJ). TJ do not represent a rigid barrier but rather a fence, regulating the passage of ions, water and other molecules as well as immune cells, through the paracellular pathway. TJ consists of several physiologically regulated proteins forming the circumferential seals around

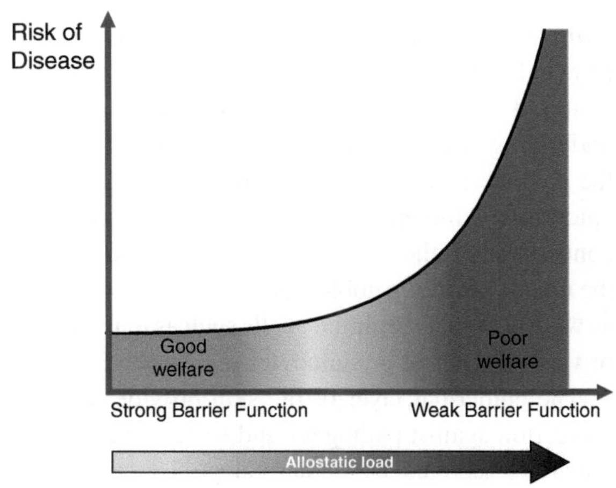

Fig. 2 The graph visualizes the connection between increasing allostatic load/stress, reduced barrier function and increased risk of disease, which from a function-based point of view means a reduced welfare

adjacent epithelial cells. These proteins are connected to the actin ring of the cytoskeleton, creating a continuous structure facing the apical side of the intestinal epithelium (Schneeberger and Lynch 2004).

The intestinal epithelial layer is protected by an extrinsic mucus layer. This is created by the apical release of mucin glycoproteins from goblet cells scattered in the intestinal epithelium (Shephard 1994). An important role of the mucus layer is to prevent uncontrolled attachment of bacteria to the intestinal epithelial cells. This is achieved by the glycoproteins forming a matrix to physically establish an effective diffusion barrier for the bacteria. The mucus layer further contains antimicrobial factors such as antibacterial peptides, lysozyme and proteases (Ellis 2001), which actively neutralize intraluminal pathogens. Moreover, the mucus is constantly replenished by the goblet cells, which results in a continuous flushing of the intestinal lumen, carrying harmful substances and pathogens away and out of the fish.

Within the gut, pathogens such as viruses and bacteria, along with bacterial toxins and other harmful substances, are mixed with nutrients and water. Although the extrinsic and physical barriers together are highly effective in preventing harmful substances from entering the internal milieu, total exclusion is impossible. Pathogens that breach the intestinal barrier will encounter the immunological barrier, *i.e.* the gut-associated lymphoid tissue (GALT).

Main characteristics of the skin primary barrier The physical skin barrier comprises an epidermis made of a stratified squamous epithelium, together with underlying dermis containing the scales and finally the hypodermis. The basic cellular element of the epidermis is the epithelial cell. Additional cell types contributing to the barrier function of the skin include the mucus-secreting goblet cells, ion-transporting cells as well as intrusive immune cells such as macrophages or various types of granulocytes.

The epidermal layer of the skin provides physical protection against pathogens and secretes mucus. The mucus is secreted by epidermal goblet cells and is composed of water and glycoproteins. Similar to the situation in the intestine, the skin mucus has dual roles: to prevent pathogen adherence by being continuously produced and sloughed off and to act as a repository of numerous innate immune factors such as lysozyme, immunoglobulins, complement proteins, lectins, C-reactive protein, proteolytic enzymes and various other antibacterial proteins and peptides (Shephard 1994). Immune cells are also present and active in the epidermal barrier of the skin, and their number greatly increases in response to skin injury or infection with pathogens.

Farming situations adding allostatic load on the primary barriers The severity of stress elicited by the different husbandry conditions tested and their impact on the functioning of the primary barriers can vary with number of stressors, their intensity and duration as well as their nature. Short-term, acute stress under fish-farming conditions is represented for instance by netting or short-term transportation. Such treatments have been shown to result in goblet cell depletion in the skin barrier, in detachment of the enterocytes from the basement membrane in the intestinal barrier (e.g. Szakolczai 1997) and in a reduction in the thickness of the protective mucus layer, giving way to increased epithelial–bacteria interaction (Szakolczai 1997; Olsen et al. 2002, 2005). Social stress resulted in flattening of the mucosal folds and loosening of the cell-to-cell contacts between enterocytes in the intestine of European eel (Peters 1982). Changes in paracellular barrier permeability due to alterations in intercellular junctions as a consequence of acute stress have been described for Atlantic salmon and Atlantic cod (Olsen et al. 2002, 2008). Similar intercellular damage has been observed in rainbow trout subjected to 15-min acute stress. Moreover, the acute stress resulted in intestinal barrier dysfunction measured as increased paracellular permeability to mannitol (Olsen et al. 2005). Cortisol is believed to be one mediator behind these changes in epithelial permeability. In fact, in rainbow trout, intraperitoneal-located slow-release cortisol implants resulted in increased paracellular permeability to mannitol concomitant with a decrease in transepithelial electrical resistance (TER) (Sundell, unpublished results). In addition, also the most initial and rapid primary stress response, adrenaline, has been shown to modulate the integrity of the intestinal barrier by increasing the permeability to Cl^- through the tight junctions (Bakker et al. 1993).

A common, often more long-term stressor in fish culture is periods of low levels of dissolved oxygen

(DO). Indeed, the DO levels have been suggested as a key limiting factor for salmon aquaculture (Ellis et al. 2002). This is supported by data from sea cage environmental studies showing that DO levels can be highly variable in both space and time and are affected by factors such as stocking density, degree of stratification, water currents, position of the sea cages and seasonal variations. Moreover, tidal cycles exhibit a major influence on the DO levels in farms situated in fjords, as these farms are sheltered from other causes of water movement like winds, waves and strong currents (Johansson et al. 2007). Cyclic drops in DO levels, reaching as low as 50% (<5 mg ml^{-1}) for several hours have been observed when the tidal current changes direction and the water current is close to zero (slack water). Not surprisingly, in late summer, the combination of slack water and high water temperatures led to the lowest and most critical DO levels within the sea cages (Johansson et al. 2007). Subjecting Atlantic salmon to such common husbandry conditions, constituting low and fluctuating DO levels in combination with high temperatures, under a series of controlled experiments led to an induction of the primary stress response reflected in elevated plasma cortisol levels (Sundh et al. 2010). Experimental analyses based solely on plasma cortisol levels would conclude that the allostatic load decreases with time, as the cortisol levels returned to basal values after 30–60 days, depending on the type of stressor. However, the impact of the stressors as assessed by measuring intestinal barrier function revealed a continued impairment during prolonged periods, even after full recovery of normal cortisol levels. Consequently, several different husbandry conditions creating constantly low and fluctuating DO levels and poor water quality result in elevated plasma cortisol levels up to 4 weeks after onset of the treatment, but not at later time points, whereas the intestinal barrier functions were disturbed throughout the experimental period. Moreover, the detrimental effects of the experimental environment on the intestinal barrier were more severe at higher temperatures, which suggest that elevated temperature per se acts as an additional stressor creating a higher allostatic load. In addition to disturbing the physical barrier, as shown by increased paracellular permeability, many of these husbandry conditions also affected the immunological properties of the intestinal barrier. An increased infiltration of neutrophils and altered gene expression patterns of important pro-inflammatory and anti-inflammatory cytokines in the intestinal mucosa was demonstrated as a result of both decreased DO levels and increased temperature (Sundh et al. 2010; Sundh et al., in preparation; Niklasson et al., submitted). An immunosuppressive effect of the stressors was suggested as IFN-γ was down-regulated by low DO, and this effect was even stronger with higher water temperature. Taken together, environmental situations commonly observed in sea cages during rearing of Atlantic salmon are apparently stressful to the fish, and this leads, in addition to transient changes in the classical stress response parameter, cortisol, to persistent alterations in the fish's defence system, in particular the immunology and permeability of the intestinal barrier.

The mucus layer in the skin has also been shown to be affected by unfavourable environmental conditions. Mucus production rates are known to increase during times of stress. For example, the mucification of acid-exposed fish during short-term exposures (within the first day of exposure) is correlated with a reduction in mucus cell numbers due to the exhausted secretion of the mucus cells. In long-term exposures to acidic water, the mucification is thought to be due to hyperplasia of mucus cells in skin and gills (Segner et al. 1988). Changes in salinity also have an influence on skin mucus production. In general, the abundance of mucus cells on fish surfaces decreases as salinity increases (Shephard 1994). A reduction in epithelium height and numbers of cell layers is also a feature of sea water-adapted fishes (Wendelaar Bonga and Meis 1981). Recently, in a study with smoltifying Atlantic salmon, it has been shown that intermixing of sea water (20‰) in combination with intensive rearing decreases mucus cell numbers and epithelium thickness, changes mucus quality and subsequently increases the susceptibility to infection with *Moritella viscosa* in Atlantic salmon after transfer to full-strength sea water (Toften et al., unpublished results). Alternatively, infestation of farmed salmon, even with low numbers of sea louse, *Lepeophtheirus salmonis*, can also reduce the number of mucous cells and results in stress-related changes to the skin epithelium (Nolan et al. 1999).

Not only low DO can be detrimental, hyperoxygenation of fresh water can also affect both intestinal and skin barrier functions. The intensification of Atlantic salmon smolt production includes moves towards stocking at higher fish densities. As water flow and DO become limiting, oxygen is added to maintain adequate oxygen levels in the holding tanks (Wedemeyer 1996). DO levels of 200–300% saturation are commonly used in aquaculture in order to be able to reduce the specific water flow (L kg^{-1} min^{-1}) (Brun 2003). This practice can have negative impacts on water quality in terms of increased levels of metabolic wastes such as carbon dioxide and ammonia together with decreased pH (Fivelstad et al. 1999, 2004; Ellis et al. 2002). Such an environment in the FW stage of Atlantic salmon farming has been hypothesized to be a contributing factor to the increased disease susceptibility to fish pathogens observed after SW transfer. Several studies indicate that Atlantic salmon smolts have increased susceptibility to infectious pancreatic necrosis virus (IPNV) and *Moritella viscosa* after exposure to reduced specific water flow and hyper-oxygenation (Fridell et al. 2007; Sundh et al. 2009; Toften et al., unpublished results).

Subjecting Atlantic salmon parr to hyperoxic conditions for 26 days resulted in chronically elevated plasma cortisol levels (Fridell et al. 2007). An increased allostatic load was also manifested as a disturbed intestinal barrier function resulting in increased paracellular permeability and increased translocation rate of *Aeromonas salmonicida* in the distal intestine (Sundh et al.2009). These observed decreases in intestinal barrier function are thought to be responsible for an increased risk of infection. Indeed, fish subjected to a similar experimental protocol revealed increased disease susceptibility to an IPNV challenge after subsequent transfer to SW (Fridell et al. 2007).

Moreover, the IPNV challenge per se can serve as a stressor. This was shown by elevated levels of plasma cortisol after IPNV co-habitant challenge (Fridell et al. 2007). This was also reflected in the cortisol release rate into the water (Sundh 2009). Interestingly, fish subjected to stress in FW responded with a stronger primary stress response to the IPNV challenge compared with unstressed fish. This reveals that stress in one life stage may affect the outcome of additional stressors in a subsequent life stages. The IPNV challenge per se also decreased the intestinal barrier function, as shown by both an increased paracellular permeability and increased translocation of *A. salmonicida.* Thus, the disturbed intestinal barrier function may be one explanation behind the increased risk of disease from secondary bacterial infections observed during acute IPNV infections (Johansen and Sommer 2001). In summary, the commonly used husbandry practice of hyperoxygenating FW is likely to reduce the ability of Atlantic salmon post-smolts to cope with additional stressors and represents therefore a major risk to the welfare of the fish.

Another typical stressor is farming under conditions where fish densities are too high (Gytre 2004; Huntingford et al. 2006). As fish density increases, water quality parameters are affected in a complex manner yielding decreased levels of dissolved oxygen (DO) and increased levels of carbon dioxide, nitrates and ammonia as well as a decreased pH (Ellis et al. 2002). Several studies have shown that high stocking density per se can be stressful to the fish (Ellis et al. 2002; Adams et al. 2007; Schram et al. 2010). For instance, Atlantic salmon kept at a density of 70 kg m^{-3} displayed elevated plasma cortisol levels, while this was not seen in fish kept at the intermediate densities of 30 and 50 kg m^{-3} (Sundh et al., in preparation). The elevated plasma cortisol levels decreased with time, suggesting habituation. The intestinal barrier function decreased in a dose-dependent manner in response to the severity of the stress, i.e. the fish density. Again, this was observed at time points when no differences in plasma cortisol levels were detected. Increased fish density resulted in decreased TER as well as increased paracellular permeability to mannitol. The threshold density for mediating a decreased barrier function was between 30 and 50 kg m^{-3}. Additionally, local signs of inflammation were more evident in fish kept at 70 kg m^{-3} as compared to 10 kg m^{-3}, which could be an effect of increased leakage of luminal content that attracts immune cells to the mucosa. Also, the expression of cytokines in the intestinal mucosa was affected by density. IFN-γ was down-regulated in the 70 kg m^{-3} group compared to 10 kg m^{-3}, suggesting a suppression of the mucosal immune barrier. In conclusion, high stocking density most frequently also in association with poor water quality chronically stressful to the fish as eliciting a primary stress response followed by a disturbance of physical and immunological barrier properties of the intestinal epithelium. Thus, this type of environment should be regarded as a threat to the welfare and health of the fish.

What is the association between stress and disease in fish?

As described previously, husbandry-induced stress can modulate immune function and disease susceptibility of fish (Yin et al. 1995; Davis et al. 2003; Iguchi et al. 2003; Pruett 2003; Binuramesh et al. 2005; Small and Bilodeau 2005; Welker et al. 2007). Both immunosuppressive and immunoenhancing effects of stress have been described (e.g. Demers and Bayne 1997, Vazzana et al. 2002). These discrepancies are most probably related to differences in study design, stress intensity and duration, nature of stressor etc. However, despite many examples of impaired immune status and enhanced disease incidences in stressed fish, the relationship between stress and disease must not be generalized (Weyts et al. 1999; Final Report of WEALTH project no 501984, 2008). Various genetic, developmental and environmental factors modulate magnitude, duration and consequences of the stress response and may obscure the relationship between stress and immune/disease status. A brief overview of several studies performed during the last decade, as given in Table 1, may illustrate our still fragmentary knowledge on how stress modulates immune parameters and disease susceptibility. Even if significant alterations in immunological functions occur following exposure to stressors, the consequences for the overall host resistance may not always be deleterious (Pruett 2003). For example, channel catfish subjected to low water stress displayed a higher lysozyme activity than un-stressed individuals, while disease resistance to *Edwardsiella ictaluri* was significantly lower (Small and Bilodeau 2005).

Stressor effects on the immune function and disease susceptibility have mainly been attributed to elevated cortisol levels (Pottinger and Carrick 1999a; Weyts et al. 1999). Indeed, cortisol has immunomodulatory effects in fish as reviewed by Weyts et al. (1999), Yada and Nakanishi (2002) and Segner et al. (2006). Cortisol acts as a regulator inhibiting certain constituents of the immune system and enhancing others (e.g., Espelid et al. 1996; Pruett 2003; Esteban et al. 2004). Also, the regulatory effect varies with cortisol concentration, duration of stress impact (acute versus chronic stress) or species differences in stress sensitivity (Weyts et al. 1999; Ruane et al. 1999; Barton 2002; Pruett 2003; Small and Bilodeau 2005; Ruane et al. 2007). However, it is clear that cortisol is but one amongst several players in mediating the immunosuppressive effects of stressful husbandry conditions on fish. Unfortunately, the differential role of other stress-induced endocrine processes in modulating immune functions of fish is little understood to date (Pruett 2003). Also stress-induced paracrine hormone secretion in the immune system itself must not be overlooked in this context (Arnold and Rice 2000). Thus, instead of keeping a rather one-dimensional focus on cortisol, we should try to develop a more general view on how (neuro)endocrine and paracrine processes mediate the effects of stressful husbandry conditions on the fish immune system (see Verburg-van Kemenade et al. 1999, 2009).

The utility of health/disease parameters as welfare indicators

The pros and cons of using health-related parameters as welfare indicators

While the previous chapter addressed how poor welfare status, as associated, for instance, with poor husbandry conditions, translates into poor health status of fish, the focus of the present chapter is on the use of health-related parameters as indicators of fish welfare. It is a strength of the health and biological function concept that it has potential to provide such welfare indicators (WIs) for fish culture. Welfare indicators are defined as *operational welfare indicators* (OWIs) when they can be corrected instantly, whereas they are designated *strategic* when they cannot be corrected instantly. Both types of indicators are referred to as *direct* if they are related to the fish and as *indirect* if they refer to the system or rearing conditions. As discussed previously, the basis for using health/disease parameters as WI or OWI is that poor welfare conditions can result in reduced fish health and increased disease incidence. In addition, disease per se can implicate reduced well-being of the animal.

Advantages of biological function and health-related parameters such as growth, body indices or feed utilization are that they are straightforward to measure. Physical health is the one of the most universally used indicators of welfare, as is relatively

Table 1 Brief overview of 38 studies related to common stressors in aquaculture: type of stressor employed, fish species, health and function indicators, bacterial challenge tests performed

Authors	Stressor(s)	Species	Physiological indicators	Immune parameters	Bacterial challenge
Acerete et al. (2004)	Transport loading	*Perca fluviatilis*	Cortisol, glucose, osmolality, RBC, haematocrit, lactates, haemoglobin	Not evaluated	Not performed
Barcellos et al. (2004)	Tank draining, transfer, crowding	*Rhamdia quelen*	Cortisol, RBC, haemoglobin	WBC	Not performed
Barcellos et al. (2006)	Handling air emersion	*Rhamdia quelen* (fingerlings)	Cortisol	Not evaluated	Not performed
Barnett and Pankurst (1998)	Crowding, chasing, exercise	*Rhombosolea tapirina*	Cortisol, lactates	Not evaluated	Not performed
Binuramesh et al. (2005)	Confinement, sex ratio	*Oreochromis mossambicus*	Cortisol	ROS production lysozyme immunoglobulin	*Aeromonas hydrophila*
Biswas et al. (2006)	Photoperiod manipulation, handling, confinement	*Pagrus major*	Cortisol, glucose, cholesterol, total protein, Ions, SGR, haematocrit, osmolarity	Not evaluated	Not performed
Cairns et al. (2008)	Handling confinement	*Onocrhynchus mykiss*	Expression of stress-related genes	Not evaluated	Not performed
Davis et al. (2003)	Dietary cortisol	*Ictalurus punctatus*	Cortisol	Not evaluated	*Ichthyophtirius multifiliis* Channel Catfish Virus (CCV)
Demers and Bayne (1997)	Handling	*Oncorhynchus mykiss*	Cortisol, adrenaline	Lysozyme	Not performed
Dror et al. (2006)	Handling	*Carassius auratus*	Glucose, cortisol	Interleukin-10	*Aeromonas salmonicida*
Espelid et al. (1996)	Handling, cortisol injection	*Salmo salar*	Cortisol	WBC	*Aeromonas salmonicida*
Fanouraki et al. (2007)	Anaesthesia, chasing, air exposure, density, temperature	*Pagrus pagrus*	Cortisol, glucose, lactates, norepinephrine, epinephrine, electrolytes, prothrombin, ions	Not evaluated	Not performed
Gornati et al. (2004)	Density	*Dicentrarchus labrax*	HSP, metallothioneins, stress-related genes	Not evaluated	Not performed
Hosoya et al. (2007)	Handling	*Melanogrammus aeglefinus*	Cortisol, glucose, HSP70	Not evaluated	Not performed
Iguchi et al. (2003)	Density	*Plecoglossus altivelis*	Cortisol	IgM	*Flavobacterium psychophilum*
Jentoft et al. (2005)	Handling	*Perca fluviatilis*	Cortisol, glucose, growth	Not evaluated	Not performed

Table 1 continued

Authors	Stressor(s)	Species	Physiological indicators	Immune parameters	Bacterial challenge
Lankford et al. (2003)	Air immersion, temperature, daytime	*Acipencer medirostris*	Cortisol, glucose, lactates	Not evaluated	Not performed
Mazur and Iwama (1993)	Density	*Oncorynchus tshawytsha*	Cortisol, haematocrit	Not evaluated	Not performed
McCormick et al. (1998)	Handling	*Salmo salar*	Cortisol, IGF-1, growth	Not evaluated	Not performed
Ortuno et al. (2002)	Complex stressful event	*Sparus aurata*	Cortisol, glucose	ROS production complement	Not performed
Pottinger and Carrick (1999a)	Confinement	*Oncorhynchus mykiss*	Cortisol, glucose	Not evaluated	Not performed
Pottinger and Carrick (1999b)	Confinement	*Oncorhynchus mykiss*	Cortisol	Not evaluated	Not performed
Pottinger et al. (2000)	Confinement	*Leuciscus cephalus*	Cortisol, GR, glucose, lactates, steroids	Not evaluated	Not performed
Rotllant et al. (1997)	Crowding	*Pagrus pagrus*	Cortisol, glucose, haematology	Complement, Ig lysozyme, WBC, haemagglutination	Not performed
Ruane and Komen (2003)	Density	*Cyprinus carpio*	Cortisol, glucose, free fatty acids, growth Food conversion	Not evaluated	Not performed
Saeij et al. (2003)	Handling, cortisol treatment	*Cyprinus carpio*	Not evaluated	Immune gene expression	*Trypanoplasma borreli*
Small and Bilodeau (2005)	Low water stress, dietary cortisol	*Ictalurus punctatus*	Cortisol	Lysozyme	*Edwardsiella ictaluri*
Strand et al. (2007)	Shadows movements, tank cleaning	*Perca fluviatilis*	Weight, feed intake, SGR	Not evaluated	Not performed
Terova et al. (2005)	Density	*Dicentrarchus labrax*	GRmRNA	Not evaluated	Not performed
Trenzado et al. (2003)	Confinement	*Oncorhynchus mykiss*	Cortisol, glucose, lactates, amino acids, glycogen aminotransferase	Not evaluated	Not performed
Trenzado et al. (2006)	Crowding	*Oncorhynchus mykiss*	Glucose, weight, SGR, feed intake, feed efficiency, haematocrit, haemoglobin, glycogen, liver enzymes	Not evaluated	Not performed
Vazzana et al. (2002)	Density	*Dicentrarchus labrax*	Cortisol, glucose, osmolarity	ROS production cytotoxic cells activity	Not performed

Table 1 continued

Authors	Stressor(s)	Species	Physiological indicators	Immune parameters	Bacterial challenge
Vijayan et al. (1997)	Confinement, cortisol treatment	*Oreochromis mossambicus*	Cortisol, glucose, lactates, free amino acids, glycogen, PEPCK, LDH	Not evaluated	Not performed
Welker et al. (2007)	Hypoxia	*Ictalurus punctatus*	Cortisol, glucose	Lysozyme, CH50, bactericidal activity, antibody response	*Edwardsiella ictaluri*
Wilkinson et al. (2006)	Confinement, hypoosmotic stress	*Salmo salar* *Oncorhynchus mykiss*	Cortisol, GH, IGF-1, IGF-2	Not evaluated	Not performed
Yin et al. (1995)	Crowding	*Cyprinus carpio*	Cortisol, glucose, chlorides	Proteins, lysozyme, complement, phagocytic cells activity	*Aeromonas hydrophila*

Captions: RBC, red blood cells; WBC, white blood cells; ROS, reactive oxygen species; SGR, specific growth rate; HSP, heat shock protein; IgM, type M immunoglobulin; IGF-1 and -2, insulin-like growth factor 1 and 2; GH, growth hormone; GR, glucocorticoid receptor; PEPCK, phosphoenolpyruvate carboxykinase; LDH, lactate dehydrogenase; CH50, total haemolytic complement

straightforward to recognize and to measure. In comparison with primary stress parameters such as cortisol, health-related parameters are less transient but integrate the adverse consequences of low welfare conditions over time. As they are more directly related to the defence mechanisms of the fish, such as indicators of immune system and primary barrier functioning, they are reliable indicators of disturbed function and decreased welfare. Finally, as promoting health and preventing or reducing disease support the economical success of a fish farm, this is an avenue through which farm owners can be motivated to take measures for improving fish welfare.

Although health and disease parameters are objective and can often be quantified supporting their use as OWIs in practical aquaculture, they can be difficult to interpret, and the caveats in the welfare–health relationship must not be neglected (Dawkins 2006). A disadvantage is that health-related OWIs may not be very specific. For instance, parameters such as condition factor or feed conversion are not influenced by fish welfare alone but also by, e.g., feed quality or water temperature. Thus, such parameters should not be used as stand-alone OWIs but in concert with other indicators. Another critical point is the question to what extent the presence of disease in a farm system is indeed an indicator of poor welfare? As pointed out by Wolffrom (2004) and Broom and Corke (2002), presence of disease normally indicates impaired welfare. Diseases result in poor welfare through various clinical and subclinical effects, although Huntingford et al. (2006) pointed out that for specific diseases, we need further research to fully understand their welfare implications. More difficult to answer is the question whether the presence of disease in a farm is a consequence of low welfare conditions. Disease may develop as a consequence of bad husbandry and stressful conditions and indicate an underlying problem in the farming facility. As discussed earlier, stressful husbandry conditions are likely to impair defence capabilities such as the immune system or the primary barriers of fish, and this eventually leads to disease. Similarly, poor hygiene in the husbandry system can lead to an increased pathogen load in the water eventually overloading the defence capability of the fish and causing disease. However, the occurrence of disease is not necessarily caused by poor welfare conditions, but can occur also under good welfare conditions. For instance, if the farm system is connected with a river system or in the case of cages is embedded in the natural environment, even a farmed stock kept in good welfare may show disease due to transmission of pathogens form wild stocks. Generally, with respect to welfare conditions, diseases arising from opportunistic pathogens may be more informative than diseases arising from infections with obligatory pathogens.

Alternatively, the absence of disease is not necessarily an indicator of good welfare. For instance, if a farmed fish stock is vaccinated, this may protect the fish from disease outbreaks despite poor welfare conditions. Thus, it is not so much the absence/presence alone that is meaningful as a direct welfare indicator instead, to be able to more conclusively predict the disease indicator should be embedded in a coherent scheme along with other indicators. For instance, it may be combined with a disease challenge test that informs on the immune capacity of the fish, or with measurements of stress indicators such as cortisol or barrier leakage (see WEALTH project no 501984, 2008). In addition, one should discriminate between pathogenic and non-pathogenic (i.e. nephrocalcinosis, gas bubble disease, deformations, fin damage) origins of disease, with the latter being directly indicative of poor environmental quality. Moreover, a distinction is needed between obligate and opportunistic pathogenic disease. Obligate pathogenic disease will occur even if the fish is in prime condition, while opportunistic pathogens will become a problem mainly under poor welfare conditions as the equilibrium between the fish and the opportunistic pathogen becomes imbalanced (WEALTH project no 501984, 2008).

Practical application of health-related parameters as welfare indicators in aquaculture

Scientific understanding of the processes and mechanisms relating husbandry conditions, fish health and welfare, as discussed in the previous chapter, is paramount to support the use of health parameters as welfare indicators, but practical use and applicability of these indicators is still another issue. In the following sections, we will discuss the use of OWIs for the example of rainbow trout, although the same general principles could be applied to any teleost being domesticated and introduced into artificial systems where conflicts arise between the demands of the species in question and the technology and economy of the husbandry conditions.

When addressing OWIs in fish production, it is important to focus on freshwater and marine systems separately. Furthermore, the production technologies of fry and fingerlings are so different from those for growth production of larger fish, making generalizations and conclusions difficult. Therefore, it is necessary to consider separately the different rearing systems such as flow-through, recirculation, concrete raceways, fibre-glass tanks, earth ponds and net-pen-systems both in fresh and in salt water.

In practice, a series of inspections and observations should be taken in order to determine the state of welfare on the farm (Table 2). These should refer to the fish, e.g., fish appearance (size, condition, skin, fin, eye and gill integrity and colour), fish behaviour (feed intake, location in water column, air-gasping, balance, activity), but also the system (fish density, unit size), water quality (oxygen, carbon dioxide, ammonia, BOD) (Table 3) or feed quality (essential nutrients, oxidation state of lipids, contaminants) (Table 4). Health inspections should be conducted and include virology, bacteriology and parasitology (Table 5). Serological parameters may be included if possible due to the fact that factors such as serum cortisol, glucose, lactic acid and others may reveal

Table 2 Initial observation of parameters related directly to the fish in the farms and the possibility for instant correction

Observation	Instant correction possible
Mortality	±
Condition factor	–
Epaxial musculature	–
Hypaxial musculature	–
Exophthalmia	–
Colour/pigmentation	–
Skin	–
Fins	–
Gills	–
Ulcers	–
Balance disturbances	±
Air-gasping	±
Scratching	–
Lethargy	±
Visible parasites	±

Table 3 Water quality parameters and their possible use as operational welfare indicators

Parameter	Instant correction possible
Temperature	±
Salinity	±
pH	±
Water flow	±
Turbidity	+
BOD	+
Ammonia	+
Nitrite	+
Nitrate	+
Phosphate	+

Table 4 Nutritional parameters and their possible use for OWI

Nutritional element	Instant correction possible
Proteins	
Amino acids	+
Lipids	
Fatty acids	+
Carbohydrates	+
Vitamins	+
Minerals	+
Antioxidants	+
Immunostimulants	+
Contaminants	±

both acute and chronic welfare problems. For instance, antibody titres can reflect both prior and current infections of specific pathogens. The physicochemical conditions in the production unit can in some cases be corrected. For instance, in recirculation systems operating with limited water volumes, it is possible to replenish fish tank water (adjusting BOD, salinity), re-oxygenate water, adjust salinity by adding sodium chloride or to adjust pH by adding hydrochloric acid or sodium hydroxide. In larger enterprises such as net-pen culture in the sea, it is not always feasible to correct the problems instantly.

The issue of health conditions is a good example of a strategic OWI. If poor values of direct, i.e. fish health-related OWIs are observed, this can be difficult to correct. For instance, health inspection on the farm site may reveal that clinical signs of the fish could be caused by excessive abundance of ectoparasites/ectocommensals on the gills, fins and skin of the fish. The problems associated with this may be corrected instantly through water treatment using formaldehyde, hydrogen peroxide, sodium percarbonate, sodium chloride or other disinfectants. If the health problem of the fish is not related to superficial colonization with ectocommensals or parasites, which can easily be managed as described above, it is necessary to make a plan for correction of the adverse conditions. This may include stamping out (if the disease is not treatable), drug treatment or vaccination (if available), radical change in water source or feed source. Here, a major welfare issue in fish culture comes into play and that is the serious lack of veterinary medicines (Wall 2008) and effective vaccines (Biering et al. 2005) to treat fish diseases.

How can fish farmers, farm inspectors and authorities use health-related OWIs? Producers are required to maintain strict health management plans incorporating a number of areas such as biosecurity controls, recording of movements and mortalities. It can be envisaged that in the near future, such management plans may include a range of health-related OWIs that the individual producers are required to monitor in order to show that the highest possible standards of welfare can be maintained. The systematic observations of important farm and fish parameters may be powerful tools for a future description of the welfare systems in a particular farm environment. The inevitably robust nature of OWIs makes them useful guidelines for farmers.

If one is to assume that maintaining a good health status of farmed fish is a cornerstone of good welfare, then practical health-related OWIs allowing the producer to rapidly and reliably measure welfare is a key factor. There are currently a number of resources available to producers to allow them to maintain a good health status on the farm. These strategies range from the legislative (EU Directives, national legislation) to agreed codes of practice in various industries and down to individual management plans on each farm. From a legislative point of view, EU Council Directive 2006/88/EC consolidates and updates aquaculture health controls for all aquatic animals within the EU (Breed 2008). There have been numerous studies on the use and application of risk analysis in managing health issues in aquaculture (Murray and Peeler 2005; Peeler et al.

Table 5 Some infections with various pathogens in rainbow trout farming (freshwater/marine) and relevance as operational welfare indicators

Pathogen	Control method available	Instant correction possible	Instant correction not possible
Virus			
VHSV	+	−	+
IHNV	+	−	+
IPNV	+	−	+
Bacteria			
Flavobacterium psychrophilum	+	−	+
Yersinia ruckeri	+	−	+
Aeromonas salmonicida	+	−	+
Vibrio anguillarum	+	−	+
Parasites			
Ectocommensals			
Apiosoma	+	+	−
Ambiphrya	+	+	−
Capriniana	+	+	−
Epistylis	+	+	−
Ectoparasites			
Trichodina	+	+	−
Chilodonella	+	+	−
Skin parasites			
Ichthyobodo	+	+	−
Tetrahymena	+	−	+
Ichthyophthirius	+	−	+
Gyrodactylus	+	+	−
Argulus	+	−	+
Lepeophtheirus	+	−	+
Caligus	+	−	+
Intestinal parasites			
Spironucleus	+	−	+
Eubothrium	+	−	+
Crepidostomum	+	−	+
Eye parasites			
Diplostomum	+	−	+
Tylodelphys	+	−	+
Kidney parasites			
Tetracapsuloides bryosalmonae	+	−	+
Cartilage parasites			
Myxobolus cerebralis	+	−	+

Possibility for control indicated and if this correction can be performed instantly

2007), and this risk-based approach is reflected in the new Directive. The Directive also allows each member state to address new and emerging disease issues within their country. This may take the form of listing a particular disease, thus making it notifiable to the authorities or developing codes of practice for the management of the disease in a particular country. Codes of practice have been developed in many countries for aquaculture such as the 'Code of good practice for Scottish finfish aquaculture' and the

'Code of conduct for European aquaculture' issued by the Federation of European Aquaculture Producers. These codes provide a voluntary framework and guidelines for the industry allowing them to improve practices and health of the farmed stock and thus, by default, improving welfare of their farmed fish. In fact, the RSPCA in conjunction with Freedom Foods in the UK has produced a welfare standards document for farmed Atlantic salmon. Producers who sign up to agree with the standards are audited and if accepted are allowed to produce labelled fish, which may demand a higher premium in the market place.

One aspect that is yet to address, but which is vital to sustainability and general acceptance of aquaculture, is that health and welfare of farmed fish need to be considered in the context of the complete ecosystem to which they belong. Or, in other words, what is the cost of maintaining welfare in the box to the cost of welfare outside, including the immediate and the networked environment? The 'ecological footprint' extends well beyond the farming system itself and includes the socio-economic and ecological impacts of feed production to supply the farm, as well as 'down-stream' effects, all of which can have severe consequences on welfare for many species, including humans. This is especially dramatic for shrimp farming, which has spawned efforts towards integrated production rather than monoculture (i.e. Azad et al. 2009). Indeed, to make a real step towards improved welfare, the question arises whether rather than focusing on the consequences of technologies permitting increasing fish densities in monoculture (as we discussed previously), an adaption of integrated multi-trophic aquaculture (IMTA) may be a better way to go (Chopin et al. 2001; Neori 2008)? Of course, this would require a wide diversity of solutions, each tailored to specific ecosystem economies and climates (i.e. Pigneguy 2008). It would be interesting to investigate how the 'five freedoms' criteria for welfare in IMTA compare with conventional monoculture. As far as the authors are aware, this has not been done nor have OWIs, as outlined earlier for salmonids in monoculture, been applied to IMTA. An obvious difficulty is how to establish reliable criteria for comparison between such diverse systems. Here, the measurement of barrier function, as described previously, could be one such approach. Poor welfare conditions lead to heavy use of antibiotics and other treatment strategies to maintain health. These loads are passed onto the environment in the form of toxic wastes, high levels of organic materials and antibiotic-resistant microbial strains (Boerlin and Reid-Smith 2008). This increased risk and pressure on health and welfare outside the box is seldom assessed and does not yet figure in the OWIs for the farm, although it is essential this be done.

The seriousness of this problem and the extent of the knowledge gaps can be exemplified by several recent studies. Even to understand how the wild fish population outside the box will react to welfare measures applied on the farm, we need to better understand their welfare (Berg 2007). For example, although the diversity of potentially pathogenic bacteria was found to be very high in wild Chinook salmon fry (Evans and Neff 2009), the prevalence of individual bacterial strains tends to be low, as was also found in other juvenile salmonids (Dionne et al. 2009). The potential for overloading this natural diversity with high concentrations of pathogens released from intensive aquaculture is a very real concern. To understand the potential harmful effects of aquaculture on the welfare in the immediate environment, we therefore need more long-term studies of aquaculture localities, preferably starting even before farms are established, but in the least as monitoring programmes to follow the effects of farming strategies. An example of how valuable this can be is a 23-year monitoring programme of the emergence of columnaris disease (*Flavobacterium columnare*) in salmon fingerlings on a fish farm in northern Finland (Pulkkinen et al. 2010). The authors show how intensive farming is coupled with the evolution of more virulent pathogens, with higher transmission amongst homogeneous subsets of fish. That major antibiotics used to treat fish are also, in part, front-line reagents in treating microbial infections of other animals, including humans, demonstrates how the selection for resistant and virulent strains can have wide-ranging consequences (Boerlin and Reid-Smith 2008).

Conclusions

In this brief review, we focus on which processes and mechanisms lay the basis for maintaining biological functions and good health under the pressures imposed by the economics of successful aquaculture production, and how this again relates to good welfare in aquaculture. This focus is chosen as well founded, and measurable health parameters are arguably the most readily adopted as operational welfare indicators for use on the farm. There is quite some effort being invested in deriving putative feeling-based and nature-based parameters and the resultant OWIs topics of other articles in this volume. One of the great challenges will be integrating these data with function- and health-based OWIs to develop comprehensive and robust welfare plans.

A definite advantage of function- and health-based OWIs is that they remain valid and applicable, regardless of whether fish can be regarded as sentient or not. For all sentient beings, protracted suffering will inevitably lead to poorer health, a measurable commodity across species boundaries. For non-sentient species, measuring biological function and health parameters or processes is probably one of the few indicators we have available as markers of welfare. This focus places a strong reliance on the reliability of parameters currently under consideration and in use and argues strongly for the development of validation based on methods measuring secondary and tertiary response indicators. For this reason, we have outlined in this review the strengths and shortcomings of the current methodologies and approaches. We have drawn attention to one of the major front lines in the fish allostatic response to environmental stressors, the primary barriers provided by skin, intestine and gills. Injury of these barriers, whether mechanically or chemically or biologically, inevitably leads to a displaced homeostasis and threatens health. Primary barrier responses, including the ensuing systemic immune responses, are areas of urgently needing further research. In addition to characterizing physiological responses, the application of system biology approaches, including functional genomics and proteomics (see for example the article by Prunet, this volume), should be at the forefront of such efforts. This is necessary for robust scientific backing of health- and function-related OWIs. Indicators based on sound physiological understanding are critical if health and, as a consequence, welfare considerations are to receive the necessary acceptance for legislation to protect a sustainable aquaculture industry where good welfare is an integral and central component.

References

Acerete L, Balasch JC, Espinosa E, Josa A, Tort L (2004) Physiological responses in Eurasian perch (Perca fluviatilis L.) subjected to stress by transport and handling. Aquaculture 237:167–178

Adams CE, Turnbull JF, Bell A, Bron JE, Huntingford FA (2007) Multiple determinants of welfare in farmed fish: stocking density, disturbance, and aggression in Atlantic salmon (Salmo salar). Can J Fish Aquat Sci 64:336–344

Anderson PA, Berzins IK, Fogarty F, Hamlin HJ, Gillette LJ (2011) Sound stress, and sea horses: the consequences of a noisy environment to animal health. Aquaculture 311:129–138

Arnold RE, Rice CD (2000) Channel catfish, *Ictalurus punctatus*, leukocytes secrete immunoreactive adrenal corticotropin hormone (ACTH). Fish Physiol Biochem 22:303–310

Ashley PJ (2007) Fish welfare: current issues in aquaculture. Appl Anim Behav Sci 104:199–235

Azad AK, Jensen KR, Lin CK (2009) Coastal aquaculture development in Bangladesh: unsustainable and sustainable experiences. Environ Manage 44:800–809

Bakker R, Dekker K, De Jonge HR, Groot JA (1993) VIP, serotonin, and epinephrine modulate the ion selectivity of tight junctions of goldfish intestine. Am J Physiol 264:R362–R368

Barcellos LJG, Kreutz LC, de Souza C, Rodrigues LB, Fioreze I, Quevedo RM, Cericato LM, Soso AB, Fagundes M, Conrad J (2004) Hematological changes in jundia (Rhamdia quelen Quoy and Gaimard Pimelodidae) after acute and chronic stress caused by usual aquacultural management, with emphasis on immunosuppressive effects. Aquaculture 237:229–236

Barcellos LJG, Kreutz LC, Quevedo LM (2006) Previous chronic stress does not alter the cortisol response to an additional acute stressor in jundia (Rhamdia quelen, Quoy and Gaimard) fingerlings. Aquaculture 253:317–321

Barnett CW, Pankurst NW (1998) The effects of common laboratory and husbandry practices on the stress reponse of greenback flounder Rhombosolea tapirina (Günther, 1862). Aquaculture 162:313–329

Barton BA (2002) Stress in fishes: a diversity of responses with particular reference to changes in circulating corticosteroids. Integr Comp Biol 42:517–525

Berg O (2007) The dual myths of the healthy wild fish and the unhealthy farmed fish. Dis Aquat Org 75:159–164

Biering E, Villoin S, Sommerset I, Christie KE (2005) Update on viral vaccines for fish. Dev Biol 121:97–113

Binuramesh C, Prabakaran M, Steinhagen D, Michael RD (2005) Effect of chronic confinement stress on the immune responses in different sex ration groups of *Oreochromis mossambicus* (Peters). Aquaculture 250:47–59

Biswas AK, Seoka M, Takii K, Maita M, Kumai H (2006) Stress response of red sea bream Pagrus major to acute handling and chronic photoperiod manipulation. Aquaculture 252:566–572

Boerlin P, Reid-Smith RJ (2008) Antimicrobial resistance: its emergence and transmission. Anim Health Res Rev 9:115–126

Breed AJ (2008) New fish health directive: Council Directive 2006/88/EC of 24th October 2006 on animal health requirements for aquaculture animals and products thereof, and on the prevention and control of certain diseases of aquatic animals. Fish Vet J 10:21–28

Broom DM (2007) Welfare in relation to feelings, stress and health. REDVET: 2007, vol VIII, No 12B, http://www.veterinaria.org/revistas/redvet/n121207B/BA018ing.pdf. Accessed 7 April 2011

Broom DM, Corke MJ (2002) Effects of disease on farm animal welfare. Acta vet Brno 71:133–136

Brun E (2003) Epidemiology. In: Evensen O, Rimstad E, Stagg R, Brun E, Midtlyng P, Johansen LH, Jensen I (eds) IPN in salmonids: a review. FHL & VESO, Trondheim (2003), pp 51–67

Cairns MT, Johnson MC, Talbot AT, Pemmasani JK, McNeill RE, Houeix B, Sangrador-Vegas A, Pottinger TG (2008) A cDNA microarray assessment of gene expression in the liver of rainbow trout (*Oncorhynchus mykiss*) in response to a handling and confinement stressor. Compar Biochem Physiol D Genom Proteom 3:51–66

Chandroo KP, Duncan IJH, Moccia RD (2004) Can fish suffer? Perspectives on sentience, pain, fear and stress. Appl Anim Behav Sci 86:225–250

Chopin T, Buschmann AH, Halling C, Troell M, Kautsky N, Neori A, Kraemer G, Zertuche-Gonzalez J, Yarish C, Neefus C (2001) Integrating seaweeds into aquaculture systems: a key towards sustainability. J Phycol 37:975–986

Conte FS (2004) Stress and the welfare of cultured fish. Appl Anim Behav Sci 86:205–223

Davis KB, Griffin BB, Gray WL (2003) Effect of dietary cortisol on resistance of channel catfish to infection by *Ichthyopthirius multifiliis* and channel catfish virus disease. Aquaculture 218:121–130

Dawkins MS (2006) A user's guide to animal welfare science. TREE 2:77–82

Demers NE, Bayne CJ (1997) The immediate effects of stress on hormones and plasma lysozyme in rainbow trout. Develop Comp Immunol 21:363–373

Dionne M, Miller KM, Dodson JJ, Bernatchez L (2009) MHC standing genetic variation and pathogen resistance in wild Atlantic salmon. Philos Trans R Soc B 364:1555–1565

Dror M, Sinyakov MS, Okun E, Dym M, Sredni B, Avtalion RR (2006) Experimental handling stress as infection-facilitating factor for the goldfish ulcerative disease. Vet Immunol Immunopathol 109:279–287

Duncan IJ (2005) Science-based assessment of animal welfare: farm animals. Rev Sci Tech 24:483–492

Ellis AE (2001) Innate host defense mechanisms of fish against viruses and bacteria. Develop Comp Immunol 25:827–839

Ellis T, North B, Scott AP, Bromage NR, Porter M, Gadd D (2002) The relationships between stocking density and welfare in farmed rainbow trout. J Fish Biol 61:493–531

Espelid S, Lokken GB, Steiro K, Bogwald J (1996) Effects of cortisol and stress on the immune system in Atlantic Salmon (*Salmo salar L.*). Fish Shellfish Immunol 6:95–110

Esteban MA, Rodriguez A, Ayala G, Meseguer J (2004) Effects of high doses of cortisol on innate cellular immune response of seabream (*Sparus aurata* L.). Gen Comp Endocrinol 137:89–98

Evans ML, Neff BD (2009) Major histocompatibility complex heterozygote advantage and widespread bacterial infections in populations of Chook salmon (Oncorhynchus tshawytscha). Mol Ecol 18:4716–4729

Evans DH, Piermarini PM, Choe KP (2005) The multifunctional fish gill: dominant site of gas exchange, osmoregulation, acid base regulation and excretion of nitrogenous waste. Physiol Rev 85:97–177

Fanouraki E, Divanach P, Pavlidis M (2007) Baseline values for acute and chronic stress indicators in sexually immature red porgy (*Pagrus pagrus*). Aquaculture 265:294–304

Fivelstad S, Olsen AB, Kloften H, Ski H, Stefansson S (1999) Effects of carbon dioxide on Atlantic salmon (Salmo salar L.) smolts at constant pH in bicarbonate rich freshwater. Aquaculture 178:171–187

Fivelstad S, Bergheim A, Hølland PM, Fjermedal AB (2004) Water flow requirements in the intensive production of Atlantic salmon (*Salmo salar* L.) parr-smolt at two salinity levels. Aquaculture 231:263–277

Fridell F, Gadan K, Sundh H, Taranger GL, Glette J, Olsen RE, Sundell K, Evensen Ø (2007) Effect of hyperoxygenation and low water flow on the primary stress response and susceptibility of Atlantic salmon *Salmo salar* L. to experimental challenge with IPN virus. Aquaculture 270:23–35

Gornati R, Papis E, Rimoldi S, Terova S, Saroglia M, Bernardini G (2004) Rearing density influences the expression of stress-related genes in sea bass (*Dicentrarchus labrax* L.). Gene 341:111–118

Gytre T (2004) What is life like for a farmed fish? Marine Research News, Institute of Marine Research, Bergen, Norway, http://www.imr.no/english/__data/page/6335/What_is_Life_Like_for_a_Farmed_Fish.pdf

Hosoya S, Johnson SC, Iwama GK, Gamperl AK, Afonso LOB (2007) Changes in free and total plasma cortisol levels in juvenile haddock (*Melanogrammus aeglefinus*) exposed to long-term handling stress. Comp Biochem Physiol A Mol Integrat Physiol 146A:78–86

Huntingford FA, Adams C, Braithwaite VA, Kadri S, Pottinger TG, Sandoe P, Turnbull JF (2006) Current issues in fish welfare. J Fish Biol 68:332–372

Iguchi K, Ogawa K, Nagae M, Ito F (2003) The influence of rearing density on stress response and disease susceptibility of ayu (*Plecoglossus altivelis*). Aquaculture 220: 515–523

Iwama GK, Pickering AD, Sumpter JP, Schreck CB (1997) Fish stress and health in aquaculture. Cambridge University Press, Cambridge

Jentoft S, Aastveit AH, Torjesen A, Andersen O (2005) Effects of stress on growth, cortisol and glucose levels in non-domesticated Eurasian perch (*Perca fluviatilis*) and domesticated rainbow trout (*Oncorhynchus mykiss*). Comp Biochem Physiol 141A:353–358

Johansen LH, Sommer AI (2001) Infectious pancreatic necrosis virus infection in Atlantic salmon *Salmo salar* post-smolts affects the outcome of secondary infections with infectious salmon anaemia virus or Vibrio salmonicida. Dis Aquat Organ 47:109–117

Johansson D, Juell J-E, Oppedal F, Stiansen J-E, Ruohonen K (2007) The influence of the pycnocline and cage resistance on current flow oxygen flux and swimming behaviour of Atlantic salmon (*Salmo salar* L.) in production cages. Aquaculture 265:271–287

Korte SM, Olivier B, Koolhaas JM (2007) A new animal welfare concept based on allostasis. Physiol Behav 92:422–428

Landys MM, Ramenofsky M, Wingfield JC (2006) Actions of glucocorticoids at a seasonal baseline as compared to stress-related levels in the regulation of periodic life processes. Gen Comp Endocrinol 148:132–149

Lankford SE, Adams TE, Cech JJJ (2003) Time of day and water temperature modify the physiological stress response in green sturgeon, *Acipenser medirostris*. Comp Biochem Physiol 135A:291–302

Lupatsch I, Santos GA, Schrama JW, Verreth JAJ (2010) Effect of stocking density and feeding level on energy expenditure and stress responsiveness in European sea bass, Dicentarchus labrax. Aquaculture 298:245–250

Mazur CF, Iwama GKK (1993) Handling and crowding stress reduces the number of plaque-forming cells in Atlantic salmon. J Aquat Animal Health 5:98–101

McCormick SD, Shrimpton JM, Carey JB, O'Dea MF, Sloan KE, Moriyama S, Bjornsson BT (1998) Repeated acute stress reduces growth rate of Atlantic salmon parr and alters plasma levels of growth hormone, insulin-like growth factor I and cortisol. Aquaculture 168:221–235

McEwen BS, Wingfield JC (2003) The concept of allostasis in biology and biomedicine. Horm Behav 43:2–15

Moberg GP (2000) Biological responses to stress: implications for animal welfare. In: Moberg GP, Mench JA (eds) The biology of animal stress: basic principles and implications for animal welfare. CAB International, Wallingford, pp 1–21

Mommsen T, Vijayan M, Moon T (1999) Cortisol in teleost: dynamics, mechanisms of action and metabolic regulation. Rev Fish Biol Fish 9:211–268

Murray AG, Peeler EJ (2005) A framework for understanding the potential for emerging diseases in aquaculture. Prev Vet Med 67:223–235

Neori A (2008) Essential role of seaweed cultivation in integrated multi-trophic aquaculture farms for global expansion of mariculture: an analysis. J Appl Phycol 20:567–570

Niklasson L, Sundh H, Fridell F, Taranger GL, Sundell K (2011) Disturbance of the intestinal mucosal immune system caused by prolonged husbandry stress in farmed Atlantic salmon (*Salmo salar*). Fish Shellfish Immunol (submitted)

Nolan DT, Reilly P, Wendelaar Bonga SE (1999) Infection with low numbers of the sea louse *Lepeophtheirus salmonis* induces stress-related effects in post-smolt Atlantic salmon (*Salmo salar*). Can J Fish Aquat Sci 56:947–959

Olsen RE, Sundell K, Hansen T, Hemre GI, Myklebust R, Mayhew TM, Ringø E (2002) Acute stress alters the intestinal lining of Atlantic salmon, *Salmo salar* L.: an electron microscopical study. Fish Physiol Biochem 26: 211–221

Olsen RE, Sundell K, Mayhew TM, Myklebust R, Ringø E (2005) Acute stress alters intestinal function of rainbow trout, *Oncorhynchus mykiss* (Walbaum). Aquaculture 250: 480–495

Olsen RE, Sundell K, Ringø E, Myklebust R, Hemre G-I, Hansen T, Karlsen Ø (2008) The acute stress response in fed and food deprived Atlantic cod, *Gadus morhua* L. Aquaculture 280:232–241

Ortuno J, Esteban MA, Meseguer J (2002) Effects of phenoxyethanol on the innate immune system of gilthead seabream (*Sparus aurata* L.) exposed to crowding stress. Vet Immunol Immunopathol 89:29–36

Peeler EJ, Murray AG, Thebault A, Brun E, Giovaninni A, Thrush MA (2007) The application of risk analysis in aquatic animal health management. Prev Vet Med 81: 3–20

Peters G (1982) The effect of stress on the stomach of the European eel, *Anguilla anguilla* L. J Fish Biol 21:497–512

Pigneguy M (2008) Tapping into sustainability. NZ Aquaculture 22:12–14

Pottinger TG, Carrick TR (1999a) A comparison of plasma glucose and plasma cortisol as selection markers for high and low stress-responsiveness in female rainbow trout. Aquaculture 175:351–363

Pottinger TG, Carrick TR (1999b) Modification of the plasma cortisol response to stress in rainbow trout by selective breeding. Gen Comp Endocrinol 116:122–132

Pottinger TG, Carrick TR, Appleby A, Yeomans WE (2000) High blood cortisol levels and low cortisol receptor affinity: Is the chub, *Leuciscus cephalus*, a cortisol-resistant teleost? Gen Comp Endocrinol 120:108–117

Press CM, Evensen Ø (1999) The morphology of the immune system in teleost fishes. Fish Shellfish Immunol 9: 309–318

Pruett SB (2003) Stress and the immune system. Pathophysiology 9:133–153

Pulkkinen K, Suomalainen LR, Read AF, Ebert D, Rintamäki P, Valtonen ET (2010) Intensive fish farming and the evolution of pathogen virulence:the case of columnaris disease in Finland. Proc R Soc B 277:593–600

Roger PA (2008) The impact of disease and disease prevention on sheep welfare. Small Rumin Res 76:104–111

Rotllant J, Pavlidis M, Kentouri M, Abad ME, Tort L (1997) Non-specific immune responses in the red porgy *Pagrus pagrus* after crowding stress. Aquaculture 156:279–290

Ruane NM, Komen H (2003) Measuring cortisol in the water as an indicator of stress caused by increased loading density in common carp (*Cyprinus carpio*). Aquaculture 218:685–693

Ruane NM, Wendelaar Bonga SE, Balm PHM (1999) Differences between rainbow trout and brown trout in the regulation of the pituitary-interrenal axis and physiological performance during confinement. Gen Comp Endocrinol 115:210–219

Ruane NM, HJTh Goos, Komen J (2007) Stress-induced facilitation of the cortisol response in 17α-hydroxylase deficient XX mas-1/mas-1 carp (*Cyprinus carpio*). Gen Comp Endocrinol 150:473–479

Saeij JPJ, Verburg-van Kemenade LBM, van Muiswinkel WB, Wiegertjes GF (2003) Daily handling stress reduces resistance of carp to *Trypanoplasma borreli*: in vitro modulatory effects of cortisol on leukocyte function and apoptosis. Develop Comp Immunol 27:233–245

Schneeberger EE, Lynch RD (2004) The tight junction: a multifunctional complex. Am J Physiol 286:C1213–C1228

Schram E, Roques JAC, Abbink W, Spnings T, de Vries P, Biernan S, van de Vis H, Flik G (2010) The impact of elevated water ammonia concentration on physiology, growth and feed intake of African catfish (*Clarias gariepinus*). Aquaculture 306:108–115

Segner H, Marthaler R, Linnenbach M (1988) Growth, aluminium uptake and mucous cell morphometrics of brown trout, *Salmo trutta*, early life stages in low pH water. Environl Biol Fish 21:153–159

Segner H, Eppler E, Reinecke M (2006) The impact of environmental hormonally active substances on the endocrine and immune systems of fish. In: Reinecke M, Zaccone G, Kapoor BG (eds) Fish endocrinology. Science Publishers, Enfield (NH), pp 809–865

Shephard KL (1994) Functions for fish mucus. Rev Fish Biol Fish 4:401–429

Small BC, Bilodeau AL (2005) Effects of cortisol and stress on channel catfish (*Ictalurus punctatus*) pathogen susceptibility and lysozyme activity following exposure to *Edwardsiella ictaluri*. Gen Comp Endocrinol 142:256–262

Stott GH (1981) What is animal stress and how is it measured? J Anim Sci 52:150–153

Strand A, Magnhagen C, Alanara A (2007) Effects of repeated disturbances on feed intake, growth rates and energy expenditures of juvenile perch, *Perca fluviatilis*. Aquaculture 265:163–168

Sturm A, Segner H (2005) P-glycoprotein and xenobiotic efflux transport in fish. In: Mommsen TP, Moon TW (eds) Biochemistry and molecular biology of fishes, vol VI. Environmental toxicology. Elsevier, Amsterdam, pp 496–533

Sundh H (2009) Chronic stress and intestinal barrier function: Implications for infection and inflammation in intensive salmon aquaculture. Doctoral thesis, University of Gothenburg, Gothenburg, Sweden, http://hdl.handle.net/2077/20826

Sundh H, Olsen RE, Fridell F, Gadan K, Evensen Ø, Glette J, Taranger GL, Sundell K (2009) The intestinal primary barrier in Atlantic salmon (*Salmo salar*): effect of hyperoxygenation and reduced flow in freshwater and subsequent IPN virus challenge in sea water. J Fish Dis 32: 687–698

Sundh H, Kvamme BO, Fridell F, Olsen RE, Ellis T, Taranger GL, Sundell K (2010) Intestinal barrier function of Atlantic salmon (*Salmo salar* L.) post smolts is reduced by common sea cage environments and suggested as a possible physiological welfare indicator. BMC Physiol 10:22. doi:10.1186/1472-6793-10-22

Sundh H, Fridell F, Ellis T, Niklasson L, Taranger GL, Pettersen EF, Wergeland HI, Sundell K (2011) High stocking density and poor water quality disturb physical and immunological barriers in the Atlantic salmon (*Salmo salar*, L.) intestine (in preparation)

Szakolczai J (1997) Histopathological changes induced by environmental stress in common carp, Japanese coloured carp, European eel, and African catfish. Acta Vet Hung 45:1–10

Terova G, Gornati R, Rimoldi S, Bernardini G, Saroglia M (2005) Quantification of glucocorticoid receptor in sea bass (Dicentrarchus labrax) reared at high stocking density. Gene 357:144–151

Thomas S, Kinkead R, Walsh PJ, Wood CM, Perry SF (1991) Desensitization of adrenaline-induced red blood cell H^+ extrusion in vitro after chronic exposure of rainbow trout to moderate environmental hypoxia. J Exp Biol 156:233–248

Trenzado C, Carrick ETR, Pottinger TG (2003) Divergence of endocrine and metabolic responses to stress in two rainbow trout lines selected for differing cortisol responsiveness to stress. Gener Compe Endocrinol 133:332–340

Trenzado CE, Morales AC, de la Higuera M (2006) Physiological effects of crowding in rainbow trout, *Oncorhynchus mykiss*, selected for low and high stress responsiveness. Aquaculture 258:583–593

Turnbull JF, Kadri S (2007) Safeguarding the many guises of farmed fish welfare. Dis Aquat Org 75:173–182

Vazzana M, Cammarata M, Cooper EL, Parrinello DN (2002) Confinement stress in sea bass (Dicentrarchus labrax) depresses peritoneal leukocyte cytotoxicity. Aquaculture 210:231–243

Verburg-Van kemenade BML, Nowak B, Engelsma MY, Weyts FAA (1999) Differential effects of cortisol on apoptosis and proliferation of carp B-lymphocytes from head kidney, spleen and blood. Fish Shellfish Immunol 9:405–415

Verburg-Van Kemenade BML, Stolte EH, Metz JR, Chadzinska M (2009) Neuroendocrine-immune interactions in teleost fish. In: Bernier NJ, van der Kraak G, Farrell AP, Brauner CJ (eds) Fish neuroendocrinology. Elsevier, Amsterdam

Vijayan M, Pereira C, Grau EG, Iwama GK (1997) Metabolic responses associated with confinement stress in Tilapia: The role of cortisol. Comp Biochem Physiol 116C:89–95

Volpato GL, Goncalves-de-Freitas E, Fernandes-de-Castilho M (2007) Insights into the concept of fish welfare. Dis Aquat Org 75:165–171

Wall T (2008) Disease and medicines—the welfare implications. In: Branson EJ (ed) Fish welfare. Blackwell, Oxford, pp 195–201

WEALTH project no 501984 (2008) Welfare and Health in sustainable aquaculture

Wedemeyer GA (1996) Physiology of fish in intensive culture systems. Chapman & Hall, ITP, New York

Welker TL, McNulty ST, Klesius PH (2007) Effect of sublethal hypoxia on the immune response and susceptibility of channel catfish, *Ictalurus punctatus*, to enteric septicaemia. J World Aquacult Soc 38:12–23

Wendelaar Bonga SE (1997) The stress response in fish. Physiol Rev 77:591–630

Wendelaar Bonga SE, Meis S (1981) Effects of external osmolality, calcium and prolactin on growth and differentiation of the epidermal cells of the cichlid teleost Sarotherodon mossambicus. Cell Tissue Res 221:109–123

Weyts FA, Cohen AN, Flik G, Verburg-van Kemenade BML (1999) Interactions between the immune system and the hypothalamo-pituitary-interrenal axis in fish. Fish Shellfish Immunol 9:1–20

Wilkinson RJ, Porter M, Woolcott H, Longland R, Carragher JF (2006) Effects of aquaculture related stressors and nutritional restriction on circulating growth factors (GH, IGF-I and IGF-II) in Atlantic salmon and rainbow trout. Comp Biochem Physiol Part A Mol Integr Physiol 145:214–224

Wolffrom T (2004) Farmed fish and welfare EU commission DG general for fisheries. Research and Scientific Analysis Unit, Brussels

Yada T, Nakanishi T (2002) Interaction between endocrine and immune systems in fish. Int Rev Cytol 220:35–92

Yin Z, Lam TJ, Sin YM (1995) The effects of crowding stress on the non-specific immuneresponse in fancy carp (*Cyprinus carpio* L.). Fish Shellfish Immunol 5:519–529

Demand Feeding and Welfare in Farmed Fish – Summary

Joël Attia, Sandie Millot, Carole Di-Poï, Marie-Laure Bégout, Chris Noble, Javier Sanchez-Vazquez, Genciana Terova, Marco Saroglia and Børge Damsgård

Demand-feeding systems can optimise feed management and production performance without compromising welfare. They are suitable tools for commercial farms. However, feed must be provided in sufficient quantity and quality to allow fish to express their normal feeding behaviour. Feeding to demand may be very different from one individual to another within the same group subjected to the same conditions. It depends on multiple parameters including density, social organisation, genetics, individual learning ability and boldness.
A benefit of self-feeding technology is its ability to test the individual's capacity for differentiating between different types of feed. When fish can freely access two or more self-feeders containing complementary feeds, they exhibit an ability to select nutritionally balanced diets from sub-optimal feeds. This is of particular interest when it comes to developing alternative plant-based feed sources, using fish memory and knowledge of feed to optimise feeding and thus welfare.

Background

Over the last few decades, many experiments have shown that fish do have a memory that lasts more than 3 seconds (no offence to the sea bream Dory...). Fish are able to learn relatively difficult tasks, and in particular to obtain food from a distributor by pushing or pulling a device. Such feeding systems, called self-feeders or demand feeders, are increasingly used in fish farms, as they can improve welfare by allowing fish to eat when they want and because they reduce food waste. But is it clear that these kinds of feeding systems improve fish welfare? Many experiments have been conducted to investigate feeding motivation and feed intake in fish reared with demand feeding systems. This paper reviews a selection of studies in the field, focusing on the key factors related to demand feeding and fish welfare.

Scientific Significance

These studies show the potential value of using demand-feeding systems for investigating fish behaviour in captivity. As appetite levels can be used as an operational welfare indicator (OWI) in farmed fish, providing the fish with enough food and feeding fish to appetite can, therefore, improve their welfare.

Practical Application

In this paper we argue that the use of demand-feeding systems can improve welfare of fish. In our view, this is in line with the demands of an increasing number of European consumers who require more healthy and ethical products.

Contact

J. Attia
ENES, CNPS, UMR CNRS 8195, University
of Saint-Etienne, 42000 Saint-Etienne, France
e-mail: attia@univ-st-etienne.fr

Fish Physiol Biochem (2012) 38:107–118
DOI 10.1007/s10695-011-9538-4

Demand feeding and welfare in farmed fish

Joël Attia · Sandie Millot · Carole Di-Poï · Marie-Laure Bégout · Chris Noble · F. Javier Sanchez-Vazquez · Genciana Terova · Marco Saroglia · Børge Damsgård

Received: 6 April 2010 / Accepted: 20 June 2011 / Published online: 5 July 2011

Abstract Following the development of demand-feeding systems, many experiments have been conducted to explore feeding motivation and feed intake in farmed fish. This work aims to review a selection of studies in the field, focusing on three key factors, related to demand feeding and fish welfare. Firstly, we outline how demand feeders should be considered when developing feed management strategies for improving welfare in production conditions. Secondly, via laboratory demand-feeding experiments, we show self-feeding activities depend not only on feeding motivation and social organisation, but also on individual learning capacity and risk-taking behaviour. Thirdly, we report encouraging results demonstrating that when presented with two or more self-feeders containing complementary foods, fish select a diet according to their specific nutritional requirements, suggesting that demand feeders could be used to improve welfare by allowing fish to meet their nutritional needs.

Keywords Demand-feeding systems · Food selection · Self-feeder · Domestication · Selection · Stress · Welfare

J. Attia (✉)
ENES, CNPS, UMR CNRS 8195, University of Saint-Etienne, 42000 Saint-Etienne, France
e-mail: attia@univ-st-etienne.fr

S. Millot
Centro de Ciências do Mar do Algarve (CCMAR), University of Algarve, Campus de Gambelas, 8005-139 Faro, Portugal

C. Di-Poï
Groupe Mémoire et Plasticité comportementale (GMPc) - EA 4259, University of Caen, Esplanade de la paix, BP 5186, 14032 Caen, France

M.-L. Bégout
Ifremer, Place Gaby Coll, BP 5, 17137 L'Houmeau, France

C. Noble · B. Damsgård
Nofima Marin, PO Box 6122, 9291 Tromsø, Norway

F. J. Sanchez-Vazquez
Department of Physiology and Pharmacology, Faculty of Biology, University of Murcia, 30100 Espinardo, Murcia, Spain

G. Terova · M. Saroglia
Department of Biotechnology and Molecular Sciences (DBSM), University of Insubria, Via J.H. Dunant, 3-21100 Varese, Italy

Introduction

Feeding has been and still is the subject of numerous aquacultural studies. In particular, in the last three decades there has been an increase in the number of experiments that use demand-feeding systems. For example, from an experimental perspective demand-feeding systems allow researchers to explore the feeding motivation of fish held in captivity. From an applied point of view, demand-feeding systems allow fish to dictate the timing of feed delivery and the quantity of feed delivered, which in turn can minimise feed wastage. Demand-feeding systems can be separated into two basic types: self-feeders and interactive feedback systems. In the first type, fish obtain food by actuating a trigger, which in turn initiates the release of feed as a reward (Jobling et al. 2001). The feed delivery can be linked mechanically or electrically to the trigger actuation. In general, fish have to push or pull a rod to trigger the feed delivery (Covès et al. 2006). In interactive feedback systems, the amount of feed delivered is regulated by the amount of feed waste and the number of uneaten pellets. Feed waste can be estimated by using an infra-red pellet sensor (Blyth et al. 1993) or a camera-based pellet detection (Ang and Petrell 1997). In this review, we present the results of representative studies that use demand feeders in both laboratory and production conditions, outlining the aspects that could be interesting for those who want to improve welfare in farmed fish. This paper draws extensively on the work of others but does include data from our own studies, including previously unpublished recent works.

In "Feed delivery strategies and fish behaviour", we highlight a number of problems that need to be addressed when developing operational feed management strategies for production conditions, and we give the arguments in favour of demand-feeding systems with regard to balancing production performance and fish welfare. In "Behavioural motivation and state dependency of self-feeding activities" we note that, besides hunger, other factors influence the triggering activity in small groups of fish using demand-feeding systems. In "Feed selection and fish welfare", we explore the feed selection abilities of fish placed in front of two or more self-feeders containing different feed stuffs.

Feed delivery strategies and fish behaviour

In production conditions, feed management strategies can be used to regulate and control growth performance in farmed fish. Feed management strategies govern how much feed is fed to the fish, the frequency of the feed delivery (the number of meals per day), the timing of the meals, and the spatial distribution of the feed as well as feed type (Talbot et al. 1999). It is well established that these variables can affect production performance, e.g. fish growth and feeding efficiency (Noble et al. 2008), but they can also have a marked effect upon fish behaviour by stimulating competition between conspecifics. This competition can take two forms, scramble competition or contest competition (see Ward et al. 2006 for an in-depth review). Scramble competition occurs when a forager is non-aggressively excluded from feeding by a faster reacting competitor during a feeding bout (Gill and Hart 1996), whereas contest competition occurs when an individual is aggressively thwarted by a conspecific in its attempts to feed (Ryer and Olla 1998).

From a behavioural viewpoint, the amount of feed delivered to fish can affect competition in different ways. For example, juvenile Atlantic salmon (*Salmo salar*) can utilise contest competition and become more aggressive when underfed (Symons 1968; Cañon Jones et al. 2010). In contrast, when gilthead sea bream (*Sparus aurata*) are underfed, they increase both their swimming speed and the complexity of their swimming manoeuvres during feeding, but do not become more aggressive (Andrew et al. 2004). Hence, they may primarily utilise scramble, rather than contest competition to gain access to the feed resource.

The manner with which feed is dispersed also has a major influence upon the equality of access to the resource amongst group-held fish. When feed is delivered in a spatially localised fashion, highly competitive individuals can exclude less-competitive conspecifics, which may lead to resource monopolisation, disproportionate feed acquisition (Ryer and Olla 1996) and increased size heterogeneity (Johansen and Jobling 1998). This has been demonstrated in a commercial environment, where feed monopolisation amongst cage-held adult Atlantic salmon was reduced when feed was spatially dispersed (Thorpe et al. 1990). Numerous other studies have shown that spatially increasing feed dispersal can reduce aggression (e.g. Grant and Guha 1993; Ryer and Olla 1995) and competition for a feed resource.

When feed is delivered in a temporally localised fashion and fish are given their feed over a relatively short period of time, fish become less aggressive compared to when feed is delivered in a temporally uniform manner and feed is available all day (Grant and Kramer 1992). This suggests that a feed resource becomes indefensible when a high number of pellets are delivered at the same time, reducing the benefits of being aggressive (Ryer and Olla 1996). However, two recent studies on salmonids using on-demand feeders have reported that temporally clumping feed to one period each day increases aggression (Brännäs et al. 2005; Noble et al. 2007a). In the study of Brännäs et al. (2005), aggression increased 1 h prior to, but not during, the temporal clumping of feeding demand compared to when the fish had unrestricted access. In the study of Noble et al. (2007a), aggression increased during both feeding and non-feeding periods when a satiation ration was temporally clumped to a single demand-feeding period each day. This suggests fish can also become more competitive when given a relatively short feeding window, even if fish are allowed to feed to satiation.

Delivery rate is another aspect of feed management that dictates feed availability, which in turn can influence competition for a feed resource. Delivery rate is defined as the number of pellets delivered in a given unit of time. A low feed delivery rate allows feed items to become defensible and can increase aggression in a number of salmonids (Jobling 1983; Hakoyama and Iguchi 1997), whereas a high feed delivery rate can lead to a feed resource becoming indefensible, minimising competition and reducing aggression (Hakoyama and Iguchi 1997). However, there is also evidence that feed delivery rate has no influence upon aggression or swimming speeds, in gilthead sea bream for example (Andrew et al. 2004).

Both aggression and scramble competition have implications for the health and welfare of farmed fish. Scramble competition can lead to a portion of the group being underfed or starving, if individual fish are repeatedly excluded from feeding by faster competitors (Gill and Hart 1996). Contest competition may also lead to underfeeding, as fish may refrain from feeding in the presence of a competitor, due to intimidation (Huntingford et al. 1993). Aggression can lead to increased mortality (Sakakura et al. 1998) and it can be detrimental to health by increasing the prevalence of injuries, primarily fin damage in Atlantic salmon (Turnbull et al. 1996; Cañon Jones et al. 2010) and rainbow trout (Abbott and Dill 1985). This direct damage may also encourage pathogenic infection by *Aeromonas salmonicida*, which causes furunculosis (Turnbull et al. 1996). The recipients of aggression may become stressed and exhibit elevated cortisol levels (Øverli et al. 1999; Dunlap et al. 2002), which may have additional detrimental effects upon immune–competence (Barton et al. 1987; Peters et al. 1988), tissue repair (Gregory and Wood 1999) and survival (Pickering and Pottinger 1989).

From a production perspective, feed management strategies can affect numerous productivity indicators and can have a marked effect upon the profitability of a farm. Feed costs are a major constituent of a farms' total running costs, and can account for up to 60% of a farms' expenditure (Sveier and Lied 1998). Therefore, selecting a correct ration is particularly important as a farmer can easily over- or under-feed its fish. Underfeeding can be particularly detrimental to a number of important production parameters. For example, underfeeding can limit the growth potential of fish, decrease feed conversion ratio (Cho 1992; Bureau et al. 2006), and increase inter-individual growth variation and size heterogeneity (Johansen and Jobling 1998). All of these factors can reduce the value of a farmer's livestock and are detrimental to productivity. Further, underfeeding is detrimental to welfare as fish may starve and may also become injured as competition for feed increases. For example, fish fed with reduced rations can exhibit greater fin damage than those fed to satiation (Damsgård et al. 1997; Moutou et al. 1998; Ytrestøyl et al. 2005; Cañon Jones et al. 2010). Overfeeding, where fish receive more feed than they can consume, is also detrimental to production and environmental sustainability of aquaculture, as feed wastage increases feed conversion ratio (Talbot et al. 1999), potentially increases environmental degradation (Cho and Bureau 1998), and can be a burden on filtration systems in re-circulating aquaculture systems.

A farmer must be careful about addressing each of these factors when designing welfare friendly and economically sustainable feed management strategies. A way to circumvent many of these problems would be to feed fish with a spatially dispersed ration to satiation, and to make this daily ration available at times that the fish desire it. However, this cannot always be achieved in an operational context, as the appetite levels of fish are known to vary within and between days (Blyth et al. 1993), and also between groups of the same species (Noble et al. 2005). In addition, it can be difficult to judge satiation levels when feeding a population using current techniques, such as a cessation in surface activity (Paspatis et al. 1999).

To limit the problems of poor feed management e.g. aggression or fin damage, farmers could consider the use of demand-feeding technology as feed management tools. Demand feeders have been demonstrated to help a farmer improve growth and feed conversion ratio (Alanärä 1992; Azzaydi et al. 1998; Noble et al. 2007b, 2008) by synchronising feed delivery with appetite. But they also have been shown to reduce aggression (Almazán–Rueda et al. 2004), scramble competition (Andrew et al. 2002; Noble et al. 2007b, c), stress (Endo et al. 2002) and fin damage (Noble et al. 2007b; Suzuki et al. 2008). Thus, demand feeders are a potentially effective technology for optimising both growth performance and welfare in farmed fish.

To apply demand-feeding technology in commercial aquaculture, farmers must first consider the behaviour of their specific species, utilising knowledge from their previous husbandry experience and also from scientific to industrial literature sources. For example, with regarding to self-feeding systems, the fish must rapidly learn and utilise the feeding systems for effective on-farm deployment. Another key point is to understand how each feeding system affects, and is affected by, the behaviour of each species e.g. will the efficacy and effectiveness of the self-feeding system be affected by potential dominance hierarchies? If so, does its efficacy rely on the establishment of spatial territories? There are also some other useful traits to consider when using self-feeding systems, e.g. whether certain fish will aggressively defend the self-feeding trigger from usage by conspecifics, or whether self-feeding utilisation is affected by the level of boldness or curiosity of the species, or by the quality of the rearing environment.

Some of these characteristics have already been identified for a number of commercial fish species, in both the scientific literature and also directly by the farmers in their day to day husbandry practises. In addition, several behavioural characteristics can change throughout the rearing cycle and these characteristics can be affected by a number of production parameters, e.g. levels of aggression can evolve and adapt in response to changes in husbandry practises, such as size grading, changes in numerical density or feed availability. Therefore, a combination of theoretical knowledge and also rapid observation of changes in fish behaviour are required to constantly adapt and optimise feed delivery. Fish behaviour, especially aggression and swimming activity can be monitored and quantified by commercial devices including video camera and image analysis software. Where feed systems allow fish to adjust feed delivery in response to changes in appetite, the data from these systems can also be a useful monitoring tool.

In circumstances where fish are known to be aggressive, or to become aggressive in relation to feed delivery parameters, e.g. feed distribution or ration size, feed should be delivered in a spatially dispersed manner and in amounts that satisfy the appetite of the fish. This can be achieved by providing production facilities with feed distribution technologies that deliver a widely dispersed ration over the majority of the surface of the rearing unit. If these technologies are expensive or unavailable, a farmer could increase the number of feed dispensers per rearing unit to increase spatial distribution of feed. In addition, the use of multiple low cost self-feeders can provide fish with a choice of different diets, allowing fish to select a diet according to their specific nutritional requirements (see Sect., "Feed selection and fish welfare"). Future research should focus on increasing co-operation between farmers and scientists, to develop feeding systems and feed management strategies for commercial scale aquaculture, thus providing customised solutions for the cost-effective optimisation of growth and welfare for each specific farmed species.

Behavioural motivation and state dependency of self-feeding activities

The different phases of feed intake (from feeding motivation to effective feed intake) of farmed fish are generally dependent upon the capacity to find, handle and ingest feed pellets, and the physiological capacity to digest them. These phases are dependent upon a range of environmental factors, which might control or limit any stage of the process. In addition, they vary according to time or from one individual to another and these variations depend upon complex relationships between e.g. the individual state of the fish and the social relationships between the individuals.

The development of demand-feeding systems has represented an interesting way to explore the impact of different factors on the feeding motivation and feed intake in farmed fish. Obviously, studies using self-feeding systems, where individual fish actuate a trigger may have limited values as absolute appetite measurements in groups of fish, as such measurements do not necessarily reflect the actual appetite of the group, e.g. competition, intimidation, dominance hierarchies may prevent some individuals to actuate the trigger. As an equivalent to a Skinner box for terrestrial animals, self-feeding experiments in fish may however provide important insights into inter-individual differences, the flexibility of the feeding motivation and the state dependency of feed intake.

Before going further, an important parameter to keep in mind is that self- feeding experiments in laboratory conditions often include a relatively low number of fish and a low stocking density. It is well known that many farmed fish species compete for feed, and aggression is well documented in domesticated fish (e.g. Huntingford et al. 1990). The level and intensity of the aggression tend to balance between the advantages and disadvantages for the individuals concerned (Grant 1997), which take into account the energetic investment to defend the resource, and the relative importance of feed intake and growth. Such a model has implications for the numbers of competitors, and it generally tends to be more aggression in systems with low density or a low number of fish. For example, African catfish, *Clarias gariepinus*, fight less beyond a threshold stocking density (Kaiser et al. 1995). In addition, this implies that the mechanisms causing feeding hierarchies enable some individuals to consistently feed and grow well at the expense of others, described in many fish species such as Atlantic salmon (e.g. Adams et al. 1998), rainbow trout (e.g. McCarthy et al. 1992) and Arctic charr (e.g. Winberg et al. 1992; Jobling and Baardvik 1994). The aggression of individual fish in such hierarchies was state-dependent and changed on a day-to-day basis, dependent on e.g. water current and feed availability (Damsgård et al. 1997).

In small groups of fish fed using self-feeders, only one or a few fish trigger the system whilst other fish have access to the feed according to the feeding demand of these triggering fish. The feed intake is as in the producer–scrounger model in which the scroungers (low- or null-triggering fish) benefit from the resources found by the producers (high-triggering fish). In rainbow trout (*Oncorhynchus mykiss*) and Arctic charr (*Salvelinus alpinus*) (5–15 young individuals; 2–6 kg m^{-3}), a dominance hierarchy develops and a good correlation is observed between the social rank and the number of self-feeder actuations: the one or two dominant individuals displayed the highest number of actuations, whilst sub-dominants and subordinates showed a medium and low number of actuations, respectively (Alanärä and Brännäs 1993, 1996; Brännäs and Alanärä 1993, 1994). Moreover, Alanärä et al. (1998) found a strong negative correlation between feeding demand (or growth rate) and brain serotonergic activity in groups of Arctic charr: high-triggering fish showed a higher specific growth rate and a lower brain serotonergic activity than low- or null-triggering fish. Serotonergic activity is commonly appraised by the 5-HIAA/5-HT ratio, 5-HIAA being the major metabolite of serotonin (5-HT). This ratio is a reliable neurochemical marker of stress in fish (Winberg and Nilsson 1993; Winberg et al. 1997; Winberg and Lepage 1998). In young European sea bass (*Dicentrarchus labrax*), no agonistic behaviours or signs of them (biting, scale loss or fin damage) were observed in groups of 50 individuals (20 kg m^{-3}) using self-feeder (Covès et al. 2006; Di–Poï et al. 2007, 2008; Millot et al. 2008). The absence of agonistic behaviours or dominance hierarchy in European sea bass, when compared to Arctic charr or rainbow trout, could be due to the difference of densities: 6 kg m^{-3} in Arctic charr and rainbow trout versus 20 kg m^{-3} in European sea bass. Nevertheless, as in Arctic charr and rainbow trout, only a few fish strongly actuated the trigger: 2–4 individuals were responsible for more than 80% of the triggering activity (Di–Poï et al. 2007; Millot et al. 2008; Millot and Bégout 2009). Sometimes two high-triggering fish actuated the trigger during the same period, but in general the high-triggering fish followed each other (Fig. 1).

They displayed a higher specific growth rate than the other fish but only during the time they occupied the position of high-triggering fish (Millot et al. 2008). As in Arctic charr and rainbow trout, high-triggering fish exhibit the lowest brain serotonergic activity, suggesting a lower stress level than low- and null-triggering fish (Di–Poï et al. 2007).

To conclude, in general, the high-triggering fish are regarded as the most dominant or high-ranked fish. There are however, several alternative explanations as to why a single fish triggers a feeder, and these behavioural mechanisms are not mutually exclusive.

First of all a triggering behaviour must be based upon the learning capacity of individual fish, and

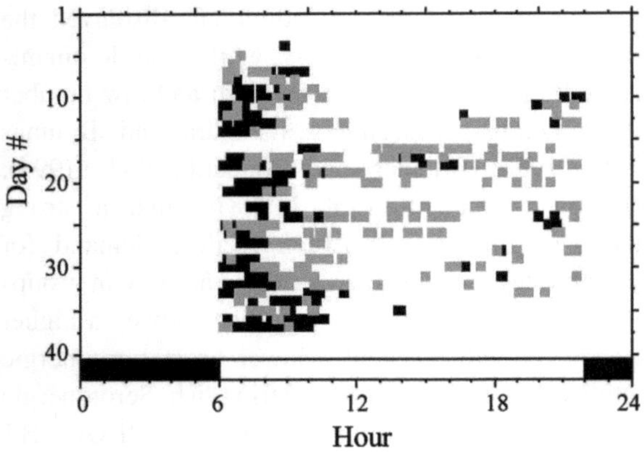

Fig. 1 Feeding demand of high-triggering fish. In a group of 50 juvenile European sea bass, the two highest-triggering fish have different temporal patterns: the first one mainly actuates the trigger during the first 6 h of the light phase whilst the activity of the second ranges over the whole light phase. *Black bar* first high-triggering fish. *Grey bars* second high-triggering fish. Each *black* or *grey bar* corresponds to one actuation (Attia and Beauchaud, unpublished)

most experiments using self-feeders includes a preconditioning period. This learning period includes the time lag before the first fish learn to trigger the system, and typically the period before reaching a stable level is between 1 and 2 weeks (e.g. Landless 1976; Boujard and Leatherland 1992). After learning to use the system, several experiments shows that fish might even learn to discriminate between several triggers (e.g. Adron et al. 1973). Several sequential experiments also show that fish might become more adept with the feeding system, as for example the overall triggering activity increases over time (Alanärä and Brännäs 1996). Landless (1976) suggested that the reward level could be fine-tuned to optimise growth and feed efficiency. There are examples demonstrating relationships between hunger levels and triggering activity, but there are no reasons to believe that an individual fish would trigger more based on the feed intake of the whole group, but purely its own hunger level. However, in experiments using self-feeders, competition, stress level and feed waste decrease, when compared to automatic feeders, encouraging their use in production conditions. In general, the speed of the learning process in the triggering system demonstrates the capacity for naive fish to adapt in culture.

Secondly, a fish may continue to trigger the system if it has an individual gain, e.g. first access to feed, and the fish may control the area close to the trigger or to the feeder. Territorial fish may even control the access to feed for the other fish, even after it has fulfilled its own need for feed. For example, subordinate Arctic charr had a lower swimming activity more than a week after the dominant fish were relocated (Øverli et al. 1998), indicating the prolonged effects of dominance. A comparative study of feeding demand in rainbow trout and Arctic charr on different stocking densities demonstrated that the number of triggering fish increased at higher densities, probably because fish had decreasing possibilities to monopolise the triggering area (Alanärä and Brännäs 1996). A long-term study of individual feeding behaviour in European sea bass revealed that it is not necessarily the same individual fish that trigger the system, and the shift in triggering fish seems abrupt and not related to other described factors (Millot and Bégout 2009). European sea bass that triggered the system often had a higher specific growth rate, so it was clearly an individual gain of triggering, but there were few other physiological differences between triggering and non-triggering fish (Millot et al. 2008). On the contrary, in other experiments triggering European sea bass had a lowered specific growth rate compared to non-triggering fish (Covès et al. 2006), indicating that the triggering behaviour also comes with a cost. A study of the numbers of meal in rainbow trout showed that size heterogeneity increases over time in all self-feeding regimes, and the level of aggression was highest in the groups receiving one meal per day (Noble et al. 2007a). In addition, several self-feeding experiments report considerable growth variation between replicate groups (e.g. Boujard and Leatherland 1992; Brännäs and Alanärä 1993; Alanärä and Brännäs 1996), also indicating the existence of dominance hierarchies. A detailed study of individual triggering behaviour using PIT-tags (Brännäs and Alanärä 1993) did show a large variation between groups, but also that several fish trigger the system during the first days, whilst increasingly more of the triggering was conducted by fewer and fewer fish. The triggering fish is however not necessarily the largest one in the group (e.g. Brännäs and Alanärä 1994), and dominance may not be directly linked to fish size (Huntingford et al. 1990). A positive relationship between numbers of bites and specific growth rates was demonstrated in Arctic charr, but not in rainbow trout (Brännäs and Alanärä 1994), indicating possible differences between species.

Thirdly, the triggering behaviour may be a part of risk-taking or boldness behaviour. Bold individuals, often termed pro-active individuals, tend to be more willing to expose themselves to new environments and are more eager or faster to learn new ways of feeding (see e.g. review by Øverli et al. 2007). Learning capacity, aggression and boldness are however not independent of each other, but are state-dependent and affected by various environmental and biological factors. Learning ability is probably most important during the first days. Thereafter, dominance might play an increasing role, and after some time, risk-taking behaviour may become more important due to low feed intake in non-triggering fish. For example, after a period of low feed availability in a group of coho salmon (*Oncorhynchus kisutch*), the fish both resume feeding faster and are more aggressive towards each other in order to catch up with the reduced feed intake (Damsgård and Dill 1998). Such compensatory behaviour is an important part of the so-called recovery growth or catch-up growth, and fish that undergoes periods of feed restrictions have an ability to catch up for the lost growth by initiating a temporary growth spurt or compensatory response, described in many fish species, such as rainbow trout (e.g. Weatherley and Gill 1981) and Arctic charr (e.g. Miglavs and Jobling 1989). In the wild, periods of feed restriction are not surprising, as carnivorous fish generally inhabit environments in which feed availability fluctuates or where encounters with appropriate feed items are rare and unpredictable. Thus, many fish species are accustomed to differences measured in days or weeks between meals. The eel, as an extreme instance, may migrate several thousand miles for several months without feeding, and even, at the stage of *leptocephalus*, may survive years without feeding. At the physiological level, feed deprivation does not lead very quickly to lipolysis or protein turnover in fish, as it is the case in mammalians. For instance, Wang et al. (2006) reported that genes involved in lipolysis do not show significant expression in carp (*Cyprinus carpio*) until at least 16 days of feed deprivation, whilst being activated after only 24–48 h in mammalians.

The mechanisms of compensatory growth are as yet not fully understood, but it is suggested that animals follow a predetermined 'ideal' growth trajectory under conditions of adequate feed supply. In a self-feeding system, where the total amount of feed is sufficient to feed all fish, such compensatory behaviour is expected to regulate the feeding motivation, and such behaviour may include a combination of becoming more aggressive, or taking larger risks, e.g. spending more time out in open areas in the tank. The relationships between dominance, boldness and learning are however complex, and a lot of the observed variation between species, experiments, groups and individual fish may probably be explained in terms of behavioural mechanisms.

Fourthly, triggering activity could depend on the genetic origin of the fish. Millot et al. (2010) showed that different strains of European sea bass behave differently: *Domesticated*, *Selected for growth Prosper* and *Selected for growth* showed a significantly higher feeding demand at dawn whilst the *Wild* strain did not exhibit this peak. In response to a first period of repeated acute stress, all groups modified their feeding rhythm and highly increased their feed intake whilst their growth rate decreased (Domesticated and Selected fish groups) or remained stable (Wild). In response to a second period of acute stress, fish continued to modify their feeding rhythm (being more and more diurnal) and increased again their feed intake associated this time with an improvement of the growth rate of all populations. During the whole experiment, Selected and Domesticated fish were always characterised by a higher body mass, specific growth rate and body condition factor than Wild fish. Besides these differences, this study demonstrated that self-feeding systems could be used to measure stress responsiveness, as triggering activity increased during repeated acute stress, suggesting that feeding demand can constitute a relevant operational welfare indicator (OWI). Millot et al. (2010) conclude in this study that a first generation of domestication or selection improves fish growth performance but at this early stage, does not modify behavioural responses to repeated acute stress exposure. Furthermore, selection did not place its focus on stress tolerance traits in the tested strains.

To conclude, the intermixed influence of dominance, genetics, boldness and learning have to be kept in mind when it comes to estimate the advantages of the demand-feeding systems to improve welfare.

Feed selection and fish welfare

When fish have the possibility to freely access two or more self-feeders containing complementary feed stuffs, they show their capacity to select a diet according to their specific nutritional requirements. Indeed, dietary selection can be seen as a step forward in studying feeding behaviour taking advantage of the 'nutritional wisdom' of fish (Simpson and Raubenheimer 2001). Macronutrient selection experiments revealed the ability of fish, given the choice to feed separately on protein, fat and carbohydrate, to select amongst them and compose their own balanced diet (Aranda et al. 2000). Nutritional requirements, however, may change during growing, spawning or under stress in captivity. Another nutritional challenge in intensive aquaculture could be starvation, as fish may experience periods of feed deprivation (e.g. before grading or transportation). After a fasting period, fish usually exhibit compensatory hyperphagia. During re-feeding, fish select a balanced combination of macronutrients rather than a high-energy (fat) selection (Aranda et al. 2001). When fasting is not complete (i.e. protein or fat fasting), the response is complex and species-specific. For instance, European sea bass deprived of protein, but with free access to lipids and carbohydrates, refused to feed and actually starved (Vivas et al. 2003). However, sharpsnout sea bream challenged with protein dilutions, compensated efficiently for a protein deficiency (Vivas et al. 2006).

In intensive aquaculture, fish diets are designed to be cost-effective and include low cost, sustainable feed sources. However, low protein quality may lead to inappropriate aminoacid profiles, which may compromise fish welfare. Fish can detect the lack of a single aminoacid in the diet, such that a methionine deficient diet was avoided by European sea bass using self-feeders (Hidalgo et al. 1988). Furthermore, in laboratory conditions, an unbalanced combination of gelatin/casein (5:1) was detected by trout, which quickly learned to self-feed on the right combination (1:5) (Yamamoto et al. 2000).

Fish, like all aerobic organisms, are susceptible to oxidative stress, though antioxidants compounds (e.g. vitamins C, E) and internal systems (e.g. CAT, SOD, GpX) prevent the damage caused by reactive oxygen radicals. Diet oxidation may occur under intensive farming conditions, so that when fish are fed with oxidised oils, their antioxidants enzymes are boosted to minimise damage (Mourente et al. 2002). Such antioxidant defences in fish are dependent on nutritional factors and most interestingly on feeding behaviour (see for review Martínez–Álvarez et al. 2005). As a matter of fact, ongoing research revealed the ability of gilthead sea bream to detect and avoid a diet containing oxidised lipids. When sea bream were given the choice between two self-feeders containing either a control diet or an oxidised diet, fish quickly learned within a few days to select the control diet. Furthermore, when the diets were shifted over, after a few days sea bream resume their previous preference for the control non-oxidised diet (Fig. 2).

To lower costs and encourage sustainable aquaculture, plant protein sources such as soybean are often used to replace fish meals. However, plant protein may contain anti-nutritional factors such as phytate, which interferes with protein digestibility and phosphorous absorption, seriously compromising fish performance and health. To avoid these undesirable effects, phytase (an enzyme that hydrolyses non-digestible phytate), can be added to plant-based diets to improve digestibility. Recent dietary selection trials revealed that fish could detect the supplementation of phytase to the diet. Nile tilapia self-selected a diet containing 1,500 U kg^{-1} of phytase rather than the control diet without added phytase. Moreover, adding phytate to the diets resulted in an increased selection for the phytase diet (Fortes–Silva and Sánchez–Vázquez, unpublished).

Dietary selection is not reliant on taste or flavour. Fish fed with different gelatine capsules containing separate macronutrients are still able to select amongst them, provided they have a clue (different colours) to help differentiate the diets, and self-compose a nutritionally balanced diet (Rubio et al. 2003). Thus, the regulation of feed selection in fish is not driven by oral sensory inputs, but by post-absorptive signal that produce a nutritional reward. Ultimately, the use of multiple self-feeders has demonstrated the abilities of fish to select a diet according to their specific nutritional requirements, suggesting the possibility to improve welfare by allowing fish to meet their nutritional needs.

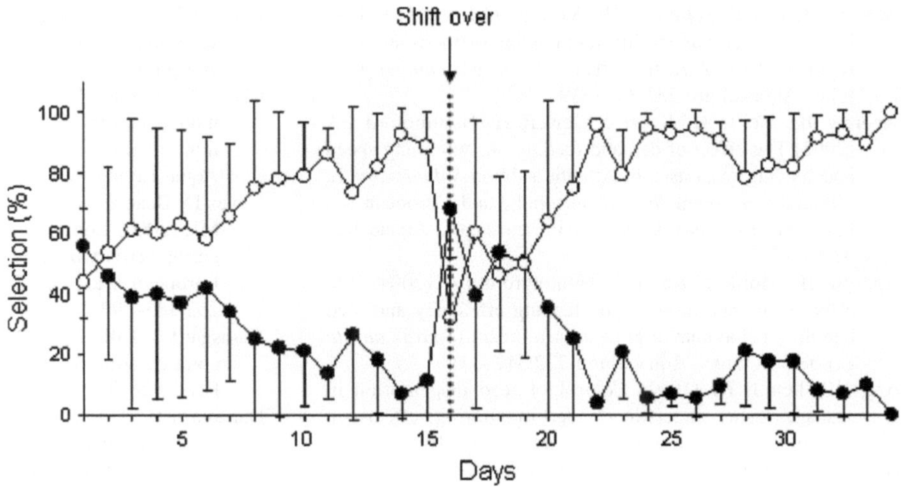

Fig. 2 Food selection in self-fed gilthead sea bream using two diets (46% protein, 34% carbohydrate and 20% fat) with two different levels of lipid oxidation (*open circle* control, POV = 6 meq/kg; *filled circle* oxidised, POV = 100 meq/kg). Results are the mean of six groups of four sea bream of 254 g maintained under LD 12:12 at 17°C. On day 16, the diets were shifted over between the self-feeders (modified from Montoya et al. 2010)

Concluding remarks

The different works presented here show different applications of demand-feeding systems. With regard to feed management, demand-feeding systems can optimise production performance without compromising welfare, and are a suitable tool for commercial farms. However, feed must be provided in sufficient quantity and quality to allow fish expressing their normal feeding behaviour. Feeding demand may be very different from one individual to another within the same group subjected to the same conditions. It depends on multiple parameters including density, social organisation, genetics, individual learning ability and boldness. A benefit of self-feeding technology is its ability to test the individual's capacity for differentiating between different types of feed. When fish can freely access two or more self-feeders containing complementary feeds, they exhibit an ability to select nutritionally balanced diets from sub-optimal feeds. This is of particular interest when it comes developing alternative plant-based feed sources, using fish memory and knowledge of feed to optimise feeding and thus welfare. Ultimately, all these studies show the strong potential of the demand-feeding systems to investigate fish behaviour in captivity, and that feeding demand can be used as an operational welfare indicator (OWI) in farmed fish.

Acknowledgments Most of the direct exchanges between the contributors to this review were supported by the COST action 867: Welfare of fish in European aquaculture (Chair: Anders Kiessling). We would like to thank Nadia Aubin-Horth (Université Laval, Québec) for her attentive re-reading of the manuscript.

References

Abbott JC, Dill LM (1985) Patterns of aggressive attack in juvenile steelhead trout (*Salmo gairdneri*). Can J Fish Aquat Sci 42:1702–1706

Adams CE, Huntingford FA, Turnbull JF, Beattie C (1998) Alternative competitive strategies and the cost of food acquisition in juvenile Atlantic salmon (*Salmo salar*). Aquaculture 167:17–26

Adron JW, Grant PT, Cowey CB (1973) A system for the quantitative study of the learning capacity of rainbow trout and its application to the study of food preference and behaviour. J Fish Biol 5:625–636

Alanärä A (1992) Demand feeding as a self-regulating system for rainbow trout (*Oncorhynchus mykiss*) in net-pens. Aquaculture 108:347–356

Alanärä A, Brännäs E (1993) A test of individual feeding activity and food size preference in rainbow trout using demand feeders. Aquac Int 1:47–54

Alanärä A, Brännäs E (1996) Dominance in demand-feeding behaviour in Arctic charr and rainbow trout: the effect of stocking density. J Fish Biol 48:242–254

Alanärä A, Winberg S, Brännäs E, Kiessling A, Höglund E, Elofsson U (1998) Feeding behaviour, brain serotonergic activity and energy reserves of Arctic char (*Salvelinus alpinus*) within a dominance hierarchy. Can J Zool 76:212–220

Almazán-Rueda P, Schrama JW, Verreth JAJ (2004) Behavioural responses under different feeding methods and light regimes of the African catfish (*Clarias gariepinus*) juveniles. Aquaculture 231:347–359

Andrew JE, Noble C, Kadri S, Jewell H, Huntingford FA (2002) The effect of demand feeding on swimming speed and feeding responses in Atlantic salmon *Salmo salar* L. gilthead sea bream *Sparus aurata* L. and European sea bass *Dicentrarchus labrax* L. in sea cages. Aquac Res 33:501–507

Andrew JE, Holm J, Kadri S, Huntingford FA (2004) The effect of competition on the feeding efficiency and feed handling behaviour in gilthead sea bream (*Sparus aurata* L.) held in tanks. Aquaculture 232:317–331

Ang KP, Petrell RJ (1997) Control of feed dispensation in seacages using underwater video monitoring: effects on growth and food conversion. Aquacult Eng 16:45–62

Aranda A, Sanchez-Vazquez FJ, Zamora S, Madrid JA (2000) Self-design of fish diets by means of self-feeders: validation of procedures. J Physiol Biochem 56:155–166

Aranda A, Sánchez-Vázquez FJ, Madrid JA (2001) Effect of short-term fasting on macronutrient self-selection in sea bass. Physiol Behav 73:1–5

Azzaydi M, Madrid JA, Zamora S, Sánchez-Vázquez FJ, Martínez FJ (1998) Effect of three feeding strategies (automatic, ad libitum demand feeding, and time restricted demand feeding) on feeding rhythms and growth in European sea bass (*Dicentrarchus labrax* L.). Aquaculture 163:285–296

Barton BA, Schreck CB, Barton LD (1987) Effects of chronic cortisol administration and daily acute stress on growth, physiological conditions, and stress responses in juvenile rainbow trout. Dis Aquat Organ 2:173–185

Blyth PJ, Purser GJ, Russell JF (1993) Detection of feeding rhythms in sea caged Atlantic salmon using new feeder technology. In: Reinertsen H, Dahle LA, Jørgensen L, Tvinnereim K (eds) Fish farming technology. Balkema, Rotterdam, pp 209–216

Boujard T, Leatherland JF (1992) Demand-feeding behaviour and diel pattern of feeding activity in Oncorhynchus mykiss held under different photoperiod regimes. J Fish Biol 40:535–544

Brännäs E, Alanärä A (1993) Monitoring the feeding activity of individual fish with a demand-feeding system. J Fish Biol 42:209–215

Brännäs E, Alanärä A (1994) Effect of reward level on individual variability in demand-feeding activity and growth rate in Arctic charr and rainbow trout. J Fish Biol 45: 423–434

Brännäs E, Berglund U, Eriksson LO (2005) Time learning and anticipatory activity in groups of Arctic charr. Ethology 111:681–692

Bureau DP, Hua K, Cho CY (2006) Effect of feeding level on growth and nutrient deposition in rainbow trout (*Oncorhynchus mykiss* Walbaum) growing from 150 to 600 g. Aquac Res 37:1090–1098

Cañon Jones HA, Hansen L, Noble C, Damsgård B, Broom DM, Pearce GP (2010) Social network analysis of behavioural interactions influencing fin damage development in Atlantic salmon (*Salmo salar*) during feed-restriction. Appl Anim Behav Sci 127:139–151

Cho CY (1992) Feeding systems for rainbow trout and other salmonids with reference to current estimates of energy and protein requirements. Aquaculture 100:107–123

Cho CY, Bureau DP (1998) Development of bioenergetic models and the Fish-PrFEQ software to estimate production, feeding ration and waste output in aquaculture. Aquat Living Resour 11:199–210

Covès D, Beauchaud M, Attia J, Dutto G, Bouchut C, Bégout ML (2006) Long-term monitoring of individual fish triggering activity on a self-feeding system: an example using European sea bass (*Dicentrarchus labrax*). Aquaculture 253:385–392

Damsgård B, Dill LM (1998) Risk-taking behaviour in weight-compensating coho salmon, *Oncorhynchus kisutch*. Behav Ecol 9:26–32

Damsgård B, Arnesen AM, Baardvik BM, Jobling M (1997) State-dependent feed acquisition among two strains of hatchery-reared Arctic charr. J Fish Biol 50:859–869

Di-Poï C, Attia J, Bouchut C, Dutto G, Covès D, Beauchaud M (2007) Behavioral and neurophysiological responses of European sea bass groups reared under food constraint. Physiol Behav 90:559–566

Di-Poï C, Beauchaud M, Bouchut C, Dutto G, Covès D, Attia J (2008) Effects of high food-demand fish removal in groups of juvenile sea bass (*Dicentrarchus labrax*). Can J Zool 86:1015–1023

Dunlap KD, Pelczar PL, Knapp R (2002) Social interactions and cortisol treatment increase the production of aggressive electrocommunication signals in male electric fish, *Apteronotus leptorhynchus*. Horm Behav 42:97–108

Endo M, Kumahara C, Yoshida T, Tabata M (2002) Reduced stress and increased immune responses in Nile Tilapia kept under self-feeding conditions. Fish Sci 68:253–257

Gill AB, Hart PJB (1996) Unequal competition between three-spined stickleback, *Gasterosteus aculeatus*, L. encountering sequential prey. Anim Behav 51:689–698

Grant JWA (1997) Territoriality. In: Godin JGJ (ed) Behavioural ecology of teleost fishes. University Press, Oxford, pp 81–103

Grant JWA, Guha RT (1993) Spatial clumping of food increases its monopolisation and defense by convict cichlids, *Cichlasoma nigrofasciatum*. Behav Ecol 4:293–296

Grant JWA, Kramer DL (1992) Temporal clumping of food arrival reduces its monopolisation and defence by zebrafish, *Brachydanio rerio*. Anim Behav 44:101–110

Gregory TR, Wood CM (1999) The effects of chronic plasma cortisol elevation on the feeding behaviour, growth, competitive ability, and swimming performance of juvenile rainbow trout. Physiol Biochem Zool 72:286–295

Hakoyama H, Iguchi K (1997) Why is competition more intense if food is supplied more slowly? Behav Ecol Sociobiol 40:159–168

Hidalgo F, Kentouri M, Divanach P (1988) The utilisation of a self-feeder as a tool for the nutritional study of sea bass, *Dicentrarchus labrax*—preliminary results with methionine. Aquaculture 68:177–190

Huntingford FA, Metcalfe NB, Thorpe JE, Graham WD, Adams CE (1990) Social dominance and body size in Atlantic salmon Parr, *Salmo salar* L. J Fish Biol 36:877–881

Huntingford FA, Metcalfe NB, Thorpe JE (1993) Social status and feeding in Atlantic salmon *Salmo salar* Parr: the effect of visual exposure to a dominant. Ethology 94:201–206

Jobling M (1983) Growth studies with fish—overcoming the problems of size variation. J Fish Biol 22:153–157

Jobling M, Baardvik BM (1994) The influence of environmental manipulation on inter- and intra-individual variation in food acquisition and growth performance of Arctic charr, *Salvelinus alpinus* L. J Fish Biol 44:1069–1087

Jobling M, Covès D, Damsgard B, Kristiansen HR, Koskela J, Petursdottir TE, Kadri S, Gudmundsson O (2001) Techniques for measuring feed intake. In: Houlihan D, Boujard T, Jobling M (eds) Food intake in fish. Blackwell Science, Oxford, pp 131–156

Johansen SJS, Jobling M (1998) The influence of feeding regime on growth and slaughter traits of cage-reared Atlantic salmon. Aquac Int 6:1–17

Kaiser H, Weyl O, Hecht T (1995) The effects of stocking density on growth, survival and agonistic behaviour in African catfish. Aquacult Int 3:217–225

Landless PJ (1976) Demand-feeding behaviour of rainbow trout. Aquaculture 7:11–25

Martínez-Álvarez RM, Morales AE, Sanz A (2005) Antioxidant defenses in fish: biotic and abiotic factors. Rev Fish Biol Fish 15:75–88

McCarthy ID, Carter CG, Houlihan DF (1992) The effect of feeding hierarchy on individual variability in daily feeding of rainbow trout, *Oncorhynchus mykiss* (Walbaum). J Fish Biol 41:257–263

Miglavs I, Jobling M (1989) Effects of feeding regime on food consumption, growth rates and tissue nucleic acids in juvenile Arctic charr, *Salvelinus alpinus*, with particular respect to compensatory growth. J Fish Biol 34: 947–957

Millot S, Bégout ML (2009) Individual fish rhythm directs group feeding: a case study with sea bass juveniles (*Dicentrarchus labrax*) under self-demand feeding conditions. Aquat Living Resour 22:363–370

Millot S, Bégout ML, Person-Le Ruyet J, Breuil G, Di-Poï C, Fievet J, Pineau P, Roue M, Severe A (2008) Feed demand behaviour in sea bass juveniles: effects on individual specific growth rate variation and health (inter-individual and inter-group variation). Aquaculture 274:87–95

Millot S, Péan S, Leguay D, Vergnet A, Chatain B, Bégout ML (2010) Evaluation of behavioral changes induced by a first step of domestication or selection for growth in the European sea bass (*Dicentrarchus labrax*): a self-feeding approach under repeated acute stress. Aquaculture 306:211–217

Montoya A, Alves Martins D, Yúfera M, Sánchez-Vázquez FJ (2010) Self-selection of diets with different oil oxidation levels in gilthead seabream (*Sparus aurata*). Aquaculture, in revision

Mourente G, Díez-Salvago E, Bell JG, Tocher DG (2002) Increased activities of hepatic antioxidant defence enzymes in juvenile githead seabream (*Sparus aurata* L.) fed dietary oxidised oil: attenuation by dietary vitamin E. Aquaculture 214:343–361

Moutou KA, McCarthy ID, Houlihan DF (1998) The effect of ration level and social rank on the development of fin damage in juvenile rainbow trout. J Fish Biol 52:756–770

Noble C, Mizusawa K, Tabata M (2005) Does light intensity affect self-feeding and food wastage in group-held rainbow trout and white-spotted charr? J Fish Biol 66:1387–1399

Noble C, Mizusawa K, Suzuki K, Tabata M (2007a) The effect of differing self-feeding regimes on the growth, behaviour and fin damages of rainbow trout held in groups. Aquaculture 264:214–222

Noble C, Kadri S, Mitchell DF, Huntingford FA (2007b) Influence of feeding regime on intraspecific competition, fin damage and growth in 1+ Atlantic salmon parr (*Salmo salar* L.) held in freshwater production cages. Aquac Res 38:1137–1143

Noble C, Kadri S, Mitchell DF, Huntingford FA (2007c) The effect of feed regime on the growth and behaviour of 1+ Atlantic salmon post-smolts (*Salmo salar* L.) in semi-commercial sea cages. Aquac Res 38:1686–1691

Noble C, Kadri S, Mitchell DF, Huntingford FA (2008) Growth, production and fin damage in cage-held 0+ Atlantic salmon pre-smolts (*Salmo salar* L.) fed either a) on-demand, or b) to a fixed satiation-restriction regime: data from a commercial farm. Aquaculture 275:163–168

Øverli Ø, Winberg S, Damsgård B, Jobling M (1998) Food intake and spontaneous swimming activity in Arctic charr (*Salvelinus alpinus*): role of brain serotonergic activity and social interactions. Can J Zool 76:1366–1370

Øverli Ø, Harris CA, Winberg S (1999) Short-term effects of fights for social dominance and the establishment of dominant-subordinate relationships on brain monoamines and cortisol in rainbow trout. Brain Behav Evol 54:263–275

Øverli Ø, Sørensen C, Pulman KGT, Pottinger TG, Korzan W, Summers CH, Nilsson GE (2007) Evolutionary background for stress-coping styles: relationship between physiological, behavioral, and cognitive traits in non-mammalian vertebrates. Neurosci Biobehav Rev 31:396–412

Paspatis M, Batarias C, Tiangos P, Kentouri M (1999) Feeding and growth response of sea bass (*Dicentrarchus labrax*) reared by four feeding methods. Aquaculture 175:293–305

Peters G, Faisal M, Lang T, Ahmed I (1988) Stress caused by social interaction and its effect on susceptibility to *Aeromonas hydrophila* infection in rainbow trout *Salmo gairdneri*. Dis Aquat Organ 4:83–89

Pickering AD, Pottinger TG (1989) Stress responses and disease resistance in salmonid fish: Effects of chronic elevation of plasma cortisol. Fish Physiol Biochem 7:253–258

Rubio VC, Sánchez-Vázquez FJ, Madrid JA (2003) Macronutrient selection through post-ingestive signals in sea bass fed gelatin capsules. Physiol Behav 78:795–803

Ryer CH, Olla BL (1995) The influence of food distribution upon the development of aggressive and competitive behaviour in juvenile chum salmon, *Oncorhynchus keta*. J Fish Biol 46:264–272

Ryer CH, Olla BL (1996) Growth dispensation and aggression in laboratory reared coho salmon: the effect of food distribution and ration size. J Fish Biol 48:686–694

Ryer CH, Olla BL (1998) Shifting the balance between foraging and predator avoidance: the importance of food distribution for a schooling pelagic forager. Environ Biol Fish 52:467–475

Sakakura Y, Tagawa M, Tsukamoto K (1998) Whole-body cortisol concentrations and ontogeny of aggressive behaviour in yellowtail (*Seriola quinqueradiata* Temminck and Schlegel: Carangidae). Gen Comp Endocr 109:286–292

Simpson SJ, Raubenheimer D (2001) A framework for the study of macronutrient intake in fish. Aquac Res 32:421–432

Suzuki K, Mizusawa K, Noble C, Tabata M (2008) The growth, feed conversion ratio and fin damage of rainbow trout *Oncorhynchus mykiss* under self-feeding and hand-feeding regimes. Fish Sci 74:941–943

Sveier H, Lied E (1998) The effect of feeding regime on growth, feed utilisation and weight dispersion in large Atlantic salmon (*Salmo salar*) reared in seawater. Aquaculture 165:333–345

Symons PEK (1968) Increase in aggression and in strength of the social hierarchy among juvenile Atlantic salmon deprived food. J Fish Res Board Can 25:2387–2401

Talbot C, Corneillie S, Korsøen O (1999) Pattern of feed intake in four species of fish under commercial farming conditions: implications for feeding management. Aquac Res 30:509–518

Thorpe JE, Talbot C, Miles MS, Rawlings C, Keay DS (1990) Food consumption in 24 hours by Atlantic salmon (*Salmo salar* L.) in sea cage. Aquaculture 90:41–47

Turnbull JF, Richards RH, Robertson DA (1996) Gross, histological and scanning electron-microscopic appearance of dorsal fin rot in farmed Atlantic salmon, *Salmo salar* L. parr. J Fish Dis 19:415–427

Vivas M, Sanchez-Vazquez FJ, Garcia–Garcia B, Madrid JA (2003) Macronutrient self-selection in European sea bass in response to dietary protein or fat restriction. Aquacult Res 34:271–280

Vivas M, Rubio VC, Sánchez-Vázquez FJ, Mena C, García-García B, Madrid JA (2006) Dietary self-selection in sharpsnout seabream (*Diplodus puntazzo*) fed paired macronutrient feeds and challenged with protein dilution. Aquaculture 251:430–437

Wang T, Hung CC, Randall DJ (2006) The comparative physiology of food deprivation: from feast to famine. Annu Rev Physiol 68:223–251

Ward AJW, Webster MM, Hart PJB (2006) Intraspecific food competition in fishes. Fish Fish 7:231–261

Weatherley AH, Gill HS (1981) Recovery growth following periods of restricted rations and starvation in rainbow trout *Salmo gairdneri* Richardson. J Fish Biol 18:195–208

Winberg S, Lepage O (1998) Elevation of brain 5-HT activity, POMC expression, and plasma cortisol in socially subordinate rainbow trout. Am J Physiol Regul Integr Comp Physiol 274:645–654

Winberg S, Nilsson GE (1993) Roles of brain monoamine neurotransmitters in agonistic behaviour and stress reactions, with particular reference to fish. Comp Biochem Physiol C 106:597–614

Winberg S, Nilsson GE, Olsén KH (1992) Changes in brain serotonergic activity during hierarchic behaviour in Arctic charr (*Salvelinus alpinus* L.) are socially induced. J Comp Physiol A 170:93–99

Winberg S, Nilsson A, Hylland P, Söderström V, Nilsson GE (1997) Serotonin as a regulator of hypothalamic-pituitary-interrenal activity in teleost fish. Neurosci Lett 230: 113–116

Yamamoto T, Shima T, Furuita H, Shiraishi M, Sánchez-Vázquez FJ, Tabata M (2000) Self-selection of diets with different aminoacid profiles by rainbow trout (*Oncorhynchus mykiss*). Aquaculture 187:375–386

Ytrestøyl T, Struksnæs G, Koppe W, Bjerkeng B (2005) Effects of temperature and feed intake on astaxanthin digestibility and metabolism in Atlantic salmon, *Salmo salar*. Comp Biochem Physiol B 142:445–455

Dietary Nitrogen and Fish Welfare – Summary

Luis Conceição, Cláudia Aragão, Jorge Dias, Benjamín Costas, Genciana Terova, Catarina Martins and Lluis Tort

Under some stress situations an increase in dietary needs of some of amino acids may occur. Therefore, supplementation of fish diets in periods of predictable stress such as transport or capture may reduce the negative effects of stress on fish welfare. Indispensable amino acids such as arginine and leucine have an important role in the immune system. However, more work is needed to understand the precise effects of each particular amino acid in immune response and disease resistance. Moreover, supplementation of tryptophan in diets has been shown to reduce aggressiveness in fish. Still, before supplementation of this amino acid can be used as a "therapeutic" agent in fish, further studies on factors such as dosages and duration of treatment are needed. Some peptides (polymers of a few amino acids) which are present in some common ingredients (e.g. milk and fish proteins) have also been found to have bioactive properties; they may provide help in disease control of cultured fish. Fish meal replacement by plant protein ingredients has been a growing trend in the fish feed industry. However, this replacement needs to be done carefully to avoid negative effects on fish welfare. Plant protein ingredients may be deficient in some amino acids and phosphorus, which are normally supplied by Fish meal, and some require technological treatments for removal or inactivation of some compounds (anti-nutritional factors) that reduce the nutritional value of the feed.

Background

Fish welfare is recognized as an important issue in aquaculture, due both to ethical issues and production advantages. Good welfare leads to gains in growth, and less susceptibility to stress and diseases. Fish welfare can be improved, and some welfare issues can be minimized, through better nutrition. The effects of vitamins and polyunsaturated fatty acids on fish health and welfare are well known. However, less reported studies are available on the relationship between by the protein fraction of the diet and fish welfare. It is known that the 20 amino acids,which are the building blocks of all proteins, have also some very other important roles in growth, aggressive behaviour, reproduction and immune responses.

Scientific Significance

This review shows that amino acid metabolism is affected when fish are under stress and this, together with sub-optimal protein formulations, may affect fish welfare. Consequently, improved dietary protein supply may help fish coping with predictable stressors.

Practical Application

Fish welfare and socio-economic sustainability of the aquaculture industry may be improved by feeding fish with diets with well balanced amino acid contents.

Contact

L. E. C. Conceicão
CCMAR-CIMAR L.A., Centro de Ciências do Mar
do Algarve, Universidade do Algarve, Campus
de Gambelas, 8005-139 Faro, Portugal
e-mail: lconcei@ualg.pt

Fish Physiol Biochem (2012) 38:119–141
DOI 10.1007/s10695-011-9592-y

Dietary nitrogen and fish welfare

Luis E. C. Conceição · Cláudia Aragão · Jorge Dias · Benjamín Costas · Genciana Terova · Catarina Martins · Lluis Tort

Received: 1 November 2011 / Accepted: 16 December 2011 / Published online: 3 January 2012

Abstract Little research has been done in optimizing the nitrogenous fraction of the fish diets in order to minimize welfare problems. The purpose of this review is to give an overview on how amino acid (AA) metabolism may be affected when fish are under stress and the possible effects on fish welfare when sub-optimal dietary nitrogen formulations are used to feed fish. In addition, it intends to evaluate the current possibilities, and future prospects, of using improved dietary nitrogen formulations to help fish coping with predictable stressful periods. Both metabolomic and genomic evidence show that stressful husbandry conditions affect AA metabolism in fish and may bring an increase in the requirement of indispensable AA. Supplementation in arginine and leucine, but also eventually in lysine, methionine, threonine and glutamine, may have an important role in enhancing the innate immune system. Tryptophan, as precursor for serotonin, modulates aggressive behaviour and feed intake in fish. Bioactive peptides may bring important advances in immunocompetence, disease control and other aspects of welfare of cultured fish. Fishmeal replacement may reduce immune competence, and the full nutritional potential of plant-protein ingredients is attained only after the removal or inactivation of some antinutritional factors. This review shows that AA metabolism is affected when fish are under stress, and this together with sub-optimal dietary nitrogen formulations may affect fish welfare. Furthermore, improved dietary nitrogen formulations may help fish coping with predictable stressful events.

Keywords Fish welfare · Protein nutrition · Amino acid requirements · Tryptophan · Bioactive peptides · Antinutritional factors

L. E. C. Conceição (✉) · C. Aragão · J. Dias · B. Costas · C. Martins
CCMAR-CIMAR L.A., Centro de Ciências do Mar do Algarve, Universidade do Algarve, Campus de Gambelas, 8005-139 Faro, Portugal
e-mail: lconcei@ualg.pt

B. Costas
CIIMAR-CIMAR L.A., Centro Interdisciplinar de Investigação Marinha e Ambiental, Universidade do Porto, Rua dos Bragas 289, 4050-123 Porto, Portugal

G. Terova
Department of Biotechnology and Molecular Sciences, University of Insubria, Via J.H. Dunant, 3, 21100 Varese, Italy

C. Martins
Aquaculture and Fisheries Group, Wageningen University, P.O. Box 338, 6700 AH Wageningen, The Netherlands

L. Tort
Department of Cell Biology, Physiology and Immunology, Universitat Autonoma Barcelona, 08193 Cerdanyola, Spain

Introduction

Welfare is an important issue in aquaculture, since animals reared under good welfare conditions have better growth performances and fecundity and are less susceptible to develop diseases and infections (Dawkins 2001). Furthermore, welfare of animals reared in captivity is increasingly becoming a relevant and significant societal topic (Dawkins 1990). Farmed fish are submitted to various kinds of stressors including grading, transport, crowding and vaccination, which can compromise their welfare status (Ashley 2007). In this review, we will use a *function-based* definition of welfare, in which welfare is correlated to biological functions and physiological stress responses (Duncan 2005).

The stress response is vitally important for all living organisms. In response to a stressor, a fish will undergo a series of biochemical and physiological changes in an attempt to compensate for the challenge imposed upon it and thereby cope with the stress (Wendelaar Bonga 1997). The stress response diverts energy from normal metabolic processes, including growth, to deliver energy to the physiological systems activated to adapt to the stressor. The several endocrine and physiological alterations often result in changes in the ability of the fish to survive, decrease the immune response, increase the incidence of diseases or limit growth (Pickering 1998). Although these changes may not be lethal, they can impair the health and appearance of the fish, which may result in welfare-, environmental- and/or aquaculture-related problems.

Therefore, nutritional strategies to modulate fish immune system and consequently disease resistance seem a valuable tool. Nutrition has an influence on health and immune responses of fish (Blazer 1992). In contrast to what is commonly found in mammals, in fish, non-specific immunity is considered as the first line of defence and represents a considerable part of the immune response. Generally, values of alternative complement pathway activity obtained for fish are considerably higher than those for mammals, suggesting that this pathway is of particular relevance for fish as non-specific immune mechanism (Montero et al. 1998; Ai et al. 2007).

The major fate of amino acids (AA) is towards protein synthesis, but emerging evidence shows that many AA regulate key metabolic pathways that are crucial to maintenance, growth, reproduction and immune responses (Li et al. 2009). Stressful husbandry conditions affect AA metabolism in fish (Pinto et al. 2007; Aragão et al. 2008; Costas et al. 2008), and under some stress situations, an increase in the requirement of certain indispensable AA may occur, which is probably related with the synthesis of proteins and other compounds related with the stress response (Aragão et al. 2008). Furthermore, some AA may assume a special importance in fish welfare. This seems to be the case of tryptophan as precursor for serotonin, which modulates aggressive behaviour and feed intake in fish. It is also important to evaluate the effects on fish welfare with the emerging use of alternative protein sources in the fish feed industry.

The purpose of this review is to give an overview on how AA metabolism may be affected when fish are under stress and the possible effects on fish welfare when sub-optimal dietary nitrogen formulations are used to feed fish. Due to space limitations, citations in this review are not exhaustive. In addition, this review intends to evaluate the current possibilities, and future prospects, of using improved dietary nitrogen formulations to help fish coping with predictable stressful periods under culture conditions.

How are amino acid metabolism and protein turnover affected when fish are under stress?

Stress affects amino acid metabolism in fish

Stress has been shown to affect several metabolic pathways in fish, and some studies have analysed the relation between the most used stress indicator—cortisol—and AA (mostly free AA) concentrations in fish. The results are sometimes contradictory, but, as reviewed by Mommsen et al. (1999), several factors as the mode of cortisol application (e.g., single or repeated injection, intraperitoneal implants or mini-pumps) or treatment time make complex the analyses of the results. Nevertheless, cortisol is known to affect protein and AA metabolism.

The results of stress experiments indicate that changes in free AA concentrations may be dependent on the type of stressor imposed to the fish. For instance, handling stress does not seem to clearly affect plasma total free AA levels in fish. These concentrations were unaffected in common dentex (*Dentex dentex*) submitted to an acute handling stress

(Morales et al. 2005) and in Senegalese sole (*Solea senegalensis*) submitted to weekly handling stress during four and 9 weeks (Aragão et al. 2008; Costas et al. 2011a), even though in all situations handling procedures evoked a stress response in fish compared to control groups. In the same way, plasma total free AA concentrations in Senegalese sole chronically exposed to exogenous ammonia were unaffected compared to control groups (Pinto et al. 2007). However, other stressful husbandry conditions as confinement, high stocking density or temperature changes clearly affected plasma levels of free AA. Plasma total free AA concentrations were significantly higher in tilapia (*Oreochromis mossambicus*) and rainbow trout (*Oncorhynchus mykiss*) confined for 24 h (Vijayan et al. 1997; Trenzado et al. 2003) and Senegalese sole after 24 h following acute hypoxia (Costas et al. 2011b). In addition, an inverse linear relationship between total free AA concentrations in plasma and stocking density was found in Senegalese sole: a reduction of 40% in the plasma concentration of total free AA was observed in fish held at high stocking density during 9 weeks compared to fish held at low stocking density (Costas et al. 2008). Furthermore, also for Senegalese sole, feed deprivation and acclimation to low and high temperatures increased total free AA levels in plasma (Costas et al. 2011c, d).

As described for the total free AA concentrations, several studies have shown that stressful husbandry conditions affect the plasma individual AA concentrations in fish. In some cases, though total free AA concentrations in plasma were unaffected, significant changes in plasma individual free AA levels were found (Pinto et al. 2007; Aragão et al. 2008; Costas et al. 2011a). These changes probably reflect the synthesis of proteins or other specific compounds related to the stress response. It is interesting to note that as reported for total free AA concentrations in plasma, changes in plasma levels of individual free AA also appear to be dependent on the type of stressor imposed to the fish. Senegalese sole submitted for 4 and 9 weeks to repeated handling stress both presented lower plasma concentrations of arginine, lysine and ornithine (Aragão et al. 2008; Costas et al. 2011a), while an increase in the stocking density led to a general decrease in plasma levels of all indispensable AA (with the exception of threonine and methionine), plus serine, glutamine and ornithine. Although most of these studies focused on the changes of AA levels in plasma, stress and cortisol have been shown to affect differently the free AA concentrations in different organs (Saha et al. 2002). For instance, Milligan (1997) showed that exhaustive exercise affected significantly alanine, glutamine, glutamate and branched-chain AA (BCAA; leucine, isoleucine and valine) levels in rainbow trout, but the responses were different in plasma, muscle and liver. This led the authors to suggest that cortisol stimulated proteolysis in liver, releasing BCAA to the plasma, which were then taken up and utilized in skeletal muscle, the major site of BCAA oxidation (Fig. 1).

Recent studies using functional transcriptomics and comparative proteomics also evidence that stress affects AA metabolism. An acute stressor has been shown to bring significant changes in the expression of genes involved in gluconeogenesis, energy metabolism, protein degradation, and the immune response in trout liver (Wiseman et al. 2007). All this may mean that under stress liver proteins may serve as transient sources of energy and building blocks for immune-related proteins. The substrates for gluconeogenesis are predominantly AA in trout and other species (Mommsen et al. 1999). Protein degradation (or turnover) may function to mobilize AA for energy production and/or for the synthesis of proteins involved in immune or stress responses. In a comparative proteomics study, Alves et al. (2010) identified glutamine synthetase (GSase) as a down-regulated protein in liver of gilthead seabream (*Sparus aurata*) exposed to chronic stress. This enzyme catalyses glutamine formation from ammonia and glutamate, an ammonia elimination/detoxification mechanism in liver. The down-regulation of GSase expression in stressed fish was suggested to result from an eventual increase in GSase early on stress exposure, followed by a feedback mechanism to reduce this accumulation after some time (Alves et al. 2010). This hypothesis is supported by the observed up-regulation of GSase expression in trout liver after acute stress (Wiseman et al. 2007).

In summary, stressful conditions have profound effects on AA metabolism, and these effects go beyond the increased need of AA as energy substrates to fuel the allostatic response. Some studies suggest that the stress response in fish may even alter their AA requirements and affect health status (e.g., Aragão et al. 2008; Costas et al. 2008; see also section "Amino

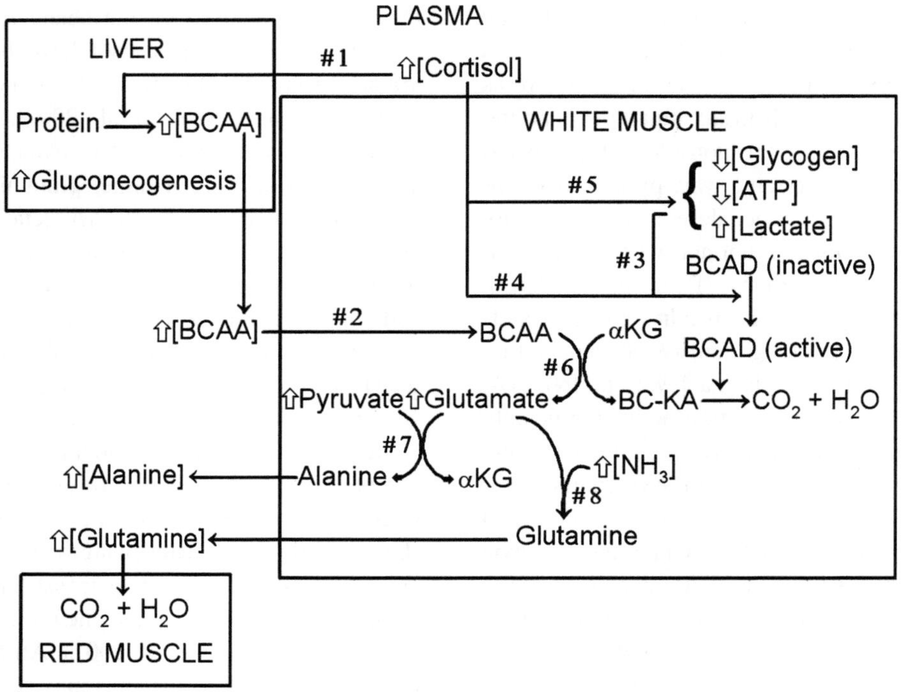

Fig. 1 Model proposed to explain the role of cortisol in amino acid mobilization following exhaustive exercise in *Oncorhynchus mykiss*. *BCAA* branched-chain amino acids (leucine, isoleucine, valine), *BCAD* branched-chain keto acid dehydrogenase, *αKG* α-ketoglutarate, *BC-KA* branched-chain keto acid, *ATP* adenosine triphosphate (With kind permission from Springer Science + Business Media: Fish Physiology and Biochemistry, The role of cortisol in amino acid mobilization and metabolism following exhaustive exercise in rainbow trout *Oncorhynchus mykiss Walbaum,* vol. 16, 1997, pp. 119–128, C.L. Milligan, Fig. 5)

acid requirements are affected by stress and immune responses" in this review).

Stress and protein turnover

Stress has been shown to affect not only the metabolism of individual AA, but also the rates of protein synthesis and protein turnover (or degradation). This has consequences in terms of protein deposition, even at identical protein intake, once protein deposition (or growth) is dependent on the balance between these two rates. Protein growth may result from both an increase in the rate of protein synthesis and from a lower rate of protein turnover. Furthermore, the synthesis of proteins has a high energy cost (Houlihan 1991), and increased protein synthesis leads to higher energy expenditure. If indispensable AA are used as energy substrates, then the efficiency of protein utilization will be reduced.

Normally, as protein intake increases protein synthesis also increases (Houlihan 1991). In adult fish and mammals (Houlihan et al. 1994), protein turnover increases when protein synthesis augments. However, this is not the case for fish larvae, probably as a result of a strong selective pressure for fast growth and thereby in the efficiency of protein deposition (Conceição et al. 1997; Conceição et al. 2008). It has been proposed that once protein turnover has high energy costs, there may be a trade-off between fast growth and stress resistance/survival in fish larvae (Conceição et al. 2001). This may also happen under some circumstances in juvenile and adult fish (Houlihan 1991). A high protein turnover probably favours a rapid adaptation to stressful conditions including changes in the environment (Kiørboe et al. 1987). In fact, environmental stress led to higher protein turnover in dab (*Limanda limanda*) and rainbow trout (Houlihan et al. 1994; Wilson et al. 1996). Recent transcriptomics studies also suggest that genes for proteins involved in protein degradation, such as cathepsin D, ubiquitin and cystatin, may be upregulated in stressed fish (Wiseman et al. 2007; Prunet et al. 2008). Moreover, protein turnover has been shown to increase when imbalanced AA diets were fed to European seabass (*Dicentrarchus labrax*) (Langar et al. 1993). This probably reflects a mobilization of AA in deficiency for synthesis of proteins required for homeostasis.

It will be interesting to verify in future studies to what extent protein intake, including AA balance, affects protein turnover, and to what extent this interacts with stress resistance and fish welfare.

How do sub-optimal dietary nitrogen formulations affect fish welfare?

Nutritional stress and transcriptomic response

Nutritional stress is a negative physiological state resulting from suboptimal quantity or quality of food available to an animal, and the dietary nitrogen plays here a major role. Whereas much is known about the physiological and biochemical response of fish to nutritional stress, only little information is available at the molecular level. In particular, very little is known about the response of genes at the transcriptional level to starvation and refeeding, although this knowledge could provide a better understanding on the mechanisms by which the expression of genes regulates fish growth and affects fish welfare.

Fasting in fish elicits dramatic alterations in the metabolic pathways responsible for the release, transport, and utilization of energy reserves, including protein. Many of these adjustments depend to a large extent on changes in the expression of genes that encode diverse groups of physiologically relevant proteins. For this reason, the fasting–refeeding paradigm has been used in fish to investigate the molecular mechanisms underlying the transition from a catabolic to an anabolic state (Terova et al. 2006). Several candidate genes whose expression is involved in response to nutritional stress in European seabass have been identified (Terova et al. 2006, 2007). The results of these studies have demonstrated that the nutritional state of the fish influences the levels of expression of several genes, including positive and negative growth factors.

It is already known that the brain processes and responds to nutritional stimuli with appropriate modifications in growth through hormonally mediated pathways (Duan 1998). A central step in this endocrine pathway is the growth hormone (GH)—insulin-like growth factors' (IGFs) axis. The majority of the growth-promoting effects of the growth hormone are believed to be mediated by insulin-like growth factor-I (Duan 1998; Leroith et al. 2001), which is structurally related to IGF-II (Shamblott and Chen 1992). IGF-I and IGF-II are ubiquitous small peptides, well known for stimulating many anabolic responses on a variety of target cells in both endocrine and autocrine/paracrine fashions (Wood et al. 2005). IGFs (also known as somatomedins) are unique among the growth factors, having the property of stimulating both proliferation and differentiation of muscle precursor cells (myoblasts or satellite cells), as well as myocyte hypertrophy during muscle regeneration (Musaro et al. 1999). The evidence that IGF-I and IGF-II mRNAs are also expressed in other tissues (Wood et al. 2005) and the presence of receptors in these tissues suggests that in fish, paracrine and autocrine actions of IGFs are also involved in organ-specific functions.

Nutrient availability modulates the rate of muscle growth in fish. The proliferation activity of myogenic precursor cells in vitro and in vivo depends largely on the feeding status in fish (Fauconneau and Paboeuf 2000; Brodeur et al. 2002), suggesting that the cellular aspects of muscle growth are influenced by the feeding status of the fishes. Under fasting, both hypertrophic and hyperplasic components of muscle growth are stopped (Kiessling et al. 1990). This is thought to be associated with changes in the expression of growth factors in muscle. Furthermore, in various fish species, it has been shown that hepatic and muscular IGF-I and IGF-II mRNA levels depend on feeding status. For instance, Terova et al. (2007) have analysed the pattern of IGF-I and -II gene expression in liver and myotomal muscle of European seabass and reported that these were down-regulated during fasting and up-regulated during recovery from fasting. Therefore, IGF-I and -II participate in response to nutritional stress.

The complex processes of myogenesis and muscle growth are precisely regulated. Growth factors such as fibroblast growth factor 6 (FGF6) and myostatin (MSTN), a member of the transforming growth factor-β (TGF-β) superfamily, are involved. Whereas FGF6 is involved both in the proliferation and in the differentiation of the myogenic lineage (Armand et al.

2003) by stimulating myoblast proliferation, myostatin is a negative regulator of the number of myoblasts and, hence, number of muscle fibres. MSTN is the main factor involved in depressing both the number of myoblasts that reach terminal differentiation divisions and, to some extent, the degree of fibre hipertrophy. As a consequence, the lack of MSTN results in increased muscle mass (Bass et al. 1999). Nutritional stress influenced myostatin abundance levels in fish muscle, inducing an up-regulation during fasting and a down-regulation during the recovery from fasting (Terova et al. 2006). Therefore, transcriptomic studies have demonstrated that the nutritional state of the fish influences the levels of expression of positive and negative growth factors involved in the nutritional stress response, which are closely related to protein accretion and metabolism.

Nutritional stress response and intestinal oligopeptide transporters

During digestion, dietary proteins in both teleosts and mammals are subject to hydrolysis by a range of proteases and peptidases that generate a mixture of free AA and small peptides, which are efficiently absorbed across the apical membranes of enterocytes (Clements and Raubenheimer 2006; Taylor et al. 2011). Cellular transport of AA and small peptides is thus a key final step in intestinal processing. Distinct types of enterocyte transport systems are involved in this process: (1) AA transport which may or may not be sodium dependent; (2) transport of small peptides (2–3 AA) coupled to a H^+ gradient on an abundant high-capacity, low-specificity peptide transporter (PepT1) or through endocytosis. Intestinal peptide transport is of major nutritional significance in that the intraluminal products of protein digestion are predominantly di- and tripeptides (up to 85% of the digesta in mammals) and not AA as previously believed (Taylor et al. 2011; Verri et al. 2011). In addition, many AA are more rapidly and efficiently absorbed in peptide form (for a review see Gilbert et al. 2008).

Terova et al. (2009) have shown that fasting nutritional stress significantly influenced PepT1 mRNA copy number in the proximal intestine (Fig. 2), inducing a down-regulation during prolonged fasting (35 days) and an up-regulation during the recovery from fasting. These findings offer new information about the nutritional regulation of PepT1 mRNA level and support a role of this membrane transporter protein in the nutritional stress response of fish. In

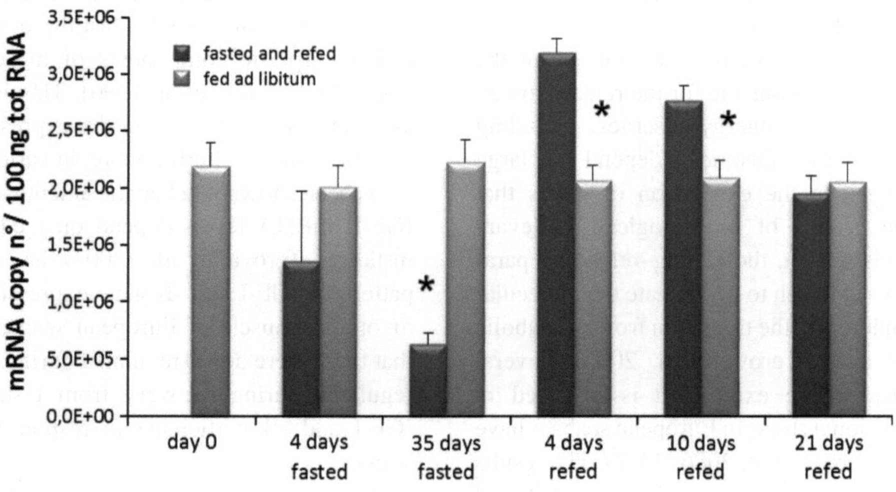

Fig. 2 Expression levels of PepT1 in *D. labrax* proximal intestine in the course of the experiment. PepT1 mRNA copy number was normalized as a ratio to 100 ng total RNA. Fish were sampled before fasting (day 0), 4 days after fasting (4 days fasted), at the end of fasting (35 days fasted), and then sequentially at 4, 14 and 21 days following refeeding. The means of five animals in each group are shown. *Bars* indicate standard error of the mean. Differences were determined by one-way analysis of variance (ANOVA). (*asterisk*) indicates significantly different means from controls, for each time point tested ($P < 0.05$). This figure has been adapted from Terova et al. (2009)

addition, PepT1 gene expression is up-regulated with refeeding and correlates with fish compensatory growth (Verri et al. 2011). Moreover, Bakke et al. (2010) have shown that dietary protein hydrolysates and free AA affect the expression of PepT1 along the digestive tract of Atlantic cod (*Gadus morhua*), indicating an adaptive response to keep a maximal efficiency of protein absorption at the intestinal level. Still, while the importance of this peptide transporter to protein utilization in fish remains to be quantified, it may be particularly important for the uptake of bioactive peptides and their role in fish welfare (see section later in this paper).

Amino acid requirements are affected by stress and immune responses

Stress response in fish may alter their AA requirements and affect health status. Still, few studies concerning the effects of dietary AA and protein levels on disease resistance and welfare in fish have been conducted. Diet can have significant effects on disease resistance mechanisms, and the influence that dietary factors may have on disease outbreaks in cultured fish has been recognized for many years (reviewed by Blazer 1992). Still, the consequences of AA deficiencies have not so far been correlated with major disease outbreaks, severe reduction in infection resistance nor important mortality. However, marginal dietary deficiencies are difficult to identify and may have been overlooked. This gain especially importance when replacing fishmeal by plant protein sources (see section "Plant protein sources, antinutritional factors and fish welfare"). Furthermore, the additional requirements under stressful conditions have been poorly studied. For the purpose of this review, we will try to elucidate mechanisms by which the requirements of individual AA may increase in fish under stressful situations.

As in other vertebrates, a balanced diet is a key for an adequate performance including immune response and disease resistance. Thus, diets have to gather at least three conditions: the components have to bear enough energy for the biological processes involved in the immune defence; the components have to be balanced in order to avoid unwanted effects; and they have to include key components that are relevant for specific physiological functions related to the immune response. Therefore, an adequate balanced diet plays a crucial role in the management of stress and immunocompetence, and nutrition seems to be a critical determinant in the interactions between the CNS, the endocrine system and the immune system, when fish have to build a response under alarm or stress conditions (Romeo et al. 2008).

In terms of proteins and AA as major nutrient requirements, it is known that most fish species do not require an absolute protein level but 10 indispensable AA. Clearly, AA metabolism may undergo dramatic changes when fish are in stressful situations, what will result in altered AA requirements and may affect stress resistance and growth potential. Moreover, different stressful husbandry conditions seem to affect differently AA metabolism. Nevertheless, which of these AA are relevant for the adequate function of immune responses remains to be assessed and only partial information is available for several species and specific AA. In addition, the influences of either the deficiency or the supplementation of each one of these AA are still poorly understood.

Regarding the effects of particular AA on the immune competence in fish, little work has been done up to present and only partial information can be obtained. Table 1 summarizes the effects of several stressful conditions on plasma (also muscle for histidine) AA levels, as well as some health-related compounds derived from indispensable AA which may have important roles during stress and immune responses in fish. Besides its importance as dietary indispensable AA for fish, some of AA have been shown to be particularly versatile. For instance, arginine is involved in a number of metabolic pathways (Mommsen 2001). Hydrolysis by arginase leads to production of ornithine, which can be used for the synthesis of urea and proline, glutamate, glutamine or polyamines (Gouillou-Coustans et al. 2002; Taylor et al. 2011). Positive effects of arginine-enriched diets on the resistance of channel catfish and Senegalese sole to infection with *Edwardsiella ictaluri* and *Photobacterium damselae* subsp. *piscicida*, respectively, have been demonstrated (Buentello and Gatlin 2001; Costas et al. 2011e). Furthermore, when juvenile channel catfish were injected intraperitoneally with lipopolysaccharide and fed high dietary levels of arginine, significantly higher levels of haemoglobin, haematocrit, and circulating erythrocytes but not the number of circulating leucocytes, lymphocytes,

Table 1 Overview of some important health-related substances which need indispensable amino acids to be synthesized and may be utilized during immune and stress responses

Amino acid	Fish species	Stressor	Effect on plasma/tissue levels	Function observed in fish	Main health-related substance
Arginine	Senegalese sole *Solea senegalensis*	Acute hypoxia, chronic handling and high density	Decrease	Dietary arginine supplementation enhanced respiratory burst and nitric oxide production in Senegalese sole, and disease resistance in channel catfish and sole.[1,2]	Nitric oxide Polyamines
	Tilapia *Oreochromis mossambicus*	24 h Confinement	Increase	Dietary arginine in sole and polyamines incubation in seabream leucocytes increased the level of expression of hepcidin antimicrobial peptide.[2,3]	
Valine Leucine Isoleucine	Senegalese sole *Solea senegalensis*	Acute hypoxia, chronic handling and high density	Decrease	Dietary β-hydroxyl-β-methyl-butyrate (HMB; a leucine metabolite) supplementation resulted in non-specific immune enhancement in several fish species as rainbow trout, pikeperch (*Sander lucioperca*) or tench (*Tinca tinca*).[4,5,6]	HMB
		Chronic ammonia exposure and feed deprivation	Increase		
	Tilapia *Oreochromis mossambicus*	24 h confinement	Increase		
	Rainbow trout *Oncorhynchus mykiss*	Exhaustive exercise	Increase		
Tryptophan[a]	Senegalese sole *Solea senegalensis*	Acute hypoxia	Increase	Dietary tryptophan supplementation can prevent a stress-induced cortisol surge[7], reduce stress-induced anorexia and cannibalism[8] and inhibit aggression in fish.[9]	Serotonin melatonin
		Chronic handling, high density and ammonia exposure	Decrease		
Histidine	Senegalese sole *Solea senegalensis*	Acute hypoxia, chronic high density and feed deprivation	Decrease	Histidine is a major player in the biochemical changes that occur during inflammation, and which exert a metabolic demand on AA metabolism.[10]	Histamine
	Milkfish *Chanos chanos* Rainbow trout *Oncorhynchus mykiss* Japanese eel *Anguilla japonica* Japanese dace *Leuciscus hakonensis*	Feed deprivation			
Phenylalanine tyrosine	Senegalese sole *Solea senegalensis*	Acute hypoxia	Increase	Tyrosine is converted into neurotransmitters that modulate stress responses. Dietary phenylalanine increased the synthesis and release of brain dopamine as well as plasma phenylalanine and tyrosine in Senegalese sole.[11]	Catecholamines thyroid hormones
	Tilapia *Oreochromis mossambicus*	2 h Confinement	Decrease		
		24 h confinement	Increase		
Lysine	Senegalese sole *Solea senegalensis*	Acute hypoxia, chronic handling and high density	Decrease	Lysine is one of the substrates for the synthesis of L-carnitine, which in fish is required for the transport of long-chain fatty acids to mitochondria, where they are oxidized.[13]	Carnitine
	Tilapia *Oreochromis mossambicus*	24 h confinement	Increase		

Table 1 continued

Amino acid	Fish species	Stressor	Effect on plasma/tissue levels	Function observed in fish	Main health-related substance
Methionine cysteine	Senegalese sole *Solea senegalensis*	Acute hypoxia and chronic handling and ammonia exposure	Decrease	Methionine is involved in protein and DNA methylation as well as in the synthesis of carnitine. Carp head kidney phagocytes are highly resistant to nitrosative stress by increasing total intracellular glutathione,[12] and a similar outcome has been hypothesized for Senegalese sole leucocytes,[2] suggesting a role for methionine	Carnitine Glutathione Taurine
	Tilapia *Oreochromis mossambicus*	24 h confinement	Increase		
Threonine	Senegalese sole *Solea senegalensis*	Chronic ammonia exposure	Decrease	Threonine has been shown to modulate immune function in mammals. For instance, it is needed for the synthesis of the mucin protein that is required for maintaining intestinal immune function and increases antibody production.[14] Therefore, the relation between threonine and stress response in fish seems a matter to deserve further attention	Mucin
		Acute hypoxia	Increase		
	Tilapia *Oreochromis mossambicus*	24 h confinement	Increase		

[a] Further details on the relation between tryptophan and fish welfare are given in section "Tryptophan and fish welfare"

[1] Buentello and Gatlin (2001); [2] Costas et al. (2011e); [3]Reyes-Becerril et al. (2011); [4,5,6] Siwicki et al. (2003, 2005, 2006); [7] Lepage et al. (2002); [8] Höglund et al. (2007); [9] Hseu et al. (2003); [10] Grimble and Grimble (1998); [11] Costas et al. (2011a); [12] Saeij et al. (2003); [13] Harpaz (2005); [14] Li et al. (2007)

eosinophils, and monocytes were found (Buentello et al. 2007). An in vitro model showed that a moderate level of arginine in the culture media significantly enhance phagocytosis in channel catfish peritoneal macrophages (Buentello et al. 2007). Moreover, the level of expression of g-type lysozyme, hypoxia-inducible factor, hepcidin antimicrobial peptide (HAMP-1) and macrophage inflammatory protein 1α increased in Senegalese sole fed increased levels of arginine (Costas et al. 2011e). The up-regulation of HAMP-1 gene transcripts could be related to an increased availability of polyamines. Arginine is also an essential component in endogenous production of polyamines, which are positively linked to translation and RNA and DNA polymerase activity and therefore are essential for cell proliferation, differentiation and growth (Taylor et al. 2011).

Glutamine is one of the most versatile AA in cellular metabolism and physiology, and even though it is not indispensable, it deserves special attention. Glutamine not only acts as a precursor for protein synthesis, but is also an important intermediate in a large number of metabolic pathways. Glutamine synthesis is an important pathway for ammonia detoxification in several fish species (Ip et al. 2001), and an increase in plasma glutamine levels in chronically stressed Senegalese sole has been previously observed (Pinto et al. 2007; Aragão et al. 2010; Costas et al. 2011c). Therefore, this pathway may assume a special importance in fish exposed to exogenous ammonia, prolonged starvation and osmotic challenge. Supporting the idea that glutamine formation is linked to removal and detoxification of ammonia is the observation that ammonia exposure increases abundance of glutamine synthetase transcripts in intestinal tissue (Anderson et al. 2002) and glutamine synthetase activity is increased by cortisol (Mommsen et al. 2003). Augmented cortisol levels are usually associated with increased nitrogen turnover in fish and are known to increase the availability of amino acids through activation of endogenous proteolysis (Mommsen et al. 1999). Glutamine presents also important roles in improving gastrointestinal tract maturation and health, maintaining mucosal integrity and the gut-associated lymphoid tissues, and enhancing the synthesis of secretary immunoglobulin (Wu et al. 2007). As the precursor for the synthesis of glutathione, glutamine is also essential for defending cells from oxidative stress (Wu et al. 2007).

Costas et al. (2008) reported that fish reared at high stocking densities presented high plasma cortisol levels, changes in AA metabolism, several disease outbreaks and mortality. This study contributes to confirm that chronic stress situations increase the susceptibility of fish to infectious diseases and suggests a role of indispensable AA on disease resistance. Interestingly, neutrophilic granulocytes are protected by stress. Cortisol can induce apoptosis in activated B lymphocytes but not in neutrophils (Verburg-van Kemenade et al. 2009). As neutrophilic granulocytes are of great importance to the first line of defence, it may be beneficial in situations of acute stress (and possible injury) to prolong their lifespan and maintain higher numbers in circulation. However, little attention has been paid to the effects of acute stress and catecholamines, which seem to have both stimulatory and inhibitory actions on immune functions in fish (Wendelaar Bonga 1997). In fact, an acute increase in catecholamines and cortisol levels may signal the immune system to prepare for possible consequences of a stressor and thus serve as an adaptive function (Verburg-van Kemenade et al. 2009). For instance, daily handling stress for 14 days induced a higher disease resistance in stressed Senegalese sole than in undisturbed specimens, as well as increased expression values of immune-related genes at the time of bacterial infection (Costas et al. 2011e). However, the exact mechanisms by which stressed Senegalese sole resisted to bacterial infection remain to be elucidated.

More work is clearly needed to understand the effects of particular AA and stressful rearing conditions in the mechanisms related to the immune response.

Can improved dietary nitrogen formulations help fish coping with stress?

Tryptophan and fish welfare

Tryptophan (TRP) is an indispensable AA supplied in animal diets to ensure maximum growth. Besides being used for protein synthesis, TRP has been shown to have a "calmative" or "therapeutic" effect on terrestrial farm animals and fish when supplied in excess. Such therapeutic effect is a consequence of TRP being the precursor of the neurotransmitter serotonin (5-hydroxytryptamine, 5-HT). In fact, an increased concentration of TRP in the feed has been shown to be strongly correlated with plasma TRP levels as well as TRP and 5-HT levels in the brain (e.g., Johnston et al. 1990). An increase in serotonin level in the brain has been shown to reduce aggressive behaviour, cannibalism, self-injury, depressive mood, fearfulness, pain sensitivity, size heterogeneity within groups and stress response in mammals (see Table 2). Therefore, it is not surprising the interest on feeding animals with a surplus of dietary TRP as a strategy to improve the welfare of farmed animals.

Serotonin does not cross the blood–brain barrier; therefore, its effect within the brain is dependent on the transport of its precursor, TRP, across that barrier. In teleost fish, as in mammals, TRP uptake from the blood into the brain is mediated by active transport, sharing a carrier protein with other large neutral AA (LNAA) such as tyrosine, phenylalanine, leucine, isoleucine and valine (Fernstrom and Wurtman 1997; Aldegunde et al. 1998). Consequently, TRP concentrations in the brain can either be increased by increasing plasma TRP levels or by decreasing plasma concentrations of LNAA (and therefore increasing the TRP:LNAA ratio). Despite the importance of TRP:LNAA ratio to increase the uptake of TRP into the brain, the competition between TRP and LNAA seems to be less relevant in fish than in mammals as TRP is mostly in the free state in the plasma (and not bound to albumin as in mammals) and therefore directly available for uptake into the brain (Aldegunde et al. 2000). In the brain, the first step in the synthesis of 5-HT is the hydroxylation of tryptophan to 5-hydroxytryptophan (5-HTP) by tryptophan hydroxylase (TPH, reviewed by Boadle-Biber 1993). This rate-limiting enzyme is unsaturated with TRP under normal physiological conditions, and no negative feedback mechanism from the end product of the biosynthesis pathway, that is, 5-HT, is known. The second step in the synthesis of 5-HT is the decarboxylation of 5-HTP to serotonin which is stored in vesicles of the nerve cell until release. Once in the synaptic cleft, 5-HT can bind to different receptors types (Norton et al. 2008) in different parts of the brain, influencing a variety of behaviour and physiological functions.

The work of Lepage et al. (2002, 2003) reporting plasma cortisol levels in non-stressed and stressed fish fed TRP-supplemented diets showed that TRP-supplemented diets increased plasma cortisol levels in non-stressed fish while reducing plasma cortisol levels in stressed fish. The authors suggested that TRP-supplemented diets stimulate the brain 5-HT system which consequently stimulates the hypothalamic–pituitary–inter-renal (HPI) axis. This stimulatory effect of 5-HT on the HPI axis in fish is supported by a variety of studies showing a correlation between serotonergic activity (5-HIAA:5-HT) and plasma cortisol levels (e.g., Winberg and Lepage 1998; Øverli et al. 2001). Furthermore, using cannulated rainbow trout, Winberg et al. (1997) showed that injecting the selective 5-HT_{1A} receptor agonist in the dorsal aorta resulted in elevated plasma cortisol levels. The influence of the serotonergic activation (and TRP administration) on plasma cortisol levels seems to be not only stimulatory but also inhibitory (Höglund et al. 2002). These effects seem to be dependent on which areas of the brain are stimulated by an elevated 5-HT release. Moreover, TRP supplementation during 14 days significantly increased brain 5-HT levels and decreased plasma lactate levels both in unstressed and in stressed fish when compared to fish fed a control diet (Costas et al. 2011a), supporting a modulatory role of TRP on the stress response in fish.

Winberg et al. (2001) were the first to use L-tryptophan-supplemented diets to suppress aggression in rainbow trout, but such effects depend on the treatment duration. Both short supplementation periods (i.e., 3 days) and long supplementation periods seem to fail in improving fish welfare. Lepage et al. (2003) suggested that 28 days of supplementation period may activate compensatory mechanisms such as the activation of catabolising Trp enzyme (tryptophan pyrrolase) in the liver. Inhibition of aggression after TRP supplementation was also demonstrated by Höglund et al. (2005) with Atlantic cod. A reduction in aggression was associated with an increased 5-HT activity, measured as the ratio between the main 5-HT metabolite, that is, 5-hydroxyindoleacetic (5-HIAA) and 5-HT, results that corroborate the findings of Winberg et al. (2001). Dietary TRP supplementation has also been shown to be effective in reducing a more extreme form of aggression, that is, cannibalism in juvenile grouper (Hseu et al. 2003). The possibility to reduce aggression by TRP supplementation is in line with previous observations that chronically increased brain 5-HT activity (such as in subordinate individuals) is linked with a suppression of aggressive behaviour in fish (Summers and Winberg 2006).

Table 2 Summary of studies linking dietary tryptophan (TRP) supplementation with welfare issues in fish

Welfare issue	Fish species	Dietary TRP supplementation		Results	Reference
		Concentration (g/kg dry feed)	Duration (days)		
Aggression	Rainbow trout *Oncorhynchus mykiss*	8.38 and 74.30	3 and 7	↓ Number of aggressive acts (7 days); 0 attack latency	Winberg et al. (2001)
	Atlantic cod *Gadus morhua*	27.6 and 28.1	7	↓ Number aggressive acts; ↑ Latency to first aggressive act	Höglund et al. (2005)
	Grouper *Epinephelus coioides*	2.5, 5.0 and 10	10	↓ Cannibalism	Hseu et al. (2003)
	Fighting fish *Betta splendens*	2.8 and 26.6[a]	7	0 Duration of opercular displays towards a mirror image	Clotfelter et al. (2007)
Stress (cortisol) response	Rainbow trout *Oncorhynchus mykiss*	8.38 and 74.30	3 and 7	0 Plasma cortisol	Winberg et al. (2001)
		0.44, 0.95, 1.84 and 3.57	7	↑ Basal levels of plasma cortisol in non-stressed fish; ↓ Plasma cortisol in stressed fish	Lepage et al. (2002)
		3.57	3, 7 and 28	↓ Plasma cortisol in stressed fish (7 days)	Lepage et al. (2003, 2005)
	European seabass *Dicentrarchus labrax*	5	7	↓ Plasma cortisol in stressed fish	Herrero et al. (2007)
Growth, feed intake and feed efficiency	Rainbow trout *Oncorhynchus mykiss*	8.38 and 74.30	3 and 7	0 Feed intake	Winberg et al. (2001)
		0.44, 0.95, 1.84 and 3.57	7	0 Feed intake	Lepage et al. (2002)
		3.57	3, 7 and 28	0 Feed intake	Lepage et al. (2003, 2005)
		26.9	77	↓ Growth (SGR); ↑ Feed efficiency (FCR); ↑ Feed Intake	Papoutsoglou et al. (2005)
		3.9 and 5.9	98	0 growth and feed intake	Johnston et al. (1990)
	Grouper *Epinephelus coioides*	2.5, 5.0 and 10	10	↓ Growth	Hseu et al. (2003)
	Brown trout *Salmo trutta*	TRP:LNAA of 0.22[b]	7	↑ Feed Intake recovery after stress	Höglund et al. (2007)

LNAA large neutral amino acids

↑ Means increase, ↓ means decrease, 0 means no effect; [a]units in g/kg wet feed, [b] no unit

TRP-supplemented diets have also been claimed to attenuate stress-induced anorexia in brown trout (*Salmo trutta*) (Höglund et al. 2007). These authors suggested a potential link between TRP supplementation and the release of neuropeptides (such as melanocorticortins and corticotrophin-releasing hormone-like peptides, Bernier 2006) that play an important role in mediating the anorectic effects after stress. Despite these results and suggestion for underlying mechanism, the effect of dietary TRP supplementation on feed intake remains contradictory. Winberg et al. (2001) and Lepage et al. (2002, 2003) reported no significant effect on feed intake. However, Hseu et al. (2003) reported lower growth rates for fish fed the TRP-supplemented diets and suggested a reduction in appetite as a consequence of increased 5-HT activity. Also Papoutsoglou et al. (2005) showed that dietary TRP supplementation resulted in significantly lower growth and feed efficiency, despite increased feed intake. An increase in brain 5-HT has also been shown to inhibit feed intake in fish (de Pedro et al. 1998).

The limited number of available studies indicates that dietary TRP supplementation may be used as a tool to decrease aggression and increase stress resistance in fish. Despite these promising results, many questions remain unclear and need further research to fully understand the potential of dietary TRP supplementation as a "therapeutic" agent in fish. Among this research needs, we emphasize the determination of which factors influence the response to TRP supplementation. Factors such as age, gender and social status have been shown to interfere with the response to TRP supplementation in mammals (Henry et al. 1996; Peremans et al. 2002). How far these factors influence the response of fish to TRP supplementation is still unknown. Another topic of relevance for the aquaculture industry is the possibility of using short-term TRP supplementation before slaughter to improve flesh quality in fish. Pre-slaughter conditions have been shown to strongly impair the quality of fillets (e.g., Lefèvre et al. 2008). Henry et al. (1996) showed that TRP supplementation improved muscular pH and reduced the incidence of PSE (Pale Soft and Exudative) meat in pigs. Whether these effects extend to fish species has still to be determined. Finally, considering the cost of TRP-supplemented diets, it is imperative to determine what is the best time window to apply such diets and to which extend must the TRP:LNAA be increased to induce behavioural and physiological responses that enhance fish welfare.

Bioactive peptides and fish welfare

The identification of peptides in feed ingredients with bioactive properties besides their nutritional value as AA donors for metabolism is currently an important research topic in farm animal and human nutrition. Feed-derived bioactive peptides have been defined as protein fragments with specific sequences that have, in addition to their nutritional value, a positive impact on body functions and ultimately influence health (Kitts and Weiler 2003). This concept of feed-derived peptides possessing biological activity brings a completely new dimension to dietary protein quality. Bioactive peptides may bring different health beneficial effects such as antihypertensive, cholesterol-lowering, antithrombotic, opioid, immunomodulating, antimicrobial, and antioxidant activities, and also modulate nutrient digestion and absorption, and enhance mineral absorption and/or bioavailability (Kitts and Weiler 2003; Korhonen and Pihlanto 2006; Hartmann and Meisel 2007; Madureira et al. 2010). Still, most available studies are on humans and farm animals and derive from dairy products, while information in fish is scarce.

Bioactive peptides may be present naturally in the feed (Bauchart et al. 2007) or may be released by proteolysis, during feed processing (i.e., protein hydrolysates) or gastrointestinal digestion. Some bioactive peptides may also be produced by cells of the organism, and such is the case for antimicrobial peptides that are part of the innate immune system of fish (Gómez and Balcázar 2008). Most dietary proteins contain bioactive sequences, but the corresponding peptides must be released by proteolysis if biological activity is to be expressed. Furthermore, even if bioactive peptides are present in natural and processed ingredients, to exert their physiological role in vivo, they must reach their target tissue intact, enduring the action of proteases in the gastrointestinal tract (Madureira et al. 2010). Thereby, most identified bioactive peptides range from two to twenty AA residues (Korhonen and Pihlanto 2006, Hartmann and Meisel 2007), with many of them presenting seven or less residues.

A number of works have been carried out with protein hydrolysates, which suggest a role of bioactive peptides on improved immunocompetence (Duarte et al. 2006). Dietary protein hydrolysates have been shown to influence European seabass larvae resistance to *Vibrio anguillarum* experimental infection, thus showing significantly lower mortality. These hydrolysates, as other oligo-polypeptides, are abundant in live food for larvae, such as *Artemia*, and their composition may help the immune system against pathogen infection. Although the precise mechanisms are not known, other studies have also shown that certain biologically active peptides show immunostimulating and antibacterial properties, for instance a high stimulatory effect on head kidney leucocytes of Atlantic salmon (*Salmo salar*) (Gildberg et al. 1996). Large yellow croaker (*Pseudosciaena crocea*) fed with fish protein hydrolysates significantly increased growth but also improved immune parameters. Thus, lysozyme activity, complement proteins C3 and C4 and IgM significantly increased in fish fed with supplemented diets (Tang et al. 2008). In mice, it has been demonstrated that fish protein hydrolysates may enhance immunocompetence without inducing any inflammatory immune reaction in the intestine. Phagocytic activity of peritoneal macrophages was enhanced, the number of IgA+ cells increased, and IL-4, IL-6 and IL-10 were all significantly increased in the lamina propria of the small intestine of animals that received fish protein hydrolysates. At the same time, some pro-inflammatory cytokines such as IFNγ and TNFα also increased, but the intestinal homoeostasis was maintained and no tissue damage was observed (Duarte et al. 2006). In lambs fed with diets containing fishmeal, an enhancement of some immune responses and higher resistance to infection has been shown (Israf et al. 1996). Bioactive colostral peptides have also been shown to differentially modulate the innate immune response of intestinal epithelial cells (Jørgensen et al. 2010).

The identification of peptides with bioactive properties in protein hydrolysates and the evaluation of bioactive sequences in protein ingredients, which may reproducibly arise from gastric and intestinal digestion, will most likely be an important research focus in fish nutrition in the near future. This may bring important advances in immunocompetence, control of oxidative stress, disease control, and other aspects of welfare of cultured fish. The same holds for an eventual role of bioactive peptides in modulation of nutrient digestion and absorption, in particular in what concerns mineral absorption and/or bioavailability.

Plant protein sources, antinutritional factors and fish welfare

The move towards a lower usage of finite marine-harvested resources in fish feeds is a major trend in the aquaculture industry. It is now consensual that vegetable protein sources are valid ingredients in fish feeds. Over recent years, substantial effort is being invested in exploring the use of plant proteins as substitutes for fishmeal (FM) (Gatlin et al. 2007; Tacon and Metian 2008). Although the existence of some variability between species, fish fed plant-protein-rich diets (up to 75% replacement of FM) tend to show diminished growth and a reduction on feed utilization. This poor growth performance is in most cases related to a reduction on the voluntary feed intake and generally associated with the lower biological value (indispensable AA imbalance, impaired phosphorus availability, presence of antinutritional factors) of the plant-protein sources.

Most studies aiming at developing low-FM diets have targeted an optimal growth performance, high feed efficiency and the enhanced metabolic use of nutrients, without compromising the safety, quality and the health-promoting properties of the final product to the consumer (Sitjà-Bobadilla et al. 2005). Far less studied is the effect of FM replacement by plant-protein sources on the immune status, stress response and overall welfare criteria in fish. Still, the suppression of the non-specific immune capacity by high levels (60–70%) of dietary soybean proteins has been reported in rainbow trout (Burrells et al. 1999). Similarly, 75 and 100% replacement of fishmeal by a mixture of plant-protein sources have been suggested to be immunosuppressive in gilthead seabream (Sitjà-Bobadilla et al. 2005). Moreover, Neji et al. (1993) showed that complete dietary protein replacement by plant sources resulted in higher susceptibility to infection by *Aeromonas salmonicida.* In this regard, the level of replacement appears to be highly relevant. Subhadra et al. (2006) did not found adverse effects on largemouth bass (*Micropterus salmoides*) when fishmeal substitution was only partial (25%). In the work by Bransden et al. (2001), the effects of alternative protein sources (dehulled lupin meal and hydrolysed proteins) did not show significant effects on the immune function when the replacement was up to 40% of the fishmeal protein.

Besides their role as building blocks for protein synthesis and gluconeogenic substrates, AA have also been associated with a multitude of physiological functions (Wu 2009). Plant proteins, commonly used as alternative protein sources for FM replacement, are not well balanced in indispensable AA according to fish requirements (Tacon and Jackson 1985). FM is considered a highly valuable dietary source of indispensable AA to fish, but the indispensable AA composition of plant-protein sources may differ significantly from that of FM (Fig. 3). Consequently no single ingredient can totally replace FM, and sensible blending of different vegetable protein sources is necessary to balance the indispensable AA profile (e.g., Espe et al. 2006).

Several studies show that supplementing taurine to low-FM diets can promote growth and efficiency of feed utilization in rainbow trout (Gaylord et al. 2007), dentex (Chatzifotis et al. 2008) and Japanese flounder (*Paralichthys olivaceus*) (Kim et al. 2003), but not in common carp (*Cyprinus carpio*) (Kim et al. 2008). Taurine levels in plant ingredients are very low. Carnosine is also abundant in animal proteins but absent from plants ingredients. Supplementing carnosine to a FM-free diet failed to enhance growth or feed efficiency in rainbow trout (Snyder et al. 2008); however, given its antioxidant, antiglycator and metal chelator properties, it would be interesting to assess the potential of carnosine and other histidine-derived peptides on stress-associated responses in fish. Glutamine is one of the most abundant free AA in fish plasma and muscle, whereas glutamate and its decarboxylation product (gamma-amino butyric acid, GABA) are important neurotransmitters present at high concentrations in the brain. Glutamine has been proven essential for the synthesis of purine and pyrimidine nucleotides in immune cells, a major energy substrate for leucocytes and a key modulator of cytokine and nitric oxide production (Li et al. 2007). There are also strong evidences associating glutamate and GABA as modulators of altered neurological states and aggressive behaviour in human models (Miczek and Fish 2005). Despite still controversial, glutamate has also been linked to the regulation of the maturation of osteoblasts and osteoclastogenesis, and consequently skeletal deformities in rodents (Lin et al. 2008). Given that plant-proteins present considerable higher levels of glutamate than FM, it would be important to assess its effects on the immune response, skeletal metabolism and behaviour of fish fed FM-free diets. Hydroxyproline content in FM is much higher than that in plants-proteins. In rodents, hydroxyproline

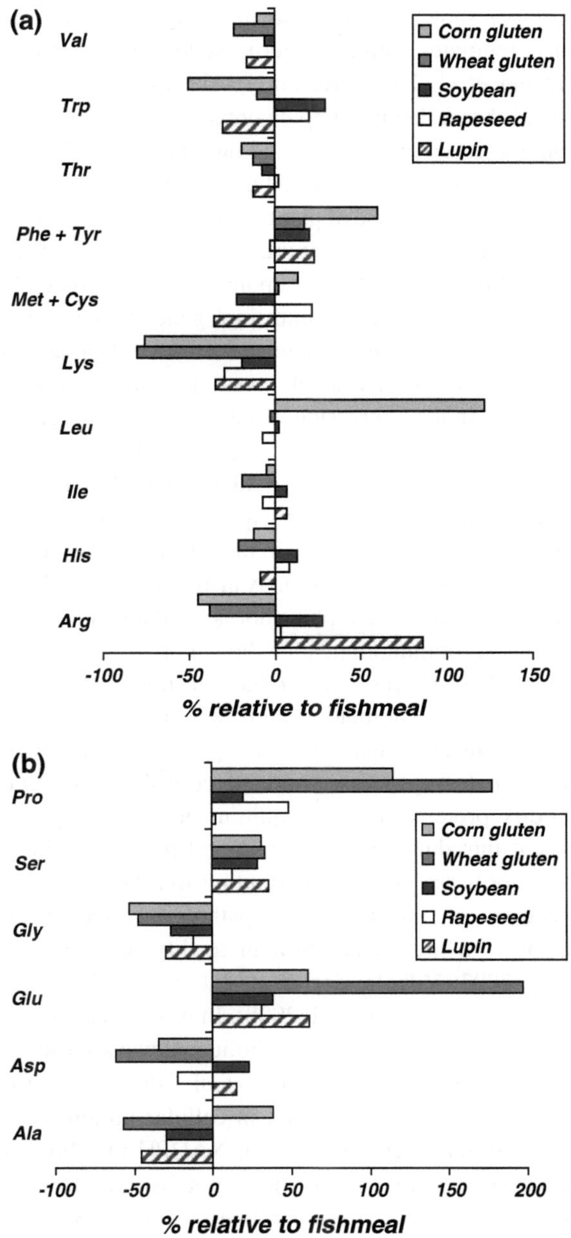

Fig. 3 **a** Indispensable amino acid profile (expressed as % of IAA profile of fishmeal) in selected ingredients commonly used in fish feeds. Data from Sauvant et al. (2004) and Storebakken et al. (2000). **b** Dispensable amino acid profile (expressed as % of DAA profile of fishmeal) in selected ingredients commonly used in fish feeds. Data from Sauvant et al. (2004) and Storebakken et al. (2000)

has been shown to participate in the signalling cascade that regulates cellular redox state and apoptosis (Phang et al. 2008). Recently, Aksnes et al. (2008) found that a dietary supplementation of hydroxyproline increased growth rate and modified bone composition of Atlantic salmon fed high plant-protein diets.

Moreover, the use of plant-protein sources may also condition dietary phosphorus availability. Kirchner et al. (2007) using a 16-K salmonid microarray recently identified a number of intestinal genes that reliably monitor phosphorus deficiency. Thus, a decrease in dietary phosphorus levels was associated with a decrease in the expression of interferon-related genes implicated in the antiviral response (Gig2, VHSV-induced proteins), in particular after 20 days of feeding, that could bear consequences by jeopardizing the defence against viruses in the intestine. Therefore, alternative protein sources may also affect fish welfare indirectly via reduced phosphorus availability.

During immunological challenge or stress, AA are diverted from protein production towards tissues involved in inflammation and immune response. As a consequence, metabolic changes associated with inflammatory processes and infectious diseases can modify protein and AA requirements. Similarly, the recent introduction of genomic and post-transcriptomics tools are greatly contributing to a better understanding of the metabolic pathways affected by changes in the dietary protein quality (e.g., Martin et al. 2003; Vilhelmsson et al. 2004; Sitjà-Bobadilla et al. 2005; Panserat et al. 2008). In brief, such studies highlighted a number of metabolic pathways sensitive to plant-protein feeding in several fish species, for example, pathways involved in cellular protein degradation, fatty acid breakdown, NADPH metabolism and the expression of heat-shock proteins, which could be indicative of a stress/immune-mediated response. Moreover, since most of these studies were performed under dietary indispensable AA balanced conditions, the effects have been solely attributed to the presence of anti-nutritional factors in plant proteins. Studies assessing the specific effects of various AA balances in terms of fish welfare are lacking. Furthermore, current trends in aquaculture feeds involve the replacement of both FM and fish oil. Therefore, the interactions between these two dietary changes need to be assessed.

Besides the inherent sulphur AA and phosphorus deficiencies, which can be readily rectified by a dietary supplementation, the full nutritional potential of plant-protein ingredients is attained only after the removal or inactivation of some antinutritional factors (ANFs—Table 3). Significant advances have been made in the removal or inactivation of ANFs in legume seeds, either by selective plant breeding or through improved extraction and processing techniques (Melcion and van der Poel 1993; Anderson and Wolf 1995), and in particular with the widespread use of extrusion in fish feeds (Romarheim et al. 2005).

Given that low-FM diets are generally achieved by a blend of several plant-protein ingredients, the type and levels of the various ANFs can be diverse. The specific effects of individual ANFs as modulators of fish welfare status are scarcely known. Despite scarce, species-specific and conditioned by fish size, studies have associated the dietary levels of ANFs to: (1) Disturbances on the absorption of nutrients through the inhibition of digestive enzymes by protease inhibitors (Krogdahl et al. 1994; Santigosa et al. 2008), alterations in digesta viscosity and consequently of feed transit time and binding of bile salts in the gut by carbohydrate fractions (Refstie et al. 1999; Leenhouwers et al. 2007), chelating effect of phytic acid forming protein and mineral complexes (Vielma et al. 2004); (2) Alteration in metabolic pathways, namely in relation to protein and AA metabolism (Martin et al. 2003; Panserat et al. 2008), primary energy metabolism (production of NADPH, ATP) (Vilhelmsson et al. 2004), somatotropic axis (Gómez-Requeni et al. 2004); (3) quantitative and qualitative changes in intestinal microflora (Refstie et al. 2006); (4) Inflammatory and histopathological lesions in the gastrointestinal tract (Urán et al. 2008); and (5) activation of non-specific immune response, stress indicators and oxidative status (Rumsey et al. 1994; Sitjà-Bobadilla et al. 2005). The full nature of ANFs effects in fish clearly requires further research.

Conclusions

While welfare is recognized as an important issue in aquaculture, with implications in growth performances, susceptibility to stress and diseases, and public image of the industry, little research has been done in optimizing the nitrogenous fraction of the diet in order to minimize welfare problems.

Table 3 Major antinutritional factors (ANFs) present in raw materials used in fish feeds

ANFs type	Ingredient	Detrimental effects	Elimination
Protease inhibitors	Soybean, rapeseed, sunflower, lupin, peas	Growth reduction, inhibition of proteolytic enzymes	Dry and moist heating
Lectins	Soybean, peas	Growth reduction, changes in gut microflora	Moist heating
Allergens	Soybean	Gastro-intestinal hypersensitivity	Solvent extraction
Goitrogens	Soybean, rapeseed	Growth reduction, thyroid hyperplasia, changes in T3 and T4 levels	Water and solvent extraction, extrusion
Cyanogens	Peas	Growth and feed efficiency reduction	Water extraction Heat treatments
Saponins	Soybean, sunflower, lupin, peas	Growth and feed efficiency reduction, absorption of lipids, cholesterol, vit A and E	Water and solvent extraction
Oestrogen	Soybean, lupin	Growth reduction, induction of vitellogenin secretion	Solvent extraction
Alkaloids	Rapeseed, sunflower, lupin	Growth reduction, reduced feed palatability, liver abnormalities	Water extraction
Phytic acid	Soybean, rapeseed, peas	Reduced mineral (Ca, P and Zn) availability, reduced protein digestibility	Autoclaving, phytase supplementation
Anti-vitamins	Soybean, peas	Reduced vitamin availability	–
Toxic fatty acids	Rapeseed	Growth reduction, mortality	–

Based on reviews by Melcion and van der Poel (1993) and Francis et al. (2001)

The major fate of AA is towards protein synthesis, but emerging evidence shows that many AA regulate key metabolic pathways that are crucial to maintenance, growth, reproduction and immune responses.

Both metabolomic and genomic evidence show that stressful husbandry conditions affect AA metabolism in fish, and under some stress situations, an increase in the requirement of certain indispensable AA may occur. This increase in requirements is probably related with an increased demand of AA for energy and/or the synthesis of proteins and other compounds related with the stress response. These additional requirements still need to be properly quantified, and the cost-effectiveness of supplying AA-supplemented diets in periods of forecastable stress (e.g., grading, capture) evaluated. Supplementation in arginine and BCAA (in particular leucine) may have an important role in enhancing the innate immune system. Lysine, methionine, threonine and glutamine have also been shown to have immune-stimulatory effects in higher vertebrates. However, more work is clearly needed to understand the effects of the particular AA in the mechanisms related to the immune response. In addition, beneficial effects of all these AA on the immune system and disease resistance, including short- vs. long-term effects, still need to be carefully evaluated for most species of commercial importance. The importance of dietary histidine levels on allergic and inflammatory reactions, vasoactive effects on smooth muscle and control of appetite also deserves further study.

Furthermore, some AA assume already a special importance in fish welfare. This is the case of tryptophan as precursor for serotonin, which modulates aggressive behaviour and feed intake in fish. Despite the limited number of available studies, it seems that dietary tryptophan supplementation can be used as a tool to decrease aggression and increase stress resistance in fish. Despite these promising results, many questions remain unclear and need further research to fully understand the potential of dietary tryptophan supplementation as a "therapeutic" agent in fish. How far factors such as age, gender and social status influence the response of fish to tryptophan supplementation is still unknown. The possibility of using short-term tryptophan supplementation before slaughter to improve flesh quality in fish also deserves future study. Finally, considering the cost of tryptophan-supplemented diets, it is imperative to determine what is the best time window to apply such diets and what dosages are needed to induce behavioural and physiological responses that enhance fish welfare.

Stress has been shown to affect not only metabolism of individual AA, but also the rates of protein synthesis and protein turnover. There is evidence that a high protein turnover probably favours a rapid adaptation to stressful conditions including changes in the environment. It will be interesting to verify in future studies to what extent protein intake, including AA balance, affects protein turnover, and to what extent this interacts with stress resistance and fish welfare.

Nutritional stress is a negative physiological state resulting from suboptimal quantity or quality of food available to an animal, and the dietary nitrogen plays here a major role. Fasting in fish elicits dramatic alterations in the metabolic pathways responsible for the release, transport, and utilization of energy reserves, including protein. Transcriptomic studies have demonstrated that the nutritional state of the fish influences the levels of expression of positive and negative growth factors involved in the nutritional stress response, which are closely related to protein accretion and metabolism. These include the GH-IGFs axis hormones, FGF6, myostatin and PepT1.

The identification of bioactive peptides in protein hydrolysates and the evaluation of bioactive sequences in protein ingredients, which may reproducibly arise from gastric and intestinal digestion, will most likely be an important research focus in fish nutrition in the near future. This may bring important advances in immunocompetence, disease control and other aspects of welfare of cultured fish.

Fishmeal replacement by plant protein ingredients has been a growing trend in the fish feed industry. Besides the inherent sulphur AA and phosphorus deficiencies, which can be readily rectified by a dietary supplementation, the replacement of FM by plant proteins undoubtedly implies an altered dietary AA supply and its implications on welfare aspects needs to be further assessed. Additionally, the full nutritional value of plant-protein ingredients is attained only after the removal or inactivation of some antinutritional factors (ANFs). The detrimental effects of several endogenous plant ANFs have been documented, but there is still little information on the physiological and metabolic mechanisms mediating such effects. The full nature of ANFs effects in fish clearly requires further research, if fish welfare is to be ensured.

In short, this review shows that AA metabolism is affected when fish are under stress, and this together with sub-optimal dietary nitrogen formulations may affect fish welfare. Furthermore, improved dietary nitrogen formulations may help fish coping with predictable stressful events.

Acknowledgments This publication benefits from participation of authors in COST action 867 "Welfare of fish in European aquaculture." LC participated in this review in the framework of project HYDRAA-PTDC/MAR/71685/2006, granted by Fundação para a Ciência e a Tecnologia (FCT), Portugal, with the support of FEDER. LT is a member of the XRAq and was founded by SGR2009-554 (Generalitat de Catalunya). C. Aragão, B. Costas and C. Martins acknowledge financial support by Fundação para a Ciência e Tecnologia, Portugal, through grants SFRH/BPD/37197/2007, SFRH/BD/38697/2007 and SFRH/BPD/42015/2007.

References

Ai QH, Mai KS, Zhang L, Tan BP, Zhang WB, Xu W, Li HT (2007) Effects of dietary beta-1, 3 glucan on innate immune response of large yellow croaker, Pseudosciaena crocea. Fish Shellfish Immunol 22:394–402

Aksnes A, Mundheim H, Toppe J, Albrektsen S (2008) The effect of dietary hydroxyproline supplementation on salmon (*Salmo salar* L.) fed high plant protein diets. Aquaculture 275:242–249

Aldegunde M, Garcia J, Soengas JL, Rozas G (1998) Uptake of tryptophan into brain of rainbow trout (*Oncorhynchus mykiss*). J Exp Zool 282:285–289

Aldegunde M, Soengas JL, Rozas G (2000) Acute effects of L-tryptophan hydroxylation rate in brain regions (hypothalamus and medulla) of rainbow trout (*Oncorhynchus mykiss*). J Exp Zool 286:131–135

Alves RN, Cordeiro O, Silva TS, Richard N, de Vareilles M, Marino G, Di Marco P, Rodrigues PM, Conceição LEC (2010) Metabolic molecular indicators of chronic stress in gilthead seabream (*Sparus aurata*) using comparative proteomics. Aquaculture 299:57–66

Anderson RL, Wolf WJ (1995) Compositional changes in trypsin inhibitors, phytic acid, saponins and isoflavones related to soybean processing. J Nutr 125:581S–588S

Anderson WG, Takei Y, Hazon N (2002) Osmotic and volaemic effects on drinking rate in elasmobranch fish. J Exp Biol 205:1115–1122

Aragão C, Corte-Real J, Costas B, Dinis MT, Conceição LEC (2008) Stress response and changes in amino acid requirements in Senegalese sole (*Solea senegalensis* Kaup 1858). Amino Acids 34:143–148

Aragão C, Costas B, Vargas-Chacoff L, Ruiz-Jarabo I, Dinis MT, Mancera JM, Conceição LEC (2010) Changes in plasma amino acid levels in a euryhaline fish exposed to different environmental salinities. Amino Acids 38:311–317

Armand A, Launay T, Pariset C, Della Gaspera B, Charbonnier F, Chanoine C (2003) Injection of FGF6 accelerates regeneration of the soleus muscle in adult mice. Bioch Biophys Acta 1642:97–105

Ashley PJ (2007) Fish welfare: current issues in aquaculture. Appl Anim Behav Sci 104:199–235

Bakke S, Jordal A-EO, Gomez-Requeni P, Verri T, Kousoulaki K, Aksnes A, Rønnestad I (2010) Dietary protein hydrolysates and free amino acids affect the spatial expression of peptide transporter PepT1 in the digestive tract of Atlantic cod (*Gadus morhua*). Comp Biochem Physiol B-Biochem Mol Biol 156:48–55

Bass J, Oldham J, Sharma M, Kambadur R (1999) Growth factors controlling muscle development. Domest Anim Endocrin 17:191–197

Bauchart C, Morzel M, Chambon C, Mirand PP, Reynes C, Buffiere C, Redmond D (2007) Peptides reproducibly released by in vivo digestion of beef meat and trout flesh in pigs. Br J Nutr 98:1187–1195

Bernier NJ (2006) The corticotropin-releasing factor system as a mediator of the appetite-suppressing effects of stress in fish. Gen Comp Endocrinol 146:45–55

Blazer VS (1992) Nutrition and disease resistance in fish. Annu Rev Fish Dis 2:309–323

Boadle-Biber MC (1993) Regulation of serotonin synthesis. Prog Biophys Mol Biol 60:1–15

Bransden MP, Carter CG, Nowak BF (2001) Effects of dietary protein source on growth, immune function, blood chemistry and disease resistance of Atlantic salmon (Salmo salar L.) parr. Anim Sci 73:105–113

Brodeur JC, Peck LS, Johnston LA (2002) Feeding increases MyoD and PCNA expression in myogenic progenitor cells of *Notothenia coriiceps*. J Fish Biol 60:1475–1485

Buentello JA, Gatlin DM (2001) Effects of elevated dietary arginine on resistance of channel catfish to exposure to *Edwardsiella ictaluri*. J Aquat Anim Health 13:194–201

Buentello JA, Reyes-Becerril M, Romero-Geraldo Mde J, Ascencio-Valle Fde J (2007) Effects of dietary arginine on hematological parameters and innate immune function of channel catfish. J Aquat Anim Health 19:195–203

Burrells C, Williams PD, Southgate PJ, Crampton VO (1999) Immunological, physiological and pathological responses of rainbow trout (*Oncorhynchus mykiss*) to increasing dietary concentrations of soybean proteins. Vet Immunol Immunopathol 72:277–288

Chatzifotis S, Polemitou I, Divanach P, Antonopoulou E (2008) Effect of dietary taurine supplementation on growth performance and bile salt activated lipase activity of common dentex, *Dentex dentex*, fed a fish meal/soy protein concentrate-based diet. Aquaculture 275:201–208

Clements KD, Raubenheimer D (2006) Feeding and nutrition. In: Evans DH, Claiborne JB (eds) The Physiology of Fishes, 3rd edn. CRC Press, Boca Raton, pp 47–81

Clotfelter ED, O'Hare EP, McNitt MM, Carpenter RE, Summers CH (2007) Serotonin decreases aggression via 5-HT_{1A} receptors in the fighting fish *Betta splendens*. Pharmacol Biochem Behav 87:222–231

Conceição LEC, Houlihan DF, Verreth JAJ (1997) Fast growth, protein turnover and costs of protein metabolism in yolk-sac larvae of the African catfish *(Clarias gariepinus)*. Fish Physiol Biochem 16:291–302

Conceição LEC, Skjermo J, Skjåk-Bræk G, Verreth JAJ (2001) Effect of an immunostimulating alginate on protein turnover of turbot (*Scophthalmus maximus* L.) larvae. Fish Physiol Biochem 24:207–212

Conceição LEC, Morais SJ, Dinis MT, Rønnestad I (2008) Tracer studies in fish larvae. In: Cyrino JEP, Bureau D, Kapoor BG (eds) Feeding and digestive functions in fishes. Science Publishers, Enfield, New Hampshire, pp 349–392

Costas B, Aragão C, Mancera JM, Dinis MT, Conceição LEC (2008) High stocking density induces crowding stress and affects amino acid metabolism in Senegalese sole *Solea senegalensis* (Kaup 1858) juveniles. Aquacult Res 39:1–9

Costas B, Aragão C, Soengas JL, Míguez JM, Rema P, Dias J, Afonso A, Conceição LEC (2011a) Effects of dietary amino acids and repeated handling on stress response and brain monoaminergic neurotransmitters in Senegalese sole (*Solea senegalensis* Kaup, 1858) juveniles. Comp Biochem Physiol A. doi:10.1016/j.cbpa.2011.08.014

Costas B, Conceição LEC, Aragão C, Martos JA, Ruiz-Jarabo I, Mancera JM, Afonso A (2011b) Physiological responses of Senegalese sole (*Solea senegalensis* Kaup, 1858) after stress challenge: effects on non-specific immune parameters, plasma free amino acids and energy metabolism. Aquaculture 316:68–76

Costas B, Aragão C, Ruiz-Jarabo I, Vargas-Chacoff L, Arjona FJ, Dinis MT, Mancera JM, Conceição LEC (2011c) Feed deprivation in Senegalese sole *(Solea senegalensis*, Kaup 1858) juveniles: effects on blood plasma metabolites and free amino acid levels. Fish Physiol Biochem 37:495–504

Costas B, Aragão C, Ruiz-Jarabo I, Vargas-Chacoff L, Arjona FJ, Mancera JM, Dinis MT, Conceição LEC (2011d) Different environmental temperatures affect amino acid metabolism in the eurytherm teleost Senegalese sole (*Solea senegalensis* Kaup, 1858) as indicated by changes in plasma metabolites. Amino Acids. doi:10.1007/s00726-011-1082-0

Costas B, Conceição LEC, Dias J, Novoa B, Figueras A, Afonso A (2011e). Dietary arginine and repeated handling increase disease resistance and modulate innate immune mechanisms of Senegalese sole (*Solea senegalensis* Kaup, 1858). Fish Shellfish Immunol. doi:10.1016/j.fsi.2011.07.024

Dawkins MS (1990) From an animal's point of view: motivation, fitness and animal welfare. Behav Brain Sci 13:1–161

Dawkins MS (2001) Who needs animal suffering? Anim Welf 10:S19–S31

de Pedro N, Pinillos ML, Valenciano AI, Alonso-Bedate M, Delgado MJ (1998) Inhibitory effect of serotonin on feeding behavior in goldfish: involvement of CRF. Peptides 19:505–511

Duan C (1998) Nutritional and developmental regulation of insulin-like growth factors in fish. J Nutr 128:306S–314S

Duarte J, Vinderola G, Ritz B, Perdigo G, Matar C (2006) Immunomodulating capacity of commercial fish protein hydrolysate for diet supplementation. Immunobiology 211:341–350

Duncan IJH (2005) Science-based assessment of animal welfare: Farm animals. Rev Sci Tech Off Int Epiz 24:483–492

Espe M, Lemme A, Petri A, El-Mowafi A (2006) Can Atlantic salmon grow on diets devoid of fish meal? Aquaculture 255:255–262

Fauconneau B, Paboeuf G (2000) Effect of fasting and refeeding on in vitro muscle cell proliferation in rainbow trout (*Oncorhynchus mykiss*). Cell Tissue Res 3101:459–463

Fernstrom JD, Wurtman RJ (1997) Brain serotonin content: physiological regulation by plasma neutral amino acids (reprinted). Obes Res 5:377–380

Francis G, Makkar HPS, Becker K (2001) Antinutritional factors present in plant-derived alternate fish feed ingredients and their effects in fish. Aquaculture 199:197–227

Gatlin DM, Barrows FT, Brown P, Dabrowski K, Gaylord TG, Hardy RW, Herman E, Hu G, Krogdahl A, Nelson R, Overturf K, Rust M, Sealey W, Skonberg D, Souza EJ, Stone D, Wilson R, Wurtele E (2007) Expanding the utilization of sustainable plant products in aquafeeds: a review. Aquac Res 38:551–579

Gaylord TG, Barrows FT, Teague AM, Johansen KA, Overturf KE, Shephed B (2007) Supplementation of taurine and methionine to all-plant protein diets for rainbow trout (*Oncorhynchus mykiss*). Aquaculture 269:514–524

Gilbert ER, Wong EA, Webb KE Jr (2008) Peptide absorption and utilization: implications for animal nutrition and health. J Anim Sci 86:2135–2155

Gildberg A, Bogwald J, Johansen A, Stenberg E (1996) Isolation of acid peptide fractions from a fish protein hydrolysate with strong stimulatory effect on Atlantic Salmon (*Salmo salar*) head kidney leucocytes. Comp Biochem Physiol B 114:97–101

Gómez GD, Balcázar JL (2008) A review on the interactions between gut microbiota and innate immunity of fish. FEMS Immunol Med Microbiol 52:145–154

Gómez-Requeni P, Mingarro M, Calduch-Giner JA, Médale F, Martin SAM, Houlihan DF, Kaushik S, Pérez-Sánchez J (2004) Protein growth performance, amino acid utilisation and somatotropic axis responsiveness to fish meal replacement by plant protein sources in gilthead sea bream (*Sparus aurata*). Aquaculture 232:493–510

Gouillou-Coustans MF, Fournier V, Métailler R, Vachot C, Desbruyères E, Huelvan C, Moriceau J, Le Delliou H, Kaushik SJ (2002) Dietary arginine degradation is a major pathway in ureagenesis in turbot (*Psetta Maxima*). Comp Biochem Physiol A 132:305–319

Grimble RF, Grimble GK (1998) Immunonutrition: role of sulfur amino acids, related amino acids, and polyamines. Nutrition 14:605–610

Harpaz S (2005) L-carnitine and its attributed functions in fish culture and nutrition—a review. Aquaculture 249:3–21

Hartmann R, Meisel H (2007) Food-derived peptides with biological activity: from research to food applications. Curr Opin Biotechnol 18:163–169. doi:10.1016/j.copbio.2007.01.013

Henry Y, Sève B, Mounier A, Ganier P (1996) Growth performance and brain neurotransmitters in pigs as affected by tryptophan, protein, and sex. J Anim Sci 74:2700–2710

Herrero MJ, Martínez FJ, Míguez JM, Madrid JA (2007) Response of plasma and gastrointestinal melatonin, plasma cortisol and activity rhythms of European sea bass (*Dicentrarchus labrax*) to dietary supplementation with tryptophan and melatonin. J Comp Physiol B 177:319–326

Höglund E, Balm PH, Winberg S (2002) Stimulatory and inhibitory effects of 5-HT1A receptors on adrenocorticotropic hormone and cortisol secretion in a teleost fish, the Arctic charr (*Salvelinus alpinus*). Neurosci Lett 324:193–196

Höglund E, Bakke MJ, Øverli Ø, Winberg S, Nilsson GE (2005) Suppression of aggressive behaviour in juvenile Atlantic cod (*Gadus morhua*), by L-tryptophan supplementation. Aquaculture 249:525–531

Höglund E, Sorensen C, Bakke MJ, Nilsson GE, Øverli Ø (2007) Attenuation of stress-induced anorexia in brown trout (*Salmo trutta*) by pre-treatment with dietary L-tryptophan. Br J Nutr 97:786–789

Houlihan DF (1991) Protein turnover in ectotherms and implications for energetics. In: Gilles R (ed) Advances in comparative and environmental biology 7. Springer, Berlin, pp 1–43

Houlihan DF, Costello MJ, Secombes CJ, Stagg R, Brechin J (1994) Effects of sewage sludge exposure on growth, feeding and protein synthesis of dab *Limanda limanda* L. Mar Environ Res 37:331–353

Hseu JR, Lu FI, Su HM, Wang LS, Tsai CL, Hwang PP (2003) Effect of exogenous tryptophan on cannibalism, survival and growth in juvenile grouper, *Epinephelus coioides*. Aquaculture 218:251–263

Ip YK, Chew SF, Randall DJ (2001) Ammonia toxicity, tolerance, and excretion. In: Wright PA, Anderson AJ (eds) Nitrogen excretion. Fish physiology, vol 20. Academic Press Inc., San Diego, pp 109–148

Israf DA, Coop RL, Stevenson LM, Jones DG, Jackson F, Jackson E, MacKellar A, Huntley JF (1996) Dietary protein influences upon immunity to *Nematodirus battus* infection in lambs. Vet Parasitol 61:273–286

Johnston W, Atkinson JL, Hilton JW, Were KE (1990) Effect of dietary tryptophan on plasma and brain tryptophan, brain serotonin and brain 5-hydroxyindoleacetic acid in rainbow trout. J Nut Biochem 1:49–54

Jørgensen ALW, Juul-Madsen HR, Stagsted J (2010) Colostrum and bioactive, colostral peptides differentially modulate the innate immune response of intestinal epithelial cells. J Pept Sci 16:21–30. doi:10.1002/psc.1190

Kiessling AH, Johansson L, Kiessling JH (1990) Effect of starvation on rainbow trout muscle. Acta Agr Scand 40: 309–324

Kim SK, Takeuchi T, Yokoyama M, Murata Y (2003) Effect of dietary supplementation with taurine, β-alanine, and GABA on the growth of juvenile and fingerling Japanese flounder *Paralichthys olivaceus*. Fish Sci 69:242–248

Kim SK, Matsunari H, Takeuchi T, Yokoyama M, Furuita H, Murata Y, Goto T (2008) Comparison of taurine biosynthesis ability between juveniles of Japanese flounder and common carp. Amino Acids 35:161–168

Kiørboe T, Munk P, Richardson K (1987) Respiration and growth of larval herring *Clupea harengus*: relation between specific dynamic action and growth efficiency. Mar Ecol Prog Ser 40:1–10

Kirchner S, McDaniel NK, Sugiura SH, Soteropoulos P, Tian B, Fletcher J, Ferraris RP (2007) Salmonid microarrays identify intestinal genes that reliably monitor P deficiency in rainbow trout aquaculture. Anim Genet 38:319–331

Kitts DD, Weiler K (2003) Bioactive proteins and peptides from food sources. Applications of bioprocesses used in isolation and recovery. Curr Pharm Des 9:1309–1323

Korhonen H, Pihlanto A (2006) Bioactive peptides: production and functionality. Int Dairy J 16:945–960. doi:10.1016/j.idairyj.2005.10.015

Krogdahl A, Lee TB, Olli JJ (1994) Soybean protease inhibitors affect intestinal trypsin activities and amino acid digestibilities in rainbow trout (*Oncorhynchus mykiss*). Comp Biochem Physiol A 107:215–219

Langar H, Guillaume J, Metailler R, Fauconneau B (1993) Augmentation of protein-synthesis and degradation by poor dietary amino-acid balance in European sea bass (*Dicentrarchus labrax*). J Nutr 123:1754–1761

Leenhouwers JI, Ortega RC, Verreth JAJ, Schrama JW (2007) Digesta characteristics in relation to nutrient digestibility and mineral absorption in Nile tilapia (*Oreochromis niloticus* L.) fed cereal grains of increasing viscosity. Aquaculture 273:556–565

Lefèvre F, Bugeon J, Aupérin B, Aubin J (2008) Rearing oxygen level and slaughter stress effects on rainbow trout flesh quality. Aquaculture 284:81–89

Lepage O, Tottmar O, Winberg S (2002) Elevated dietary intake of L-tryptophan counteracts the stress-induced elevation of plasma cortisol in rainbow trout (*Oncorhynchus mykiss*). J Exp Biol 205:3679–3687

Lepage O, Vílchez IM, Pottinger TG, Winberg S (2003) Time-course of the effect of dietary L-tryptophan on plasma cortisol levels in rainbow trout *Oncorhynchus mykiss*. J Exp Biol 206:3589–3599

Lepage O, Larson ET, Mayer I, Winberg S (2005) Tryptophan affects both gastrointestinal melatonin production and interrenal activity in stressed and nonstressed rainbow trout. J Pineal Res 38:264–271

Leroith D, Bondy C, Yakar S, Liu JL, Butler A (2001) The somatomedin hypothesis. Endocr Rev 22:53–74

Li P, Yin Y, Li D, Kim WK, Wu G (2007) Amino acids and immune function. Br J Nutr 98:237–252

Li P, Mai K, Trushenski J, Wu G (2009) New developments in fish amino acid nutrition: towards functional and environmentally oriented aquafeeds. Amino Acids 37:43–53

Lin TH, Yang RS, Tang CH, Wu MY, Fu WM (2008) Regulation of the maturation of osteoblasts and osteoclastogenesis by glutamate. Eur J Pharmacol 589:37–44

Madureira AR, Tavares T, Gomes AMP, Pintado ME, Malcata FX (2010) Invited review: physiological properties of bioactive peptides obtained from whey proteins. J Dairy Sci 93:437–455

Martin SAM, Vilhelmsson O, Médale F, Watt P, Kaushik S, Houlihan DF (2003) Proteomic sensitivity to dietary manipulations in rainbow trout. Biochim Biophys Acta 1651:17–29

Melcion JP, Van der Poel AFB (1993) Process technology and antinutritional factors: principles, adequacy and process optimization. In: Van der Poel AFB, Huisman J, Saini HS (eds) Recent advances in antinutritional factors in legume seeds. EAAP Publication, Wageningen, The Netherlands, pp 419–434

Miczek KA, Fish EW (2005) Monoamines, GABA, glutamate and aggression. In: Nelson RJ (ed) Biology of aggression. Oxford University Press, Oxford

Milligan CL (1997) The role of cortisol in amino acid mobilization and metabolism following exhaustive exercise in rainbow trout (*Oncorhynchus mykiss* Walbaum). Fish Physiol Biochem 16:1119–1128

Mommsen TP (2001) Paradigms of growth in fish. Comp Biochem Physiol B 129:207–219

Mommsen TP, Vijayan MM, Moon TW (1999) Cortisol in teleosts: dynamics, mechanisms of action, and metabolic regulation. Rev Fish Biol Fish 9:211–268

Mommsen TP, Busby ER, von Schalburg KR, Evans JC, Osachoff HL, Elliott ME (2003) Glutamine synthetase in tilapia gastrointestinal tract: zonation, cDNA and induction by cortisol. J Comp Physiol B 173:419–427

Montero D, Tort L, Izquierdo MS, Robaina L, Vergara JM (1998) Depletion of serum alternative complement pathway activity in gilthead seabream caused by alpha-tocopherol and n-3 HUFA dietary deficiencies. Fish Physiol Biochem 18:399–407

Morales AE, Cardenete G, Abellán E, García-Rejón L (2005) Stress-related physiological responses to handling in common dentex (*Dentex dentex* Linnaeus, 1758). Aquacult Res 36:33–40

Musaro A, McCullagh KJ, Naya FJ, Olson EN, Rosenthal N (1999) IGF-I induces skeletal myocyte hypertrophy through calcineurin in association with GATA-2 and NF-ATcl. Nature 400:81–585

Neji H, Naimi N, Lallier R, de la Noue J (1993) Relationships between feeding, hypoxia, digestibility and experimentally induced furunculosis in rainbow trout. In: Kaushik SJ, Luquet P (eds) Fish nutrition in practice. Institut National de la Recherche Agronomique, Paris, pp 187–197

Norton WH, Folchert A, Bally-Cuif L (2008) Comparative analysis of serotonin receptor (HTR1A/HTR1B families) and transporter (slc6a4a/b) gene expression in the zebrafish brain. J Comp Neurol 511:521–542

Øverli Ø, Pottinger TG, Carrick TR, Øverli E, Winberg S (2001) Brain monoaminergic activity in rainbow trout selected for high and low stress responsiveness. Brain Behav Evol 57:214–224

Panserat S, Kolditz C, Richard N, Plagnes-Juan E, Piumi F, Esquerré D, Médale F, Corraze G, Kaushik S (2008) Hepatic gene expression profiles in juvenile rainbow trout (*Oncorhynchus mykiss*) fed fishmeal or fish oil-free diets. Br J Nutr 28:1–15

Papoutsoglou SE, Karakatsouli N, Chiras G (2005) Dietary L-tryptophan and tank colour effects on growth performance of rainbow trout (*Oncorhynchus mykiss*) juveniles reared in a recirculating water system. Aquac Eng 32:277–284

Peremans K, Audenaert K, Blanckart P, Jacobs F, Coopman F, Verschooten F, Van Bree H, Van Heeringen C, Mertens J, Slegers G, Dierckx R (2002) Effects of aging on brain perfusion and serotonin-2A receptor binding in the normal canine brain measured with single photon emission tomography. Prog Neuropsychopharmacol Biol Psychiatry 26:1393–1404

Phang JM, Donald SP, Pandhare J, Liu Y (2008) The metabolism of proline, as a stress substrate, modulates carcinogenic pathways. Amino Acids 35:681–690

Pickering AD (1998) Stress responses of farmed fish. In: Black K, Pickering A (eds) Biology of farmed fish. Sheffield Academic Press, Sheffield, pp 222–238

Pinto W, Aragão C, Soares F, Dinis MT, Conceição LEC (2007) Growth, stress response and free amino acid levels in

Senegalese sole (*Solea senegalensis* Kaup 1858) chronically exposed to exogenous ammonia. Aquacult Res 38: 1198–1204

Prunet P, Cairns MT, Winberg S, Pottinger TG (2008) Functional genomics of stress responses in fish. Rev Fish Sci 16:157–166

Refstie S, Svihus B, Shearer KD, Storebakken T (1999) Nutrient digestibility in Atlantic salmon and broiler chickens related to viscosity and non-starch polysaccharide content in different soybean products. Anim Feed Sci Technol 79: 331–345

Refstie S, Landsverk T, Bakke-McKellep AM, Ringø E, Sundby A, Shearer CD, Krogdahl Å (2006) Digestive capacity, intestinal morphology, and microflora of 1-year and 2-year old Atlantic cod (*Gadus morhua*) fed standard or bioprocessed soybean meal. Aquaculture 261:269–284

Reyes-Becerril M, Ascencio-Valle F, Tovar-Ramírez D, Meseguer J, Esteban MA (2011) Effects of polyamines on cellular innate immune response and the expression of immune-relevant genes in gilthead seabream leucocytes. Fish Shellfish Immunol 30:248–254

Romarheim OH, Aslaksen MA, Storebakken T, Krogdahl A, Skrede A (2005) Effect of extrusion on trypsin inhibitor activity and nutrient digestibility of diets based on fish meal, soybean meal and white flakes. Arch Anim Nutr 59:365–375

Romeo J, Warnberg J, Gomez-Martinez S, Diaz LE, Marcos A (2008) Neuroimmunomodulation by nutrition in stress situations. Neuroimmunomodulation 15:165–169

Rumsey GL, Siwicki AK, Anderson DP, Bowser PR (1994) Effect of soybean protein on serological response, non-specific defense mechanisms, growth, and protein utilization in rainbow trout. Vet Immunol Immunopathol 41:323–339

Saeij JPJ, van Muiswinkel WB, van de Meent M, Amaral C, Wiegertjes GF (2003) Different capacities of carp leukocytes to encounter nitric oxide-mediated stress: a role for the intracellular reduced glutathione pool. Dev Comp Immunol 27:555–568

Saha N, Dutta S, Bhattacharjee A (2002) Role of amino acid metabolism in an air-breathing catfish, *Clarias batrachus* in response to exposure to a high concentration of exogenous ammonia. Comp Biochem Physiol B 133:235–250

Santigosa E, Sánchez J, Médale F, Kaushik S, Pérez-Sánchez J, Gallardo MA (2008) Modifications of digestive enzymes in trout (Oncorhynchus mykiss) and sea bream (*Sparus aurata*) in response to dietary fish meal replacement by plant protein sources. Aquaculture 282:68–74

Sauvant D, Perez J-M, Tran G (2004) Tables of composition and nutritive value of feed materials. Pigs, poultry, cattle, sheep, goats, rabbits, horses, fish. INRA Editions Versailles

Shamblott MJ, Chen TT (1992) Identification of a second insulin-like growth factor in a fish species. PNAS 89: 8913–8917

Sitjà-Bobadilla A, Peña-Llopis S, Gómez-Requeni P, Médale F, Kaushik S, Pérez-Sánchez J (2005) Effect of fish meal replacement by plant protein sources on non-specific defence mechanisms and oxidative stress in gilthead sea bream (*Sparus aurata*). Aquaculture 249:387–400

Siwicki AK, Morand M, Fuller J Jr, Nissen S, Goryczko K, Ostaszewski P, Kazun K, Glombski E (2003) Influence of feeding the leucine metabolite β-hydroxy-β-methylbutyrate (HMB) on the non-specific cellular and humoral defence mechanisms of rainbow trout (*Oncorhynchus mykiss*). J Appl Ichthyol 19:44–48

Siwicki AK, Zakes Z, Fuller JC Jr, Nissen S, Trapkowska S, Glabski E, Kazun K, Kowalska A, Terech-Majewska E (2005) The effect of feeding the leucine metabolite β-hydroxy-β-methylbutyrate (HMB) on cell-mediated immunity and protection against *Yersinia ruckeri* in pikeperch (*Sander lucioperca*). Aquacult Res 36:16–21

Siwicki AK, Zakes Z, Fuller JC Jr, Nissen S, Kazun K, Glabski E (2006) The influence of β-hydroxy-β-methylbutyrate (HMB) on cell-mediated immunity in tench *Tinca tinca* (*L.*): in vitro and in vivo study. Aquacult Int 14:153–161

Snyder GS, Gaylord TG, Barrows FT, Hardy RW (2008) Carnosine supplementation on an all-plant protein diet for rainbow trout Oncorhynchus mykiss. Book of Abstracts of Aquaculture America 08, February 9–12, Orlando, Florida, USA, p 369

Storebakken T, Shearer KD, Baeverfjord G, Nielsen BG, Åsgård T, Scott T, De Laporte A (2000) Digestibility of macronutrients, energy and amino acids, absorption of elements and absence of intestinal enteritis in Atlantic salmon, *Salmo salar*, fed diets with wheat gluten. Aquaculture 184:115–132

Subhadra B, Lochmann R, Rawles S, Chen R (2006) Effect of dietary lipid source on the growth, tissue composition and hematological parameters of largemouth bass (*Micropterus salmoides*). Aquaculture 255:210–222

Summers CH, Winberg S (2006) Interactions between the neural regulation of stress and aggression. J Exp Biol 209: 4581–4589

Tacon AGJ, Jackson AJ (1985) Utilization of conventional and unconventional protein sources in practical fish feeds. In: Cowey CB, Mackie AMM, Bell JG (eds) Nutrition and feeding in fish. Academic Press, London, pp 119–145

Tacon AGJ, Metian M (2008) Global overview on the use of fish meal and fish oil in industrially compounded aquafeeds: trends and future prospects. Aquaculture 285:146–158

Tang HG, Wu TX, Zhao ZY, Pan XD (2008) Effects of fish protein hydrolysate on growth performance and humoral immune response in large yellow croaker (*Pseudosciaena crocea* R.). J Zhejiang Univ Sci B 9:684–690

Taylor JR, Cooper CA, Mommsen TP (2011) Implications of GI function for gas exchange, acid-base balance and nitrogen metabolism. In: Grosell M, Farrell AP, Brauner CJ (eds) The multifunctional gut of fish. Fish physiology, vol 30. Elsevier Science, San Diego, pp 213–259

Terova G, Bernardini G, Binelli G, Gornati R, Saroglia M (2006) cDNA encoding sequences for myostatin and FGF6 in sea bass (*Dicentrarchus labrax* L.) and the effect of fasting and refeeding on their expression levels. Domest Anim Endocrin 30:304–319

Terova G, Rimoldi S, Chini V, Gornati R, Bernardini G, Saroglia M (2007) Cloning and expression analysis of insulin-like growth factor I and II in liver and muscle of sea bass (*Dicentrarchus labrax* L.) during long-term fasting and refeeding. J Fish Biol 70(SB):219–233

Terova G, Corà S, Verri T, Rimoldi S, Bernardini G, Saroglia M (2009) Impact of feed availability on PepT1 mRNA expression levels in sea bass (*Dicentrarchus labrax*). Aquaculture 294:288–299

Trenzado CE, Carrick TR, Pottinger TG (2003) Divergence of endocrine and metabolic responses to stress in two rainbow trout lines selected for differing cortisol responsiveness to stress. Gen Comp Endocrinol 133:332–340

Urán PA, Gonçalves AA, Taverne-Thiele JJ, Schrama JW, Verreth JAJ, Rombout JHWM (2008) Soybean meal induces intestinal inflammation in common carp (*Cyprinus carpio* L.). Fish Shellfish Immunol 25:751–760

Verburg-van Kemenade BML, Stolte EH, Metz JR, Chadzinska M (2009) Neuroendocrine-immune interactions in teleost fish. In: Bernier NJ, Van Der Kraak G, Farrell AP, Brauner CJ (eds) Fish Neuroendocrinology. Fish physiology, vol 28. Academic Press Inc., San Diego, pp 313–364

Verri T, Terova G, Dabrowski K, Saroglia M (2011) Peptide transport and animal growth: the fish paradigm. Biol Lett 7:597–600

Vielma J, Ruohonen K, Gabaudan J, Vogel K (2004) Top-spraying soybean meal-based diets with phytase improves protein and mineral digestibilities but not lysine utilization in rainbow trout, *Oncorhynchus mykiss* (Walbaum). Aquac Res 35:955–964

Vijayan MM, Pereira C, Gordon Grau E, Iwama GK (1997) Metabolic response associated with confinement stress in tilapia: the role of cortisol. Comp Biochem Physiol C 116:89–95

Vilhelmsson OT, Martin SAM, Médale F, Kaushik SJ, Houlihan DF (2004) Dietary plant-protein substitution affects hepatic metabolism in rainbow trout (*Oncorhynchus mykiss*). Br J Nutr 92:71–80

Wendelaar Bonga SE (1997) The stress response in fish. Physiol Rev 77:591–625

Wilson RW, Wood CM, Houlihan DF (1996) Growth and protein turnover during acclimation to acid and aluminium in juvenile rainbow trout *Oncorhynchus mykiss*. Can J Fish Aquat Sci 53:802–811

Winberg S, Lepage O (1998) Elevation of brain 5-HT activity, POMC expression, and plasma cortisol in socially subordinate rainbow trout. Am J Physiol Regul Integr Comp Physiol 274:R645–R654

Winberg S, Nilsson A, Hylland P, Söderstöm V, Nilsson GE (1997) Serotonin as a regulator of hypothalamic-pituitary-interrenal activity in teleost fish. Neurosci Lett 230: 113–116

Winberg S, Øverli Ø, Lepage O (2001) Suppression of aggression in rainbow trout (*Oncorhynchus mykiss*) by dietary L-tryptophan. J Exp Biol 204:3867–3876

Wiseman S, Osachoff H, Bassett E, Malhotra J, Bruno J, VanAggelen G, Mommsen TP, Vijayan MM (2007) Gene expression pattern in the liver during recovery from an acute stressor in rainbow trout. Comp Biochem Physiol D 2:234–244

Wood AW, Duan CM, Bern HA (2005) Insulin-like growth factor signaling in fish. Int Rev Cytol Surv Cell Biol 243:215–285

Wu G (2009) Amino acids: metabolism, functions, and nutrition. Amino Acids 37:1–17

Wu G, Bazer FW, Davis TA, Jaeger LA, Johnson GA, Kim SW, Knabe DA, Meininger CJ, Spencer TE, Yin YL (2007) Important roles of the arginine family amino acids in swine nutrition and production. Livest Sci 112:8–22

Does Feeding Time Affect Fish Welfare? – Summary

Jose Fernando López-Olmeda, Chris Noble and Javier Sánchez-Vázquez

Meal timing is crucial for fish welfare. Fish fed at random times display higher stress levels when compared with fish fed at scheduled times. Besides classic physiological indicators of stress (plasma cortisol and glucose), fish behaviour and growth performance can be seriously compromised by the feeding protocol. If fish are not able to predict the arrival of the meal, they are forced to increase their activity to be in a constant state of alertness. However, when meal time is fixed, food availability can be predicted. In this case fish display increased activity in anticipation of the forthcoming meal several hours before mealtime, known as food anticipatory activity. Furthermore, feeding strategies should be flexible, ready to meet possible changes in the appetite of fish along the year. For instance, sea bass shifts to nocturnal feeding in winter, and so night meals proved most useful to improve growth and food efficiency during this season. Finally, inappropriate feeding strategies lead to poor welfare because fish compete for food and display aggressive behaviour. As result a detrimental effect upon fin damage occurs.

Background

When do fish feed? One may think fish eat whatever food item they encounter at any time, which is wrong. Actually, food available in the wild is usually restricted to particular times of the day/night, when prey abundance is highest and the risk of predators is lowest. Therefore, it should not be surprising that fish have biological clocks to anticipate those convenient feeding times. Feeding is one of the largest production cost in fish farming and can be one of the biggest stressors for fish if meal timing is not tuned according to fish appetite. This review shows the impact of feeding time upon fish physiology and behaviour, which show circadian rhythms as well.

Scientific Significance

Feeding time influences the stress (both physiological and behavioural) responses of fish, depending on their species-specific daily pattern of feeding behaviour. As a result, a number of validated and operational welfare indicators, which may also show daily rhythms, are modulated by meal timing. Furthermore, this time factor can have an interactive and cumulative effect upon other welfare indicators e.g. the effect of meal timing upon aggression and the additional role aggression plays in increasing injury and mortality.

Practical Application

Feeding is not only the largest production cost in a fish farm, but also one of the biggest stressors for fish. Under farming conditions, fish are challenged with artificial diets and feeding regimes. Inadequate feeding conditions cause stress, alteration of normal behavioural patterns, poor performance and eventually diseases and death, which are by no means acceptable both economically and ethically.

Contact

F. J. Sánchez-Vázquez
Department of Physiology, Faculty of, Biology of Murcia, University
30100 Murcia, Spain
e-mail: javisan@um.es

Fish Physiol Biochem (2012) 38:143–152
DOI 10.1007/s10695-011-9523-y

Does feeding time affect fish welfare?

**J. F. López-Olmeda · C. Noble ·
F. J. Sánchez-Vázquez**

Received: 21 October 2010 / Accepted: 3 June 2011 / Published online: 14 June 2011

Abstract Increased aquaculture production has raised concerns about managing protocols to safeguard the welfare of farmed fish, as consumers demand responsible aquaculture practices to provide 'welfare friendly' products. Feeding is one of the largest production cost in a fish farm and can be one of the biggest stressors for fish. Under farming conditions, fish are challenged with artificial diets and feeding regimes, and inadequate feeding conditions cause stress, alteration of normal behavioural patterns, poor performance and eventually diseases and death, which are by no means acceptable neither economically nor ethically. This review aims to highlight the impact of feeding rhythms and feeding time upon physiological and behavioural welfare indicators, which show circadian rhythms as well. Therefore, all these variables should be considered when designing feeding strategies in farming conditions and assessing the welfare state of cultured fish.

Keywords Feeding schedule · Feeding strategy · Behaviour · Fish welfare · Operational welfare indicators · Biological rhythms

J. F. López-Olmeda · F. J. Sánchez-Vázquez (✉)
Department of Physiology, Faculty of Biology, University of Murcia, 30100 Murcia, Spain
e-mail: javisan@um.es

C. Noble
Nofima Marine, Muninbakken 9-13, P.O. Box 6122, 9291 Tromsø, Norway

Feeding rhythms in fish

A simple observation of fish feeding behaviour will reveal its rhythmic nature: fish do not feed constantly, but feed at certain times of the day or night. This is because living organisms have evolved under a cyclic environment (governed by the Earth's rotation on its axis and around the Sun) and food is not constantly available in the wild but may be restricted to particular times of the day and/or season. In addition, fish must cope with the presence of predators, whose activities can be also restricted to certain times of the day. Therefore, fish have evolved time-keeping mechanisms to anticipate such cyclic, predictable events. Fish are active either during the day or during the night, but rarely throughout a 24 h period. Thus, the diurnal or nocturnal daily pattern of behaviour has been fixed genetically in most species according to their time niche (Daan 1981). Fish, however, exhibit unique rhythmic characteristics because they posses a flexible circadian system that show dual phasing: they can shift from diurnal to nocturnal behaviour, and vice versa, depending on seasonal factors or other factors such as feeding time (Madrid et al. 2001).

In aquaculture, fish are usually challenged with different forms of environmental and social stressors such as crowding, handling, aggression, artificial diets and feeding regimes. These factors compromise fish welfare by inhibiting feed intake and altering feeding behaviour, although the mechanisms that control and affect appetite are complex (Kulckzykowska and

Sánchez-Vázquez 2010). A better knowledge of fish feeding rhythms, however, may provide valuable data to improve feeding schedules that match feeding behaviour and optimise fish welfare.

Feeding rhythms have been reported in a wide variety of fish species. Most of them have been characterised as daily rhythms, but also tidal, lunar and seasonal rhythms have been observed (Reebs 2002). Early research on daily feeding rhythms in fish (Hoar 1942) reported that Atlantic salmon (*Salmo salar*) and brook trout (*Salvelinus fontinalis*) displayed preferences for feeding during the light phase in summer. Some years later, the first self-feeding device was designed for fish and became an important tool for the study of feeding rhythms in fishes (Rozin and Mayer 1961). Interestingly, the classification of daily behavioural patterns in teleost fish can be difficult since individuals from the same species can display a high degree of flexibility in their daily feeding and locomotor activity patterns. These individuals can display either diurnal or nocturnal behaviour, shifting from one type of phasing to the other along their life cycle. This ability is known as dualism and was first reported in Arctic fish species inhabiting high latitudes, which are forced to change their behaviour as they have to cope with extreme changes in the photoperiod between seasons (Eriksson 1978). Later, this dualism was reported in temperate fish species and is ultimately thought to be a common feature amongst many fish species. The mechanisms that drive dualism and allow fish to adapt quickly to environmental changes are unknown, although such flexibility seems to be related to the highly flexible circadian system and entraining mechanisms of the molecular clock of fish (López-Olmeda and Sánchez-Vázquez 2010).

In addition to daily feeding rhythms, seasonal rhythms have also been reported in fish. As ectothermic animals, the food intake of fish is directly influenced by the seasonal variations in water temperature (Higgins and Talbot 1985). The seasonal changes in photoperiod can also decisively influence the phase of feeding rhythms. Jørgensen and Jobling (1989, 1990) reported seasonal phase inversions in the daily feeding patterns of Arctic charr (*Salvelinus alpinus*), a behaviour that was later observed in other species such as European sea bass (*Dicentrarchus labrax*) (Sánchez-Vázquez et al. 1998). This species displayed nocturnal feeding behaviour under natural environmental conditions during winter, while exhibiting diurnal behaviour during the rest of the year (Sánchez-Vázquez et al. 1998). Current research aims to clarify the role of clock genes in the control of feeding rhythms and the dualism in this species (Sánchez et al. 2009b).

Finally, tidal and lunar feeding rhythms may also exist and are often synchronised not only to geophysical cycles but also to biological cycles such as the vertical and horizontal movements of prey items that use tides to migrate and synchronise their reproductive rhythms. The reports of lunar influence on feeding rhythms are scarce to date, with only few studies reporting such rhythms in brown trout (*Salmo trutta*), coho salmon (*Oncorhynchus kisutch*), rainbow trout (*Oncorhynchus mykiss*) and Arctic charr (Farbridge and Leatherland 1987a, b; Dabrowski et al. 1992).

Coping with scheduled-meals: food anticipatory activity

When food is available in a periodic and predictable manner, meals can be predicted by the endogenous clocks of animals. Feeding cycles actually act as a potent synchroniser of many molecular, physiological and behavioural rhythms in animals. When subjected to scheduled-feeding, most animals display an increase in locomotor activity in anticipation of the forthcoming meal, and this can occur several hours before mealtime. This is known as food anticipatory activity (FAA) (Mistlberger 1994) and has been reported in a wide variety of animals, from bees to mammals (Stephan 2002) and also in fish (Sánchez-Vázquez and Madrid 2001; López-Olmeda and Sánchez-Vázquez 2010). To the authors knowledge, the first evidence of FAA in fish was observed in the bluegill (*Lepomis macrochirus*) and largemouth bass (*Micropterus salmoides*), which displayed pre-feeding activity consisting of increases in swimming activity 1–3 h before the scheduled mealtime (Davis 1963). Some species such as goldfish (*Carassius auratus*) and golden shiner (*Notemigonus crysoleucas*) show FAA either during the light or during the dark phase of the LD cycle. Although the occurrence of FAA in these species is independent of the phase of the LD cycle in which food is provided, the lighting conditions influence the behavioural patterns observed. For instance, when animals are fed near dusk or dawn, FAA is usually shortened (Reebs and Lague 2000; López-Olmeda et al. 2009a). Nevertheless, fish fed at a fixed mealtime in the absence of photoperiodic information (constant light or darkness) exhibit FAA, thus indicating that food alone acts as the external signal that synchronises FAA (Aranda et al. 2001). Under these constant lighting conditions, if mealtime is suppressed and fish are fasted for several days, FAA persists with a periodicity close to 24 h, which indicates that this behaviour is generated by an endogenous pacemaker synchronised to feeding time (Sánchez-Vázquez and Madrid 2001; López-Olmeda and Sánchez-Vázquez 2010).

Scheduled-meals can reset the flexible internal clock of fish which in turn may change their daily pattern of behaviour accordingly (i.e. fish may become diurnal when fed during daytime and vice-versa). However, some fish species with more rigid feeding rhythms may not modify their behaviour in response to the feeding schedule. For example, tench (*Tinca tinca*) is a strict nocturnal species which feeds exclusively at night, and refuses to change their nocturnal locomotor activity pattern when food is supplied during daytime (Herrero et al. 2005).

Besides locomotor activity, physiological variables can display anticipation to feeding time. In mammals, there is also an increase in corticosterone secretion, body temperature, intestinal motility, hormones involved in the regulation of appetite such as ghrelin and enzymes involved in digestion and metabolism (Mendoza 2007). In fish, the anticipation of metabolic parameters was studied in goldfish, showing increases in the plasma levels of neuropeptide Y (NPY) and in the enzymatic activity of gut amylase 2 h before the scheduled-feeding (Vera et al. 2007). Recently, the anticipation in the activity of digestive enzymes such as the amylase and alkaline protease has also been observed in the gilthead sea bream (*Sparus aurata*) (Montoya et al. 2010b). Taken together, these data highlight the adaptive advantages of exhibiting anticipation to a periodic meal, preparing the organism and increasing the optimisation and utilisation of nutrients.

Growth performance and meal timing

Fish display feeding rhythms that can vary between days and also between seasons. These rhythmic changes in appetite are of great interest to aquaculturists as they can determine when food is eaten and also when it is wasted. An area that has attracted additional attention is the timing of feed delivery and consumption, as numerous studies have demonstrated that meal timing can improve the growth performance of many commercially cultured species, such as rainbow trout (Reddy et al. 1994; Boujard et al. 1995), European sea bass (Azzaydi et al. 1999) and channel catfish (*Ictalurus punctatus*) (Noeske-Hallin et al. 1985). The number of meals that fish are offered on any given day can also affect the growth of a number of farmed fish including channel catfish (Jarboe and Grant 1997), Atlantic halibut (*Hippoglossus hippoglossus*) (Schnaittacher et al. 2005), rainbow trout (Ruohonen et al. 1998) and tiger puffer (*Takifugu rubripes*) (Kikuchi et al. 2006).

The effect of feeding time on fish growth seems to be tightly related with the daily patterns of behaviour of the species studied. For instance, in the nocturnal African catfish (*Heterobranchus longifilis*), feeding the same amount of food (3% b.w.) during the dark phase increases growth when compared with fish fed the same amount during the light phase (Kerdchuen and Legendre 1991). In other siluriforms considered to be nocturnal, such as the Indian catfish (*Heteropneustes fossilis*) and the African catfish (*Clarias lazera*), a similar positive effect of nocturnal feeding has been reported (Hogendoorn 1981; Sundaraj et al. 1982). On the other hand, diurnal fish such as the rainbow trout display higher growth rates when they are fed at dawn or in the middle of the day (Reddy et al. 1994; Boujard et al. 1995), which coincides with the feeding activity of rainbow trout allowed to self-feed, which show the highest feeding activity at dawn or during the light phase (Sánchez-Vázquez and Tabata 1998).

The seasonal rhythms of feeding behaviour should also be considered when programming the feeding time for fish. As commented above, some fish are able to display changes in their daily feeding habits, changing from diurnal to nocturnal or vice versa (Eriksson 1978). These changes are related to seasonal variations in animals maintained under natural environmental conditions, as has been observed in the Arctic charr, Atlantic salmon or the European sea bass (Jørgensen and Jobling 1989, 1990; Fraser et al. 1995; Sánchez-Vázquez et al. 1998). In the case of sea bass, these fish display diurnal feeding behaviour during the second half of spring, summer and autumn, and shift to nocturnal behaviour in winter, before returning to diurnal behaviour during spring (Sánchez-Vázquez et al. 1998). Therefore, we may hypothesise that feed efficiency at a fixed time of the day would change with season, in accordance with the feeding behaviour during that season. Research by Azzaydi et al. (2000) was performed to elucidate this issue. Sea bass were fed by means of three different feeding strategies: nocturnal automatic-feeding (NF), diurnal automatic-feeding (DF) and free access to self-feeders (SF). The experiment was performed during winter and spring, with sea bass maintained under natural conditions so fish from the SF group display nocturnal self-feeding activity in most of the groups during the first part of the experiment, shifting to diurnal self-feeding activity at the end of winter and at the beginning of spring (Azzaydi et al. 2000). In contrast, NF and DF fish were forced to feed during night or day, respectively, and therefore they only displayed nocturnal or diurnal feeding activity. Feeding strategies affected specific growth rate (SGR) and feed conversion ratio (FCR), which also varied along the time (Fig. 1). In the first part of the study, performed during winter, NF fish showed high SGR and low FCR, in contrast to DF fish. However, in the middle and in the last part of the study, coinciding with the behavioural shift in daily patterns from nocturnal to diurnal, SGR increased in DF and decreased in NF fish, and FCR decreased in DF and increased in NF fish (Fig. 1), demonstrating that food efficiency and growth can be affected by the behavioural patterns of the species, which should be considered when designing the feeding strategies and meal timing.

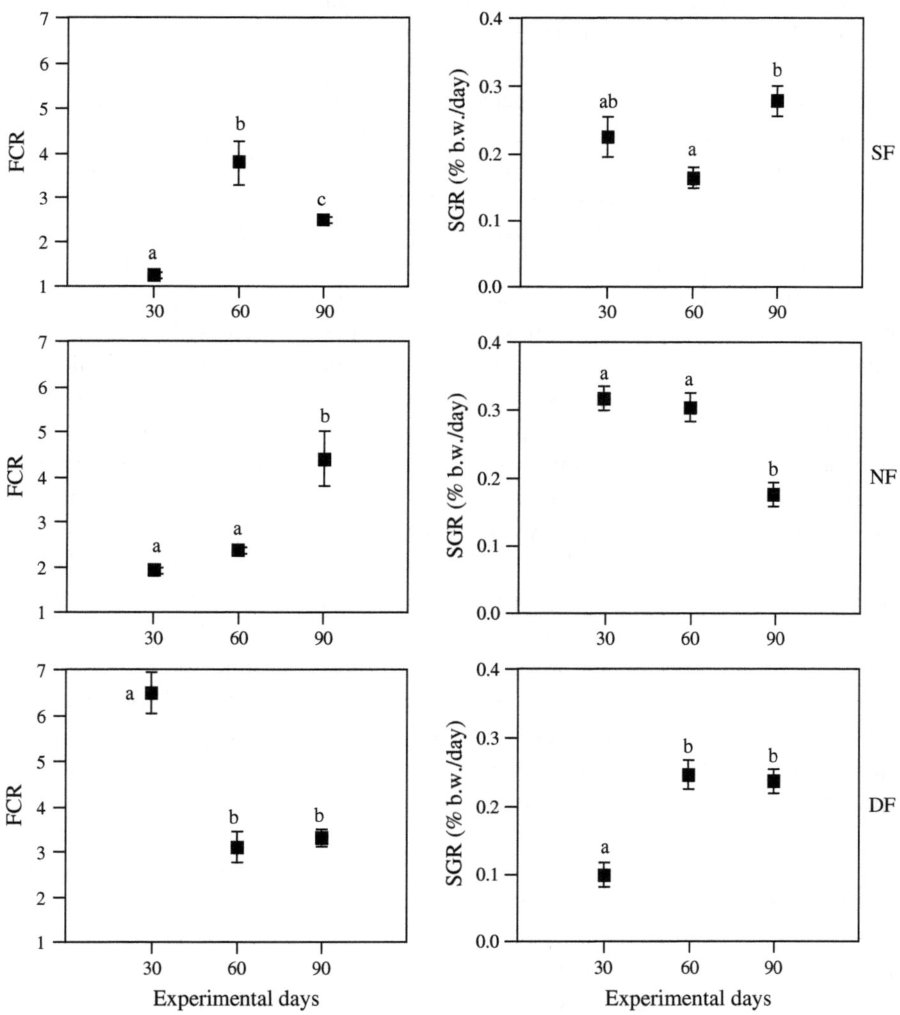

Fig. 1 The evolution of SGR and FCR during three experimental periods in sea bass fed with three feeding regimes: SF, where fish have free access to the self-feeder; NF, nocturnal automatic-feeding; DF, diurnal automatic-feeding. Sea bass were divided into groups of 15 fish each and maintained under natural environmental conditions in 460 l tanks. The experiments began on 15 January, and the sampling times when fish were weighted correspond to 30, 60 and 90 days after the beginning, thus finishing in April. SF fish displayed nocturnal feeding activity during the beginning of the experiment and shifted to diurnal feeding activity between day 30 and day 60, depending on the group. Means with *different letters* indicate statistically significant differences (ANOVA, Tukey HSD, $P < 0.05$, mean ± SEM) (From Azzaydi et al. 2000)

Behavioural responses

The welfare of farmed fish can be assessed using a number of behavioural, physiological and morphological indicators in tandem or in isolation (Huntingford et al. 2006). Other indicators can include increased mortality, reduced appetite and feed intake, and increased susceptibility to disease. Two common behavioural indicators of poor welfare in farmed fish are aggression and swimming activity. Increased aggression can have additional consequences for other welfare indicators as it can lead to increased mortality (Sakakura et al. 1998) and it can increase the incidence of injuries such as fin damage (Turnbull et al. 1996). If individual fish are frequently prevented from feeding by more successful competitors during scramble competition (Gill and Hart 1996), this may have a consequential effect upon their overall feed intake. Meal timing has a documented effect upon aggression in salmonids. When self-fed salmonids receive a single meal vs. either (1) unrestricted access (Brännäs et al. 2005) or (2) unrestricted light phase access (Noble et al. 2007), aggression levels significantly increase, suggesting competition levels increase even though fish are fed to satiation. In addition, a low feeding frequency can increase aggression in salmonids (Brännäs et al. 2005) and also foraging activity in yellowtail flounder (*Limanda ferruginea*) (Dwyer et al. 2002), and it is a potential indicator of increased scramble competition or underfeeding.

Two primary morphological indicators of poor welfare are injury and deformity. Fin damage is increasingly used as an indicator of poor welfare (Ellis et al. 2008) and both meal time and frequency have been implicated as risk factors for its occurrence. When self-fed rainbow trout receive a single daily meal at the start of the light phase, their recovery from historical dorsal fin erosion is impeded in comparison with trout given free light phase access to self-feeders (Noble et al. 2007). On the other hand, when fish receive a fixed satiation ration, *increasing* feeding frequency from 1 to 3 meals per day has a demonstrated detrimental effect upon pectoral fin damage in rainbow trout (Rasmussen et al. 2007). The authors' hypothesised that it was due to increased competition around repeated feeding events, as fish received their daily ration in a number of small meals, rather than a single large one. However, when fish receive their daily satiation ration from self-feeders, *reducing* feeding frequency from 3 to 1 meal per day hinders the recovery from historical dorsal fin erosion in rainbow trout (Noble et al. 2007). To the authors' knowledge, no studies investigated any potential interactions between meal time and the incidence of deformities in fish, and only one study has investigated the effect of feeding frequency upon the prevalence of deformities and reported no effect in Senegalese sole (*Solea senegalensis*) (Engrola et al. 2005).

In summary, when designing feeding strategies, it should be carefully considered that the feeding regime can compromise the welfare of fish by influencing their behaviour, increasing the risk of aggression and the occurrence of fin damage.

Physiological stress responses

From an array of physiological parameters, elevated plasma cortisol levels emerge as a commonly used and robust indicator for poor fish welfare, as it has a well-documented detrimental effect upon fish (Huntingford et al. 2006). As regards cortisol and feeding, the factors from the HPI (hypothalamic-pituitary-interrenal) axis involved in the stress response have been described to act as anorexigenic signals, inhibiting food intake in fish (Bernier and Peter 2001; Volkoff et al. 2005). This anorectic effect is very clear for the corticotropin-releasing factor (CRF) produced in the hypothalamus, which is a precursor of cortisol (Bernier 2006). In the case of cortisol, results from several papers also suggest an anorexigenic effect of this hormone in fish (Bernier and Peter 2001; Bernier et al. 2004).

Recent research has revealed that meal timing can influence plasma cortisol and glucose levels (Sánchez et al. 2009a), with fish exhibiting increased levels of both stress indicators when fed a single meal randomly during the day in comparison with those fish fed a single meal at a fixed time. When sea bream were allowed to self-feed, restricting food-rewards to either diurnal or nocturnal periods meant little differences were observed in both cortisol and glucose levels (López-Olmeda et al. 2009b). In addition, feeding frequency did not affect plasma cortisol levels in Nile tilapia (*Oreochromis niloticus*) (García and Villarroel 2009), but an increase in feeding frequency did affect serum glucose levels in ayu (*Plecoglossus altivelis*) (Nakagawa et al. 1999).

Besides feeding frequency and meal timing, the feeding strategy (scheduled vs. random) is another factor that deserves consideration. In a recent experiment, sea bream fed once a day at a random time (RF) displayed higher levels in several factors used as welfare indicators when compared with animals fed once a day at a fixed time (FF) (Fig. 2) (Sánchez et al. 2009a). In that study, FF fish showed FAA, in contrast to RF fish that displayed sustained activity during the whole light phase and higher levels of total daily

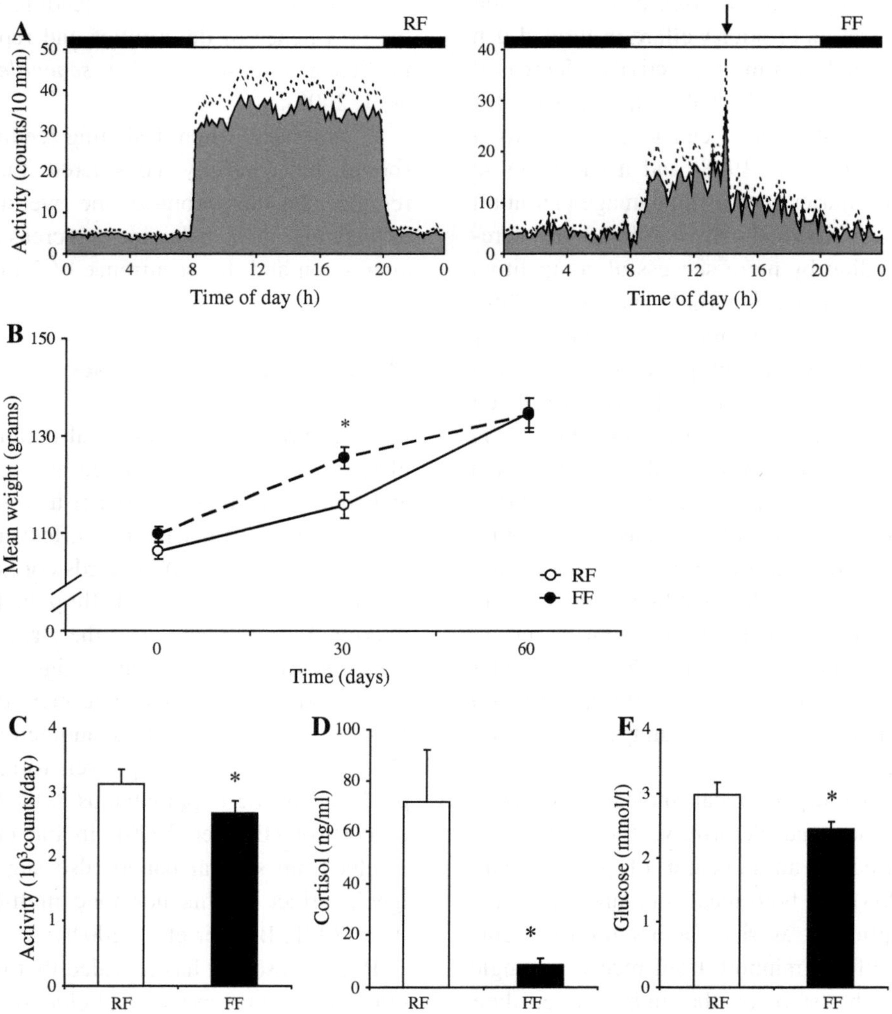

Fig. 2 Influence of the feeding strategy on the daily patterns of locomotor activity (**a**) and several physiological variables: fish performance (**b**), total daily activity (**c**), cortisol (**d**) and glucose (**e**). Gilthead seabream were maintained under two different feeding strategies: fed once a day at a random time (RF) or fed once a day at a fixed time (FF). Daily activity patterns have been represented as mean waveforms (**a**) each point has been calculated as the mean ± SD from 10-min binned data across all the experimental days. The *continuous line* and *grey area* indicate the mean light-beam interruptions and the *dotted line* indicates the SD. *Bars* above meanwaves represent light (*open bars*) and dark (*black bars*) phases of the LD cycle. Mealtime in FF is represented by an *arrow*. Fish performance (**b**) (mean ± SEM, $n = 24$) was evaluated at the beginning of the experiment (day 0), at day 30 and at the end of the experiment (day 60) for both RF (*white circles*) and FF (*black circles*) groups. Total daily activity (**c**) was calculated with data collected from all the experimental days. Cortisol (**d**) and glucose (**e**) (mean ± SEM, $n = 24$) were analysed from samples collected at the end of the experiment (day 60) for both RF (*white bars*) and FF (*black bars*). Blood samples were collected immediately before feeding time for FF and fish from both groups had received their last meal 24 h before sampling. *Asterisks* indicate statistically significant differences between groups (*t* test, $P < 0.05$) (From Sánchez et al. 2009a)

activity than FF (Fig. 2a, c). This finding suggests that the RF group maintained higher activity as they were not able to predict the arrival of the meal and hence were in a constant state of alertness. Fish growth was similar between groups at the end of the experimental phase, although scheduled-fed fish showed higher growth in the middle part of the trial (Fig. 2b). This catch-up growth could be explained as food provided was the same for both groups and no food waste could be observed. However, when humoral indicators of fish welfare were analysed, both cortisol and glucose in RF groups showed higher levels than FF groups (Fig. 2d, e), indicating the higher stress state of the animals fed randomly when compared with the scheduled-fed animals. These results also suggested that fish performance could be misleading when it is used as a welfare indicator and should be interpreted with caution (Sánchez et al. 2009a).

Both cortisol and glucose display daily variations in most vertebrates studied, including many fish species (Dickmeis 2009; López-Olmeda et al. 2009b). Therefore, the time of day should be considered when sampling fish for the evaluation of cortisol and/or glucose to analyse the stress state of the animals. In addition, food can modify both parameters (Small 2005; Polakof et al. 2007) and, in goldfish maintained under constant lighting conditions, feeding time is able to synchronise the rhythms of cortisol secretion (Spieler and Noeske 1984). Previous research in gilthead seabream reported that the cortisol rhythm peaked around the beginning of the light phase, but its phase was shifted by 4 h in animals allowed to self-feed during the night when compared with animals allowed to self-feed during day (López-Olmeda et al. 2009b). Further in an experiment performed on gilthead seabream, food was restricted to a fixed time of the day, with two different scheduled times being evaluated: mid-light (ML) and mid-dark (MD) (Fig. 3) (Montoya et al. 2010a). Under these conditions, cortisol levels exhibited daily variations in both groups, but the shift in the phase was almost 12 h, with the peak of cortisol in the animals fed at ML being found during the light phase, and the peak of the animals fed at MD being found during the dark phase (Fig. 3). In both cases, the elevation in cortisol levels was found after mealtime. These data highlight the role of food and mealtime as synchroniser and modulator of plasma cortisol that, as stated earlier, is one of the most commonly used welfare indicators. In addition, a third

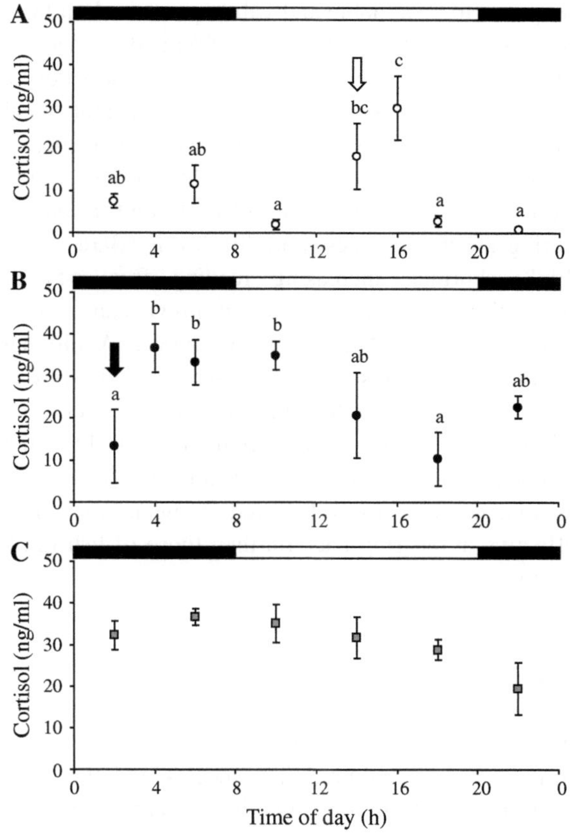

Fig. 3 Daily rhythms of cortisol in gilthead seabream fed three different strategies: once a day in the middle of the light phase (**a**), once a day in the middle of the dark phase (**b**) and once a day at a random time (**c**). *White* and *black bars* above the graphs represent the light and dark phases of the LD cycle, respectively. *White* and *black arrows* represent feeding time for fish fed at mid-light phase and at mid-dark phase, respectively. When sampling coincided with feeding time values represent the mean ± SEM (n = 5/time point). *Different letters* indicate statistically significant differences between time points within the same group (ANOVA, $P < 0.05$) (From Montoya et al. 2010a and unpublished data)

group of fish was fed once a day at a random time. In this group, cortisol levels remained elevated compared with the scheduled-fed animals, in accordance with that was previously reported by Sánchez et al. (2009a). Moreover, cortisol rhythms were attenuated in this group, displaying lower amplitude (Fig. 3c).

Other operational welfare indicators

Other indicators can also be affected by meal timing and feeding frequency. Changes in mortality are proposed to be a viable and valid operational on farm welfare indicator (North et al. 2008) and farmers do their upmost to reduce its intensity and prevalence. However, some studies have reported that meal timing and feeding frequency can have a detrimental effect upon mortality. For example, feeding vundu catfish (*Heterobranchus longifilis*) at night instead of during the day reduces mortality levels (Baras et al. 1998). Further, increasing feeding frequency in Australian snapper (*Pagrus auratus*) can reduce mortality rates (Tucker et al. 2006). Additional welfare indicators such as disease susceptibility can also be affected by feeding frequency. García and Villarroel (2009) have reported that increased feeding frequency improves disease resistance in Nile tilapia, as macrophages from fish fed 2 meals were less effective at clearing bacteria than those of fish fed 8 meals per day.

Conclusions

In summary, it is difficult to precise a specific feeding strategy for all cultured fish species and the most appropriate seems to be studying different strategies and evaluating a set of parameters and welfare indicators to select the optimal. Amongst these strategies, in general, the self-feeding system emerges as a good option as seems to lower the levels of some welfare indicators and allows fish to choose their feeding time and food ratio. When dealing with automatic-feeding times, fixed times are more recommended than random feeding times. In this case, the effect of meal frequency on fish welfare seems to be species-dependent, and therefore, this should be investigated before applying the feeding protocol. In addition, when imposing a scheduled meal timing, this should be located preferentially during the active feeding phase of the species (diurnal or nocturnal). In the case of dual species that display seasonal changes in their feeding patterns, such as European sea bass or Atlantic salmon, the season of the year should also be considered for selecting the time of day in which fish will be fed.

On the other hand, since fish exhibit distinct feeding rhythms, feeding time influences the stress (both physiological and behavioural) responses of fish, depending on the species-specific daily pattern of feeding behaviour. As a result, a number of validated and operational welfare indicators, which may also show daily rhythms, are modulated by meal timing. Furthermore, this time factor can have an interactive and additive effect upon other welfare indicators, e.g., the effect of meal timing upon aggression and the additional role aggression plays in increasing injury and mortality. Therefore, when assessing the welfare state of the animals, it is strongly recommended (1) to analyse several operational welfare indicators and (2) to consider meal timing and the time of day when such indicators are being measured.

Acknowledgments This paper was partially funded by national (MCINN) 'Aquagenomics' and 'Circasole', and regional (SENECA) 'Circabass' projects to FJSV. Part of the results and discussions were presented at the COST ACTION 867 'Welfish' at the Krakow and Iceland meetings.

References

Aranda A, Madrid JA, Sánchez-Vázquez FJ (2001) Influence of light on feeding anticipatory activity in goldfish. J Biol Rhythms 16:50–57

Azzaydi M, Martínez FJ, Zamora S, Sánchez-Vázquez FJ, Madrid JA (1999) Effect of meal size modulation on growth performance and feeding rhythms in European sea bass (*Dicentrarchus labrax*, L.). Aquaculture 170:253–266

Azzaydi M, Martínez FJ, Zamora S, Sánchez-Vázquez FJ, Madrid JA (2000) The influence of nocturnal vs. diurnal feeding under winter conditions on growth and feed conversion of European sea bass (*Dicentrarchus labrax*, L.). Aquaculture 182:329–338

Baras E, Tissier F, Westerloppe L, Mélard C, Philippart JC (1998) Feeding in darkness alleviates density-dependent growth of juvenile vundu catfish *Heterobranchus longifilis* (Clariidae). Aquat Living Resour 11:335–340

Bernier NJ (2006) The corticotropin-releasing factor system as a mediator of the appetite-suppressing effects of stress in fish. Gen Comp Endocrinol 146:45–55

Bernier NJ, Peter RE (2001) The hypothalamic-pituitary-interrenal axis and the control of food intake in teleost fish. Comp Biochem Physiol B 129:639–644

Bernier NJ, Bedard N, Peter RE (2004) Effects of cortisol on food intake, growth, and forebrain neuropeptide Y and corticotropin-releasing factor gene expression in goldfish. Gen Comp Endocrinol 135:230–240

Boujard T, Gélineau A, Corraze G (1995) Time of a single daily meal influences growth performance in rainbow trout (*Oncorhynchus mykiss*). Aquac Res 26:341–349

Brännäs E, Berglund U, Eriksson L-O (2005) Time learning and anticipatory activity in groups of Arctic charr. Ethology 111:681–692

Daan S (1981) Adaptive daily strategies in behavior. In: Aschoff J (ed) Handbook of behavioral neurobiology 4: biological rhythms. Plenum Press, New York, pp 275–298

Dabrowski K, Krumschnabel G, Paukku M, Labanowski J (1992) Cyclic growth and activity of pancreatic enzymes in alevins of Arctic charr (*Salvelinus alpinus* L.). J Fish Biol 40:511–521

Davis RE (1963) Daily 'predawn' peak of locomotion in fish. Anim Behav 12:272–283

Dickmeis T (2009) Glucocorticoids and the circadian clock. J Endocrinol 200:3–22

Dwyer KS, Brown JA, Parrish C, Lall SP (2002) Feeding frequency affects food consumption, feeding pattern and growth of juvenile yellowtail flounder (*Limanda ferruginea*). Aquaculture 213:279–292

Ellis T, Oidtmann B, St-Hilaire S, Turnbull J, North B, MacIntyre C, Nikolaidis J, Hoyle I, Kestin S, Knowles T (2008) Fin erosion in farmed fish. In: Branson E (ed) Fish welfare. Blackwell, Oxford, pp 121–149

Engrola S, Conceição LEC, Gavaia PJ, Cancela ML, Dinis MT (2005) Effects of pre-weaning feeding frequency on growth, survival, and deformation of Senegalese sole, *Solea senegalensis* (Kaup, 1858). Isr J Aquac Bamidgeh 57:10–18

Eriksson LO (1978) Nocturnalism versus diurnalism–dualism within individuals. In: Thorpe JE (ed) Rhythmic activity of fishes. Academic Press, London, pp 69–89

Farbridge KF, Leatherland JF (1987a) Lunar cycles of coho salmon, *Oncorhynchus kisutch*. I. Growth and feeding. J Exp Biol 128:165–178

Farbridge KF, Leatherland JF (1987b) Lunar periodicity of growth cycles in rainbow trout, *Salmo gairdneri* Richardson. J Interdiscip Cycle Res 18:169–177

Fraser NHC, Heggenes J, Metcalfe NB, Thorpe JE (1995) Low summer temperatures cause juvenile Atlantic salmon to become nocturnal. Can J Zool 73:446–451

García JA, Villarroel M (2009) Effect of feed type and feeding frequency on macrophage functions in tilapia (*Oreochromis niloticus* L.). Fish Shellfish Immunol 27:325–329

Gill AB, Hart PJB (1996) Unequal competition between three-spined stickleback, *Gasterosteus aculeatus*, L. encountering sequential prey. Anim Behav 51:689–698

Herrero MJ, Pascual M, Madrid JA, Sánchez-Vázquez FJ (2005) Demand-feeding rhythms and feeding-entrainment of locomotor activity rhythms in tench (*Tinca tinca*). Physiol Behav 84:595–605

Higgins PJ, Talbot C (1985) Growth and feeding in juvenile Atlantic salmon (*Salmo salar*, L.). In: Cowey CB, Mackie AM, Bell JG (eds) Nutrition and feeding in fish. Academic Press, London, pp 243–263

Hoar WS (1942) Diurnal variations in feeding activity of young salmon and trout. J Fish Res Board Can 34:1655–1669

Hogendoorn H (1981) Controlled propagation of the African catfish, *Clarias lazera* (C. & V.). IV. Effect of feeding regime on fingerling culture. Aquaculture 24:123–131

Huntingford FA, Adams C, Braithwaite VA, Kadri S, Pottinger TG, Sandøe P, Turnbull JF (2006) Current understanding on fish welfare: a broad overview. J Fish Biol 68:332–372

Jarboe HH, Grant WJ (1997) The influence of feeding time and frequency on the growth, survival, feed conversion, and body composition of channel catfish, *Ictalurus punctatus*, cultured in a three-tier, closed, recirculating raceway system. J Appl Aquac 7:43–52

Jørgensen EH, Jobling M (1989) Patterns of food intake in Arctic charr, *Salvelinus alpinus*, monitored by radiography. Aquaculture 81:155–160

Jørgensen EH, Jobling M (1990) Feeding modes in Arctic charr, *Salvelinus alpinus* L.: the importance of bottom feeding for the maintenance and growth. Aquaculture 86:379–385

Kerdchuen N, Legendre M (1991) Influence de la fréquence et de la période de nourrissage sur la croissance et l'efficacité alimentaire d'un silure africain, *Heterobranchus longifilis*. Aquat Living Resour 4:241–248

Kikuchi K, Iwata N, Kawabata T, Yanagawa T (2006) Effect of feeding frequency, water temperature, and stocking density on the growth of tiger puffer, *Takifugu rubripes*. J World Aquac Soc 37:12–20

Kulckzykowska E, Sánchez-Vázquez, FJ (2010) Neurohormonal regulation of feed intake and response to nutrients in fish: aspects of feeding rhythm and stress. Aquac Res 41:654–667. doi:10.1111/j.1365-2109.2009.02350.x

López-Olmeda JF, Sánchez-Vázquez FJ (2010) Feeding rhythms in fish: from behavioural to molecular approach. In: Kulczykowska E (ed) Biological clock in fish. Science Publishers, Enfield, pp 155–184

López-Olmeda JF, Egea-Álvarez M, Sánchez-Vázquez FJ (2009a) Glucose tolerance in fish: is the daily feeding time important? Physiol Behav 96:631–636

López-Olmeda JF, Montoya A, Oliveira C, Sánchez-Vázquez FJ (2009b) Synchronization to light and restricted-feeding schedules of behavioural and humoral daily rhythms in gilthead sea bream (*Sparus aurata*). Chronobiol Int 26:1389–1408

Madrid JA, Boujard T, Sánchez-Vázquez FJ (2001) Feeding rhythms. In: Houlihan DF, Boujard T, Jobling M (eds) Food intake in fish. Blackwell Science, Oxford, pp 189–215

Mendoza J (2007) Circadian clocks: setting time by food. J Neuroendocrinol 19:127–137

Mistlberger R (1994) Circadian food-anticipatory activity: formal models and physiological mechanisms. Neurosci Biobehav R 18:171–195

Montoya A, López-Olmeda JF, Garayzar ABS, Sánchez-Vázquez FJ (2010a) Synchronization of daily rhythms of locomotor activity and plasma glucose, cortisol and thyroid hormones to feeding in Gilthead seabream (*Sparus aurata*) under a light-dark cycle. Physiol Behav 101:101–107

Montoya A, López-Olmeda JF, Yúfera M, Sánchez-Muros MJ, Sánchez-Vázquez FJ (2010b) Feeding time synchronises daily rhythms of behaviour and digestive physiology in gilthead seabream (*Sparus aurata*). Aquaculture 306:315–321

Nakagawa H, Umino T, Mizogami H (1999) Effect of feeding regime on biological and biochemical parameters in Ayu. Fisheries Sci 65:259–263

Noble C, Mizusawa K, Suzuki K, Tabata M (2007) The effect of differing self-feeding regimes on the growth, behaviour and fin damage of rainbow trout held in groups. Aquaculture 264:214–222

Noeske-Hallin TA, Spieler RE, Parker NC, Suttle MA (1985) Feeding time differentially affects fattening and growth of channel catfish. J Nutr 115:1228–1232

North BP, Ellis T, Knowles T, Bron JE, Turnbull JF (2008) The use of stakeholder focus groups to identify indicators for the on-farm assessment of trout welfare. In: Branson E (ed) Fish welfare. Blackwell, Oxford, pp 243–267

Polakof S, Ceinos RM, Fernández-Durán B, Míguez JM, Soengas JL (2007) Daily changes in parameters of energy metabolism in brain of rainbow trout: dependence of feeding. Comp Biochem Physiol A 146:265–273

Rasmussen RS, Larsen FH, Jensen S (2007) Fin condition and growth among rainbow trout reared at different sizes, densities and feeding frequencies in high-temperature re-circulated water. Aquacult Int 15:97–107

Reddy PK, Leatherland JF, Khan MN, Boujard T (1994) Effect of daily meal time on the growth of rainbow trout fed different ration levels. Aquacult Int 2:1–15

Reebs SG (2002) Plasticity of diel and circadian activity rhythms in fishes. Rev Fish Biol Fish 12:349–371

Reebs SG, Lague M (2000) Daily food-anticipatory activity in golden shiners: a test of endogenous timing mechanisms. Physiol Behav 70:35–43

Rozin P, Mayer J (1961) Regulation of food intake in the goldfish. Am J Physiol 201:968–974

Ruohonen K, Vielma J, Grove DJ (1998) Effects of feeding frequency on growth and food utilisation of rainbow trout (*Oncorhynchus mykiss*) fed low-fat herring or dry pellets. Aquaculture 165:111–121

Sakakura Y, Tagawa M, Tsukamoto K (1998) Whole-body cortisol concentrations and ontogeny of aggressive behaviour in yellowtail (*Seriola quinqueradiata* Temminck and Schlegel: Carangidae). Gen Comp Endocrinol 109:286–292

Sánchez JA, López-Olmeda JF, Blanco-Vives B, Sánchez-Vázquez FJ (2009a) Effects of feeding schedule on locomotor activity rhythms and stress response in sea bream. Physiol Behav 98:125–129

Sánchez JA, Madrid JA, Sánchez-Vázquez FJ (2009b) Molecular cloning, tissue distribution and daily rhythms of expresión of *per1* gene in European sea bass (*Dicentrarchus labrax*). Chronobiol Int 26:1120–1135

Sánchez-Vázquez FJ, Madrid JA (2001) Feeding anticipatory activity in fish. In: Houlihan DF, Boujard T, Jobling M (eds) Food intake in fish. Blackwell Science, Oxford, pp 216–232

Sánchez-Vázquez FJ, Tabata M (1998) Circadian rhythms of demand-feeding and locomotor activity in rainbow trout. J Fish Biol 52:255–267

Sánchez-Vázquez FJ, Azzaydi M, Martínez FJ, Zamora S, Madrid JA (1998) Annual rhythms of demand-feeding activity in sea bass: evidence of a seasonal phase inversion of the diel feeding pattern. Chronobiol Int 15:607–622

Schnaittacher G, King W, Berlinsky DL (2005) The effects of feeding frequency on growth of juvenile Atlantic halibut, *Hippoglossus hippoglossus* L. Aquacult Res 36:370–377

Small BC (2005) Effect of fasting on nychthemeral concentrations of plasma growth hormone (GH), insulin-like growth factor I (IGF-I), and cortisol in channel catfish (*Ictalurus punctatus*). Comp Biochem Physiol B 142:217–223

Spieler RE, Noeske TA (1984) Effects of photoperiod and feeding schedule on diel variations of locomotor activity, cortisol, and thyroxine in goldfish. Trans Am Fish Soc 113:528–539

Stephan FK (2002) The "other" circadian system: food as a *zeitgeber*. J Biol Rhythms 17:284–292

Sundaraj BI, Nath P, Halberg F (1982) Circadian meal timing in relation to lighting schedule optimizes catfish body weight gain. J Nutr 112:1085–1097

Tucker BJ, Booth MA, Allan GL, Booth D, Fielder DS (2006) Effects of photoperiod and feeding frequency on performance of newly weaned Australian snapper *Pagrus auratus*. Aquaculture 258:514–520

Turnbull JF, Richards RH, Robertson DA (1996) Gross, histological and scanning electron-microscopic appearance of dorsal fin rot in farmed Atlantic salmon, *Salmo salar* L. parr. J Fish Dis 19:415–427

Vera LM, De Pedro N, Gómez-Milán E, Delgado MJ, Sánchez-Muros MJ, Madrid JA, Sánchez-Vázquez FJ (2007) Feeding entrainment of locomotor activity rhythms, digestive enzymes and neuroendocrine factors in goldfish. Physiol Behav 90:518–524

Volkoff H, Canosa LF, Unniappan S, Cerdá-Reverter JM, Bernier NJ, Kelly SP, Peter RE (2005) Neuropeptides and the control of food intake in fish. Gen Comp Endocrinol 142:3–19

Assessment of Welfare in Fish

Safeguarding the Welfare of Farmed Fish at Harvest – Summary

Jeff Lines and Jade Spence

Food withdrawal prior to harvest is necessary to limit the build-up of ammonia and carbon dioxide in the water and to ensure a long shelf life for the product. However, traditional fasting periods are longer than is necessary to empty the gut. There is little evidence that this extended period provides any quality benefit. Following fasting, fish are crowded for collection. This should take place slowly and progressively. Oxygen should be monitored and situations which result in vigorous activity avoided. Poor crowding increases the risk of exhausting the fish. This compromises both welfare and product quality. Removal of fish from the water using a dry brail should be avoided. Whenever possible, wet brailing is the best option. Pumping can be used, but it requires high density crowding. Pumped fish are frequently transported for some distance through pipelines. Their welfare can be considered compromised if they are unable to swim normally following this transit. Poorly designed and built pipelines can cause injuries. Only three methods are recommended for slaughtering of fish destined for human consumption: percussive stunning, electric stunning or euginol; however the last option is not widely permitted by national authorities. Electric and percussive stunning result in high standards of welfare if properly carried out. If used inappropriately (e.g. insufficient or poorly targeted current or percussive force), the outcome is likely to be very poor welfare. For humane stunning, it is necessary to achieve immediate and prolonged insensibility. Electric stunning must be followed by killing while fish are insensible. Appropriate, practical methods include bleeding and/or chilling in ice. The following killing methods are still routinely used for fish without prior stunning and none are considered humane: asphyxia in air, immersion in ice slurry, use of carbon dioxide, salt or ammonia baths, bleeding or decapitation.

Background

At harvest, fish welfare is frequently compromised by poor choice of handling and killing methods and unnecessary adherence to farming traditions. Farmed fish experience frequent, intense handling operations at this time. Particular care must therefore be taken by the stockmen to avoid causing distress or suffering.

Scientific Significance

Carcass quality cannot be used as the sole indicator of welfare at slaughter. There remains a need for greater understanding of fish welfare requirements, particularly their behavioural and physiological responses to fasting and handling.

Practical Application

We foresee a wider adoption of current recommendations for fish welfare at harvest. We recommend that industry adapts traditional practices in order to continually improve welfare and to deliver a product considered acceptable by the consumer.

Contact

J. A. Lines
Silsoe Livestock Systems, Wrest Park, Silsoe, Bedford
MK45 4HS, UK
e-mail: Jeff.Lines@silsoeresearch.org.uk

Fish Physiol Biochem (2012) 38:153–162
DOI 10.1007/s10695-011-9561-5

Safeguarding the welfare of farmed fish at harvest

J. A. Lines · J. Spence

Received: 11 August 2010 / Accepted: 21 September 2011 / Published online: 12 October 2011

Abstract Fish welfare at harvest is easily compromised by poor choice of handling and slaughter methods, lack of attention to detail and by unnecessary adherence to fish farming traditions. The harvest process comprises fasting the fish to empty the gut, crowding the fish, gathering and moving the fish using brails, fish pumps, and sometimes also road or boat transport and finally stunning and killing the fish. The harvesting processes commonly used for bass, bream, carp, catfish, cod, eel, halibut, pangasius, salmon, tilapia, trout, tuna and turbot are outlined. These harvesting processes are discussed; the consequences for fish welfare identified and practical tests which can be made at the harvest site highlighted. Welfare at harvest for the majority of farmed fish species can be improved by adopting and adapting existing procedures already known to be beneficial for fish welfare through their use in other fish farming systems or with other species. It is seldom necessary to develop completely new concepts or methods.

J. A. Lines (✉)
Silsoe Livestock Systems, Wrest Park, Silsoe, Bedford MK45 4HS, UK
e-mail: Jeff.Lines@silsoeresearch.org.uk

J. Spence
The Humane Slaughter Association, The Old School, Brewhouse Hill, Wheathampstead, Hertfordshire AL4 8AN, UK

Keywords Fish · Welfare · Slaughter · Stunning · Harvest · Transport

Introduction

During harvesting, farmed fish are subjected to a unique period of frequent and intense handling operations. The focus of the farm at this time is to avoid financial loss due to premature death, loss or visible injury. The longer term concerns about food conversion ratio, growth rate and disease resistance, which inform handling decisions at earlier stages are of little relevance at this point. This change in priorities could mean that aspects of fish welfare, which were previously protected for financial reasons, are downgraded. Particular attention therefore needs to be given to protecting the welfare of fish at harvest.

A general approach to fish welfare and a discussion of stress, fear and pain in fish is to be found in EFSA (2009a). Specific detailed issues related to welfare hazards at slaughter of salmon, trout, sea bass, sea bream, carp, eel, turbot and tuna are considered in detail and evaluated in a range of recent EFSA reports (2009b, c, d, e, f, g, h), which are built on the extensive and detailed literature currently available and also on expert opinion.

This paper provides a short overview of the whole subject highlighting the key points that should be focused on when assessing the welfare of a batch of farmed fish at harvest. It indicates simple tests and observations that can be made and identifies areas for discussion with the farm management. In this paper, we do not attempt to add to the detailed evaluations provided by the EFSA documents or to provide detailed instructions to guide specific farm operations.

The process of harvesting farmed fish involves three stages: a period of food withdrawal; the collection and movement of the fish to the point of slaughter and the process of stunning and killing. The impact of these processes on the welfare of the fish varies significantly between species of fish, with the methods used and with the care and attention to detail with which they are completed. We first describe the general harvest patterns for a range of farmed fish and then consider the individual processes identifying the welfare hazards potentially associated with them. While it is rarely possible to operate a harvest method that results in completely stress-free slaughter, awareness of the key potential problems enables alternative systems to be compared, mistakes avoided and fish welfare improved.

General harvest descriptions

Salmon and cod

Salmon are cultured primarily in nets at sea although a small but increasing number are grown in tanks of recirculation water. Prior to harvest, they are fasted and crowded to enable them to be pumped or brailed either into well boats or to a mobile stunning and slaughter facility. Following transport by well boat, the fish are again pumped over distances that may range from a few metres to more than a kilometre. The fish may be held in lairage following transport, but before stunning and slaughter, both to rest the fish and to optimise the use of the slaughter facility. In some systems, the fish are cooled either during well boat transport or during a holding period prior to stunning (EFSA 2009c).

The principle method for stunning and killing is automatic, semi-automatic or manual percussive stunning. Immediately after the stun, they are bled by cutting the gill arches. Again, this may be achieved manually or automatically. Alternative techniques include electrical stunning followed by bleeding and killing in water perfused with carbon dioxide or eugenol and then bleeding. Salmon which are not to be used for human consumption due to size, maturity, disease or for other farm management reasons may be stunned and killed in an electrical system, killed by asphyxia by holding them in tanks with reduced water flow or in air, killed by maceration or by pharmacological methods.

Trout

Rainbow trout can be both farmed at sea, enabling them to grow to a size of several kilograms, or held in fresh water where they tend to be slaughtered at a weight of less than 1 kg. Trout on-grown at sea are generally harvested in a similar way to salmon. The smaller freshwater trout are typically grown in tanks, raceways or in nets in lakes. At the time of harvest, they are fasted, crowded and gathered using brails or pumps. Since these fish are too small to be conveniently stunned by percussion, they are traditionally placed directly in ice to die by asphyxia and cooling. In the UK, however, many freshwater trout are electrically stunned before being placed directly in ice where they die by asphyxia before recovering sensibility. Fish not for human consumption are killed by asphyxia or pharmacological methods (EFSA 2009d).

Tilapia

Tilapia are grown to slaughter weight in tanks or ponds or floating cages. Following a period of fasting, they are netted and placed in ice, in water or in dry crates for transportation to the point of slaughter. At the point of slaughter, they are bled or cooled in ice to immobilise them prior to exsanguination or direct evisceration.

Bass, bream and meagre

Sea bass, bream and meagre (*Argyrosomus regius*) are frequently grown on the same farm—most often in nets in the sea but bass and bream may also be grown in salt water ponds, raceways and tanks. At harvest, they are fasted and then crowded before collection by brail or occasionally by pumping. The most common killing method is direct immersion in iced water where they are immobilised and eventually lose consciousness due to thermal shock and then die by asphyxia. In some operations, the fish are killed by simply removing them from the water, but on at least one farm, they are electrically stunned prior to killing in ice. Bass and bream are seldom transported more than a few tens of metres from the on-growing facility to the stunning or killing point (EFSA 2009b).

Eel

Eel are on-grown in indoor tanks. Following fasting, the eels are gathered from the tanks by crowding and netting and transported without water to the killing facility. This may be within the same building or may require transport by lorry. Eels are not kept in water for this journey or during any subsequent holding period at the point of killing. At killing, they may be electrically stunned prior to evisceration but more commonly are immobilised in ice or killed by placing them in salt or an ammonia solution prior to evisceration (EFSA 2009f).

Turbot and sole

Turbot and sole are grown in shallow covered tanks. Following fasting, they are crowded using nets or partitions in the tanks and then collected by hand brail, removed from the water and placed onto ice for transport by road to a killing and processing plant. Killing is usually by exsanguination or evisceration without further measures to ensure the fish are insensible (EFSA 2009h).

Halibut

Halibut are grown in flat-bottomed nets in the sea. Following fasting, they are crowded, collected by wet brail and placed on a working surface where they are manually percussively stunned, placed in water perfused with carbon dioxide, or bled by severing several gill arches or the caudal vein. They are then placed in ice for transport to the processing unit.

Tuna

The flesh quality of tuna is very dependant on pre-slaughter handling and so all pre-slaughter processes are controlled by this overriding requirement. Tuna are lightly crowded and then killed by underwater shooting, by shooting from the surface or by gaffing, hoisting out of the water and then spiking to directly destroy the brain (EFSA 2009g).

Pangasius and catfish

Following fasting, pangasius and catfish are crowded and netted into dry containers that are transported to a killing facility where the majority are chilled and killed by decapitation. However, in the Netherlands, around 60% of the catfish are electrically stunned prior to killing. Efforts are also being made to introduce percussive or electrical stunning for pangasius.

Carp

Carp are grown in raceways or ponds and collected by crowding and brailing. The majority of carp are sold alive and killed by the end-user in homes or restaurants. The period from leaving the farm to killing commonly involves periods out of water and periods held in small quantities of water of poor quality. During this period, they are in particular danger of physical injury by being dropped on hard surfaces or onto other fish during transfers. At killing, carp are commonly held out of water for a period to reduce struggling and then may be stunned by multiple percussive blows followed by bleeding or evisceration (EFSA 2009e).

Processes applied at harvest and potential welfare problems

Fasting

Farmed fish routinely have their food withdrawn for a period prior to handling, transport or slaughter. An empty gut during these operations benefits the fish by lowering their metabolic activity and so reducing the rate of ammonia and carbon dioxide build-up in the water during crowding and transport (Ashley 2007). It also avoids faecal contamination during processing, which can shorten the shelf life of the fish (Wall 2001). The duration of fasting necessary to empty the gut is species and water temperature dependent but may be expected to be from 1 to 5 days. Food withdrawal for this period is unlikely to cause significant welfare problems.

In commercial practice, a range of food withdrawal periods are to be found, often far longer than is necessary to simply empty the gut. This is often a matter of tradition and operational convenience, facilitating a series of partial harvests from a tank, race or net and justified because the quality of the product is not harmed. In the past, long periods of fasting were used in the mistaken belief that this could improve the flesh quality by reducing oil levels (Wall 2001). Evidence for quality benefits from prolonged fasting is weak and mixed. Trials with brown trout have indicated that prolonged pre-harvest fasting can slightly improve the flesh colour, but this is at the expense of weight loss and fillet yield (Regost et al. 2001). Work with sea bream has found a measurable but possibly insignificant increase in muscle firmness but a decrease in the shelf life (Ginés et al. 2002; Álvarez et al. 2008). Mørkøre et al. (2008) concluded that prolonged fasting improves the ability of salmon to withstand severe stress during harvesting; however, this problem is more properly handled by directly addressing the causes of stress. Periodic feeding and starving throughout the growing cycle have been shown to have some beneficial quality effects in halibut culture (Foss et al. 2009); however, there is no reason to expect that prolonged fasting just at the end of the production cycle will have the same effect.

An empathetic assessment of the effect of food withdrawal on fish welfare is difficult because fish are cold blooded and so can adjust their metabolic rate in response to the availability of nutrition. However, the motivation to eat is clearly strong and as necessary to fish as to any other animal. Excessive periods of fasting clearly infringe the principles of the five freedoms of animal welfare (Webster 2001). The critical question is whether the duration has been minimised to that simply needed to empty the gut. Robb (2008) considers that no evidence exists for additional benefits from fasting salmon for longer than 72 h.

Crowding

Crowding is the first stage in most gathering and transport operations. In sea nets, this is achieved by slowly lifting part of the net or inserting a second net into the water. In ponds, raceways and tanks, it is achieved by the use of a seine net or by moving partitions or grids. Wall (2001) comments that crowding is one of the prime causes of poor welfare during harvest. The most common problem associated with crowding is shortage of oxygen; however, even when high oxygen levels are maintained, crowding changes many other aspects of the fish physiology, the effects of which can be observed for days afterwards (Ortuño et al. 2001). Different species respond to this intervention in different ways. For species such as bass and tuna, the process needs to be carried out slowly and with great care to avoid a panic reaction in the fish, which can result in high levels of stress, injury and mortality. Cod cage depth needs to be reduced slowly prior to crowding to avoid over inflating the swim bladder (Brown et al. 2010). Other species such as turbot and eel appear to be tolerant of the crowding operation. Guidelines for crowding salmon emphasise the need to crowd progressively and to ensure the water remains deep so that the fish are not exposed to excessive sunlight. For benthic species such as halibut, a cover suspended above the crowd will reduce light levels and may encourage fish to use the full column of water, rather than bunch at the bottom of the net (HSA 2008). The surface water of a crowd should be as calm as possible, with few fish breaking the water's surface and no vigorous activity (see HSA (2005) for a guide to assessing crowds). Fish must be observed for signs of abnormal behaviour, such as moribundness, leaping out of the water, signs of asphyxia or inversion. Oxygen monitoring may be required. Unlike fasting, poor crowding technique is likely to result in external injury and mortality, so farms have a direct financial incentive to ensure that crowding is well done.

Pumping

Fish pumping is achieved using centrifugal or vacuum pumping systems. Fish pass directly through these pumps, and there is a small risk of them being trapped or injured by the moving parts. Under some circumstances, pumping is also achieved using airlift or venturi systems where the fish are not exposed to any moving parts. In the tubes through which the fish are pumped, there is a risk of injury through collisions with sharp corners or interior features in the pipe. These usually are due to poor design or manufacture. During long pumped transits, hazards can also arise from poor water quality or overcrowding, particularly during stoppages or if the pipe becomes blocked. Single chamber and, to a lesser extent, double chamber vacuum pumps result in very uneven and turbulent water flow in pipes. Struggling against this turbulence quickly exhausts the fish. Following a pumped transit to the slaughter plant, most salmon are too exhausted to swim properly (EFSA 2009c), even when the transit time is as little as 2–4 min. At the end of a pumped transfer, observations should confirm that the fish are free of injury and have sufficient energy to maintain their equilibrium (upright orientation) and to swim normally. After pumping, pipes must be checked to ensure no fish remain in them.

Wet and dry brailing

Lower levels of crowding are required for brailing than for pumping. This can benefit fish. However, very high densities of fish occur inside the brail for short periods. Brails can contain anything from a few kilograms of fish to several hundred kilograms. In dry brailing, where water is not held in the brail, there are particular dangers to the fish of bruising, crushing, puncture and abrasion injuries from contact with other fish, contact with the net and contact through the net with other hard surfaces. In wet brails, where water is collected and lifted together with the fish, the fish are largely protected from these hazards. The exit from the brail can be problematic if the fish fall onto other fish, onto a solid surface or into water from a height. Wet brails can be used to collect and move fish short distances with few welfare problems. Dry brails are frequently used by tradition and for operational convenience. However, since they represent a significant welfare hazard, the possibility of using an alternative method should always be examined.

Transport prior to slaughter

Some farmed species, most notably salmon, are transported by well boat to the slaughter point. Others such as pangasius, carp, tilapia and eel are routinely transported by land without water.

Considerable investment has been made in the design and development of well boats for transporting salmon, cod and trout from the farm to the point of slaughter. They are designed to enable water quality to be monitored and controlled, and so that the tanks can be emptied without leaving fish without water at the bottom of the tank. Under some circumstances, the fish are cooled during the journey. This may benefit fish welfare by reducing their need for oxygen and their rate of ammonia production. The RSPCA Welfare Standards for Farmed Atlantic salmon (RSPCA 2010) recommend that cooling should be no faster than 1.5°C per hour and that water temperature should not be reduced any lower than 4°C. Recent research however suggests that Atlantic salmon may cope relatively easily with much more rapid temperature drops. New data from Foss et al. (2011) indicate that temperature drops from 16 to 4°C in 1 h does not result in large increases in cortisol level or other biochemical indicators of stress; however, temperature decreases from 16 to 0°C required an adjustment period of 5 h. During and at the end of a journey, the behaviour of the fish can be observed to ensure that they are responsive and coordinated in their swimming behaviour. Records of water temperatures and oxygen levels can be kept to ensure that water quality remained at an acceptable level throughout the journey. These records should be routinely reviewed and potential problems identified and rectified.

Any fish transported without water are likely to be extremely stressed by the lack of water, lack of oxygen, physical vibration, pressure and temperature. The fact that many of these fish can survive for long periods out of water is not an indication that it is a stress-free experience. Dry transport is unlikely to be in the interests of any fish species regardless of whether they survive and recover when placed back in water.

Percussive stunning

Percussive stunning results in rapid and prolonged or permanent insensibility if the blow is strong enough and if it is applied at the correct location. When it is performed manually, these factors are very dependant on the ability, training and state of fatigue of the operator. Automated percussive stunning equipment largely overcomes these problems but remains dependant on the fish being presented to the machine in the correct orientation and on the correct adjustment of the equipment. When salmon are harvested without prior size grading, the percussive stunning equipment is unlikely to be suitable for the whole range of fish sizes that will be encountered. Some form of intervention is therefore needed to select out the extremes of size before the fish reach the percussive device. For halibut, there is the additional complication of adjusting the location of the percussive blow for sinistral and dextral fish. Some other species of farmed fish including catfish, pangasius, carp and tilapia are very resistant to percussive stunning due to the shape and protection of the skull (Marx et al. 1999; Lambooij et al. 2007). Electrical stunning may be a more suitable stunning method for these species.

The goal of percussive stunning should always be to stun the fish with the first blow. Any manual or automated system in which fish are *routinely* exposed to more than one blow should be investigated. If the reason for 'double-stunning' is due to a genuine welfare concern, then the equipment should be examined for faults and the staff assessed for their ability to stun the fish and, if necessary, retrained.

The duration that the fish are out of the water prior to stunning and the force used to grip or restrain them for stunning must be monitored. Salmon and trout should not be kept out of water for more than 10–15 s before they are stunned or killed (EFSA 2009d; HSA 2005). Manual percussive stunning can be used as a back-up method for individual fish that are ineffectively stunned by the primary stunning system. This is good practice for fish welfare. However restraint of conscious fish, by hand, for back-up stunning can be problematic, particularly if the fish are large and if they struggle. This can prolong the time taken by the slaughterman to apply an effective back-up stun. In order to reduce handling time and improve the accuracy of back-up stunning, it is recommended that slaughter teams fix a simple, appropriately shaped restraint on the bleeding table; e.g. a V-shaped block can securely hold 'round' fish of varying sizes. An operator can gather a miss-stunned fish into the restraint block and then manually percussively stun/kill the fish to rapidly alleviate any suffering. This approach is also preferable to reloading ineffectively stunned fish into the primary stunning system, which may take more time and exacerbate any distress

Following percussive stunning, any sign of controlled, coordinated movement should be considered to indicate that the fish was not adequately stunned. Signs include the rhythmic motion of the opercula (gill covers), the vestibulo-ocular reflex (commonly referred to as 'eye roll'), any struggling, swimming activity, efforts to remain upright or regain equilibrium. Occasional spasmodic convulsions can be observed in effectively stunned fish. Lambooij et al. (2010) warn that even the lack of eye roll cannot be considered as conclusive evidence that fish are insensible; however, this is probably the best indicator available in a commercial setting.

Electric stunning

Electric stunning can be achieved both in water and out of water. Determination of the correct voltage, current or electric field that needs to be applied to stun the fish is dependant on the species, the orientation of the fish and the conductivity of the water. Further details for various species can be found in a range of publications including Lambooij et al. (2006, 2007, 2008), Lines and Kestin (2004, 2005), Robb et al. (2002), Robb and Roth (2003) and Roth et al. (2004).

In assessing the effectiveness of an electric stun, it is necessary to ensure that unconsciousness is both immediate and long lasting. To test whether unconsciousness is immediate and so avoid the danger that fish are exposed to a painful period of electrical stimulation before loss of consciousness, a sample of fish can be exposed to the electricity for just 1 s, removed from the electricity and immediately assessed for indicators of consciousness as described for percussion. After a 1-s exposure to electricity, the period of unconsciousness may be very short, so the loss of eye roll can be very difficult to observe in some species.

If insensibility can be generated within 1 s of current application, the next priority is to identify how long the fish should be exposed to the electricity and how they should be treated following removal from the electricity to ensure they never recover consciousness. Unlike percussive stunning, electrical stunning without suitable post-stun treatment is seldom permanent. The duration of unconsciousness is dependant on the electrical parameters of the stun, the duration of electrical application and on the post-stun treatment.

If species such as salmon, trout, charr, bass and bream are stunned for a suitable duration and then placed in ice, they will die from hypoxia before they are able to recover consciousness. However, if these fish are returned to moving or agitated water so that a flow of oxygen to the gills is maintained, they may recover, particularly if the water is cold enough to reduce their oxygen demand. Ice, rather than ice slurry or iced water, should therefore be used in the harvest bins or dolavs to reduce the amount of water that is washed over the gills during transport from the farm to the processing factory. Long exposures to electricity can result in death for salmonids almost regardless of the subsequent treatment; however, this is only rarely practical. Other fish such as turbot, halibut, eel, tilapia, carp and catfish are tolerant of low oxygen levels and so require a follow-up killing method such as immediate bleeding to avoid recovery. Immediately after stunning, these fish must have all of the gill arches severed on at least one side of the head. It is important that slaughter personnel inspect the gills of a sample of fish during each harvest to ensure that they have effectively cut all gill arches since failure to do so could result in a delayed bleed out and some individuals recovering consciousness before death occurs. Such checks are particularly important because fish are often not inspected for a long period of time between slaughter and further processing and therefore any animal welfare issues are unlikely to be visible at this time.

The most secure method for determining unconsciousness, either after a 1-s exposure or after a complete stun, is to observe brain activity using electroencephalogram (EEG) measurements (Kestin et al. 2002). However, in practical applications, this is seldom possible. A practical field approach is to pick the fish up and assess it for the presence of eye roll, then place it in water and observe it for any rhythmic motion of the lower jaw or opercula, any signs of coordinated motion, swimming or struggling to recover posture. The presence of the first two of these indicators should be considered as signs that the fish is recovering sensibility; the presence of the latter indicates that a fish that has already recovered consciousness (Kestin et al. 2002). Occasional spasmodic convulsions or gasps can be observed in effectively stunned fish and should not be a cause for concern. Since electricity stimulates the muscles directly, all observations need to be made once the electricity has been removed.

Some advantages of electrical stunning over percussive stunning are that it can be applied in water and to large numbers of fish simultaneously. The fish do not need to be removed from the water, handled, restrained or orientated until they are insensible. Undersized, oversized or deformed individuals will be effectively stunned without special adjustment of the equipment. This can result in very high standards of welfare at slaughter and in very high throughput rates, making electrical stunning uniquely suitable for small fish with a low individual value. Where fish are removed from water and stunned 'dry', it is important to ensure that the fish are arranged so that they do not receive pre-stun shocks such as would occur if their tails made contact with the electrodes before their heads. Various means have been developed in order to achieve this, the efficacy of which should be checked by direct observation.

Use of anaesthetics

In a few countries, the use of eugenol (clove oil) is permitted as a stunning method for fish entering the food chain (Robb and Kestin 2002). For fish not destined to enter the food chain, a wider range of anaesthetics is available including MS222 and Benzocaine. These treatments can result in permanent insensibility if applied for long enough, or the fish can be killed by bleeding or asphyxiation while insensible. There is currently some welfare concern about the use of MS222 and Benzocaine as exposure to them may be stressful to fish (Kiessling et al. 2009; Zahl et al. 2009).

Physical brain destruction for tuna

Due to the large size and high individual value of tuna, it is commercially feasible to take considerable precautions to protect the welfare of individual tuna. There is little available analysis of the slaughter of tuna; however, if shooting to destroy the brain while the fish is in water can be achieved quickly, reliably and accurately, then this will result in a high standard of welfare at slaughter. In contrast, gaffing, hoisting and restraining conscious tuna out of water prior to killing will compromise welfare (EFSA 2009g).

Other stunning and killing methods

A wide range of other methods are commonly used to stun or kill fish, largely determined by historical practice and commercial convenience. None of the methods identified below can be considered to be humane.

Immersion in water infused with carbon dioxide is a method that was very well established for killing salmon. The clear behavioural indications of distress by fish subjected to this treatment and the slow onset of insensibility has resulted in it being prohibited in several countries. Trials by Erikson (2011) explored the possibility that low concentrations of carbon dioxide might have a stunning or anaesthetic effect on salmon without causing stress. The conclusion from this research was that when the concentration of CO_2 was great enough to have any useful effect on the salmon, then it also resulted in clear indications of stress. The Norwegian scientific committee for food safety has concluded that while there may be potential for other gasses to be used to humanely kill fish, no such methods have yet been found (VKM 2010).

Stunning by cold shock involves immersion in ice or iced water and has been investigated because superficially it can appear to result in little struggling. Cold shock is used widely for a range of farmed fish and is particularly defended for fish cultured in warm water because the rapid cooling results in the rapid cessation of struggling. Research shows however that bream transferred from water at 23°C to ice slurry take 5 min to lose normal patterns of brain activity (van de Vis et al. 2003). The brain is therefore still active several minutes after the fish has become immobilized (Kestin et al. 2002; Bagni et al. 2007). Roth et al. (2009) similarly investigated the industrial practice of chilling turbot. These trials showed that after 90 min in chilled seawater at about -1°C, the turbot remained alive and capable of full recovery even though their internal temperature was below 1°C and their muscles were stiff, resembling rigour. From these and other similar results, we conclude that while a fish's physical reactions may stop or slow relatively quickly after being placed in ice or ice slurry, brain activity indicates the continuation of consciousness for a substantial period. On this basis, we can conclude that cold shock initially paralyses fish rather than stunning them.

Decapitation without prior stunning is unlikely to be a humane killing method for any species of farmed animal because the brain continues to function for an appreciable time. van de Vis et al. (2003) have shown from EEG measurements that eel brain function continues for up to 13 min following decapitation. Maceration is suitable only for small fish where the whole of the fish is likely to be destroyed simultaneously. Asphyxiation, bleeding, live evisceration and the use of salt baths or ammonia baths without prior stunning are all killing methods currently used on live and conscious farmed fish. None of these are humane due to the extended duration of suffering before the fish lose consciousness.

Conclusion

Practical observations of the fish at various stages of the harvest process and simple field tests can be used by farm staff and auditors to assess the welfare of fish at harvest. For almost all species of farmed fish, changes in procedure are possible that would improve fish welfare at harvest. In most cases, techniques for gathering, moving and stunning fish without unnecessarily stressing them are already available. The practical application of these techniques however is dependent on the determination and resolve of the industry to improve fish welfare by developing these techniques into practical, safe, simple, routine and widely used procedures.

In the long term, the relative lack of scientific evidence to support firm conclusions for the welfare of fish during fasting, crowding, transport and other handling procedures must be addressed. Work is required to identify humane killing methods and parameters for species of farmed fish all over the world, which have not yet been formally studied at slaughter. The interest in this field is growing continuously, and incorporation of new techniques and technology is likely to assist this progress. As industry and science collaborate to unravel the effects of farming practices on fish species, recommendations can be made and fish welfare steadily improved.

References

Álvarez A, García García B, Garrido MD, Hernández MD (2008) The influence of starvation time prior to slaughter on the quality of commercial-sized gilthead seabream (*Sparus aurata*) during ice storage. Aquaculture 284:106–114

Ashley P (2007) Fish welfare: current issues in aquaculture. Appl Anim Behav Sci 104:199–235

Bagni M, Civitareale C, Priori A, Ballerini A, Finoia M, Brambilla G, Marino G (2007) Pre-slaughter crowding stress and killing procedures affecting quality and welfare in sea bass (*Dicentrarchus labrax*) and sea bream (*Sparus aurata*). Aquaculture 263:52–60

Brown JA, Watson J, Bourhill A, Wall A (2010) Physiological welfare of commercially reared cod and effects of crowding for harvesting. Aquaculture 298:315–324

EFSA (2009a) General approach to fish welfare and to the concept of sentience in fish. Scientific opinion of the panel on animal health and welfare. EFSA J 954:1–27

EFSA (2009b) Species-specific welfare aspects of the main systems of stunning and killing of farmed seabass and seabream. EFSA J 1010:1–52
EFSA (2009c) Species-specific welfare aspects of the main systems of stunning and killing of farmed atlantic salmon. EFSA J 2012:1–77
EFSA (2009d) Species-specific welfare aspects of the main systems of stunning and killing of farmed fish: rainbow trout. EFSA J 1013:1–55
EFSA (2009e) Species-specific welfare aspects of the main systems of stunning and killing of farmed carp. EFSA J 1013:1–37
EFSA (2009f) Species-specific welfare aspects of the main systems of stunning and killing of farmed eels (*Anguilla anguilla*). EFSA J 1014:1–42
EFSA (2009g) Species-specific welfare aspects of the main systems of stunning and killing of farmed tuna. EFSA J 1072:1–53
EFSA (2009h) Species-specific welfare aspects of the main systems of stunning and killing of farmed turbot. EFSA J 1073:1–34
Erikson U (2011) Assessment of different stunning methods and recovery of farmed Atlantic salmon: isoeugenol, nitrogen, and three levels of carbon dioxide. Anim Welf 20:365–375
Foss A et al (2009) Compensatory growth in Atlantic halibut: effect of starvation and subsequent feeding on growth, maturation, feed utilization and flesh quality. Aquaculture 290:304–310
Foss A, Grimsbø E, Vikingstad E, Nortvedt R, Slinde E, Roth B (2011) Live chilling of Atlantic salmon: physiological response to handling and temperature decrease on welfare. Fish Physiol Biochem. doi:10.1007/s10695-011-9536-6
Ginés R, Palicio M, Zamorano MJ, Argüello A, López JL, Afonso JM (2002) Starvation before slaughtering as a tool to keep freshness attributes in gilthead sea bream (*Sparus aurata*). Aquacult Int 10:379–389
HSA (2005) Humane harvesting of salmon and trout. Guidance notes no 5. The Humane Slaughter Association, Herts. http://www.hsa.org.uk. ISBN 1 871561 29 9
HSA (2008) Humane harvesting of halibut. Technical note 24 Humane Slaughter Association, Herts. Available at http://www.hsa.org.uk
Kestin SC, van de Vis JW, Robb DHF (2002) Protocol for assessing brain function in fish and the effectiveness of methods used to stun and kill them. Vet Rec 150:302–307
Kiessling A, Johansson D, Zahl IH, Samuelsen OB (2009) Pharmacokinetics, plasma cortisol and effectiveness of benzocaine, MS-222 and isoeugenol measured in individual dorsal aorta-cannulated Atlantic salmon (*Salmo salar*) following bath administration. Aquaculture 286:301–308
Lambooij E, Kloosterboer RJ, Gerritzen MA, van de Vis JW (2006) Assessment of electrical stunning in fresh water of African catfish (*Clarias gariepinus*) and chilling in ice water for loss of consciousness and sensibility. Aquaculture 254:388–395
Lambooij E, Pilarczyk M, Bialowas H, van den Boogaart JGM, van de Vis JW (2007) Electrical and percussive stunning of the common carp (*Cyprinus carpio* L.): neurological and behavioural assessment. Aquac Eng 37:171–179
Lambooij B, Gerritzen MA, Reimert H, Burggraaf D, André G, Van De Vis H (2008) Evaluation of electrical stunning of sea bass (*Dicentrarchus labrax*) in seawater and killing by chilling: welfare aspects, product quality and possibilities for implementation. Aquacult Res 39:50–58
Lambooij E, Grimbø E, van de Vis JW, Reimert H, Nortvedt R, Roth B (2010) Percussion and electrical stunning of Atlantic salmon (*Salmo salar*) after dewatering and subsequent effect on brain and heart activities. Aquaculture 300:107–112
Lines JA, Kestin SC (2004) Electrical stunning of fish: the relationship between the electric field strength and water conductivity. Aquaculture 241:219–234
Lines JA, Kestin SC (2005) Electric stunning of trout: power reduction using a two-stage stun. Aquac Eng 32:483–491
Marx H, Sengmuller-Sieber T, Hoffmann R, Stolle A (1999) Untersuchungen zu Stressbelastung und Produktqualität von Forelle, Wels und Flunder in Zusammenhang mit Betäubung und Schlachtung. Archiv für Lebensmittelhygiene 50:37–40
Mørkøre T, Mazo TP, Tahirovic V, Einen O (2008) Impact of starvation and handling stress on rigor development and quality of Atlantic salmon (*Salmon salar* L.). Aquaculture 277:231–238
Ortuño J, Esteban M, Meseguer J (2001) Effects of short-term crowding stress on the gilthead seabream (*Sparus aurata* L.) innate immune response. Fish Shellfish Immunol 11:187–197
Regost C, Arzel J, Cardinal M, Laroche M, Kaushik SJ (2001) Fat deposition and flesh quality in seawater reared, triploid brown trout (*Salmo trutta*) as affected by dietary fat levels and starvation. Aquaculture 193:325–345
Robb DHF (2008) Welfare of fish at harvest. In: Branson EJ (ed) Fish welfare. Blackwell Publishing, London, pp 217–242
Robb DHF, Kestin SC (2002) Methods used to kill fish: field observations and literature reviewed. Anim Welf 11:269–282
Robb DHF, Roth B (2003) Brain activity of Atlantic salmon (*Salmo salar*) following electrical stunning using various field strengths and pulse durations. Aquaculture 216:363–369
Robb DHF, O'Callaghan MO, Lines JA, Kestin SC (2002) Electrical stunning of rainbow trout (*oncorhynchus mykiss*): factors that affect stun duration. Aquaculture 205:359–371
Roth B, Moeller D, Slinde E (2004) Ability of electric field strength, frequency and current duration to stun farmed Atlantic salmon and pollock and relations to observed injuries using sinusoidal and square wave alternating current. N Am J Aquac 66:208–216
Roth B, Imsland AK, Foss A (2009) Live chilling of turbot and subsequent effect on behaviour, muscle stiffness, muscle quality, blood gases and chemistry. Anim Welf 2009(18):33–41
RSPCA Welfare Standards for Farmed Atlantic Salmon (2010) RSPCA, Wilberforce Way, West Sussex. ISBN 1898331685
van de Vis JH, Kestin SC, Robb DHF, Oehlenschläger J, Lambooij E, Münkner W, Kuhlmann W, Kloosterboer K, Tejada M, Huidobro A, Tejada M, Otterå H, Roth B, Sørensen NK, Aske L, Byrne H, Nesvadba P (2003) Is humane slaughter of fish possible for industry? Aquacult Res 34:211–220

VKM (2010) Risikovurdering knyttet til bruk av gass, slag mot hode og strøm til bedøving av fisk: Uttalelse fra Faggruppe for dyrehelse og dyrevelferd i Vitenskapskomiteen for mattrygghet. http://www.english.vkm.no/dav/f942fb5728.pdf. Last accessed 13 May 2011

Wall AJ (2001) Ethical considerations in the handling and slaughter of farmed fish. In: Kestin SC, Warriss PD (eds) Farmed fish quality. Fishing News Books, Blackwell science, Oxford, pp 108–115

Webster AJF (2001) Farm animal welfare: the five freedoms and the free market. Vet J 161:229–237

Zahl HI, Kiessling A, Samuelsen OB and Olsen RE (2009) Anaesthesia induces stress in Atlantic salmon (*Salmo salar*), Atlantic cod (*Gadus morhua*) and Atlantic halibut (*Hippoglossus hippoglossus*). Fish Physiol Biochem. Published online 13 August 2009. doi:10.1007/s10695-009-9346-2

Cortisol and Finfish Welfare – Summary

Tim Ellis, Hijran Yavuzcan Yildiz, Jose López-Olmeda, Maria Teresa Spedicato, Lluis Tort, Øyvind Øverli and Catarina Martins

Stress is generally accepted as an important animal welfare issue. Recent studies of fish welfare have, therefore, examined stress by assessing physiological / biochemical measures. The stress hormone cortisol is probably the best studied indicator of stress in fish. Cortisol is readily measured in the blood, responds to a wide range of acute and chronic stressors, shows a marked uni-directional response proportional to the severity of stress, and affects physiological function and behaviour. However, when viewed from a functional perspective only, the level of cortisol in the blood is not predictive of how well fish are coping with the environment. In this paper we argue that as welfare should be considered in terms of mental health and feelings, stress in fish should be viewed from a 'psychological'perspective, in addition to its traditional physiological context. We emphasize that cortisol release is triggered after perception of a threat and studies indicate that fish show cortisol responses to psychological stressors, i.e., those with no direct physicochemical action. We, therefore, suggest that researchers should attempt to use cortisol as an indicator of how fish perceive (and feel about) their environment, bodily status and physiological stress.

Background

By definition, poor animal welfare is associated with unpleasant (negative) feelings (e.g. of fear, pain, illness, hunger, discomfort, frustration) that relate to suffering in conscious animals and cause stress. Good welfare will be associated with the absence of suffering and the presence of positive feelings. Therefore, to assess welfare in fish, an indicator is needed for the presence and severity of negative (and possibly the presence of positive) feelings.

Scientific Significance

We discuss the practicalities of using cortisol "to ask the fish themselves" how they feel about the farm environment and husbandry practices. Single time point measurements of cortisol are of little value in assessing stress levels in fish, as studies need to account for diurnal and seasonal variations, and environmental and genetic factors. Areas in need of greater clarity are the separation of (physiological) stress from (psychological) distress, cortisol dynamics during chronic stress, and the interactions between feelings, cortisol, mood and behaviour.

Practical Application

Stress hormone responses can be understood by the layman, and may ultimately have value in determining the relative capacity of fish to experience feelings (and suffer) in relation to species and developmental stage. A stress hormone like cortisol can be used (with caution) as indicator to assess welfare in fish.

Contact

T. Ellis
Cefas Weymouth Laboratory, Barrack Road, The Nothe,
Weymouth, Dorset DT4 8UB, UK
e-mail: tim.ellis@cefas.co.uk

Fish Physiol Biochem (2012) 38:163–188
DOI 10.1007/s10695-011-9568-y

Cortisol and finfish welfare

Tim Ellis · Hijran Yavuzcan Yildiz · Jose López-Olmeda · Maria Teresa Spedicato · Lluis Tort · Øyvind Øverli · Catarina I. M. Martins

Received: 9 July 2010 / Accepted: 24 October 2011 / Published online: 24 November 2011

Abstract Previous reviews of stress, and the stress hormone cortisol, in fish have focussed on physiology, due to interest in impacts on aquaculture production. Here, we discuss cortisol in relation to fish welfare. Cortisol is a readily measured component of the primary (neuroendocrine) stress response and is relevant to fish welfare as it affects physiological and brain functions and modifies behaviour. However, we argue that cortisol has little value if welfare is viewed purely from a functional (or behavioural) perspective—the cortisol response itself is a natural, adaptive response and is not predictive of coping as downstream impacts on function and behaviour are dose-, time- and context-dependent and not predictable. Nevertheless, we argue that welfare should be considered in terms of mental health and feelings, and that stress in relation to welfare should be viewed as psychological, rather than physiological. We contend that cortisol can be used (with caution) as a tractable indicator of how fish perceive (and feel about) their environment, psychological stress and feelings in fish. Cortisol responses are directly triggered by the brain and fish studies do indicate cortisol responses to psychological stressors, i.e., those with no direct

T. Ellis (✉)
Cefas Weymouth Laboratory, Barrack Road, The Nothe, Weymouth, Dorset DT4 8UB, UK
e-mail: tim.ellis@cefas.co.uk

H. Y. Yildiz
Department of Fisheries and Aquaculture, Ankara University, 06110 Ankara, Turkey

J. López-Olmeda
Department of Physiology, Faculty of Biology, University of Murcia, 30100 Murcia, Spain

M. T. Spedicato
COISPA, Stazione Sperimentale per lo Studio delle Risorse del Mare, 70126 Bari, Italy

L. Tort
Unit de Fisiología Animal, Faculty of Biosciences, Department of Cell Biology, Physiology and Immunology, Universitat Autonoma de Barcelona, 08193 Bellaterra, Spain

Ø. Øverli
Department of Animal and Aquacultural Sciences, Norwegian University of Life Sciences, PO Box 5003, 1432 Aas, Norway

C. I. M. Martins
CCMAR, Centro de Ciências do Mar, Universidade do Algarve, Campus de Gambelas, 8005-139 Faro, Portugal

C. I. M. Martins
Aquaculture and Fisheries Group, Wageningen University, PO Box 338, 6700 AH Wageningen, The Netherlands

physicochemical action. We discuss the practicalities of using cortisol to ask the fish themselves how they feel about husbandry practices and the culture environment. Single time point measurements of cortisol are of little value in assessing the stress level of fish as studies need to account for diurnal and seasonal variations, and environmental and genetic factors. Areas in need of greater clarity for the use of cortisol as an indicator of fish feelings are the separation of (physiological) stress from (psychological) distress, the separation of chronic stress from acclimation, and the interactions between feelings, cortisol, mood and behaviour.

Keywords Stress · Psychological · Feelings · HPI axis · Brain

Why review cortisol and fish welfare?

The term welfare was possibly first applied to fish by Shelbourne (1975) when discussing acclimation to the captive environment of wild caught broodstock plaice *Pleuronectes platessa*. The concept of fish welfare was arguably first legitimatised in 1986 when fish were included within the UK's animal experimentation legislation (http://www.archive.official-documents.co.uk/document/hoc/321/321-xa.htm). Implicit in this legislation is that fish are deemed to have the capacity to suffer and are protected against unlicensed scientific procedures that may induce pain, suffering or lasting harm. Since then, the concept of fish welfare has spread to other countries and all areas of human interaction with fish, i.e., farming, zoos, pets, 'hunting' for food and sport (e.g. Casamitjana 2004; Huntingford et al. 2006; Metcalfe 2009) mirroring the welfare avenues for mammals and birds. This spread has occurred despite continuing dispute within the scientific community about whether fish do have the psychological capacity to suffer (e.g. see Rose 2002, 2007; Huntingford et al. 2006).

With the increasing interest in fish welfare within the scientific community, fish farming stakeholders and the European Government (having a societal responsibility to protect sentient animals), the European COST Action 867 Wellfish was initiated (http://www.fishwelfare.com). During the inaugural scientific meeting of this talking shop in 2006, it was recognised that 'stress' is a key issue in discussions of fish welfare and that cortisol is the most frequently used indicator of stress in fish. Cortisol was being used (almost ubiquitously) in fish welfare studies (e.g. Ellis et al. 2002; Turnbull et al. 2005; North et al. 2006; Varsamos et al. 2006) as a non-specific stress indicator, without questioning its value. It was therefore decided that a review of cortisol in relation to fish welfare was timely.

Stress and cortisol in fish have been reviewed many times over the last four decades (Table 1). This considerable interest in stress in fish reflects awareness of its detrimental impacts on key production issues such as disease susceptibility, growth, food conversion efficiency, flesh quality and reproduction (e.g. Pickering 1992). What we attempt to do here that is distinct from previous reviews is to restrict discussions to those pertinent to fish welfare. [Readers are referred to Mormède et al. (2007) for the discussion of cortisol in assessing terrestrial animal welfare]. The aim of our review is to discuss questions that may be asked by fish biologists interested in applying cortisol to fish welfare studies. We recognise that some views are contentious and simplified, but this is deliberate to widen understanding, thought and debate, which will ultimately move the subject of fish welfare forwards.

What is stress?

The meaning of stress is often discussed with definitions varying (Barton 1997) but often resembling '*A state of threatened homeostasis which is re-established by a complex repertoire of physiological and behavioural adaptive responses of the organism*' (Chrousos 1998). [The theoretical view of stress is evolving: please see McEwen and Wingfield (2010) for the discussion of stress in relation to homeostasis and allostasis.]

A stress response can be observed at different levels (Mazeaud and Mazeaud 1981; Donaldson 1981; Wedemeyer et al. 1990; Anderson 1990; Barton 1997):

- 1° response—the neuroendocrine responses—a perceived threat triggers release of stress hormones (catecholamines and corticosteroids) that precipitate the 2° response—and immediate behavioural changes, i.e., freeze or escape.

Table 1 Some examples of previous reviews of stress and cortisol in fish

Authors (Year)	Title
Mazeaud et al. (1977)	Primary and secondary effects of stress in fish: some new data with a general review
Barton and Toth (1980)	Physiological stress in fish: a literature review with emphasis on cortisol dynamics
Donaldson (1981)	The pituitary–interrenal axis as an indicator of stress in fish
Schreck (1982)	Stress and rearing salmonids
Carragher and Sumpter (1990)	Corticosteroid physiology in fish
Schreck (1990)	Physiological, behavioural and performance indicators of stress.
Wedemeyer et al. (1990)	Stress and acclimation
Barton and Iwama (1991)	Physiological changes in fish from stress in aquaculture with emphasis on the response and effects of corticosteroids
Pickering (1992)	Rainbow trout husbandry: management of the stress response
Sumpter (1992)	The stress response and its consequences in cultured fish
Pickering et al. (1993)	Stress and adaptation
Sumpter (1993)	The deleterious effects of stress and their significance to aquaculture
Sumpter et al. (1994)	The wide ranging effects of stress on fish
Sumpter (1997)	The endocrinology of stress
Wendelaar Bonga (1997)	The stress response in fish
Mommsen et al. (1999)	Cortisol in teleosts: dynamics, mechanisms of action and metabolic regulation
Barton et al. (2002)	Physiological and condition-related indicators of environmental stress in fish
Barton (2002)	Stress in fishes: a diversity of responses with particular reference to changes in circulating corticosteroids
Davis (2006)	Management of physiological stress in finfish aquaculture
Pottinger (2008)	The stress response in fish: mechanisms, effects and measurement
Prunet et al. (2008)	Functional genomics of stress responses in fish
Aluru and Vijayan (2009)	Stress transcriptomics in fish: a role for genomic cortisol signalling
Galhardo and Oliveira (2009)	Psychological stress and welfare in fish
Tort (2010)	Stress in farmed fish. Its consequences in health and performance
Tort et al. (2011)	Stress and welfare in sparid fishes
Vijayan (2011)	Hormone response to stress

- 2° response—the physiological changes (e.g. increased number of red blood cells, increased glucose in blood) and behavioural changes (e.g. reduced feeding activity) that enable the animal to respond to the threat. However, if these 2° responses are maintained over a period of time, 3° responses can result.
- 3° responses—the effects at the whole-animal level, e.g., decreased growth and reproductive investment, increased disease susceptibility and mortality.

This classification of stress—stress hormones precipitating physiological changes that allow response/adaptation to the stressor, but adverse whole-animal effects if extreme or sustained—parallels Selye's classic model of the generalised adaptation syndrome (Table 2), originally developed for understanding the impact of stressors on mammals. Various indicators of the three levels of stress response have been used to assess stress in fish (Table 2) and are potential stress indicators.

Where does cortisol fit in as a stress indicator?

Cortisol—identified as the primary corticosteroid hormone in fish nearly 50 years ago (Donaldson 1981)—is a stress hormone that is released into the fish's bloodstream from the interrenal cells of the head kidney. As a steroid hormone, cortisol is lipid soluble, can diffuse through cell membranes, and cannot therefore be stored. It is produced de novo from

Table 2 Stages of Selye's conceptual model of stress termed the general adaptation syndrome, alongside examples of the primary, secondary and tertiary stress response indicators (adapted from Mazeaud and Mazeaud 1981; Donaldson 1981; Wedemeyer et al. 1990; Anderson 1990; Barton 1997)

General Adaptation Syndrome stages	Stress responses in fish										
	Neuro-endocrine			Physiological				Physical		Behavioural	Performance measures
	Endocrine	Molecular	Neuro-transmitters	Bio-chemical	Haemato-logical	Hydro-mineral	Immuno-logical	Histo-pathology	Biometric		
1. Alarm	1° Stress response										
Activation of stress hormone (catecholamine and corticosteroid) pathways; the role of the stress hormones is to initiate a series of physiological compensation mechanisms to return the fish to homeostasis	Catecholamines (adrenaline, noradrenaline); ACTH Cortisol Melanocyte-stimulating hormone	Gluco-corticoid receptors	Serotonin					Interrenal cell size, number, nuclear diameter		Escape responses	
2. Resistance	2° Stress response										
The compensatory physiological processes triggered to achieve acclimation; bio-energetic cost may reduce performance capacity	Growth hormone			Acute phase proteins; plasma glucose, lactic acid, cholesterol; liver & muscle glycogen, adenylate energy charge	Haematocrit, leukocrit; erythrocyte / leukocyte numbers & ration; thrombocyte numbers; Blood clotting time Haemoglobin	Plasma chloride, sodium, potassium, protein, osmolality	Cytokine release; Phagocytic index, oxidative burst, lysozyme, pinocytosis, complement	Gastric tissue morphology; gill chloride cells; epidermal mucus cells	Skin colour	Ventilation rate;	
3. Exhaustion	3° Stress response										
Duration or severity of the stressor surpasses compensatory ability; adverse effects become apparent									Condition factor; organo-somatic indices	Food intake; Activity; Space use; Shoaling/ schooling; refuge use; Aggression Reproductive behaviour	Growth; FCR; Size variation Disease incidence, mortality Reproductive performance

cholesterol when the interrenal cells within the head kidney are stimulated by a hormonal cascade. This cascade (the hypothalamus–pituitary–interrenal (HPI) axis; Fig. 1) is triggered by higher brain centres and is one of two neuroendocrine axes constituting the 1° response to stress. The location of the interrenal cells, near the post-cardinal vein, facilitates the release of cortisol into the bloodstream. The cortisol cascade in fish is largely analogous to that in mammals and other higher vertebrates except for two differences—corticotropin releasing hormone (CRH) moves from the hypothalamus to the pituitary via direct neural contact in fish (rather than in the blood), and it is the interrenal tissue (rather than the adrenal gland) that produces cortisol (Sumpter 1997).

It must also be recognised that cortisol is not the only stress hormone in fish:

- CRH and adrenocorticotropin (ACTH): These are precursor hormones in the HPI cascade ending with the production of cortisol. It has been shown that CRH acts in tissues other than the pituitary and also that it can activate immune messengers such as cytokines and therefore induce secondary actions without cortisol (Verburg-Van Kemenade et al. 2009).
- Cortisone, another potent corticosteroid, has occasionally been recorded in fish blood at levels greater than cortisol (Pottinger and Moran 1993). However, although cortisone is undoubtedly of physiological importance, it is assumed to be a metabolic breakdown product of cortisol, meaning the latter corticosteroid is the better indicator of HPI axis activation.
- Other steroids: elasmobranchs do not produce cortisol but 1-α-hydroxy-corticosterone. Little research has focussed on this steroid, possibly due to the difficulty of obtaining antibodies.
- Catecholamines (adrenalin = epinephrine and noradrenalin = norepinephrine) (Mazeaud and Mazeaud 1981) are released from the chromaffin cells (also within the head kidney) and similarly released into the post-cardinal vein and its branches (Grassi Milano et al. 1997). The main difference between the catecholamines and cortisol is that the release of catecholamines is triggered by nerve circuitry (rather than by a cascade of chemical messengers) so the response is very quick, making it virtually impossible to obtain baseline blood concentration unaffected by the sampling itself. Catecholamines in fish have therefore received much less attention than cortisol but merit greater investigation (Pottinger 2010).

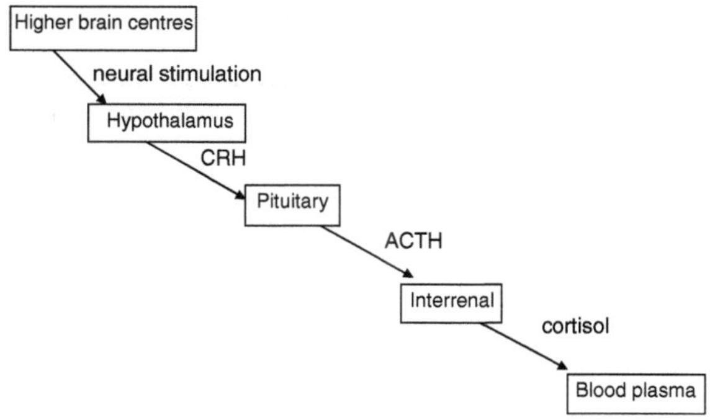

Fig. 1 Simplified hypothalamus–pituitary–interrenal axis of fish (after Sumpter 1997). *CRH* corticotropin releasing hormone, *ACTH* adrenocorticotropin

There is a variety of potential methods for assessing activation of the HPI axis in fish, which assess responses at different levels within the axis (Table 3). To gauge processes occurring in the brain, the ideal would be to measure indicators within the brain itself (i.e. CRH) or close to the origin of the HPI hormonal cascade (i.e. ACTH). However, measurements of CRH and ACTH in fish are rare, which presumably reflects the practical difficulties of measurement and obtaining baseline levels (the very rapid responses make it difficult to avoid the effect of sampling itself as for catecholamines). Although the concentration of cortisol in the blood plasma has to be recognised as *not* the ideal indicator of activation of the HPI axis, it has become the standard measure because

- as it needs to be produced, blood cortisol levels typically take a few minutes to rise in response to acute (sampling) stress and therefore are not prone to sampling effects (Molinero et al. 1997)
- blood sampling is a relatively simple, standard technique and potentially non-lethal. (For discussion of non-invasive alternatives to blood sampling see Scott et al. 2008).
- cortisol measurement using immunoassays (radioimmunoassay (RIA), enzyme-linked immunosorbent assay (ELISA)/enzyme immunoassay (EIA)) is relatively simple and enables a high throughput

Table 3 Potential parameters available for assessment of HPI axis stimulation in fish (after Donaldson 1981; Pickering and Stewart 1984; Oliveira et al. 1999)

	Biological matrix	Parameter
Lethal	Pituitary	Pituitary ACTH content
	Interrenal	Interrenal tissue volume
		Interrenal cell nuclear diameter
		Interrenal cell size
		Interrenal cell nuclear/cytoplasm ratio
		Interrenal RNA content
	Whole body	Whole body cortisol concentration
	Tissue	Tissue cortisol concentration
Lethal/non-lethal but invasive	Blood	Plasma ACTH content
		Plasma cortisol concentration
		Plasma cortisol secretion rate
		Plasma cortisol response to ACTH or stressor application
		Plasma cortisone concentration
	Bile	Glucuronated cortisol concentration
Non-invasive	Faeces	Glucuronated cortisol concentration
	Water	Free cortisol release rate
		Free + sulphated + glucoronated cortisol release rate

of samples. [For discussion of RIA versus EIA/ELISA see Sink et al. (2008).]

Further qualities of cortisol as a fish welfare indicator include the following:

- As a stress hormone, it is readily understood by the layman.
- Blood levels show a unidirectional (increasing) response that is readily interpreted
- Cortisol production in fish responds to a wide variety of events and conditions that are expected to be stressors to fish (Table 4).
- The signal (response/basal level ratio) is typically strong for both purported acute (abrupt onset, short duration, high severity inducing a rapid response) and chronic (slower onset, longer duration, less severe inducing a less marked response) stressors (Fig. 2).
- Blood levels respond relatively rapidly to the onset (within minutes) and cessation (within hours—days) of stressors, and therefore relate to the current/recent environment, and do not require extended observation periods. (Note that recovery of cortisol levels after an acute stress typically precedes 2° physiological measures, although it can succeed behavioural measures, e.g., Barton 2000; Pickering and Pottinger 1987a; Olla et al. 1992).
- It plays a regulatory role in many important physiological processes, and elevated plasma cortisol levels are thought to affect physiological functions leading to adverse impacts on traits important to aquaculture such as growth, disease resistance and reproductive output (e.g. Barton et al. 1987; Barton and Iwama 1991; Lankford and Weber 2006).
- Behavioural changes, consistent with a behavioural stress response, are apparently mediated via cortisol (see below)

How does cortisol affect physiology, behaviour and whole-animal processes?

Once synthesised, cortisol diffuses out of the interrenal cells into the blood plasma. Fishes do not appear to possess specific corticosteroid-binding proteins in the plasma (Mommsen et al. 1999), and cortisol can therefore diffuse freely out of blood vessels. Cortisol precipitates physiological and behavioural responses

Table 4 Examples of aquacultural stressors documented to elicit a cortisol response in salmonid fish

Stressor	Example reference
Net handling	Ellis et al. (2004, 2007a)
Transfer (fish pump)	Weber et al. (2002)
Confinement	Pottinger (2010), Pottinger et al. (1992)
Transportation	Barton and Peter (1982)
Grading	Flos et al. (1988)
Temperature shock	See Donaldson (1981), Barton and Peter (1982)
Disease treatments	See Donaldson (1981)
Anaesthetics	See Donaldson (1981), Tort et al. (2002)
Disease	See Ellis et al. (2007b)
Poor water quality	See Donaldson (1981)
Noise	Gilham and Baker (1985)
Social interactions	Laidley and Leatherland (1988)
Crowding	Pickering and Stewart (1984)

by binding to and activating specific cortisol receptors in target tissues. Receptors are located either (1) extracellularly on the membrane of target cells and so mediate a rapid, direct action (non-genomic pathway) or (2) intracellularly so when the receptor–hormone complex diffuses into the nucleus in the target cell, it binds and acts as a transcription factor and either activates or represses one or several genes (a slower genomic pathway). The rapid effects of cortisol on behaviour are thought to occur through non-genomic pathways (Sandi 1996; Schjolden et al. 2009; Mikics et al. 2004). There are two types of receptors for cortisol: mineralocorticoid receptors (MR) and glucocorticoid receptors (GR). MR have a higher affinity for cortisol and are believed to play a role in regulating circadian fluctuations of the hormone. GR have a lower affinity and are believed to be more important in the regulation of the response to stress. Most fishes possess MR and two GR isoforms, GR1 and GR2. GR2 appears to be more sensitive to low levels of circulating cortisol, whereas GR1 is sensitive only to higher levels, and thus acute stressors (Stolte et al. 2008).

What effects does cortisol have on physiology?

Cortisol seems to be a key controller of aerobic and anaerobic metabolism in fish, stimulating several aspects of intermediary energy metabolism, elevating oxygen uptake, increasing gluconeogenesis, inhibiting glycogen synthesis and resulting in a 'higher cost of living' (Barton et al. 1987; Morgan and Iwama 1996; Wendelaar Bonga 1997; Tort 2010). An increase in metabolic rate is thought to contribute to reduced growth: Lankford and Weber (2006) found a negative correlation between growth and cortisol responsiveness to a confinement stressor; Jentoft et al. (2005) found that domesticated fish show a lower cortisol responsiveness, a lower energy consumption and higher growth rate when exposed to repeat stressors than non-domesticated fish. Cortisol may also influence growth by affecting other steroids (Pickering 1993). Furthermore, increased plasma cortisol levels have been suggested to suppress appetite, acting through elevated plasma glucose levels (Heath 1995). Elevated cortisol levels may therefore manifest as reduced food intake, food conversion efficiency and growth (De Boeck et al. 2001; Gregory and Wood 1999).

Cortisol appears to play a pivotal role in the regulation of muscle glycogen resynthesis: a bout of high intensity exhaustive exercise in rainbow trout resulted in a near total depletion of white muscle glycogen stores and an elevation of plasma cortisol levels (Milligan 1996). Elevated plasma cortisol appears to inhibit glycogenesis, as there is no evidence of net muscle glycogen synthesis until cortisol levels begin to decline (Pagnotta et al. 1994). Allowing the fish to swim slowly after exercise prevents the post-exercise cortisol elevation and also promotes muscle glycogen resynthesis, compared to fish held in still water (Milligan et al. 2000).

Cortisol also performs an osmoregulatory function in teleosts, being a key hormone for sea water adaptation and ion uptake (McCormick 2001). Administration of cortisol to interrenalectomised and hypophysectomised fish increases plasma sodium and chloride ion levels (Chan et al. 1968, 1969; Fortner and Pickford 1982). The differential promotion of ion secretion or uptake may be partially dependent on the relative activities of growth hormone and prolactine (McCormick 2001). Cortisol is involved in salt

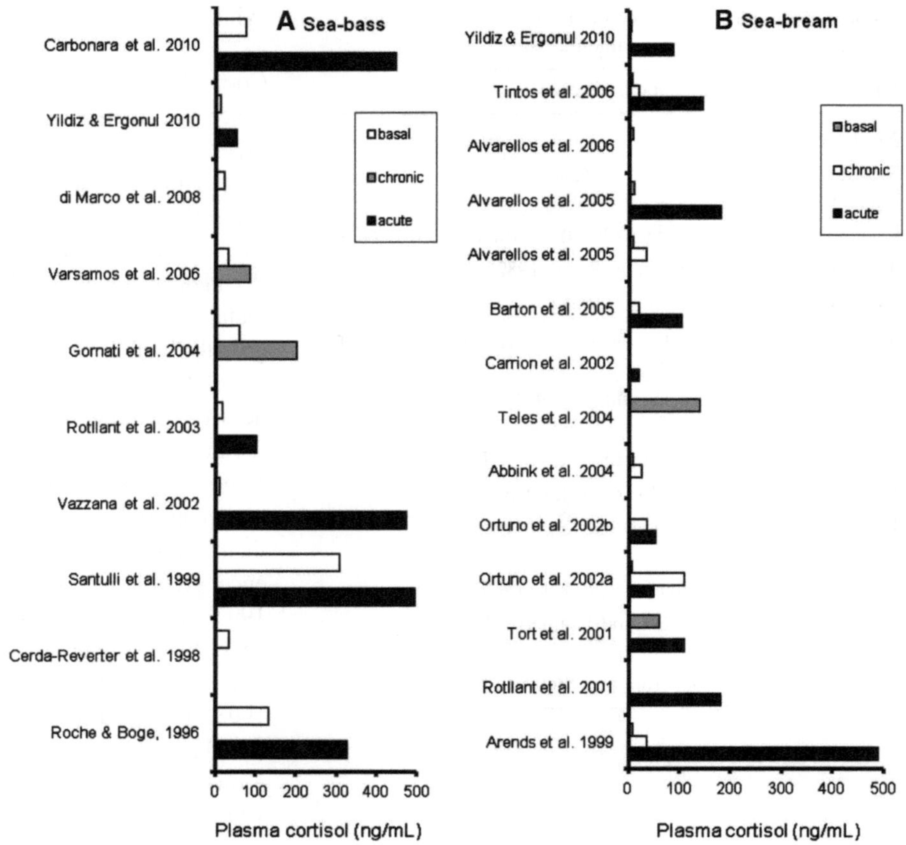

Fig. 2 Collation of recent studies on plasma cortisol concentrations in sea bass *Dicentrarchus labrax* and sea bream *Sparus aurata* indicating mean levels under basal (unstressed) and stressed (acute and chronic) conditions. Please note categorisation of basal, chronic and acute stress ascribed by the present authors

secretion, in conjunction with the growth hormone/insulin-like growth factor axis. In many salmonid species, cortisol and GH administration increases gill Na^+, K^+-ATPase activity and salinity tolerance (McCormick 1996), probably by increasing the number of Na^+, K^+-ATPase immunoreactive cells (chloride cells) in the gill (Seidelin et al. 1999), where cortisol receptors are preferentially located (Uchida et al. 1998). The number of gill cortisol receptors seems to be strongly correlated with the capacity of cortisol to stimulate gill Na^+, K^+-ATPase (Shrimpton and McCormick 1999).

Stress is thought to suppress reproductive physiology via a wide range of mechanisms (Schreck 2010), although the role of cortisol is less clear. Cortisol-implanted sexually maturing brown trout *Salmo trutta* had less gonadotropin in the pituitary gland, lower sex steroid and vitellogenin levels in the plasma, and smaller gonads than control fish (Carragher et al. 1989). Since reproduction is linked to body size, cortisol-induced impairment of growth may indirectly affect reproductive output.

What are the possible effects of cortisol on the fish brain?

Corticosteroid hormones are known to be central to mood control and emotion in a range of vertebrate species (Fietta and Delsante 2009; Wolkowitz et al. 2009). A probable mode of action is cortisol affecting neuronal death and neurogenesis and other aspects of brain structural plasticity (Cameron and Gould 1994; Radley and Morrison 2005; Wong and Herbert 2006). In mammals and birds, turnover of neurons is essential for maintaining normal cognition and emotion (reviewed by Abrous and Wojtowicz 2008; Zhao et al. 2008; Perera et al. 2008). Although our specific knowledge of how cortisol affects brain processes in fish is scant, all observed behavioural and neurobiological effects of either selection for divergent cortisol levels (Moreira et al. 2004; Øverli et al. 2001, 2005) or cortisol exposure (Gregory and Wood 1999; Øverli et al. 2002a; Bernier et al. 2004; DiBattista et al. 2005) indicate a similar role for this hormone in behavioural control and cognition in fishes as in mammals. Furthermore,

- socially subordinate rainbow trout, showing several signs of being chronically stressed, had reduced rates of cell proliferation in their forebrains (Sørensen et al. 2011a).
- brain cell proliferation, as indicated by proliferating cell nuclear antigen immuno-staining, is reduced by cortisol administration (Sørensen et al. 2011b).

Taken together, the above studies strongly suggest that varying levels of cortisol exposure over time is an important modulator of welfare in fishes. Further work is required to examine how cortisol (and stress, other hormones and environmental factors) affects cell proliferation and neurogenesis in the teleost brain, and hence mood.

What are the effects of cortisol on fish behaviour?

Many aquaculture-relevant stressors have been shown to affect various fish behaviours such as feeding, swimming and aggression (Almazán-Rueda et al. 2005; Martins et al. 2006; Ruyet et al. 2008; van de Nieuwegiessen et al. 2008; Santos et al. 2010). However, such studies do not prove cortisol causes the behavioural changes; alternative hypotheses are that behavioural changes cause cortisol changes, or both may be changed by another (triggering/confounding) factor. To isolate the role of cortisol as a modulator of behaviour, studies have increased plasma cortisol levels by treating groups with exogenous cortisol (via implants/intraperitoneal injection, food administration, bath immersion). These studies are supplemented by behavioural comparisons between strains of fish selectively bred for differential cortisol responsiveness to an acute (confinement) stress. However, as other phenotypic features may be inadvertently selected for alongside cortisol (c.f. skin pigmentation, Kittilsen et al. 2009), caution must again be used when attributing behavioural differences between selected lines to differing cortisol levels.

On cognition?

The effects of cortisol on cognition have long been recognised in mammals (e.g. de Kloet et al. 1999; Belanoff et al. 2001), particularly its inverted U-shape effect in which very low or high levels of cortisol impair, whereas moderate elevations facilitate cognition such as the acquisition and retention of memories. The differential activation of MRs and GRs is thought to be responsible for such a relationship (Mateo 2008).

In fish, the limited evidence available supports the concept that cortisol affects cognitive abilities such as learning and memory. Moreira et al. (2004) exposed selected lines of rainbow trout to a conditioning learning paradigm: a conditioned stimuli (CS) of interruption in water flow and a unconditioned stimuli (US) of a confinement stressor. Having learnt the association between CS and US, individuals of both selected lines were exposed to the CS only at weekly intervals and the plasma cortisol assessed. Low cortisol responders retained learnt responses longer than high cortisol responders, suggesting differences in cognitive function between the two lines.

Barreto et al. (2006) supported the findings of Moreira et al. (2004) using unselected rainbow trout. They administrated exogenous cortisol via intraperitoneal implants and compared the retention of a conditioned response (CS: a water jet on the surface of the tank water; US: 30 min of confinement) in cortisol-treated and sham-implant fish. The cortisol-treated fish did not retain the CR 5 days after conditioning ceased, whereas the control fish did.

On feeding behaviour?

The weight of evidence indicates that elevated cortisol levels suppress feeding behaviour in fish. Gregory and Wood (1999) showed that rainbow trout *Oncorhynchus mykiss* injected peritoneally with cortisol consumed less food and showed greater variability in the amount of food consumed from meal to meal, than control and sham-treated fish under both satiation and half-satiation feeding regimes. Barton et al. (1987) observed a loss of appetite in cortisol-fed rainbow trout. Øverli et al. (2002b) examined time to resumption of feeding after isolation in two strains of rainbow trout: low cortisol responders started feeding sooner than high cortisol responders. These three studies indicate an appetite-inhibitory effect of cortisol. A possible mechanism is that cortisol increases glucose and/or amino acids in the blood, which may suppress appetite (Andersen et al. 1991). Cortisol has also been suggested to interact with appetite regulatory pathways of the brain at the level of the expression of CRH and the neuropeptide Y (Bernier et al. 1999, 2004). CRH and some related peptides (urotensin and urocortins) have been described as potent food intake inhibitors, with effects observed when administered intracerebroventricularly (Bernier 2006). The appetite-inhibitory effect of cortisol may be dose-specific: Bernier et al. (2004) found reduced food intake in goldfish *Carassius auratus* fed high cortisol-treated diets (500 μg cortisol/g food), whereas those fed a lower dose (50 μg cortisol/g food) showed increased food intake compared to controls. It must be recognised that cortisol is only one of a complex network of hormones (produced by both brain and peripheral tissues) that regulate food intake in fish (Volkoff et al. 2009).

On aggression and social hierarchy position?

Several observational studies have shown a chronic elevation of plasma cortisol in subordinate fish (e.g. Winberg and Lepage 1998; Sloman et al. 2001, 2002), and experimental studies have confirmed a linkage. Munro and Pitcher (1985), using an immersion technique in the cichlid *Aequidens pulcher,* established a direct relationship between increased cortisol, increased submissive behaviour and decreased aggression. Gregory and Wood (1999) observed greater fin damage in cortisol-implanted fish than control and sham individuals, which was interpreted as reflecting inferior competitive ability (typical of subordinate fish) when coupled with observations on food intake and its variability. Øverli et al. (2002a) found that aggressive behaviour was inhibited in 48 h (dietary) cortisol-treated fish. More recently, DiBattista et al. (2005) found that cortisol exposure (implant, 110 mg kg^{-1} fish) predisposed juvenile rainbow trout to a low social status. They also measured brain monoamines and suggested that the effects of cortisol on social status in fish may be mediated via the modulation of central signalling systems. Schjolden et al. (2009) recently found that attack latency (but not the intensity of aggression) is affected by exogenous cortisol. In dyadic contests between individuals from rainbow trout strains with contrasting cortisol responsiveness, high cortisol responders lost more often than low cortisol responders (Pottinger and Carrick 2001 and reviewed in Øverli et al. 2005).

However, in contrast to the above studies, cortisol implants in an electric fish, *Apteronotus leptorhynchus*, were found to increase the production of electrocommunication signals interpreted as aggressive (Dunlap et al. 2002). Furthermore, in the study of Øverli et al. (2002a), short-term exposure (1 h) to dietary cortisol administration did not affect aggression. The effects of cortisol on aggression may therefore be time-dependent and explained in evolutionary terms: short-term stimulatory effects may be related to the fight-or-flight response whilst long-term inhibitory effects may be related to a conservation withdrawal response (Øverli et al. 2002a).

On swimming activity?

The impact of cortisol on swimming activity is unclear and likely to be time and context-dependent. Espmark et al. (2008) found that cortisol prenatal exposure (through implants in the mother) reduced offspring activity, by increasing 'time spent non-swimming' and 'time spent at the bottom' when exposed to a novel environment. Øverli et al. (2002a) found that long-term cortisol exposure (via food) inhibited locomotory activity, but short-term exposure stimulated locomotory activity. These impacts on swimming activity were not apparent in undisturbed fish, but only when fish were challenged by a conspecific intruder. Gregory and Wood (1999) found no effect on aerobic swimming performances in cortisol-implanted rainbow trout. They suggested that as swimming is a key behavioural response to stressors, impairment would be disadvantageous.

Cortisol, by affecting energetic metabolism, can affect swimming performance. Recovery tests applied to cortisol-implanted and unstressed sea bass (Carbonara et al. 2010) showed that treated fish were able to swim better and longer during a first bout, but they were unable to recover the energetic loss after a period of swimming at low speed (Milligan 2003).

On reproductive behaviour?

Cortisol also has a role in the regulation of reproductive behaviour. In two fish species, Gulf toadfish (*Opsanus beta*) and plainfin midshipman (*Porichthys notatus*), reproductive behaviour involves acoustic signals to attract females to lay eggs at the nest. In both species, systemic injections of cortisol increased the duration of acoustic signals within a few minutes (Remage-Healey and Bass 2004, 2006). Morgan et al. (1999) showed that cod, stressed by confinement, had increased cortisol levels, initiated fewer courtships, and were more likely to skip activities in the courtship sequence than controls. However, in fish exhibiting parental care, cortisol seems not to affect brood care behaviour (e.g. Knapp et al. 1999; Bender et al. 2008; O'Connor et al. 2009). Differences in breeding strategies are thought to explain the different roles cortisol appears to play in the regulation of behaviours in amphibians (Orchinik 1998). Similarly, in fish, the sensitivity of reproductive behaviours to modulation by cortisol is likely to depend on social and environmental conditions.

Can cortisol be used as a welfare indicator?

There are three different approaches to viewing and assessing animal welfare (Huntingford et al. 2006):

- Functional: is the animal functioning well; is it able to cope with the environmental and husbandry conditions?
- Behavioural: do the husbandry and environmental conditions allow the animal to behave naturally/normally?
- Feelings: how does the animal feel about the husbandry and environmental conditions?

Previous examinations of fish welfare (e.g. Ellis et al. 2002) have used cortisol as a non-specific stress indicator assuming that elevated levels indicate a disturbance of physiological function. We now challenge this assumption and the value of cortisol as a functional indicator of welfare.

What is important for function is whether the fish is coping or not. Cortisol measurements cannot provide this information. A cortisol response simply shows that the HPI axis, which is an integral part of normal physiology, is functional and operating to facilitate coping. A cortisol response itself is not predictive of the fish's ability to cope with a situation:

- downstream impacts on function (and behaviour) are dose-, time- and context-dependent. In contrast to the common view that corticosteroids are immunosuppressive, it has been suggested that acute surges of cortisol may enhance some components of the immune defence mechanism in fish (see Ellis et al. 2007b).
- the density of, and affinity of receptors in target tissues for stress hormones (Pottinger 2010), will affect the magnitude of the 2° responses, which will then become further moderated at the 3° level.
- variability in basal and response levels of cortisol (Fig. 2, see below) further prevent prediction of the ability to cope.

If fish welfare is viewed solely from a functional perspective, then surely cortisol (and the 1° neurohormonal responses) should be discarded in favour of the 2° and 3° responses as such measures better reflect the animal's overall response and its coping ability? However, if we adopt a feelings approach aiming to determine how fish themselves perceive the environment and feel about it, then the 1° response is the most appropriate measure, due to the direct link to the brain. Assessment of the 1° stress response should provide a means of asking the fish themselves what conditions they perceive as challenging.

Should fish welfare really address feelings?

A classic definition of animal welfare (Anon 2009) is

> the welfare of an animal is its state as regards its attempts to cope with its environment

However, such a definition provides little guidance to the naïve. In our experience, fish biologists entering the fish welfare field have a poor appreciation of what the term 'welfare' represents. They are familiar with the conjoined term 'health and welfare' in which health is typically associated with infectious disease, whilst welfare covers other non-infectious conditions caused by poor handling, husbandry or environmental conditions, i.e., injury, environmental diseases and stress.

A straightforward aid to understanding fish (animal) welfare is to view health as physical state and welfare as mental state. Mental state (welfare) refers to the *feelings* of the fish: in relation to the environment, husbandry conditions and practices to which they are exposed. There is obviously overlap between health and welfare: a poor physical state (health) can lead to a poor mental state (welfare), and vice versa where a poor mental state (welfare) can lead to a poor physical state (health). This explanation recognises that a poor mental state can occur whilst the physical state is good, and a poor mental state will co-occur with a poor physical state if the fish *feels* unwell.

That feelings are the crux of animal welfare is illustrated by the debate on whether fish are only capable of nociception or can *feel* pain, i.e., whether it is just a neurophysiological response or there is an additional mental experience (Rose 2002, 2007; Sneddon 2009). The 'Five Freedoms' (http://www.fawc.org.uk/freedoms.htm) is a well-respected foundation of animal welfare thinking that has stood the test of time (Anon 1992). These freedoms are simply a paraphrasing (repackaging) of negative feelings (i.e. freedom from feelings of hunger, thirst, discomfort, pain, sickness, frustration, boredom, loneliness, fear, anxiety, sadness, etc). These freedoms are used to guide judgements on welfare, with an infringement of a freedom indicating the presence of negative feelings and hence poor welfare.

Thinking on welfare has evolved since the conception of the 'Five Freedoms' and now, rather than being considered as a binary state (i.e. if welfare is not poor, then it is good), animal welfare is considered as a continuum (Fig. 3) extending from poor, through neutral to good welfare, the latter being associated with positive feelings (Boissy et al. 2007; Yeates and Main 2008).

Feelings may appear to be only one of the three approaches to viewing and assessing animal welfare. The two alternative approaches (function, behaviour) recognise the fact that the objective assessment of animals' (subjective) feelings is extremely difficult and are favoured by those uncomfortable with attributing feelings to non-human animals or preferring a pragmatic approach (e.g. Arlinghaus et al. 2007, 2009). However, these two alternative approaches simply represent proxy measures for feelings (Mormède et al. 2007), i.e.

- If an animal cannot function well, then negative feelings may ensue and, vice versa, if an animal has negative feelings, this may express as compromised function.
- If an animal cannot perform natural behaviour, then negative feelings may ensue and, vice versa, if an animal has negative feelings, this may express as abnormal or compromised behaviour.

WELFARE:	Poor ⟵⟶	Neutral ⟵⟶	Good
FEELINGS:	Negative	No positive or negative, or net balance	Positive
EXAMPLE FEELINGS	pain, ill, tired, hungry, hot, cold, sad (depressed), angry, scared (frightened, anxious), frustrated, bored, uncomfortable		happy, excited
ASSOCIATED STATE OF MIND/ MOOD	pessimistic		optimistic

Fig. 3 Representation of animal welfare as a continuum in relation to feelings

Animals considered to have the capacity for feelings are typically termed 'conscious' or 'sentient' (Chandroo et al. 2004a; Dawkins 2006). Only those higher taxa and developmental stages of animals that are considered to have sufficient mental capacity to experience feelings are endowed with welfare consideration. The physical state (health), function and behaviour of lower animals (and plants) and early developmental stages of higher animals do change in relation to handling, husbandry and environmental quality; however, the deemed lack of a capacity for feelings means that they are not bestowed with welfare consideration. If one accepts the concept of fish welfare, then one necessarily agrees that fish do have the capacity for at least a limited range of feelings. As animal welfare science is driven by public concern, anthropomorphism (bestowing animals with feelings as experienced by humans) needs to be recognised.

The argument for fish and tetrapods having feelings is evolutionary: presumably humans have feelings because they confer a survival advantage and have been selected for; this adaptation is unlikely to have evolved spontaneously in the recent human lineage. Implicit in the evolutionary argument is that the capacity for feelings is likely to differ substantially between species: fish are expected to have a much less advanced capacity for feelings than humans and other endothermic tetrapods; the capacity for feelings is likely to differ substantially between fish taxa due to both evolutionary lineage and the selective pressures associated with life history (Chandroo et al. 2004a); fish may experience feelings that humans do not and cannot comprehend.

Gauging the subjective feelings of animals is recognised to be problematic (hence, the alternative function and behaviour approaches). The most objective means available for assessing the presence of negative feelings in animals is generally accepted (although perhaps not explicitly) to be the measurement of 1° stress response indicators. The underlying assumption is that negative feelings in animals will manifest in a stress response, analogous to that which we recognise in humans (Korte 2001; Levine et al. 2007). [Please note that behavioural indicators provide an additional potential means of assessing fish feelings (reviewed by Martins et al. 2011). Nevertheless, this route may be compromised due to the suggestion that fish, as a common prey item, may not exhibit behavioural changes that could increase vulnerability to predators (CCAC 2005)].

Carl Schreck, a 'grandfather' of fish stress research, was an early advocate of a psychological component in fish endocrine stress responses (Schreck 1981). He interpreted the immediacy and magnitude of the cortisol response to capture and handling events as evidence for a psychological component—fright and anxiety (i.e. feelings). Schreck (1981) noted that 'the psychological component of stress in fish has received little attention in fishes'. Studies over the subsequent 30 years have largely focussed on stress responses to direct physicochemical stressors, and few authors have braved inferring psychological stress in fish until recently (Galhardo and Oliveira 2009).

Is there evidence for negative feelings and psychological stress in fish?

To date, debate around the capacity of fish for feelings has focussed on neuroanatomy, neurophysiology and behaviour (e.g. Rose 2002, 2007; Sneddon et al. 2003, Braithwaite and Huntingford 2004, Chandroo et al. 2004a, b; Huntingford et al. 2006; Broom 2007; Lund et al. 2007; Braithwaite and Boulcott 2007). In his pioneering report on fish welfare, Kestin (1994) presented the endocrinological stress response of fish alongside neurology as evidence that they did merit welfare concern. Cortisol responses in fish seem to have been overlooked, but can contribute to the debate on fish feelings. Cortisol is widely used in human psychobiological studies as an indicator of stress, anxiety and depression (e.g. Levine et al. 2007). Analogous to humans, the cortisol cascade in fish is triggered by the brain, but the key question(s) is: Does cortisol release indicate (negative) feelings and psychological stress in fish, or is it just a hard-wired physiological response?

Schreck (1981) cited two examples of cortisol responses as evidence for fright (≡fear) in fish:

- Female sockeye salmon *Oncorhynchus nerka* exposed to the physical presence of a dipnet being placed in the tank showed a five-fold cortisol elevation.
- Carp *Cyprinus carpio* showed a cortisol response to a brief (few seconds) exposure to the thrashing of another fish (Redgate 1974).

In these two studies, there was no direct physicochemical impact on the fish, but there was a cortisol response to indirectly threatening stimuli (sound/vibrations in the carp experiment; visual in the sockeye salmon experiment). Later, studies have also shown cortisol production to indirectly threatening stimuli, e.g.:

- Sea bream showed a cortisol response when other fish from the same tank were removed by dip-net (Molinero et al. 1997).
- Gulf toadfish showed a cortisol response to the acoustic signal produced by foraging dolphins (Remage-Healey et al. 2006).
- Goldfish showed a cortisol response to solely visual cues of predatory fish (Kagawa and Mugiya 2000).

These disparate studies illustrate cortisol responses in fish to stimuli with no direct noxious (physical or chemical) action, indicating that environmental information was perceived by the fish, interpreted as constituting a threat and translated (presumably within higher brain structures) to produce a cortisol response.

Habituation, acclimation and conditioning studies further indicate that there is a psychological component to the cortisol response in fish:

- Schreck (1981) illustrated that the magnitude of the cortisol response of coho salmon *Oncorhynchus kisutch* to the same handling stress diminished with exposure
- In convict cichlids *Amatitlania nigrofasciata*, the magnitude of the cortisol response to the same handling and confinement decreased with 3–4 exposures over a period of days, which was interpreted as habituation (Wong et al. 2008).

These reductions in cortisol responses may indicate psychological moderation of cortisol response, as the physicochemical stress would remain consistent, although non-psychological regulation of the HPI axis under repeated stimulation cannot be discounted.

- Pottinger and Pickering (1992) found that plasma cortisol levels were elevated in rainbow trout transferred from a communal stock held in a 1,500 L tank into individual tanks of 50 L. One week after transfer, cortisol was elevated ninefold. They concluded that the extended time of elevated cortisol level represented a 'psychological' acclimation to an unfamiliar environment as there was no direct physicochemical stressor.
- Nile tilapia *Oreochromis niloticus* were exposed to a light cue (conditioned stimulus, CS), which was coupled with a confinement stress by lowering a mesh into the tank (unconditioned stimulus, US) once daily for 9 days. When tested on day 10, the fish showed a cortisol (conditioned) response after exposure to just the light cue (Moreira and Volpato 2004)
- Carpenter and Summers (2009) developed a model for fear learning, in which fish had the opportunity to avoid social aggression from a larger conspecific (US) after the water flow was stopped (CS). They showed a 400% increase in plasma cortisol in fish that did not escape after the presentation of the CS only, i.e., no larger conspecific was presented after the water flow was stopped.

Recently, Galhardo et al. (2011) showed a reduction in cortisol response when negative events were given in a predictable way, relative to when the same events were given in an unpredictable way. The authors suggested that fish can subjectively interpret stimuli (i.e. appraise) and that the use of appraisal modulators such as predictability may become an important tool in improving fish welfare.

Schreck (1981) and others have also proposed a psychological component to fish cortisol responses due to

- anaesthetics reducing the cortisol response to handling
- the absence of visual stimuli in the dark reducing cortisol responsiveness to handling
- spatial restriction due to crowding or confinement causing a cortisol response (Li and Brocksen 1977)

However, interpretation as psychological stress from such routes is uncertain due to the possibility of alternative physicochemical explanations.

Although it is impossible to demonstrate unequivocally that fish do experience feelings, the above evidence from cortisol does indicate a psychological capacity supportive of the concept that fish do experience negative feelings.

What are likely problems in applying cortisol as a welfare indicator?

Separating stress from distress

The holy grail of stress indicators is the differentiation of physiologically induced stress from psychologically induced stress. Psychological stress implies awareness of a challenge, negative feelings, distress and hence poor welfare. Fish are not necessarily aware of a state of physiological stress; hence, it is not necessarily associated with negative feelings and poor welfare. Current information on cortisol does not allow definitive differentiation, nor does any other stress indicator of which we are aware. Therefore, it has to be recognised that assuming distress by using cortisol as an indicator of negative feelings is giving fish 'the benefit of the doubt'. Events that cause stress are inevitable within the fish farming industry (Barton 1997)—fish will need to be confined, handled, graded and transported. The onus on the fish farmer is then not to completely avoid causing distress, but to minimise it as far as possible.

Questions are often raised as to the value of cortisol in interpreting welfare. To stimulate thought, we provide provisional responses to some common questions:

Q: Isn't stress natural?

R: Yes, stress is natural, and the HPI axis has indeed evolved to cope with natural stressors. We are proposing the use of cortisol as an indicator of negative feelings. Mortality and injury are also natural and yet animal welfare science aims to reduce these to reduce negative feelings and suffering. [Please note that the theoretical view of the stress–welfare relationship is developing: the common perception of a negative relationship (↑stress = ↓welfare) has been challenged and an alternative model suggested based on a dome-shaped relationship where both too little (hypostimulation) and too much stress result in poor welfare (Korte et al. 2007)].

Q: Isn't low level stress (or pressure) good?

R: If cortisol is assumed to indicate negative feelings, then elevated cortisol level indicates negative feelings and hence poor welfare as argued above and effects on function are irrelevant. Additional information on other 1° stress response indicators (catecholamines) would however be valuable for a more holistic interpretation.

Q: Don't fish (and other animals/humans) choose to expose themselves to stressors?

R: Fish may make behavioural choices that incur endocrinological stress, e.g., dominant fish engaging in aggressive behaviour that leads to elevated cortisol levels (Noakes and Leatherland 1977). However, such choices can only be judged validly if the physiological status, motivation and range of options perceived by the fish were known.

Q: What if cortisol contradicts other welfare indicators?

R: Obviously, this will be problematic for interpretation, and further thought, discussion and research are needed to reach a consensus. An example is Ellis et al. (2005) where lightly anaesthetised rainbow trout (to Stage I, Plane 1; Tytler and Hawkins 1981) were unresponsive to external stimuli and showed no behavioural signs of distress, but highly elevated plasma cortisol levels indicated awareness and distress.

Q: Doesn't cortisol increase in relation to positive, as well as negative, feelings?

R: It has been suggested from mammal studies that elevated cortisol could simply indicate emotional arousal rather than valence, i.e., positive (e.g. excitement) as well as negative feelings. Bishop et al. (1999) suggested an increase in cortisol did not distinguish between positive and negative feelings in boar; Doyle et al. (2010) assumed positive emotions in individual sheep with an increase in cortisol. In a review, Boissy et al. (2007) suggest that both positive and negative emotions can trigger stress axes in higher vertebrates. As cortisol plays important physiological roles, we cannot dismiss the possibility that cortisol elevation may be an adaptive response for positive, as well as negative, feelings in fish.

Nevertheless, there is currently very little evidence that fish do experience positive emotional states and that cortisol levels in fish become elevated in response to assumed positive experiences. Sánchez et al. (2009) observed anticipatory activity of feeding, which could be interpreted as a response to a positive stimulus (see also review by Martins et al. 2011), but this was associated with lower plasma cortisol levels than in fish that did not show this behaviour. The multitude of studies describing cortisol elevation in fish relates to stimuli interpreted as negative stressors. The majority (but not unanimous) opinion of the authors of this review was that the assumption that cortisol elevation illustrates negative feelings (derived from current dogma that adverse stressors precipitate cortisol elevation) in fish is justified at present. Clearly, studies examining emotions in animals, especially fish, are limited to date, and suitable experimental methodologies are being developed. The assumption, and the value of cortisol in fish welfare studies, will therefore need to be reappraised in the light of future evidence.

Variability in basal and response levels

It has been suggested that basal cortisol levels in fish are generally <10 ng/mL (Sumpter 1997). However, because cortisol contributes to the maintenance of both basal and stress-related homeostasis, regulating routine metabolic functions such as glucose and ionic regulation, elevated levels cannot simply be assumed to indicate stress (Laidley and Leatherland 1988; Chrousos and Kino 2009). Both basal and stressed cortisol levels differ widely between species, and within a species (Fig. 2), being affected by factors such as temperature, gender, sexual maturity and genetics (Pottinger 2010). Barton (2002) showed that both basal and stressed cortisol levels differed between species of fish exposed to an identical procedure (30 s aerial emersion). In sea bass, control plasma cortisol concentrations (before application of stress) reported in different studies vary over two orders of magnitude (Table 5). It is currently unknown whether these major within species differences

- reflect true differences in basal cortisol concentrations associated with different fish conditions (e.g. body size, age), or
- reflect genetic/strain differences in cortisol responsiveness/coping style/personality/temperament/

behavioural syndromes, i.e., physiological–behavioural trait correlations that are consistent over time and across situation (Koolhaas et al. 1999; Øverli et al. 2007; Silva et al. 2010), or

- are due to different environmental conditions (e.g. water temperature, salinity, lightning, photoperiod, season, food composition), or
- are due to the presence of unrecognised stressors (e.g. poor husbandry conditions, disease) in some studies, or
- are a consequence of the analytical method used (RIA, EIA/ELISA, HPLC) and possible errors.

This inherent variability in plasma cortisol levels, with basal levels in some studies exceeding acute stress levels of others (Fig. 2), means that single time point measurements of cortisol are of no value in assessing the stress level of fish; studies need to compare treatment with control fish.

As cortisol plays physiological 'housekeeping' roles, basal levels change with both time of day and season, which must be considered when sampling to compare different treatments or stressors. Typically, in diurnal species, such as the human, a daily cortisol peak occurs at the beginning of the light phase, whilst in a nocturnal animal, such as the rat, the cortisol peak occurs at the end of the light phase (Dickmeis 2009). Such daily rhythms in cortisol have been described in various species of fish (Table 6). The amplitude of the variations and, more importantly the time of day when the highest values occurred (acrophase), are species-dependent and show wide differences between species. When fish are exposed to chronic stressors, the rhythm appears to be maintained with the same acrophase, but the mean values and the amplitude increase (e.g. Pickering and Pottinger 1983). Furthermore, the response to an acute stressor can depend upon the time of the day, e.g., in the green sturgeon, acute aerial exposure elicited higher cortisol values when applied at night than during the day (Lankford et al. 2003).

In addition to the daily rhythm, fish cortisol levels can display seasonal variations, which involve changes in both the amplitude and the acrophase of the cortisol rhythm, as described for brown trout and Atlantic salmon *Salmo salar* (Table 6). Water

Table 5 Plasma cortisol levels recorded in various studies of sea bass, prior to the application of stressors

Pre-stress plasma cortisol concentration (ng/mL)	Fish weight (g)	Method	Author
100–600	35	RIA	Planas et al. (1990)
84 ± 12/125 ± 8/133 ± 12	–	RIA	Roche and Bogé (1996)
36 ± 1	57	RIA	Cerdá-Reverter et al. (1998)
307 ± 25	–	ELISA	Santulli et al. (1999)
76	372	RIA	Marino et al. (2001)
10 ± 8	225	RIA	Vazzana et al. (2002)
17 ± 5	123	RIA	Rotllant et al. (2003)
234–366	14	ELISA	Teles et al. (2004)
15–23	–	HPLC	Caruso et al. (2005)
152 ± 17	1005	RIA	Peruzzi et al. (2005)
18	100	RIA	Rotllant et al. (2006)
26 ± 3	–	ELISA	Tintos et al. (2006)
335 ± 26	22	ELISA	Teles et al. (2006)
13 ± 30/186 ± 56	13	RIA	Varsamos et al. (2006)
158 ± 32/446 ± 58	83	ELISA	Herrero et al. (2007)
17 ± 3/27 ± 3	149	RIA	Di Marco et al. (2008)
136	125	ELISA	Fanouraki et al. (2008)
13 ± 2	120	ELISA	Yildiz and Ergonul (2010)
33 ± 5	120–200	Immunoassay analyser	Roque et al. (2010)
75 ± 12	325 ± 25	HPLC	Carbonara et al. (2010)

Table 6 Examples of daily rhythms of cortisol in fish and changes in rhythm in relation to season

Fish species	Cortisol (min–max)	Time of cortisol peak	Reference
Carassius auratus	10–60 ng/mL/g	Day	Singley and Chavin (1975)
Oncorhynchus mykiss	5–40 ng/mL	Midnight	Rance et al. (1982)
Heteropneustes fossilis	50–200 ng/mL (seasonal variations)	Night	Lamba et al. (1983)
Salmo trutta	5–30 ng/mL (seasonal variations)	Night	Pickering and Pottinger (1983)
Salmo salar	40–160 ng/mL (seasonal variations)	Day: June–August Night: September–May	Nichols and Weisbart (1984); Thorpe et al. (1987)
Cyprinus carpio	0.5–1.5 μmol/L	Second half of the night	Kühn et al. (1986)
Dicentrarchus labrax	24–40 ng/mL	Bimodal—Light onset and offset	Cerdá-Reverter et al. (1998)
Dentex dentex	10–40 ng/mL	Day: Long photoperiod Night: Short photoperiod	Pavlidis et al. (1999)
Salvelinus leucomaenis	5–30 ng/mL	Night	Yamada et al. (2002)
Anguilla japonica	30–80 ng/mL	Night	Li and Takei (2003)
Ictalururs punctatus	7–20 ng/mL	End of night	Small (2005)
Sparus aurata	2–30 ng/mL	Light onset	López-Olmeda et al. (2009)

temperature has been postulated to be an important factor in generating seasonal rhythms in cortisol (Kühn et al. 1986; Planas et al. 1990), but different photoperiods at the same temperature can elicit different daily cortisol profiles (Pavlidis et al. 1999), indicating that both environmental factors affect cortisol annual variations.

Interpreting the absence or loss of a cortisol response

When fish are exposed to an environmental treatment that does not trigger a cortisol response, but other functional measures indicate detrimental conditions, interpretation of the impact on welfare is difficult. For example, sub-optimal dissolved oxygen concentrations (4–5 mg L^{-1}) affect food intake, food conversion efficiency and growth of rainbow trout, but this is above the oxygen threshold ($\leq$3 mg L^{-1}) that triggers a cortisol response (see Ellis et al. 2002). The lack of a cortisol response at water oxygen concentrations $>$ 3 mg L^{-1} could be assumed to indicate that the fish were not aware that they were exposed to a challenge and therefore did not experience negative feelings. Catecholamines are also released in response to hypoxia (Reid et al. 1998), but it is unclear whether the environmental threshold differs to that for cortisol release. Furthermore, the oxygen thresholds that cause measurable behavioural responses seem to vary between different behavioural indicators, e.g., residence time in choice tanks, avoidance behaviour (Poulsen et al. 2011). Debate and scrutiny of indicators other than cortisol are therefore required to determine the acceptability of conditions in terms of fish welfare.

In some studies of assumed chronic stressors (e.g. elevated stocking density, confinement), elevated cortisol levels have been sustained for several weeks (e.g. Pickering and Stewart 1984), and the persistent alarm response may be interpreted as sustained distress. However, in other studies of chronic stressors, after an initial cortisol elevation, levels returned to basal within 1 week (e.g. Pickering and Stewart 1984; Pickering and Pottinger 1987b; Rotllant et al. 2001; Ellis et al. 2007a). It is important for the interpretation of fish welfare that the cause for the return to basal levels is established. Various hypotheses have been proposed (Pickering and Stewart 1984; Procarione et al. 1999; Pickering and Pottinger 1987a; Wendelaar Bonga 1997):

1. the HPI axis is stimulated, but cortisol secretion is impaired by interrenal exhaustion
2. the HPI axis is stimulated, but cortisol secretion is impaired by environmental factors (e.g. deterioration in water quality)
3. the HPI axis is stimulated, but the elevated cortisol production rate is masked by an increased metabolic clearance rate of cortisol from the plasma
4. the HPI axis is stimulated, but the cortisol production rate is reduced through negative feedback, e.g., inhibition of release of ACTH and CRH or down-regulation of ACTH receptors in the interrenal
5. the HPI axis is no longer being stimulated, as the fish 'acclimate' through gradual loss of awareness of the stress factor

Of the above hypotheses, only the last indicates that the fish no longer has negative feelings; the others imply that the fish still perceives the environment as threatening.

There is a lack of evidence for the first hypothesis of interrenal exhaustion. Pickering and Stewart (1984) assessed interrenal cell histology and found no differences between treatments that could account for the decrease in cortisol level in the chronic stress treatment. Evidence for the second hypothesis that cortisol secretion is impaired by environmental factors is also minimal. This concept stems largely from the study of Pickering and Pottinger (1987a) in which the cortisol response to acute confinement challenge was modulated by water chemistry. The authors themselves indicate that they found the result 'somewhat paradoxical' and suggested that the poor water quality may have reduced the 'awareness' of the fish. Endocrine disruption of the HPI axis of fish due to contaminants such as agrichemicals and metals is well documented (e.g. Cericato et al. 2008; Sandhu and Vijayan 2011), but these would need to contaminate the treatment after the initial cortisol response.

Methods are available to test the third and fourth hypotheses, i.e., the HPI axis is stimulated but plasma cortisol levels are not elevated:

- Determination of cortisol clearance rate: the hypothesis that HPI activation is not apparent from plasma cortisol levels due to increased cortisol clearance from the plasma is commonly cited. However, evidence for its validity is largely lacking as it is technically difficult to directly assess cortisol kinetics—the estimation of cortisol clearance involves injection of exogenous radiolabelled cortisol, and many steps in subsequent calculations (Schreck et al. 1985; Laidley et al. 1988)—and is therefore rarely attempted. Redding et al. (1984) found evidence for elevated plasma cortisol clearance rate in coho salmon exposed to crowding for 5 days—however, elevated cortisol levels might also be expected to still be apparent within this time scale. Schreck et al. (1985) recorded an increased cortisol clearance rate in fish exposed to a higher density in the absence of an elevated plasma cortisol level, but they were tentative in drawing a conclusion as there was no evidence from interrenal histology for HPI axis stimulation. Recent developments in measuring excreted cortisol metabolites in faeces may prove valuable in this regard, as most of the circulating cortisol is removed by the liver and excreted via the bile in a glucuronated form (Oliveira et al. 1999).
- Measurement of ACTH: ACTH is rarely measured, presumably due to the probability of a response to sampling stress. Vijayan and Leatherland (1990, cited by Wendelaar Bonga 1997) did indeed observe ACTH down-regulation, but elevated cortisol levels were still apparent. A similar result has been found in sea bream *Sparus aurata* (Rotllant et al. 2001).
- Examination of the interrenal—the volume of the interrenal tissue and the structure of the cells (interrenal cell size, nuclear size and nuclear/cytoplasm ratio) may indicate HPI axis stimulation (Pickering and Stewart 1984). Although apparently simple, the histological effort required means that such studies are few, and there do not appear to have been any studies that have convincingly demonstrated an altered interrenal structure in the absence of elevated cortisol levels (Schreck et al. 1985).
- Cortisol response to an injection of exogenous ACTH in vivo: this does not appear to have been applied to fish held under 'chronic' conditions, although has been examined in vitro (Rotllant et al. 2000).
- The magnitude of the cortisol response after exposure to an additional acute stressor (e.g.

Leatherland 1993). Fish exposed to purported chronic stressors and then exposed to an additional stressor have been found to show similar cortisol responses (Leatherland 1993; Rotllant and Tort 1997), more exaggerated cortisol responses (Ruane et al. 2002) and less exaggerated cortisol responses (Barton et al. 2005) than control fish. This diversity illustrates that interpretation of such results should be viewed with caution.

On the basis of the current lack of clear evidence to the contrary, it could seem reasonable to conclude that if plasma cortisol levels show an apparent acclimation to a chronic stress, then this indicates that the HPI axis is no longer being stimulated, the fish have acclimated, they no longer perceive the challenge as threatening and negative feelings have departed. However, more research effort is needed to illuminate this obscure area.

Cortisol and fish welfare: what next?

We hope we have encouraged readers to revise their probable view of cortisol as simply an indicator of physiological stress and affected function. We hope that the basic physiology and histology research still needed to answer some outstanding questions are not overlooked in the current wave of enthusiasm over molecular approaches to examine receptors and stress-related genes (e.g. Prunet et al. 2008); simple measurements of cortisol may ultimately prove more suitable for the assessment of fish welfare. We anticipate that the study (and acceptance) of psychological stress in fish will expand (e.g. Galhardo and Oliveira 2009). Associated areas pertinent to a better understanding of fish welfare include the following:

- further exploration of cortisol as indicator of negative, and possibly positive, feelings and identification of associated behavioural states for validation.
- how cortisol influences the sensory input (how stimuli are perceived) and the central nervous system (where behavioural integration and processing takes place) as well as the third main element of behaviour—the effector system (output) that is typically studied (Neave 2008), i.e., the interaction between cortisol and mood in fish.

We also reiterate that the stress literature for fish is skewed towards cortisol; studies on catecholamine responses in fish, and their role in physiological and behavioural responses to stressors, are inadequate (Pottinger 2010). Catecholamines are released in response to the same stressors as cortisol and precipitate metabolic responses similar to those attributed to cortisol (Reid et al. 1998; Fabbri et al. 1998). Catecholamines (as cortisol) are likely to play an important role in fish behaviour and welfare, and merit greater attention.

Acknowledgments This review paper represents an output from the COST Action 867 'Welfare of Fish in European Aquaculture' supported by the EU Commission. The text represents the authors' views. Thanks to Neil Ruane (Marine Institute, Co. Galway, Ireland) and the anonymous referees for helpful input.

References

Abbink W, Bevelander GS, Rotllant J, Canario AVM, Flik G (2004) Calcium handling in *Sparus auratus*: effects of water and dietary calcium levels on mineral composition, cortisol and PTHrP levels. J Environ Biol 207:4077–4084

Abrous DN, Wojtowicz JM (2008) Neurogenesis and hippocampal memory system. In: Gage FH, Kempermann G, Song HJ (eds) Adult neurogenesis. Cold Spring Harbour Laboratory Press, Could Spring Harbour, pp 445–461

Almazán-Rueda P, Van Helmond ATM, Verreth JAJ, Schrama JW (2005) Photoperiod affects growth, behaviour and stress variables in *Clarias gariepinus*. J Fish Biol 67:1029–1039

Aluru N, Vijayan MM (2009) Stress transcriptomics in fish: a role for genomic cortisol signalling. Gen Comp Endocrinol 164:142–150

Alvarellos S, Arjona FJ, Martin del Rio MP, Miguez JM, Mancera JM, Soengas JL (2005) Time course of osmoregulatory and metabolic changes during osmotic acclimation in *Sparus auratus*. J Exp Biol 208:4291–4304

Alvarellos SS, Polakof S, Arjona FJ, Kleszynska A, Martin del Rio PM, Miguez JM, Soengas JL, Mancera JM (2006) Osmoregulatory and metabolic changes in the gilthead sea bream *Sparus auratus* after arginine vasotocin (AVT) treatment. Gen Comp Endocrinol 148:348–358

Andersen DE, Reid SD, Moon TW, Perry SF (1991) Metabolic effects associated with chronically elevated cortisol in rainbow trout (*Oncorhynchus mykiss*). Can J Fish Aquat Sci 48:1811–1817

Anderson DP (1990) Immunological indicators: effects of environmental stress on immune protection and disease outbreaks. Am Fish Soc Symp 8:38–50

Anon (1992) FAWC updates the five freedoms. Vet Rec 131:357

Anon (2009) Scientific opinion of the panel on animal health and welfare on a request from European Commission on general approach to fish welfare and to the concept of sentience in fish. EFSA J 954:1–26

Arends RJ, Mancera JM, Munoz JL, Wendelaar Bonga SE (1999) The stress response of the gilthead seabream (*Sparus aurata* L.) to air exposure and confinement. J Endocrinol 163:149–157

Arlinghaus R, Cooke SJ, Schwab A, Cowx IG (2007) Fish welfare: a challenge to the feelings-based approach, with implications for recreational fishing. Fish Fish 8:57–71

Arlinghaus R, Schwab A, Cooke SJ, Cowx IG (2009) Contrasting pragmatic and suffering-centred approaches to fish welfare in recreational angling. J Fish Biol 75:2448–2463

Barreto RE, Volpato GL, Pottinger TG (2006) The effect of elevated blood cortisol levels on the extinction of a conditioned stress response in rainbow trout. Horm Behav 50:484–488

Barton BA (1997) Stress in finfish: past, present and future—a historical perspective. In: Iwama GK, Pickering AD, Sumpter JP, Schreck CB (eds) Fish stress and health in aquaculture. Cambridge University Press, Cambridge, pp 1–33

Barton BA (2000) Salmonid fishes differ in their cortisol and glucose responses to handling and transport stress. N Am J Aquacult 62:12–18

Barton BA (2002) Stress in fishes: a diversity of responses with particular reference to changes in circulating corticosteroids. Int Comp Biol 42:517–525

Barton BA, Iwama GK (1991) Physiological changes in fish from stress in aquaculture with emphasis on the response and effects of corticosteroids. Ann Rev Fish Dis 1:3–26

Barton BA, Peter RE (1982) Plasma cortisol response in fingerling rainbow trout, *Salmo gairdneri* Richardson, to various transport conditions, anaesthesia, and cold shock. J Fish Biol 20:39–51

Barton BA, Toth LT (1980) Physiological stress in fish: a literature review with emphasis on cortisol dynamics. Fish Res Rep Fish Wildl Div 20:1–19

Barton BA, Schreck CB, Barton LD (1987) Effects of chronic cortisol administration and daily acute stress on growth, physiological conditions, and stress responses in juvenile rainbow trout. Dis Aquat Org 2:173–185

Barton BA, Morgan JD, Vijayan MM (2002) Physiological and condition-related indicators of environmental stress in fish. In: Adams SM (ed) Biological indicators of aquatic ecosystem stress. Am Fish Soc, Bethesda, Maryland, pp 111–148

Barton BA, Ribas L, Acerete L, Tort L (2005) Effects of chronic confinement on physiological responses of juvenile gilthead sea bream, *Sparus aurata* L., to acute handling. Aqua Res 36:172–179

Belanoff JK, Gross K, Yager A, Schatzberg AF (2001) Corticosteroids and cognition. J Psych Res 35:127–145

Bender N, Heg-Bachar Z, Oliveira RF, Canario AVM, Taborsky M (2008) Hormonal control of brood care and social status in a cichlid fish with brood care helpers. Physiol Behav 94:349–358

Bernier NJ (2006) The corticotropin-releasing factor system as a mediator of the appetite-suppressing effects of stress in fish. Gen Comp Endocrinol 146:45–55

Bernier NJ, Lin X, Peter RE (1999) Differential expression of corticotropin-releasing factor (CRF) and urotensin I precursor genes, and evidence of CRF gene expression regulated by cortisol in goldfish brain. Gen Comp Endocrinol 116:461–477

Bernier NJ, Bedard N, Peter RE (2004) Effects of cortisol on food intake, growth, and forebrain neuropeptide Y and corticotropin-releasing factor gene expression in goldfish. Gen Comp Endocrinol 135:230–240

Bishop JD, Malven PV, Singleton WL, Weesner GD (1999) Hormonal and behavioral correlates of emotional states in sexually trained boars. J Animal Sci 77:3339–3345

Boissy A, Manteuffel G, Jensen MB, Moe RO, Spruijt B, Keeling LJ, Winckler C, Forkman B, Dimitrov I, Langbein J, Bakken M, Veissier I, Aubert A (2007) Assessment of positive emotions in animals to improve their welfare. Physiol Behav 92:375–397

Braithwaite VA, Boulcott P (2007) Pain perception, aversion and fear in fish. Dis Aquat Org 75:131–138

Braithwaite VA, Huntingford FA (2004) Fish and welfare: do fish have the capacity for pain perception and suffering? Animal Welfare 13:S87–S92

Broom DM (2007) Cognitive ability and sentience: which aquatic animals should be protected? Dis Aquat Org 75:99–108

Cameron HA, Gould E (1994) Adult neurogenesis is regulated by adrenal steroids in the dentate gyrus. Neuroscience 61:203–209

Carbonara P, Corsi I, Focardi S, Lembo G, Rochira S, Scolamacchia M, Spedicato MT, McKinley RS (2010) The effects of stress induced by cortisol administration on the repeatability of swimming performance tests in the European sea bass (*Dicentrarchus labrax* L.). Mar Freshwater Behav Physiol 43:283-296. doi:10.1080/10236244.2010.504046

Carpenter RE, Summers CH (2009) Learning strategies during fear conditioning. Neurobiol Learn Mem 91:415–423

Carragher JF, Sumpter JP (1990) Corticosteroid physiology in fish. In: Epple A, Scanes CG, Stetson MH (eds) Progress in comparative endocrinology. Wiley Liss, New York, pp 487–492

Carragher JF, Sumpter JP, Pottinger TG, Pickering AD (1989) The deleterious effects of cortisol implantation on reproductive function in two species of trout, *Salmo trutta* L. and Salmo gairdneri Richardson. Gen Comp Endocrinol 76:310–321

Carrion RL, Alvareollos S, Guzman JM, Martin del Rio M, Miguez JM, Soengas JL, Mancera JM (2002) Energy metabolism in fish tissues related to osmoregulation and cortisol action. Fish Physiol Biochem 27:179–188

Caruso G, Genovese L, Maracchiolo G, Mogica A (2005) Haematological biochemical and immunological parameters as stress indicators in *Dicentrarchus labrax* and *Sparus aurata* farmed in off-shore cages. Aquac Int 13:67–73

Casamitjana J (2004) Aquatic zoos: a critical study of UK public aquaria in the year 2004. Captive Animals' Protection Society, p 136

CCAC: Canadian Council on Animal Care (2005) Guidelines on: the care and use of fish in research, teaching and testing. 94 pp. (http://ccac.ca/Documents/Standards/Guidelines/Fish.pdf)

Cerdá-Reverter JM, Zanuy S, Carrillo M, Madrid JA (1998) Time-course studies on plasma glucose, insulin, and cortisol in sea bass (*Dicentrarchus labrax*) held under different photoperiodic regimes. Physiol Behav 64:245–250

Cericato L, Neto JGM, Fagundes M, Kreutz LC, Quevedo RM, Finco J, da Rosa JGS, Koakoski G, Centenaro L, Pottker E, Anziliero D, Barcellos LJG (2008) Cortisol response to acute stress in jundiá *Rhamdia quelen* acutely exposed to sub-lethal concentrations of agrichemicals. Comp Biochem Physiol C 148:281–286

Chan DKO, Chester Jones I, Mosley W (1968) Pituitary and adrenocorticol factors in the control of water and electrolyte composition of the freshwater European eel (*Anguilla anguilla* L.). J Endocrinol 42:91–98

Chan DKO, Rankin JC, Chester Jones I (1969) Influences of the adrenal cortex and the corpuscles of Stannius on osmoregulation in the European eel (*Anguilla anguilla* L.). Gen Comp Endocrinol Suppl 2:342–353

Chandroo KP, Duncan IJH, Moccia RD (2004a) Can fish suffer?: perspectives on sentience, fear and stress. App Anim Behav Sci 86:225–250

Chandroo KP, Yue S, Moccia RD (2004b) An evaluation of current perspectives on consciousness and pain in fishes. Fish Fish 5:281–295

Chrousos GP (1998) Stressors, stress, and neuroendocrine integration of the adaptive response: the 1997 Hans Selye Memorial Lecture. Ann New York Acad Sci 851:311–335

Chrousos GP, Kino T (2009) Glucocorticoid signalling in the cell. Expanding clinical implications to complex human behavioral and somatic disorders. Ann NY Acad Sci 1179:153–166

Davis KB (2006) Management of physiological stress in finfish aquaculture. N Am J Aquacult 68:116–121

Dawkins MS (2006) A user's guide to animal welfare science. Trends Ecol Evol 21:77–82

de Boeck G, Alsop D, Wood C (2001) Cortisol effect on aerobic and anaerobic metabolism, nitrogen excretion and whole-body composition in juvenile rainbow trout. Physiol Biochem Zool 74(4):858–868

de Kloet ER, Oitzl MS, Joels M (1999) Stress and cognition: are corticosteroids good or bad guys? Trends Neurosci 22:422–426

Di Marco P, Priori A, Finoia MG, Massari A, Mandich A, Marino G (2008) Physiological responses of European sea bass *Dicentrarchus labrax* to different stocking densities and acute stress challenge. Aquaculture 275:319–328. doi:10.1016/j.aquaculture.2007.12.012

DiBattista JD, Anisman H, Whitehead M, Gilmour KM (2005) The effects of cortisol administration on social status and brain monoaminergic activity in rainbow trout *Oncorhynchus mykiss*. J Exp Biol 208:2707–2718

Dickmeis Y (2009) Glucocorticoids and the circadian clock. J Endocrinol 200:3–22

Donaldson EM (1981) The pituitary-interrenal axis as an indicator of stress in fish. In: Pickering AD (ed) Stress and fish. Academic Press, New York, pp 11–47

Doyle RE, Fisher AD, Hinch GN, Boissy A, Lee C (2010) Release from restraint generates a positive judgement bias in sheep. App Anim Behav Sci 122:28–34

Dunlap KD, Pelczar PL, Knapp R (2002) Social interactions and cortisol treatment increase the production of aggressive electrocommunication signals in male electric fish, *Apteronotus leptorhynchus*. Horm Behav 42:97–108

Ellis T, North B, Scott AP, Bromage NR, Porter M, Gadd D (2002) The relationships between stocking density and the welfare of farmed rainbow trout. J Fish Biol 61:493–531

Ellis T, James JD, Stewart C, Scott AP (2004) A non-invasive stress assay based upon measuring cortisol release into the water by rainbow trout. J Fish Biol 65:1233–1252

Ellis T, James JD, Scott AP (2005) Branchial release of free cortisol and melatonin by rainbow trout. J Fish Biol 67:535–540

Ellis T, James J, Fridell F, Sundh H, Sundell K, Scott AP (2007a) Non-invasive measurement of cortisol and melatonin in seawater Atlantic salmon tanks. Aquaculture 272:707–716

Ellis T, Bagwell N, Pond M, Baynes S, Scott AP (2007b) Acute viral and bacterial infections elevate water cortisol concentrations. Aquaculture 272:698–706

Espmark AM, Eriksen MS, Salte R, Braastad BO, Bakken M (2008) A note on pre-spawning maternal cortisol exposure in farmed Atlantic salmon and its impact on the behaviour of offspring in response to a novel environment. App Anim Behav Sci 110:404–409

Fabbri E, Capuzzo A, Moon TW (1998) The role of circulating catecholamines in the regulation of fish metabolism: an overview. Comp Biochem Physiol C 120:177–192

Fanouraki E, Papandroulakis N, Ellis T, Mylonas CC, Scott AP, Pavlidis M (2008) Water cortisol is a reliable indicator of stress in European sea bass, *Dicentrarchus labrax*. Behaviour 145(SI):1267–1281

Fietta P, Delsante G (2009) Central nervous system effects of natural and synthetic glucocorticoids. Psych Clin Neurosci 63:613–622

Flos R, Reig L, Torres P, Tort L (1988) Primary and secondary stress responses to grading and hauling in rainbow trout, *Salmo gairdneri*. Aquaculture 71:99–106

Fortner NA, Pickford GE (1982) The effects of hypophysectomy and replacement therapy with prolactin, cortisone, or their combination on the blood of the black bullhead *I-catalurus melas*. Gen Comp Endocrinol 47:111–119

Galhardo L, Oliveira RF (2009) Psychological stress and welfare in fish. Ann Rev Biomed Sci 11:1–20

Galhardo L, Almeida O, Oliveira RF (2011) The role of predictability in the stress response of a cichlid fish. Physiol Behav 102:367–372

Gilham ID, Baker BI (1985) A black background facilitates the response to stress in teleosts. J Endocrinol 105:99–105

Gornati R, Papis E, Rimoldi S, Terova G, Saroglia M, Bernardini G (2004) Rearing density influences the expression of stress-related genes in sea bass (*Dicentrarchus labrax* L.) HMG-CoA reductase mRNA. Gene 341:111–118

Grassi Milano E, Basari F, Chimenti C (1997) Adrenocortical and adrenomedullary homologs in eight species of adult and developing teleosts: morphology, histology and immunohistochemistry. Gen Comp Endocrinol 108:483–496

Gregory TR, Wood CM (1999) The effects of chronic plasma cortisol elevation on the feeding behaviour, growth, competitive ability, and swimming performance of juvenile rainbow trout. Phys Biochem Zool 72:286–295

Heath AG (1995) Water pollution and fish physiology. CRC Press, VPI and State University, Blacksburg 359 pp

Herrero MJ, Martinez FJ, Miguez JM (2007) Response of plasma and gastrointestinal melatonin, plasma cortisol and activity rhythms of European sea bass (Dicentrarchus labrax) to dietary supplementation with tryptophan and melatonin. J Comp Physiol [B] 177: 319–326

Huntingford FA, Adams C, Braithwaite VA, Kadri S, Pottinger TG, Sandöe P, Turnbull JF (2006) Current issues in fish welfare. J Fish Biol 68:332–372

Jentoft S, Aastveit AH, Torjesen PA, Andersen Ø (2005) Effects of stress on growth, cortisol and glucose levels in non-domesticated Eurasian perch (*Perca fluviatilis*) and domesticated rainbow trout (*Oncorhynchus mykiss*). Comp Biochem Physiol A Mol Integr Physiol 141:353–358

Kagawa N, Mugiya Y (2000) Exposure of goldfish (*Carassius auratus*) to bluegills (*Lepomis macrochirus*) enhances expression of stress protein 70 mRNA in the brains and increases plasma cortisol levels. Zool Sci 17:1061–1066

Kestin SC (1994) Pain and stress in fish. RSPCA, Horsham, West Sussex

Kittilsen S, Schjolden J, Beitnes-Johansen I, Shawa JC, Pottinger TG, Sørensen C, Braastad BO, Bakken M, Øverli Ø (2009) Melanin-based skin spots reflect stress responsiveness in salmonid fish. Hormon Behav 56:292–298

Knapp R, Wingfield JC, Bass AH (1999) Steroid hormones and paternal care in the plainfin midshipman fish (*Porichthys notatus*). Horm Behav 35:81–89

Koolhaas JM, Korte SM, de Boer SF, van der Vegt BJ, van Reenen CG, Hopster H, de Jong IC, Ruis MAW, Blokhuis HJ (1999) Coping styles in animals: current status in behavior and stress-physiology. Neurosci Biobehav Rev 23:925–935

Korte SM (2001) Corticosteroids in relation to fear, anxiety and psychopathology. Neurosci Biobehav Rev 25:117–142

Korte SM, Olivier B, Koolhaas JM (2007) A new animal welfare concept based on allostasis. Physiol Behav 92:422–428

Kühn ER, Corneillie S, Ollevier F (1986) Circadian variations in plasma osmolality, electrolytes, glucose, and cortisol in carp (*Cyprinus carpio*). Gen Comp Endocrinol 61: 459–468

Laidley CW, Leatherland JF (1988) Cohort sampling, anaesthesia and stocking-density effects on plasma cortisol, thyroid hormone, metabolite and ion levels in rainbow trout, *Salmo gairdneri* Richardson. J Fish Biol 33:73–88

Laidley CW, Woo PTK, Leatherland JF (1988) The stress-response of rainbow trout to experimental infection with the blood parasite *Cryptobia salmositica* Katz, 1951. J Fish Biol 32:253–261

Lamba VJ, Goswami SV, Sundararaj BI (1983) Circannual and circadian variations in plasma levels of steroids (cortisol, estradiol-17b, estrone, and testosterone) correlated with the annual gonadal cycle in the catfish, *Heteropneustes fossilis* (Bloch). Gen Comp Endocrinol 50:205–225

Lankford S, Weber G (2006) Potential roles of intraovarian growth factors during follicle maturation in rainbow trout (*Oncorhynchus mykiss*). Meeting Abstract

Lankford SE, Adams TE, Cech JJ (2003) Time of day and water temperature modify the physiological stress response in green sturgeon, *Acipenser medirostris*. Comp Biochem Physiol A 135:291–302

Leatherland JF (1993) Stocking density and cohort sampling effects on endocrine interactions in rainbow trout. Aquacult Int 1:137–156

Levine A, Zagoory-Sharon O, Feldman R, Lewis JG, Weller A (2007) Measuring cortisol in human psychobiological studies. Physiol Behav 90:43–53

Li HW, Brocksen RW (1977) Approaches to the energetic costs of interspecific competition for space by rainbow trout (*Salmo gairdneri*). J Fish Biol 11:329–341

Li Y, Takei Y (2003) Ambient salinity-dependent effects of homologous natriuretic peptides (ANP, VNP, and CNP) on plasma cortisol level in the eel. Gen Comp Endocrinol 130:317–323

López-Olmeda JF, Montoya A, Oliveira C, Sánchez-Vázquez FJ (2009) Synchronization to light and restricted-feeding schedules of behavioral and humoral daily rhythms in gilthead sea bream (*Sparus aurata*). Chronobiol Int 26:1389–1408

Lund V, Mejdell CM, Röcklinsberg H, Anthony R, Håstein T (2007) Expanding the moral circle: fish as objects of moral concern. Dis Aquat Org 75:109–118

Marino G, Di Marco P, Mandich A, Finoia MG, Cataudella S (2001) Changes in serum cortisol, metabolites, osmotic pressure and electrolytes in response to different blood sampling procedures in cultured sea bass (Dicentrarchus labrax L.). J Appl Ichthyol 17:115–120

Martins CIM, Schrama JW, Verreth JAJ (2006) The effect of group composition on the welfare of African catfish (*Clarias gariepinus*). App Anim Behav Sci 97:323–333

Martins CIM, Galhardo L, Noble C, Damsgård B, Spedicato MT, Zupa W, Beauchaud M, Kulczykowska E, Massabuau J-C, Carter T, Planellas SR, Kristiansen T (2011) Behavioural indicators of welfare in farmed fish. Fish Physiol Biochem. doi:10.1007/s10695-011-9518-8

Mateo JM (2008) Inverted-U shape relationship between cortisol and learning in ground squirrels. Neurobio Learn Memory 89:582–590

Mazeaud MM, Mazeaud F (1981) Adrenergic responses to stress in fish. In: Pickering AD (ed) Stress and fish. Academic Press, London, pp 50–75

Mazeaud MM, Mazeaud F, Donaldson EM (1977) Primary and secondary effects of stress in fish: some new data with a general review. Trans Am Fish Soc 106:201–212

McCormick SD (1996) Effects of growth hormone and insulin-like growth factor I on salinity tolerance and gill Na+, K+-ATPase in Atlantic salmon (Salmo salar): Interactions with cortisol. Gen Comp Endocrinol 101:3–11

McCormick MD (2001) Endocrine control of osmoregulation in Teleost fish. Am Zool 41:781–794. doi:10.1093/icb/41.4.781

McEwen BS, Wingfield JC (2010) What is in a name? Integrating homeostasis, allostasis and stress. Horm Behav 57:105–116

Metcalfe JD (2009) Welfare in wild-capture marine fisheries. J Fish Biol 75:2855–2861

Mikics E, Kruk MR, Haller J (2004) Genomic and non-genomic effects of glucocorticoids on aggressive behaviour in male rats. Psychoneuroendocrinol 29:618–635

Milligan CL (1996) Metabolic recovery from exhaustive exercise in rainbow trout. Comp Biochem Physiol A 113:51–60. doi:10.1016/0300-9629(95)02060-8

Milligan CL (2003) A regulatory role of cortisol in muscle glycogen metabolism in rainbow trout Oncorhynchus mykiss Walbaum. J Exp Biol 206:3167–3173. doi: 10.1242/jeb.00538

Milligan CL, Hooke GB, Johnson C (2000) Sustained swimming at low velocity following a bout of exhaustive exercise enhances metabolic recovery in rainbow trout. J Exp Biol 203:921–926

Molinero A, Gómez E, Balasch J, Tort L (1997) Stress by fish removal in the sea bream *Sparus aurata*: a time course study on the remaining fish in the same tank. J Appl Aquacult 7:1–12

Mommsen TP, Vijayan MM, Monn TW (1999) Cortisol in teleosts: dynamics, mechanisms of action, and metabolic regulation. Rev Fish Biol Fish 9:211–268

Moreira PSA, Volpato GL (2004) Conditioning of stress in Nile tilapia. J Fish Biol 64:961–969

Moreira PSA, Pulman KGT, Pottinger TG (2004) Extinction of a conditioned response in rainbow trout selected for high or low responsiveness to stress. Horm Behav 46: 450–457

Morgan JD, Iwama GK (1996) Cortisol-induced changes in oxygen consumption and ionic regulation in coastal cutthroat trout (*Oncorhynchus clarki clarki*) parr. Fish Physiol Biochem 15:385–394

Morgan MJ, Wilson CE, Crim LW (1999) The effect of stress on reproduction in Atlantic cod. J Fish Biol 54:477–488

Mormède P, Andanson S, Aupérin B, Beerda B, Guémené D, Malmkvist J, Manteca X, Manteuffel G, Prunet P, van Reenen CG, Richard S, Veissier I (2007) Exploration of the hypothalamic–pituitary–adrenal function as a tool to evaluate animal welfare. Physiol Behav 92:317–339

Munro AD, Pitcher TJ (1985) Steroid hormones and agonistic behavior in a cichlid teleost, *Aequidens pulcher*. Hormones Behav 19:353–371

Neave N (2008) Hormones and behaviour: a psychological approach. University Press, Cambridge

Nichols DJ, Weisbart M (1984) Plasma cortisol concentrations in Atlantic salmon, *Salmo salar*: episodic variations, diurnal change, and short term response to adrenocorticotropic hormone. Gen Comp Endocrinol 56:169–176

Noakes DLG, Leatherland JF (1977) Social dominance and interrenal cell activity in rainbow trout, *Salmo gairdneri* (Pisces: Salmonidae). Env Biol Fish 2:131–136

North BP, Turnbull JF, Ellis T, Porter MJ, Migaud H, Bron J, Bromage NR (2006) The impact of stocking density on the welfare of rainbow trout (*Oncorhynchus mykiss*). Aquaculture 255:466–479

O'Connor CM, Gilmour KM, Arlinghaus R, van der Kraak G, Cooke SJ (2009) Stress and parental care in a wild teleost fish: insights from exogenous supraphysiological cortisol implants. Physiol Biochem Zoo 82:709–719

Oliveira RF, Canario AVM, Bshary R (1999) Hormones, behaviour and conservation of littoral fishes: current status and prospects for future research. In: Almada VC, Oliveira RF, Gonçalves EJ (eds) behaviour and conservation of littoral fishes. Instituto Superior de Psicologia Aplicada, Lisbon, pp 149–178

Olla BL, Davis MW, Schreck CB (1992) Comparison of predator avoidance capabilities with corticosteroid levels induced by stress in juvenile coho salmon. Trans Am Fish Soc 121:544–547

Orchinik M (1998) Glucocorticoids, stress and behavior: shifting the timeframe. Hormones Behav 34:320–327

Ortuno J, Esteban MA, Mesequer J (2002a) Effects of phenoxyethanol on the innate immune system of gilthead sea bream (*Sparus aurata*) exposed to crowding stress. Vet Immunol Immunopathol 89:29–36

Ortuno J, Esteban MA, Mesequer J (2002b) Lack of effect of combining different stressors on innate immune responses of sea bream (*Sparus auratus* L.). Vet Immunol Immunopathol 84:17–27

Øverli Ø, Pottinger TG, Carrick TR, Øverli E, Winberg S (2001) Brain monoaminergic activity in rainbow trout selected for high and low stress responsiveness. Brain Behav Evol 57:214–224

Øverli Ø, Kotzian S, Winberg S (2002a) Effects of cortisol on aggression and locomotor activity in rainbow trout. Hormones Behav 42:53–61

Øverli Ø, Pottinger TG, Carrick TR, Øverli E, Winberg S (2002b) Differences in behaviour between rainbow trout selected for high- and low-stress responsiveness. J Exp Biol 205:391–395

Øverli Ø, Winberg S, Pottinger TG (2005) Behavioral and neuroendocrine correlates of selection for stress responsiveness in rainbow trout—a review. Int Comp Biol 45:463–474

Øverli Ø, Sørensen C, Pulman KGT, Pottinger TG, Korzan W, Summers CH, Nilsson GE (2007) Evolutionary background for stress-coping styles: relationships between physiological, behavioral, and cognitive traits in non-mammalian vertebrates. Neurosci Biobehav Rev 31:396–412

Pagnotta A, Brooks L, Milligan L (1994) The potential regulatory role of cortisol in the recovery from exhaustive exercise in rainbow trout. Can J Zool 72:2136–2146

Pavlidis M, Greenwood L, Paalavuo M, Mölsa H, Laitinen JT (1999) The effect of photoperiod on diel rhythms in serum melatonin, cortisol, glucose, and electrolytes in the common dentex, *Dentex dentex*. Gen Comp Endocrinol 113:240–250

Perera TD, Park S, Nemirovskaya Y (2008) Cognitive role of neurogenesis in depression and antidepressant treatment. Neuroscientist 14:326–338

Peruzzi S, Varsamos S, Chatain B, Fauvel C, Menu B, Falguière Sévére A, Flik G (2005) Haematology and physiological characteristics of diploid and triploid sea bass *Dicentrarchus labrax* L. Aquaculture 244:359–367

Pickering AD (1992) Rainbow trout husbandry: management of the stress response Aquaculture 100:125–139

Pickering AD (1993) Growth and stress in fish production. Aquaculture 111:51–63

Pickering AD, Pottinger TG (1983) Seasonal and diel changes in plasma cortisol levels of the brown trout, *Salmo trutta* L. Gen Comp Endocrinol 49:232–239

Pickering AD, Pottinger TG (1987a) Poor water quality suppresses the cortisol response of salmonid fish to handling and confinement. J Fish Biol 30:363–374

Pickering AD, Pottinger TG (1987b) Crowding causes prolonged leucopenia in salmonid fish, despite interrenal acclimation. J Fish Biol 30:701–712

Pickering AD, Stewart A (1984) Acclimation of the interrenal tissue of the brown trout, *Salmo trutta* L., to chronic crowding stress. J Fish Biol 24:731–740

Pickering AD, Schreck C, Maule AG, Kaattari SL, Leatherland J, Farbridge KJ, Vijayan MM, Flett PA, Dickhoff WW (1993) In: Muir JF, Roberts RJ (eds) Stress and adaptation, recent advances in aquaculture IV. Blackwell, London, pp 153–192

Planas J, Gutiérrez J, Fernández J, Carrillo M, Canals P (1990) Annual and daily variations of plasma cortisol in sea bass, *Dicentrarchus labrax* L. Aquaculture 91:171–178

Pottinger TG (2008) The stress response in fish: mechanisms, effects and measurement. In: Branson E (ed) Fish welfare. Blackwell, Oxford, pp 32–48

Pottinger TG (2010) A multivariate comparison of the stress response in three salmonid and three cyprinid species: evidence for inter-family differences. J Fish Biol 76:601–621

Pottinger TG, Carrick TR (2001) Stress responsiveness affects dominant-subordinate relationships in rainbow trout. Hormones Behav 40:419–427

Pottinger TG, Moran TA (1993) Differences in plasma cortisol and cortisone dynamics during stress in two strains of rainbow trout (*Oncorhynchus mykiss*). J Fish Biol 43:121–130

Pottinger TG, Pickering AD (1992) The influence of social interaction on the acclimation of rainbow trout, Oncorhynchus mykiss (Walbaum) to chronic stress. J Fish Biol 41:435–447

Pottinger TG, Pickering AD, Hurley MA (1992) Consistency in the stress response of individuals of two strains of rainbow trout, *Oncorhynchus mykiss*. Aquaculture 103:275–289

Poulsen SB, Jensen LF, Nielsen KS, Malte H, Aarestrup K, Svendsen JC (2011) Behaviour of rainbow trout *Oncorhynchus mykiss* presented with a choice of normoxia and stepwise progressive hypoxia. J Fish Biol 79:969–979

Procarione LS, Barry TP, Malison JA (1999) Effects of high rearing densities and loading rates on the growth and stress responses of juvenile rainbow trout. N Am J Aquacult 61:91–96

Prunet P, Cairns MT, Winberg S, Pottinger TG (2008) Functional genomics of stress responses in fish. Rev Fish Sci 16(S1):157–166

Radley JJ, Morrison JH (2005) Repeated stress and structural plasticity in the brain. Ageing Res Rev 4:271–287

Rance TA, Baker BI, Webley G (1982) Variations in plasma cortisol concentrations over a 24-hour period in the rainbow trout *Salmo gairdneri*. Gen Comp Endocrinol 48:269–274

Redding JM, Patiño R, Schreck CB (1984) Clearance of corticosteroids in yearling coho salmon, *Oncorhynchus kisutch*, in fresh water and seawater and after stress. Gen Comp Endocrinol 54:433–443

Redgate ES (1974) Neural control of pituitary adrenal activity in *Cyprinus carpio*. Gen Comp Endocrinol 22:35–41

Reid SG, Bernier NJ, Perry SF (1998) The adrenergic stress response in fish: control of catecholamine storage and release. Comp Biochem Physiol C 120:1–27

Remage-Healey L, Bass AH (2004) Rapid, hierarchical modulation of vocal patterning by steroid hormones. J Neurosci 24:5892–5900

Remage-Healey L, Bass AH (2006) A rapid neuromodulatory role for steroid hormones in the control of reproductive behaviour. Brain Res 1126:27–35

Remage-Healey L, Nowacek DP, Bass AH (2006) Dolphin foraging sounds suppress calling and elevate stress hormone levels in a prey species, the Gulf toadfish. J Exp Biol 209:4444–4451

Roche H, Bogé G (1996) Fish blood parameters as a potential tool for identification of stress caused by environmental factors and chemical intoxication. Mar Environ Res 41:27–43

Roque A, Yavuzcan Yildiz H, Carazo I, Duncan N (2010) Physiological stress responses of sea bass (*Dicentrarchus labrax*) to hydrogen peroxide (H_2O_2) exposure. Aquaculture 304:104–107

Rose JD (2002) The neurobehavioral nature of fishes and the question of awareness and pain. Rev Fish Sci 10:1–38

Rose JD (2007) Anthropomorphism and 'mental welfare' of fishes. Dis Aquat Org 75:139–154

Rotllant J, Tort L (1997) Cortisol and glucose responses after acute stress by net handling in the sparid red porgy previously subjected to crowding stress. J Fish Biol 51:21–28

Rotllant J, Arends RJ, Mancera JM, Flik G, Wendelaar Bonga SE, Tort L (2000) Inhibition of HPI axis response to stress in gilthead sea bream (*Sparus aurata*) with physiological plasma levels of cortisol. Fish Physiol Biochem 23:13–22

Rotllant J, Balm PHM, Pérez-Sánchez J, Wendelaar-Bonga SE, Tort L (2001) Pituitary and interrenal function in gilthead sea bream (Sparus aurata L., Teleostei) after handling and confinement stress. Gen Comp Endocrinol 121:333–342

Rotllant J, Ruane NM, Caballero MJ, Montero D, Tort L (2003) Confinement stress response in sea bass (*Dicentrarchus labrax*) is characterised by an increased biosynthetic capacity of interrenal tissue, with no effect on ACTH sensitivity. Comp Biochem Physiol 136A:613–620

Rotllant J, Ruane NM, Dinis MT, Canario AVM, Power DM (2006) Intra-adrenal interactions in fish: catecholamine stimulated cortisol release in sea bass (*Dicentrarchus labrax* L.). Comp Biochem Physiol A 143:375–381

Ruane NM, Carballo EC, Komen J (2002) Increased stocking density influences the acute physiological stress response of common carp *Cyprinus carpio* (L.). Aquacult Res 33:777–784

Ruyet JP-L, Labbé L, Bayon NL, Sévère A, Roux AL, Delliou HL, Quéméner L (2008) Combined effects of water quality and stocking density on welfare and growth of rainbow trout (*Oncorhynchus mykiss*). Aquat Liv Res 21:185–195

Sánchez JA, López-Olmeda JF, Blanco-Vives B, Sánchez-Vázquez FJ (2009) Effects of feeding schedule on locomotor activity rhythms and stress response in sea bream. Physiol Behav 98:125–129

Sandhu N, Vijayan MM (2011) Cadmium-mediated disruption of cortisol biosynthesis involves suppression of corticosteroidogenic genes in rainbow trout. Aquat Toxicol 103:92–100

Sandi C (1996) Novelty-related rapid locomotor effects of corticosterone in rats. Eur J Neurosci 8:794–800

Santos GA, Schrama JW, Mamauag REP, Rombout JHWM, Verreth JAJ (2010) Chronic stress impairs performance, energy metabolism and welfare indicators in European seabass (*Dicentrarchus labrax*): the combined effects of

fish crowding and water quality deterioration. Aquaculture 299:73–80
Santulli A, Modica A, Messina L, Ceffa L, Curatolo A, Rivas G, Fabis G, D'Amelio V (1999) Biochemical responses of European sea bass (Dicentrarchus labrax L.) to the stress induced by off shore experimental seismic prospecting. Mar Pollut Bull 38:1105–1114. doi:10.1016/S0025-326X (99)00136-8
Schjolden J, Basic D, Winberg S (2009) Aggression in rainbow trout is inhibited by both MR and GR antagonists. Physiol Behav 98:625–630
Schreck CB (1981) Stress and compensation in teleostean fishes: response to social and physical factors. In: Pickering AD (ed) Stress and fish. Academic Press, London, pp 295–321
Schreck CB (1982) Stress and rearing of salmonids. Aquaculture 28:241–249
Schreck CB (1990) Physiological, behavioural and performance indicators of stress. Am Fish Soc Symp 8:29–37
Schreck CB (2010) Stress and fish reproduction: the roles of allostasis and hormesis. Gen Comp Endocrinol 165:549–556
Schreck CB, Patino R, Pring CK, Winton JR, Holway JE (1985) Effects of rearing density on indices of smoltification and performance of coho salmon, *Oncorhynchus kisutch*. Aquaculture 45:345–358
Scott AP, Hirschenhauser K, Bender N, Oliveira R, Earley RL, Sebire M, Ellis T, Pavlidis M, Hubbard PC, Huertas M, Canario A (2008) Non-invasive measurement of steroids in fish-holding water: important considerations when applying the procedure to behaviour studies. Behaviour 145(Suppl. SI):1307–1328
Seidelin M, Madsen SS, Byrialsen A, Kristiansen K (1999) Effects of insulin-like growth factor-I and cortisol on Na+, K+-ATPase expression in osmoregulatory tissues of brown trout (Salmo trutta). Gen Comp Endocrinol 113: 331–342
Shelbourne JE (1975) Marine fish cultivation: pioneering studies on the culture of the larvae of the plaice (*Pleuronectes platessa* L.) and the sole (*Solea solea* L.). MAFF Fish Investig Ser II 27(9):29
Shrimpton JM, McCormick SD (1999) Responsiveness of gill Na^+/K^+-ATPase to cortisol is related to gill corticosteroid receptor concentration in juvenile rainbow trout. J Exp Biol 202:987–995
Silva P, Martins CIM, Engrola S, Marino G, Øyvind Ø, Conceição L (2010) Individual variation in stress physiology and behaviour in the flatfish *Senegalese sole*: evidences for coping styles. Appl Anim Behav Sci 124:75–81
Singley JA, Chavin W (1975) Serum cortisol in normal goldfish (*Carassius auratus* L.). Comp Biochem Physiol 50A:77–82
Sink TD, Lochmann RT, Fecteau KA (2008) Validation, use and disadvantages of enzyme-linked immunosorbent assay kits for detection of cortisol in channel catfish, largemouth bass, red pacu, and golden shiners. Fish Physiol Biochem 34:95–101
Sloman KA, Metcalfe NB, Taylor AC, Gilmour KM (2001) Plasma cortisol concentrations before and after social stress in rainbow trout and brown trout. Physiol Biochem Zoo 74:383–389
Sloman KA, Montpetit CJ, Gilmour KM (2002) Modulation of catecholamine release and cortisol secretion by social interactions in the rainbow trout, Oncorhynchus *mykiss*. Gen Comp Endocrinol 127:136–146
Small BC (2005) Effect of fasting on nychthemeral concentrations of plasma growth hormone (GH), insulin-like growth factor I (IGF-I), and cortisol in channel catfish (*Ictalurus punctatus*). Comp Biochem Physiol 142:217–223
Sneddon LU (2009) Pain perception in fish: indicators and endpoints. ILAR J 50:338–342
Sneddon LU, Braithwaite VA, Gentle MJ (2003) Do fishes have nociceptors? Evidence for the evolution of a vertebrate sensory system. Proc R Soc Lond B Biol Sci 270:1115–1121
Sørensen C, Nilsson GE, Summers CH, Øverli Ø (2011a) Social stress reduces forebrain cell proliferation in rainbow trout (*Oncorhynchus mykiss*). Behav Brain Res (in press)
Sørensen C, Bohlin LC, Øverli Ø, Nilsson GE (2011b) Cortisol reduces cell proliferation in the telencephalon of rainbow trout (*Oncorhynchus mykiss*). Physiol Behav 102:518–523
Stolte EH, de Mazon AF, Leon-Koosterziel KM, Jesiak M, Bury NR, Sturm A, Huub FJ, Savelkoul HFJ, Verburg van Kemenade BML, Flik G (2008) Corticosteroid receptors involved in stress regulation in common carp, *Cyprinus carpio*. J Endocrinol 198:403–417
Sumpter JP (1992) The stress response and its consequences in cultured fish. Bul Inst Zool Acad Sinica Monogr 16:229–236
Sumpter JP (1993) The deleterious effects of stress and their significance to aquaculture. In: Barnabé G, Kestemont P (eds) Production, environment and quality. European Aquaculture Society Special Publication No. 18. Ghent, Belgium, pp 157–165
Sumpter JP (1997) The endocrinology of stress. In: Iwama GK, Pickering AD, Sumpter JP, Schreck CB (eds) Fish stress and health in aquaculture. Cambridge University Press, Cambridge, pp 95–118
Sumpter JP, Pottinger TG, Weaver MR, Campbell PM (1994) The wide ranging effects of stress on fish. In: Davey KG, Peter RE, Tobe SS (eds) Perspectives in comparative endocrinology. NRC, Canada, pp 535–538
Teles M, Gravato C, Pacheco M, Santos MA (2004) Juvenile sea bass biotransformation, genotoxic and endocrine responses to β-naphthoflavone, 4-nonylphenol and 17β-estradiol individual and combined exposures. Chemosphere 57:147–158
Teles M, Pacheco M, Santos MA (2006) Biotransformation, stress and genotoxic effects of 17β-estradiol in juvenile sea bass (*Dicentrarchus labrax* L.). Environ Int 32:470–477
Thorpe JE, McConway MG, Miles MS, Muir JS (1987) Diel and seasonal changes in resting plasma cortisol levels in juvenile Atlantic salmon, *Salmo salar* L. Gen Comp Endocrinol 65:19–22
Tintos A, Míguez JM, Mancera JM, Soengas JL (2006) Development of a microtitre plate indirect ELISA for measuring cortisol in teleosts, and evaluation of stress responses in rainbow trout and gilthead sea bream. J Fish Biol 68:251–263
Tort L (2010) Stress in farmed fish: its consequences in health and performance. In: Koumoundouros G (ed) Recent

advances in aquaculture research. Transworld Research Network, Trivandrum, pp 55–84

Tort L, Montero D, Robaina L, Fernández-Palacios H, Izquierdo MS (2001) Consistency of stress response to repeated handling in the gilthead sea bream, *Sparus aurata*. Aquacult Res 32:593–598

Tort L, Puigcerver M, Crespo S, Padrós F (2002) Cortisol and haematological response in sea bream and trout subjected to the anaesthetics clove oil and 2-phenoxyethanol. Aquacult Res 33:907–910

Tort L, Pavlidis M, Woo NYS (2011) Stress and welfare in sparid fishes. In: Pavlidis M, Mylonas C (eds) Sparidae. Biology and Aquaculture. Wiley-Blackwell, Oxford, pp 75–94

Turnbull J, Bell A, Adams C, Bron J, Huntingford F (2005) Stocking density and welfare of cage farmed Atlantic salmon: application of a multivariate analysis. Aquaculture 243:121–132

Tytler P, Hawkins AD (1981) Vivisection, anaesthetics and minor surgery. In: Hawkins AD (ed) Aquarium systems. Academic Press, London, pp 247–278

Uchida K, Kaneko T, Tagawa M, Hirano T (1998) Localization of cortisol receptor in branchial chloride cells in chum salmon fry. Gen Comp Endocrinol 109:175–185

van de Nieuwegiessen PG, Boerlage AS, Verreth JAJ, Schrama JW (2008) Assessing the effects of a chronic stressor, stocking density, on welfare indicators of juvenile African catfish, *Clarias gariepinus* Burchell. App Anim Behav Sci 115:233–243

Varsamos S, Flik G, Pepin JF, Wendelaar Bonga SE, Breuil G (2006) Husbandry stress during early life stages affects the stress response and health status of juvenile sea bass, *Dicentrarchus labrax*. Fish Shellfish Immunol 20:83–96

Vazzana M, Cammarata M, Cooper EL, Parrinello N (2002) Confinement stress in sea bass (*Dicentrarchus labrax*) depresses peritoneal leukocyte cytotoxicity. Aquaculture 210:231–243

Verburg-van Kemenade BML, Stolte EH, Metz JR, Chadzinska M (2009) Neuroendocrine-immune interactions in teleost fish. Fish Neuroendocrinol, Elsevier, pp 313–364

Vijayan MM (2011) Hormone response to stress. In: Farrell AP (ed) Encyclopedia of fish physiology from genome to environment. Elsevier, San Diego

Volkoff H, Xu M, MacDonald E, Hoskins L (2009) Aspects of hormonal regulation of appetite in fish with emphasis on goldfish, Atlantic cod and winter flounder: notes on actions and responses to nutritional, environmental and reproductive changes. Comp Biochem Physiol A 153:8–12

Weber ED, Borthwick SM, Helfrich LA (2002) Plasma cortisol response of juvenile chinook salmon to passage through Archimedes lifts and a Hidrostal pump. N Am J Fish Manage 22:563–570

Wedemeyer GA, Barton BA, McLeay DJ (1990) Stress and acclimation. In: Schreck CB, Moyle PB (eds) Methods for fish biology. Am Fish Soc, Bethesda, Maryland, pp 451–489

Wendelaar Bonga SE (1997) The stress response in fish. Physiol Rev 77:591–625

Winberg S, Lepage O (1998) Elevation of brain 5-HT activity, POMC expression, and plasma cortisol in socially subordinate rainbow trout. Am J Physiol Reg Int Comp Physiol 274:R645–R654

Wolkowitz OM, Burke H, Epel ES, Reus VL (2009) Glucocorticoids: mood, memory and mechanisms. Ann N Y Acad Sci 1179:19–40

Wong EYH, Herbert J (2006) Raised circulating corticosterone inhibits neuronal differentiation of progenitor cells in the adult hippocampus. Neuroscience 137:83–92

Wong S, Dykstra M, Campbell J, Earley R (2008) Measuring water-borne cortisol in convict cichlids (*Amatitlania nigrofasciata*): is the procedure a stressor? Behaviour 1245:1283–1305

Yamada H, Satoh R, Ogoh M, Takaji K, Fujimoto Y, Hakuba T, Chiba H, Kambegawa A, Iwata M (2002) Circadian changes in serum concentrations of steroids in Japanese char *Salvelinus leucomaenis* at the stage of final maturation. Zool Sci 19:891–898

Yeates JW, Main DSCJ (2008) Assessment of positive welfare: a review. Vet J 175:293–300

Yildiz YH, Ergonul MB (2010) Is prophylactic formalin exposure a stress source for gilthead sea bream (*Sparus aurata*) and sea bass (*Dicentrarchus labrax*)? Ankara Univ Vet Fak Derg 57:113–118

Zhao CM, Deng W, Gage FH (2008) Mechanisms and functional implications of adult neurogenesis. Cell 132:645–660

Mortality and Fish Welfare – Summary

Tim Ellis, Iain Berrill, Jeff Lines, James Turnbull and Toby Knowles

In general, there is a lack of robust data on mortality rates and causes within commercial fish farming. Time-series data on mortality rate are available for sea-grown Atlantic salmon, which probably reflects the more advanced state of this sector. Snapshot data are also available from case-studies which have examined specific causes of mortality. Mortality in farmed fish populations is due to a variety of biotic and abiotic causes, although it is often difficult to differentiate between underlying and immediate causes of mortality. Most mortality appears to occur during episodes associated with critical periods (in development or production) or disease outbreaks. Mortality rates can therefore vary markedly between years and locations. Mortality rates typically decrease with age, as fish develop and increase in size. Scrutiny of mortality records and determining causes may help farmers to avoid preventable mortality, improve fish welfare and increase profitability.

Background

Although animal welfare lobby groups have drawn attention to mortality in farmed fish populations, the subject has received little direct attention from fish welfare researchers. This is surprising because 1) mortality rates on commercial fish farms have been suggested to rank amongst the highest in commonly farmed vertebrate species, and 2) it is assumed that fish experience substantial suffering whilst dying.

Scientific Significance

Mortality has great potential as a welfare indicator. Fish farmers already use short-term mortality observations as an operational welfare indicator, to indicate problems and enable remedial actions. Long-term mortality rates could provide a retrospective indicator, representing overall welfare integrated over a single production stage or the whole farming cycle. To support use of mortality as a welfare indicator, researchers need to develop appropriate metrics for quantification and methods to ensure that robust data are collected; challenges are recognised to gaining such data, especially for early life stages. Documentation of current mortality rates and causes in commercial fish farms (bench-marking) would facilitate comparisons over time, between areas, between sectors, and highlight if action is needed to reduce mortality. Before targets or acceptable thresholds could be set for different stages and species, the importance of body size to susceptibility to mortality, and farmers' ability to prevent each cause of mortality, must be considered.

Practical Application

Fish farmers already monitor mortality due to commercial interests and legal requirements to promptly identify notifiable disease. As profitability in fish farming is directly linked to survival, confronting mortality should ultimately benefit both fish and farmers. Comparisons to terrestrial livestock sectors must recognise fundamental differences in i) body size, ii) production cycle duration, iii) the physiology of finfish and iv) farmers' ability to prevent mortality.

Contact

T. Ellis
Cefas Weymouth Laboratory, Barrack Road, The Nothe,
Weymouth, Dorset DT4 8UB, UK
e-mail: tim.ellis@cefas.co.uk

Fish Physiol Biochem (2012) 38:189–199
DOI 10.1007/s10695-011-9547-3

Mortality and fish welfare

Tim Ellis · Iain Berrill · Jeff Lines ·
James F. Turnbull · Toby G. Knowles

Received: 29 March 2010 / Accepted: 1 August 2011 / Published online: 16 September 2011

Abstract Mortality has received insufficient attention as a fish welfare topic. Here, we aim to prompt fish farming stakeholders to discuss fish mortalities in relation to welfare. Mortality in farmed fish populations is due to a variety of biotic and abiotic causes, although it is often difficult to differentiate between underlying and immediate causes of mortality. Most mortality appears to occur during episodes associated with disease outbreaks and critical periods (in development or production). Most causes of mortality can be assumed to be associated with suffering prior to death. As mortality rates in farmed fish populations are suspected to rank amongst the highest in commonly farmed vertebrate species, mortality should be a principal fish welfare issue. Long-term mortality rates can be used as a retrospective welfare performance indicator and short-term mortality rates as an operational welfare indicator. Scrutiny of mortality records and determining causes of death will enable action to be taken to avoid further preventable mortality. The welfare performance of fish farms should only be judged on levels of predictable and preventable mortality. Fish farmers will already be monitoring mortality due to commercial and legal requirements. As profitability in fish farming is directly linked to survival, confronting mortality should ultimately benefit both fish and farmers.

Keywords Aquaculture · Fish farming · Welfare · Indicator

T. Ellis (✉)
Cefas Weymouth Laboratory, Barrack Road, The Nothe, Weymouth, Dorset DT4 8UB, UK
e-mail: tim.ellis@cefas.co.uk

I. Berrill · J. F. Turnbull
Institute of Aquaculture, University of Stirling, Stirling FK9 4LA, Scotland, UK

J. Lines
Silsoe Livestock Systems Ltd, Wrest Park, Silsoe, Bedford MK45 4HR, UK

T. G. Knowles
School of Veterinary Science, University of Bristol, Langford, Bristol BS40 5DU, UK

Why discuss mortality of farmed fish?

In its 1996 report, the UK's Farm Animal Welfare Council (FAWC) highlighted mortality as a farmed fish welfare issue (Anon 1996). The report noted that Council members had been unable to determine typical mortality rates on commercial farms; nevertheless, the authors suggested that rates were high, likely to be associated with poor welfare, and suspected to exceed those in other livestock industries. Concern over mortality of farmed fish has since been reiterated by another animal welfare lobby group report (Stevenson 2007); in this report, the potential scale of mortality as an animal welfare issue was illustrated by linking it to the large numbers (10s of millions) of individual fish farmed. Stevenson (2007) went on to say "Such high mortality rates would rightly sound alarm bells in other branches of farming".

Although mortality has been highlighted as a welfare issue by the lobby groups, it has attracted little direct attention from the fish welfare research community. For example, mortality is conspicuous by its absence as a stand-alone topic in Branson's (2008) book entitled "Fish Welfare", although it is mentioned under other topics (e.g. Branson and Turnbull 2008; Wall 2008). In a recent paper examining mortality during the marine stage of farmed Atlantic salmon, Aunsmo et al. (2008) did not mention the word welfare. Our aim with this manuscript is to prompt stakeholders to reconsider their preconceptions and discuss mortality in relation to fish welfare.

Misperception 1: Mortality is a health, rather than welfare, issue

There is perhaps a misperception that mortality is a concern for health, rather than welfare. Mortality can be assumed to be preceded by suffering (see "Do dying fish suffer?" below) and it is this suffering associated with the process of death that makes mortality a welfare issue (Wall 2008), rather than the cause and the death itself.

Why do farmed fish die?

Misperception 2: Mortality is inevitable in farmed fish populations

Some mortality appears to be considered inevitable in populations of farmed fish and is incorporated into economic models at a set rate, e.g. 1.5% per month (James and Slaski 2006). However, fish do not just die—there has to be a cause. Researchers appear to frequently expect and record mortality whilst giving little attention to the cause of death (see Ellis et al. 2002). This is likely to be due to the perceived difficulty in determining a definitive cause of death from post-mortem examination. Although this may be valid for small fish and those at higher temperatures (due to autolysis with endogenous enzymes continuing to act at environmental temperatures), Aunsmo et al. (2008) found that cause could be determined with confidence in 92% of marine-stage salmon examined within 24 h of death in a research study. Cause of death (in these on-grown fish at late summer/autumn sea temperatures in Norway) was determined by fish health professionals from post-mortem examination, aided by additional recent and historic information. Interestingly, histopathology of dead fish and laboratory testing added little to diagnosis of cause of death (Aunsmo et al. 2008).

Aunsmo et al. (2008) found that cause of death could be classified into 22 specific categories. Additional potential sources of mortality in farmed fish populations can be gleaned from the literature and are classed in Table 1. A major issue obscuring categories of cause of death is the difficultly in separating the *immediate* from *underlying* cause of death. Aunsmo et al. (2008) provide the example of bacterial ulcers:

- although bacterial infection ultimately caused death, infection could not have occurred without an initial mechanical trauma. However, the mechanical trauma would not, on its own, cause death that only ensued after infection.

We have included "escape" as a potential cause of death (Table 1), although this is admittedly arguable as it is not an immediate cause of mortality. However, it is included as an underlying cause of mortality because: (1) reared fish are likely to die once in the open environment (Brown and Laland 2001; Reid 2006; Blanchfield et al. 2009); (2) escape represents a loss of fish from the farmed population so needs to be included if data on causes of death are to be related to metrics derived from population counts.

Disease due to infectious agents was the major cause of mortality observed in the study of Aunsmo et al. (2008). Disease also appears to be a major contributor to mortality in other case studies (see "What is the mortality rate on commercial fish farms?" below).

Misperception 3: A proportion of reared fish are destined to die due to poor genetic fitness

A seemingly common idea within both the fish farming and research communities is that a proportion of cultured fish die due to poor genetic fitness. This belief presumably originates from the assumption that

Table 1 Preliminary classification of potential causes of mortality in populations of farmed fish

Class of cause of death			Example	Reference
Biotic	Transmissible disease	Obligate and facultative pathogens	Viruses	Aunsmo et al. (2008), Rodger (2007)
			Bacteria	
			Fungi	
		Parasites		Rodger (2007)
	Compulsory slaughter for notifiable disease control			
	Culling		Moribund fish	Soares et al. (2011)
			Malformed fish	
	Cannibalism			Hecht and Pienaar (1993)
	Stress	Social		Laidley and Leatherland (1988), Pottinger and Pickering (1992)
	Predation/predation injury			Aunsmo et al. (2008)
	Toxic organisms		Jellyfish	Rodger (2007)
			Toxic insects	Savvidis et al. (2009)
			Toxic algae	Rodger et al. (2011)
	Starvation		Start-feeding	Person-Le-Ruyet et al. (1990)
			Weaning	
	Runts			Rodger (2007)
	Precocious male			Aunsmo et al. (2008)
	Morphological abnormality		Heart abnormality	Branson and Turnbull (2008)
			Uninflated swimbladder	
	Genetic abnormalities			
Abiotic	Physical injury/mechanical trauma		Trapped in cage netting	Aunsmo et al. (2008), Soares et al. (2011)
	Environmental diseases		Gas bubble disease	Wedemeyer (1996)
			Nephrocalcinosis	
	Stress	Physico-chemical	Unsmoltified	Rodger (2007), Aunsmo et al. (2008)
	Rapid temperature changes			Boyd et al. (2005)
	Episodic technical failures		Failure in oxygenation	Miller et al. (1995)
	Escape		Malicious release	Reid (2006)
Management	Culling		Excess production	Soares et al. (2011)
			Sampling	
	Chemotherapeutant treatments			Soares et al. (2011)

The classification and listing is likely be refined with further directed research; causes could be further classified in terms of predictability and preventability, depending on site

the highly fecund nature of fish will result in a proportion of genetically unfit individuals dying at some stage in the benign culture environment, which would otherwise have been predated early in development in the wild. Although some genetic abnormalities might be expected, the magnitude of this effect must be questioned. Research is increasingly showing that morphological abnormalities, rather than being due to inherent genetic defects, are induced by inappropriate environmental conditions (e.g. egg incubation temperature), nutritional deficiencies (e.g. phosphorus) and poor handling (Branson and Turnbull 2008). Although attributed to "egg quality", differences in mortality rates between batches of marine larvae are thought to be due largely to the status of the broodstock (nutrition, stress) rather than inherent genetic deficiencies (e.g. Shelbourne 1975).

Do dying fish suffer?

Misperception 4: Farmed fish are grown to be killed, so an early death is not a welfare issue

Although farmed fish are grown to be ultimately killed for the table, it is not death itself but the process of death that is key to welfare (Wall 2008). To illustrate by analogy: slaughter is only humane if the fish do not suffer, i.e. the process is without pain and fear.

Misperception 5: As all wild fish ultimately die, death of farmed fish is a natural phenomenon and therefore not a welfare issue

In order to avoid adverse welfare associated with mortality

- death must be instantaneous, or
- the fish must be at a developmental stage at which it is unaware of its state and cannot suffer [Under current UK animal experimentation law (Anon 1986), fish embryos within eggs and hatched larvae not yet developed to the stage of independent feeding are not protected], or
- the fish must be unaware of its state whilst dying, i.e. akin to anaesthetic overdose.

None of the causes of mortality identified in Table 1, apart from interventional humane culling, are humane. Even the most rapid, i.e. predation and cannibalism, are likely to be associated with periods of pursuit and ingestion, which can be expected to cause fear and pain in the prey. It is unknown how long a fish will remain aware/conscious during the ingestion process. Although wild fish may suffer during natural death and in commercial capture fisheries (Metcalfe 2009), this does not lessen the duty of care with respect to mortality in farmed fish.

Misperception 6: Mortality in captive fish is acceptable because it is orders of magnitude lower than in the wild.

One argument we have commonly encountered is that farming reduces mortality over the wild state so significantly that it should not be a welfare issue. Whilst acknowledging the fact that the protection from predators and provision of food to cultured fish does greatly reduce mortality over the wild state, this does not remove the onus on farmers to further reduce mortality if possible.

Recognising that dying fish are likely to be suffering, farmers have a welfare responsibility to intervene and humanely kill such fish. Furthermore, their removal from the population prior to death will prevent a decrease in water quality due to putrefaction and reduce the release of infectious agents if death is due to contagious disease (Sangster 1991; Wall 2008). However, the (1) identification of moribund fish that will not recover and (2) removal from the population for humane killing are areas that are likely to need further development in commercial farming.

What is the mortality rate on commercial fish farms?

As FAWC (Anon 1996) noted, there is a scarcity of documentation on mortality rates in farmed fish. The notable exception is the Atlantic salmon *Salmo salar* industry. The competent authority for fish health in Scotland publishes annually (data available for the 1990 year-class onwards) the harvested number of salmon, relative to the number that were put to sea as smolt, separated into various production areas (Anon 2001, 2009; Walker 2010). In this case, survival is for the sea production stage only (the earlier freshwater production stage is excluded), and not adjusted for duration (harvest data are presented over 3 years). This 18-year time-series shows that mortality varies greatly between year-classes and production areas, ranging around a median of 22% from as low as 1.7% to over 50% (Fig. 1). Such data sets allow comparisons between production areas and analyses of trends over time.

Soares et al. (2011) examined mortality for one Scottish salmon farming company and found an overall mortality of 24%, again for the marine stage grow-out period (averaging 89 weeks). Mortality during the comparable production stage of Atlantic salmon in Norway is similar to that in Scotland, being quoted at 10% p.a. and 15% over the marine stage

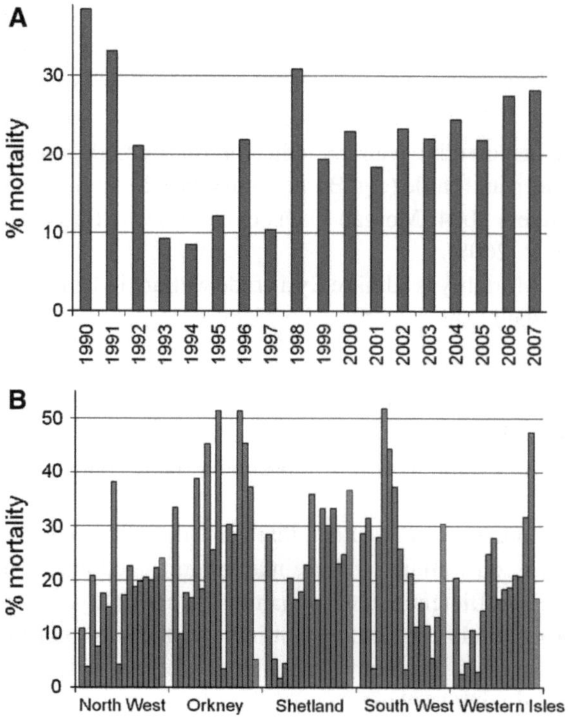

Fig. 1 Mortality of Atlantic salmon during the marine stage of farming in Scotland. Data (from Anon 2001, 2009; Walker 2010) for each smolt year-class put to sea in 1990–2007. **a** Mortality, for the whole Scottish industry (18 year-classes). **b** Mortality, by geographical area (16 year-classes)

(Aunsmo et al. 2008). Similarly, mortality during the marine stage for Irish salmon (Rodger 2007) ranges from <5 to >30% (assumed to be p.a.).

Comparable collated data for other fish farming sectors and terrestrial livestock farming industries are difficult to find (see Ortiz-Pelaez et al. 2008). Information in reports by the UK's Farm Animal Welfare Council indicates mortalities that occur in terrestrial farmed animals:

- Sheep: lamb mortality rates are generally around 10–15%, although episodes of severe weather may lead to extremely high losses (Anon 1994).
- Dairy cattle: about 170,000 calves a year die during their first month of life; scouring (diarrhoea) is the main factor contributing to their death (Anon 1997).
- Broiler hens: less than 2% mortality during rearing to 18 weeks (Anon 1998)

Other, more limited, examples are also available from recent research studies on salmonid farms:

- 12–82% mortality within 7–12 weeks during outbreaks of amoebic gill disease in seawater farmed Atlantic salmon in Norway (Steinum et al. 2008).
- 32% mortality over a 3.5 month outbreak of infectious pancreatic necrosis disease in unvaccinated seawater farmed Atlantic salmon in Norway (Ramstad et al. 2007).
- 0–16% mortality over 9 months from *Flavobacterium branchophilum* infections in salmonid batches during early rearing in Canadian Ministry hatcheries (Good et al. 2008).
- Mortalities up to 90% over the summer months in rainbow trout *Oncorhynchus mykiss* farms in Portugal and Greece due to outbreaks of the bacterium *Lactococcus garvieae* (Pereira et al. 2004; Savvidis et al. 2007).
- Mortalities of 12–33% over 3 months in rainbow trout batches during a whitespot *Ichthyophthirius multifiliis* outbreak in a UK hatchery (Shinn et al. 2009).
- "Massive" sudden mortalities thought to be due to toxic insects being washed into rivers by heavy rain and eaten by farmed rainbow trout in Greece (Savvidis et al. 2009).

These case studies provide isolated examples reflecting specific periods and do not represent the wider industry or production period. Nevertheless, they do show that episodic events can cause death of over 25% of farmed populations. Boyd et al. (2005) indicate that mortality (including episodic events) is also commonplace on commercial tilapia and catfish farms.

It must be recognised that mortality rates vary between species due to inherent differences. At hatching turbot larvae *Scophthalmus maximus* are considered less well developed than Atlantic salmon, being 100 times smaller with reserves that last for a third of the time (Person-Le Ruyet et al. 1990). These facts contribute to a higher mortality rate of turbot (60–100%) during the hatchery production stage (egg to juvenile) (Person-Le Ruyet et al. 1990). This reference is 20 years old and survival is likely to have increased, although an inherent species difference to salmon is still to be expected.

Person-Le Ruyet et al.'s (1990) generalised data for the turbot hatchery stage (Fig. 2), and more recent data for commercially farmed marine-stage Atlantic salmon (Aunsmo et al. 2008; Soares et al. 2011) illustrate key points about mortality in farmed fish populations:

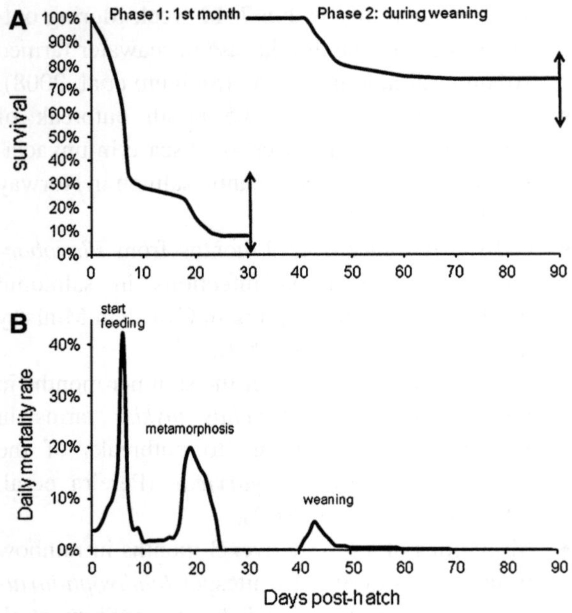

Fig. 2 **a** Mean post-hatch survival of turbot from 0 to 30 days (pre-weaning phase 1) and from 30 to 90 days (weaning phase) established over several years at IFREMER, with *vertical arrows* indicating observed ranges. Redrawn from Person-Le Ruyet (1990). **b** Derived daily mortality rate of turbot from 0 to 90 days post-hatch with critical periods (as indicated by Person-Le Ruyet 1990) *annotated*

- mortality rate typically reduces as fish develop and increase in size
- mortality tends to be episodic within
  - "critical periods" in the development/production cycle: for the hatchery stage of turbot, this is at first feeding, metamorphosis and weaning (Fig. 2); for marine-stage salmon, the first month after transfer to sea-water when "unsmoltified" fish die
  - disease outbreaks

Can mortality be used as a retrospective welfare performance indicator?

To achieve the ultimate aim of improving farm animal welfare, it is recognised that robust indicators are needed to objectively quantify welfare status (North et al. 2008; Berrill et al. 2010). Surveillance of such indicators would enable detection of trends over time, identification of sectors (or production areas) where action is needed, and allow individual farms to compare performance to the wider industry (Defra 2006). Mortality data represent a key welfare performance indicator for farmed fish, as illustrated by Fig. 1. Mortality data have similarly been recognised as a potential welfare indicator for farmed poultry, sheep and cattle (e.g. Barnett and Glatz 2004; Berg and Algers 2004; Morgan-Davis et al. 2008; Ortiz-Pelaez et al. 2008).

Mortality is admittedly a crude welfare indicator for farmed fish:

- it is only measurable at the level of the population, rather than individual
- it is an outcome variable, being too late to respond to for the benefit of those individuals contributing to the statistic
- it cannot be assumed that zero mortality indicates good welfare: welfare may be infringed but to an insufficient extent to manifest as mortality.
- mortality itself does not indicate the nature of the welfare problem (although see below on apportioning to causes)

Nevertheless, stakeholders have stated that mortality is an excellent welfare indicator (North et al. 2008). Kestin (1994) recognised that mortality indicates "something has gone seriously wrong with at least one of the Five Freedoms". Mortality will represent an integrated response of exposure to sub-optimal environmental conditions, husbandry practices, biosecurity and susceptibility to pathogens. Indeed, Aunsmo et al. (2008) implied that a mortality rate within their study populations that was lower than the industry average probably reflected good management practices at their case-study sites.

Compelling evidence of the value of mortality as a robust welfare indicator comes from its reduction with experience in culturing a novel species. This is illustrated by data within Shelbourne's (1975) report on the pioneering attempts to cultivate the early stages of marine fish (plaice *Pleuronectes platessa*). Once initial technical issues (tank design, water circulation and husbandry procedures) that resulted in very low survival had been resolved, mortality was reduced year-on-year (Fig. 3) by optimising environmental conditions (light intensity, temperature and antibiotic treatment of eggs) and "better acclimatisation and welfare of captive spawners", i.e. broodstock. Such improvements in

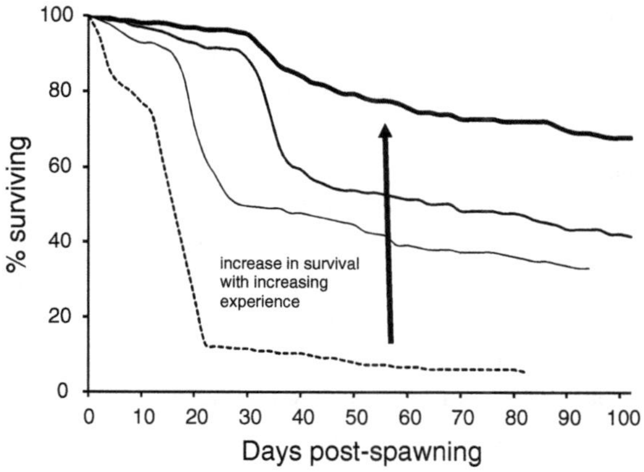

Fig. 3 The improvement in egg, larval and juvenile plaice survival (expressed as % original egg stock) during the period 1960–1962 at Port Erin (redrawn from Shelbourne 1975)

survival over time would be expected to be accompanied by improvements in other performance welfare indicators (e.g. growth, food conversion and morphological abnormality rates) as environmental requirements are better met.

A key advantage of mortality over other metrics used as welfare indicators is that it is immediately understood by a lay person or consumer. Mortality also represents a farmer-friendly means of documenting welfare: commercial pressures will drive farmers to maintain records of numbers to track the stocks of fish under their care, independently of any concern for welfare. Good practice will compel farmers to remove dead fish and record the number (e.g. Anon 2002, 2006b), and such mortality records represent a largely unexploited source of data as a key welfare performance indicator. If such records were collated and also included probable cause of death, it would enable authorities to target funding towards suspected key welfare issues. For example if, as seems probable, diseases are the dominant cause of death, then the greatest gains in fish welfare are to be made by developing preventative strategies, treatments and vaccines.

It must, however, be recognised that mortality is not a panacea for welfare indicators. Public and political pressures are changing the housing systems for laying hens, despite indications that mortality (due to infectious diseases and cannibalistic behaviour) is greater in litter-based and free range than cage systems (Fossum et al. 2009).

Can mortality be used as an operational (real-time) fish welfare indicator?

The ideal for any welfare indicator is that it is of value in real-time, i.e. it alerts farmers to a problem that can be remedied by operational intervention, to prevent further suffering. Wedemeyer (1996) pointed out that if mortality is occurring in a farmed fish population, "valuable information can sometimes be obtained by examination of the mortality pattern". He suggested that different welfare issues (i.e. oxygen depletion, exposure to acute toxins, virulent bacterial and viral pathogens, poor environmental conditions, external parasites and low virulence bacterial infections) would manifest as differing patterns of cumulative mortality (Fig. 4). Although his book largely preceded interest in fish welfare and the current terminology, Wedemeyer (1996) was clearly advocating mortality monitoring as an "operational" welfare indicator: farmers alert to changes in mortality could intervene to safeguard the remaining survivors. Admittedly, ideal operational welfare indicators alert farmers to a welfare problem before mortality of any individual occurs. Nevertheless, mortality does provide an option, in the absence of earlier warning indicators.

Wedemeyer's (1996) notional mortality curves (Fig. 4) are likely to be affected by other biotic and abiotic factors in the field (e.g. temperature, pathogenicity and variation in genetic susceptibility of the fish population). Nevertheless, the recording of mortalities represents good practice, and the scrutiny of mortality patterns to identify welfare problems remains a

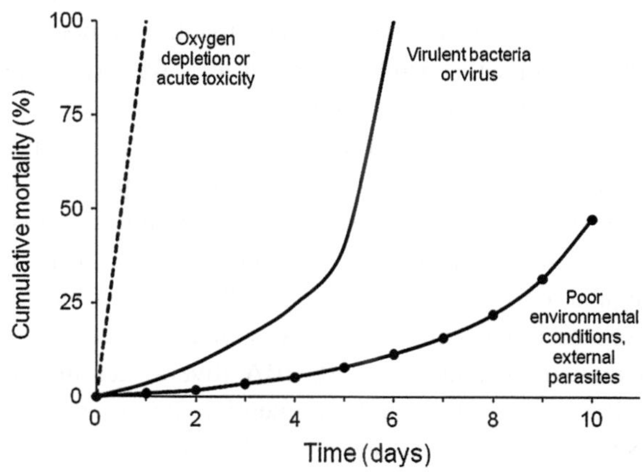

Fig. 4 Notional mortality patterns associated with various causes of death in farmed fish (redrawn from Wedemeyer 1996)

valuable idea that is already recognised in law. In Europe (under Council Directive 2006/88/EC), fish farmers are required to record mortality and notify the competent authority of abnormal mortality levels that could indicate the outbreak of a notifiable disease. If a notifiable disease is confirmed, the competent authority can take action (including humane slaughter) to prevent spread of the disease and further suffering (e.g. Anon 2006a). Research describing mortality patterns on commercial farms due to different causes may prove useful for diagnostic purposes.

Can mortality be a robust welfare indicator?

Any metric is only as robust as the data that it is derived from. Aunsmo et al. (2008) recognised the variable quality of mortality estimates for the industry at a broader scale due to inconsistencies in data collection. Nevertheless, they found that local systems for monitoring daily mortality numbers were good.

Practical problems that will be familiar to both researchers and farmers are those of accurately quantifying the numbers of (1) individuals in a population, and (2) fish that die. In commercial fish farming, operators need to know the number of animals (and their size) to ensure appropriate feeding and plan production. Population sizes are typically quantified at stocking, grading and harvest events by using fish counters, or via sample weights. Retrospective mortality can then be calculated from the difference in population size. This is the method that is used to calculate sea survival of Scottish farmed salmon (Fig. 1), and the data set illustrates that the method can provide robust, sensible values, at least for larger fish (smolt to harvest size salmon).

Regular removal of dead or dying fish from a rearing system and recording the number provides a real-time mortality metric. In some rearing systems, such as salmon sea-cages, the removal of dead fish is a regular daily process, economically justified by the need to reduce possible sources of disease. This is achieved by hand nets, SCUBA divers, hauling of collection "socks", and commercial suction devices (Sangster 1991; Soares et al. 2011). Mortality records from such collections have recently been used to retrospectively benchmark mortality rates through the production cycle (Soares et al. 2011).

However, the design of some rearing systems, e.g. large freshwater ponds, is not as conducive to removal of dead fish which may then be lost in outflows, to scavengers, or left to decompose (Boyd et al. 2005). When mortality calculated from the difference in population sizes is compared to the number of dead fish removed, anecdotal evidence provided by farmers during stakeholder meetings (North et al. 2008) suggests that the former often exceeds the latter. Such unaccounted fish have been colloquially termed "black fish" and presumably also reflect a combination of losses to predators, escapees, cannibalised individuals and counting errors. Removal of dead fish is recognised to be a time-consuming activity (Wall 2008). Research effort could be directed at determining the accuracy of mortality removal and technical methods that could improve it.

How should mortality be expressed?

Mortality, expressed as a cumulative number of dead fish, is of little value—a simple value obviously has to be related to: (1) the size of the population, and (2) a relevant time period. However, there will be a wide variety of ways of expressing mortality, and the most appropriate metric will depend upon the intended use. For example to be of value,

- as an operational welfare indicator for fish farmers, mortality needs to be responsive to changes over very short time scales and relate to the population size in a single unit, at that point in time, but be simple to calculate with clear "alert" threshold values. Figure 2 illustrates that a mortality rate gives a far clearer signal than a simple track of population size.
- as a retrospective welfare performance indicator for bench-marking and auditing batches, mortality needs to be expressed as a cumulative value over an entire stage of production, rather than over an arbitrary fixed time period. To clarify this by example, daily mortality rate may reduce at a lower temperature, but if growth rate to harvest size is reduced and the duration that mortality acts over is extended, cumulative mortality may then increase.
- as a retrospective welfare indicator for accreditation schemes, it would need to be readily

comprehensible and relate to a complete production cycle. Furthermore, it would be better expressed as survival, rather than mortality, due to the negative imagery associated with the latter term.

Fish welfare researchers, investigating mortality with the ultimate aim of reduction, also need to use appropriate methods of quantification. Research studies typically express mortality as the percentage of the initial population that died over the duration of the study (see Ellis et al. 2002). Expressing mortality in such a manner makes inter-study comparisons difficult as it has not been corrected for the duration of the study, i.e. converted to a rate. Mortality is a key metric for fisheries ecologists who typically express it as the exponential "instantaneous" mortality rate that is amenable to modelling and subdivision into categories (e.g. natural and fishing) (Pitcher and Hart 1982). Other research fields will have developed mortality metrics appropriate for their own investigations, e.g. deaths/180 calf-days (Ortiz-Pelaez et al. 2008). It seems that investigation into appropriate means of expressing mortality needs to accompany research into farmed fish mortality. Whether the pattern of mortality in farmed fish in practice is exponential, linear, or overwhelmingly episodic will be important factors in deciding on appropriate metrics and time intervals.

Misperception 7: It is too difficult to determine cause of mortality in dead fish

An important component of an on-farm mortality recording system is the development of methods for establishing cause of death (Aunsmo et al. 2008). Although often considered difficult, it has been shown to be possible: Aunsmo et al. (2008) provided mortality investigators with a list of categories of death, supplemented by key recognition points and photographs, and they concluded that the bulk of mortalities could be categorised to a cause of death. The costs (manpower) of determining cause of death would need to be determined and balanced against the benefits (to welfare and production). Employment of fish health professionals for post-mortem examination (c.f. Aunsmo et al. 2008) would represent an excessive burden on the industry. Nevertheless, tabulation of physical signs of various causes of death, which do not involve expensive and time-consuming laboratory testing, would be of practical value to farmers for simple on-farm post-mortems. Wedemeyer (1996) provides examples of simple signs for post-mortem examination that may be of value when used in conjunction with the mortality pattern, for example

- death due to oxygen depletion results in flared operculae
- some lethal toxicants (e.g. cyanide) cause massive internal haemorrhage.

Only once mortality is attributed to causes, can these be examined and classed as avoidable and unavoidable. Good management would be reflected by changes in practices and implementation of control measures, to avoid future re-occurrence of preventable causes, if the cost is balanced by the risk (Aunsmo et al. 2008). For example, early warning systems of harmful organisms can allow remedial action such as movement to alternative locations or deployment of barriers (Rodger 2007). If mortality is to be used as a welfare performance indicator, then that portion due to unforeseeable causes could legitimately be excluded from the metric.

How should farmers track mortality?

Mortality in the marine stage of farmed Atlantic salmon is considered a major economic burden for the industry (Rodger 2007; Aunsmo et al. 2008). The UK trout industry has also acknowledged that the economic losses associated with mortality and disease is unsustainable (Anon. 2007). The fish farming industry is therefore aware of mortality as a key economic issue.

Irrespective of any statutory requirements to collect and store data on mortalities, farmers have themselves emphasised the value of mortality as an on-farm welfare indicator (North et al. 2008). Farmed fish stakeholders (including individuals representing fish farmers) have also stated a need for better documentation systems for recording mortality data (Berrill et al. 2010). In practical terms, the methods used to record and archive mortality data on farms can vary greatly. Both paper and electronic methods are used. Electronic methods may use simple spreadsheet packages or more complex, dedicated farm management software. Whilst all methods are likely to be effective for storing data, time will be needed to analyse and/or interpret the data that are collected, even for those farmers that use electronic systems that include automated reporting functions.

Recording systems to collect industry wide data are less common in European aquaculture but have the potential to collate data of considerable strategic importance. In terrestrial livestock industries, such systems are being adopted more widely and often use benchmarking features as an added benefit and incentive to farmers that contribute data. There have been research efforts to address the lack of aggregated industry data in European aquaculture (e.g. http://www.sarf.org.uk/Project%20Final%20Reports/SARF028FinalReport.pdf), and although these initiatives have been supported by farmers, the outputs of such research have not yet been translated into practical systems that are used by the industry. The reasons for this are not clear, but there has been the suggestion that farmers are concerned about the security of data held in such systems, and that a breach in security could result in mortality data being used against the industry.

Misperception 8: Fish farmers do not care about mortality in their stocks

As mortality is a key determinant of the profitability of any farming operation, farmers are receptive to the introduction of better systems for monitoring and analysing mortality with a view to future reduction (Soares et al. 2011; Berrill et al. 2010). One commercial reality is that large fish represent a greater financial investment than small fish, and hence mortality of large fish represents a greater economic loss to farmers (Savvidis et al. 2007). Education may be needed to ensure that farmers appreciate their duty of care and view mortality in terms of welfare (as well as lost income) and realise that mortality of small fish should be viewed equally to that of larger fish. During stakeholder meetings (North et al. 2008; Berrill et al. 2010), farmers have voiced views that they take pride in their fish husbandry skills, with high survival demonstrating such skill.

The evolution of consensually agreed "acceptable" target mortality rates in farmed fish is a long-term goal, but one worth pursuing. All stakeholders would need to recognise inherent differences when making comparisons between developmental stages, species, and with terrestrial livestock. Categorising causes would enable the industry to defend against accusations of unacceptably high levels (e.g. Anon. 1996). Mortality is only unacceptable if it was predictable and preventable; other classes of mortality, although undesirable, need to be considered differently.

Acknowledgments This review represents the authors' thoughts. It is an output from the COST Action 867 "Welfare of Fish in European Aquaculture". Thanks to colleagues (B. Oidtmann, S. Feist, K. Denham) and industry representatives (P. Smith, N. Read) for discussions, and the referees for input.

References

Anon (1986) Guidance on the operation of the Animals (Scientific Procedures) Act 1986. http://www.archive.official-documents.co.uk/document/hoc/321/321.htm

Anon (1994) Report on the welfare of sheep. Farm Animal Welfare Council, Surbiton

Anon (1996) Report on the welfare of farmed fish. Farm Animal Welfare Council, Surbiton

Anon (1997) Report on the welfare of dairy cattle. Farm Animal Welfare Council, Surbiton

Anon (1998) Report on the welfare of broiler breeds. Farm Animal Welfare Council, Surbiton

Anon (2001) Scottish fish farms annual production survey 2000. Fisheries Research Services, Aberdeen

Anon (2002) British Trout Association code of practice. British Trout Association, Edinburgh

Anon (2006a) VHS in Yorkshire. Finfish News 2:5–8

Anon (2006b) A code of good practice for Scottish finfish aquaculture. http://www.scottishsalmon.co.uk/dlDocs/CoGp.pdf

Anon (2007) Trout health and welfare strategy for the United Kingdom. British Trout Association, Edinburgh

Anon (2009) Scottish fish farms annual production survey 2008. Marine Scotland Science

Aunsmo A, Bruheim T, Sandberg M, Skjerve E, Romstad S, Larssen RB (2008) Methods for investigating patterns of mortality and quantifying cause-specific mortality in sea-farmed Atlantic salmon *Salmo salar*. Dis Aquat Org 81:99–107

Barnett JL, Glatz PC (2004) Developing and implementing a welfare audit in the Australian chicken meat industry. In: Weeks C, Butterworth A (eds) Measuring and auditing broiler welfare. CAB International, UK, pp 231–240

Berg C, Algers B (2004) Using welfare outcomes to control intensification: the Swedish model. In: Weeks C, Butterworth A (eds) Measuring and auditing broiler welfare. CAB International, UK, pp 223–229

Berrill IK, Cooper T, MacIntyre CM, Ellis T, Knowles TG, Jones EKM, Turnbull JF (2010) Achieving consensus on current and future priorities for farmed fish welfare: a case study from the UK. Fish Physiol Biochem. doi:10.1007/s10695-010-9399-2

Blanchfield PJ, Tate LS, Podemski CL (2009) Survival and behaviour of rainbow trout (*Oncorhynchus mykiss*) released from an experimental aquaculture operation. Can J Fish Aquat Sci 66:1976–1988

Boyd CE, McNevin AA, Clay J, Johnson HM (2005) Certification issues for some common aquaculture species. Rev Fish Sci 13:231–279

Branson EJ (2008) Fish welfare, Blackwell, Oxford

Branson EJ, Turnbull T (2008) Welfare and deformities in fish. In: Branson EJ (ed) Fish welfare. Blackwell, Oxford, pp 202–216

Brown C, Laland K (2001) Social learning and life skills training for hatchery reared fish. J Fish Biol 59:471–493

Defra (2006) Delivering good animal welfare–a draft delivery strategy under the animal health and welfare strategy

Ellis T, North B, Scott AP, Bromage NR, Porter M, Gadd D (2002) The relationships between stocking density and the welfare of farmed rainbow trout. J Fish Biol 61:493–531

Fossum O, Jansson DS, Etterlin PE, Vågsholm I (2009) Causes of mortality in laying hens in different housing systems in 2001 to 2004. Acta Vet Scand 51:3. doi:10.1186/1751-0147-51-3

Good CA, Thorburn MA, Stevenson RMW (2008) Factors associated with the incidence of bacterial gill disease in salmonid lots reared in Ontario, Canada government hatcheries. Prev Vet Med 83:297–307

Hecht T, Pienaar AG (1993) A review of cannibalism and its implications in fish larviculture. J World Aqua Soc 24: 246–261

James MA, Slaski R (2006) Appraisal of the opportunity for offshore aquaculture in UK waters. Report of project FC0934, commissioned by Defra and Seafish from FRM

Kestin SC (1994) Pain and stress in fish. Royal Society for the Prevention of Cruelty to Animals

Laidley CW, Leatherland JF (1988) Cohort sampling, anaesthesia and stocking density effects on plasma cortisol, thyroid hormone, metabolite and ion levels in rainbow trout, *Salmo gairdneri* Richardson. J Fish Biol 33:73–88

Metcalfe JD (2009) Welfare in wild-capture marine fisheries. J Fish Biol 75:2855–2861

Miller SA, Wagner EJ, Bosakowski T (1995) Performance and oxygen consumption of rainbow trout reared at two densities in raceways with oxygen supplementation. Prog Fish-Cult 57:206–212

Morgan-Davis C, Waterhouse A, Pollock ML, Milner JM (2008) Body condition score as an indicator of ewe survival under extensive conditions. Anim Welf 17:71–77

North BP, Ellis T, Bron J, Knowles T, Turnbull JF (2008) The use of stakeholder focus groups to identify indicators for the on-farm assessment of trout welfare. In: Branson E (ed) Fish welfare. Blackwell, Oxford, pp 243–267

Ortiz-Pelaez A, Pritchard DG, Pfeiffer DU, Jones E, Honeyman P, Mawdsley JJ (2008) Calf mortality as a welfare indicator on British cattle farms. Vet J 176:177–181

Pereira F, Ravelo C, Toranzo AE, Romalde JL (2004) *Lactococcus garvieae*, an emerging pathogen for Portuguese trout culture. Bull Eur Assoc Fish Pathol 24:274–279

Person-Le Ruyet J, Lahaye J, Deniel C, Metailler R, Devauchelle N, Menu B, Noel T, Baudin-Laurencin F (1990) Sole and turbot culture. In: Barnabé G (ed) Aquaculture, vol 2. Ellis Horwood, New York, pp 687–734

Pitcher TJ, Hart PJB (1982) Fisheries ecology. Croom Helm, London

Pottinger TG, Pickering AD (1992) The influence of social interaction on the acclimation of rainbow trout, *Oncorhynchus mykiss* (Walbaum) to chronic stress. J Fish Biol 41:435–447

Ramstad A, Romstad AB, Knappskog DH, Midtlyng PJ (2007) Field validation of experimental challenge models for IPN vaccines. J Fish Dis 30:723–731

Reid C (2006) Activists may strike again, industry told. Fish farmer Mag 29(6):10

Rodger HD (2007) Gill disorders: an emerging problem for farmed Atlantic salmon (*Salmo salar*) in the marine environment? Fish Vet J 9:38–48

Rodger HD, Henry L, Mitchell SO (2011) Non-infectious gill disorders of marine salmonid fish. Rev Fish Biol Fisheries. doi:10.1007/s11160-010-9182-6

Sangster GI (1991) The retrieval of mortalities from salmon farm sea cages. Scott Fish Inform Pamphlet (19):15

Savvidis GK, Anatohoti SC, Kanaki Z, Vafeas G (2007) Epizootic outbreaks of Lactococcosis disease in rainbow trout, *Oncorhynchus mykiss* (Walbaum), culture in Greece. Bull Eur Assoc Fish Path 27:223–228

Savvidis G, Zartaloudis Z, Vafeas G (2009) Massive fish losses in rainbow trout cultures of Louros River (N.W. Greece) after strong summer rainfall. Implication of the sycamore lace bug *Corythucha ciliata* (Hemiptera: Tingidae). Bull Eur Assoc Fish Path 29:66–72

Shelbourne JE (1975) Marine fish cultivation: pioneering studies on the culture of the larvae of the plaice (*Pleuronectes platessa* L.) and the sole (*Solea solea* L.). MAFF Fish Investig Ser II 27(9):29

Shinn AP, Picon-Camacho SM, Bawden R, Taylor NGH (2009) Mechanical control of *Ichthyophthirius multifiliis* Fouquet, 1876 (Ciliophora) in a rainbow trout hatchery. Aquacult Eng 41:152–157

Soares S, Green DM, Turnbull JF, Crumlish M, Murray AG (2011) A baseline method for benchmarking mortality losses in Atlantic salmon (*Salmo salar*) production. Aquaculture 314:7–12

Steinum T, Kvellestad A, Ronneberg LB, Nilsen H, Asheim A, Fjell K, Nygard SMR, Olsen AB, Dale OB (2008) First cases of amoebic gill disease (AGD) in Norwegian seawater farmed Atlantic salmon, *Salmo salar* L., and phylogeny of the causative amoeba using 18S cDNA sequences. J Fish Dis 31:205–214

Stevenson P (2007) Closed waters: the welfare of farmed Atlantic salmon, rainbow trout, Atlantic cod & Atlantic halibut. Compassion in World Farming, Surrey

Walker AJ (2010) Scottish fish farm production survey 2009. Marine Scotland, The Scottish Government. http://www.scotland.gov.uk/Resource/Doc/295194/0106192.pdf

Wall T (2008) Disease and medications—the welfare implications. In: Branson EJ (ed) Fish welfare. Blackwell, Oxford, pp 195–201

Wedemeyer GA (1996) Physiology of fish in intensive culture systems. Chapman & Hall, New York

Anaesthesia of Farmed Fish: Implications for Welfare – Summary

Inger Hilde Zahl, Ole Samuelsen and Anders Kiessling

Treatment with anaesthetics is used both in fish farming and research. Its use ranges from light sedation that aims to reduce stress during confinement and handling such as netting, weighing and sorting, to full anaesthesia in order to avoid inflicting pain during e.g. surgery.
Many anaesthetics are lipid soluble and will, therefore, accumulate, in tissues with a high content of fat, such as the brain. This accumulation will result in very long wash out times and may easily yield an overdose followed by very long recovery periods and even death. It is, therefore, important not to leave fish in the anaesthetic bath longer than necessary and to observe signs of overdose in order to exchange or move the fish quickly to clean, water as a counter measure. Whereas increased water temperature generally results in quicker induction and shorter recovery times, no direct relationship was found between body weight and the effects of anaesthesia. Stress prior to anaesthesia results in quicker induction times, deeper anaesthetised fish and prolonged recovery times. It is, therefore important to reduce the dosage in stressed fish because the uptake is much larger than in unstressed fish. Therefore, the drug can be easily overdosed.

Background

Anaesthetics used for fish differ from those used in mammals due to differences in anatomy, but also because fish often are too small, too many and too difficult to catch and treat individually by injection, As consequence, we need drugs that can be dissolved in large quantities of water. Also, there are considerable differences in response between fish species. This is not surprising, considering that there are more than 27,000 described species of fish, inhabiting nearly all existing aquatic environments.

Scientific Significance

We show that differences between species are relevant for the application of anaesthetics. In addition, some of the secondary factors that might influence anaesthetic effect, i.e. body weight, water temperature and acute stress, are also discussed. Although anaesthetic agents are used to avoid stress, there are indications that anaesthesia may induce a stress response and therefore some situations become stressors by themselves. One should therefore be aware that every treatment with anaesthics will be followed by a stress reaction, but the intensity may vary between different agents.

Practical Application

In contrast to human and veterinary medicine where anaesthetic protocols include combinations of drugs, each contributing to the effects needed, such as sedation, pain relief, loss of memory etc. protocols for fish usually only refer to the use of one single agent. However, as a result of the increased focus on fish welfare there is an increased use of combinations of agents for anaesthesia.

Contact

A. Kiessling
Department of Wildlife, Fish and Environmental Studies,
Swedish University of Agricultural Sciences,
901 83 Umeå, Sweden
e-mail: Anders.kiessling@slu.se

Fish Physiol Biochem (2012) 38:201–218
DOI 10.1007/s10695-011-9565-1

Anaesthesia of farmed fish: implications for welfare

Inger Hilde Zahl · Ole Samuelsen · Anders Kiessling

Received: 30 December 2010 / Accepted: 8 October 2011 / Published online: 9 December 2011

Abstract During their life cycle as farmed animals, there are several situations in which fish are subjected to handling and confinement. Netting, weighing, sorting, vaccination, transport and, at the end, slaughter are frequent events under farming conditions. As research subjects, fish may also undergo surgical procedures that range from tagging, sampling and small incisions to invasive procedures. In these situations, treatment with anaesthetic agents may be necessary in order to ensure the welfare of the fish. The main objective of this paper is to review our knowledge of the effects of anaesthetic agents in farmed fish and their possible implications for welfare. As wide variations in response to anaesthesia have been observed both between and within species, special attention has been paid to the importance of secondary factors such as body weight, water temperature and acute stress. In this review, we have limited ourselves to the anaesthetic agents such as benzocaine, metacaine (MS-222), metomidate hydrochloride, isoeugenol, 2-phenoxyethanol and quinaldine. Anaesthetic protocols of fish usually refer to one single agent, whereas protocols of human and veterinary medicine cover combinations of several drugs, each contributing to the effects needed in the anaesthesia. As stress prior to anaesthesia may result in abnormal reactions, pre-anaesthetic sedation is regularly used in order to reduce or avoid stress and is an integral part of the veterinary protocols of higher vertebrates. Furthermore, the anaesthetic agents that are used in order to obtain general anaesthesia are combined with analgesic agents that target nociception. The increased use of such combinations in fish is therefore included as a special section. Anaesthetic agents are widely used to avoid stress during various farming procedures. While several studies report that anaesthetics are effective in reducing the stress associated with confinement and handling, there are indications that anaesthesia may in itself induce a stress response, measured by elevated levels of cortisol. MS-222 has been reported to elicit high cortisol release rates immediately following exposure, while benzocaine causes a bimodal response. Metomidate has an inhibitory effect on cortisol in fish and seems to induce the lowest release of cortisol of the agents reported in the literature. Compared to what is observed following severe stressors such as handling and confinement, the amount of cortisol released in response to anaesthesia appears to be low but may represent an extra load under otherwise stressful

I. H. Zahl · O. Samuelsen
Institute of Marine Research, P.O. Box 1870 Nordnes, 5817 Bergen, Norway
e-mail: inger.hilde.zahl@gmail.com

A. Kiessling (✉)
Department of Wildlife, Fish and Environmental Studies, Swedish University of Agricultural Sciences, 901 83 Umeå, Sweden
e-mail: Anders.kiessling@slu.se

A. Kiessling
Department of Animal and Aquacultural Sciences, Norwegian University of Life Sciences, P.O. Box 5003, 1432 Aas, Norway

circumstances. Furthermore, anaesthetics may cause secondary adverse reactions such as acidosis and osmotic stress due to respiratory arrest and insufficient exchange of gas and ions between the blood and the water. All in all, anaesthetics may reduce stress and thereby improve welfare but can also have unwanted side effects that reduce the welfare of the fish and should therefore always be used with caution. Finally, on the basis of the data reported in the literature and our own experience, we recommend that anaesthetic protocols should always be tested on a few fish under prevailing conditions in order to ensure an adequate depth of anaesthesia. This recommendation applies whether a single agent or a combination of agents is used, although it appears that protocols comprising combinations of agents provide wider safety margins. The analgesic effects of currently used agents, in spite of their proven local effects, are currently being debated as the agents are administrated to fish via inhalation rather than locally at the target site. We therefore recommend that all protocols of procedures requiring general anaesthesia should be complemented by administration of agents with analgesic effect at the site of tissue trauma.

Keywords Welfare · Anaesthesia · Sedation · Fish · Teleost

Introduction

Anaesthetic agents have been used for fish since the beginning of the last century (McFarland 1959; McFarland and Klontz 1969; Schoettger and Julin 1967). They derive partly from human and veterinary medicine and partly from other sources and were introduced to fish through trial and error. They were initially employed in order to make handling easier but improved knowledge of fish physiology and a better understanding of the importance of anaesthetic treatment to maintain welfare have led to an increased use.

Anaesthesia (from Greek: an- 'without' and aisthesis 'sensation') comprises several components, including sedation, immobilisation, unconsciousness (narcosis), amnesia (loss of memory) and analgesia (pain relief). Sedation is a reduction in sensitivity, which results in tranquillity and calmness. Narcosis (general anaesthesia) causes a state of unconsciousness and amnesia and also includes immobilisation and pain relief (analgesia). The components of anaesthesia can be obtained by various anaesthetic agents, each of which gives rise to one or several of them. In this review, the agents such as benzocaine, metacaine (MS-222), metomidate hydrochloride, isoeugenol, 2-phenoxyethanol and quinaldine are discussed as anaesthetic agents for fish. Traditionally, these agents have been used and are still used in fish to obtain different components of anaesthesia irrespective of the actual effect of the specific agent. Agents with known analgesic effect, used in human and veterinary medicine as local analgesics, such as benzocaine, are administrated to fish for the purpose of sedation, immobilisation, analgesia as well as general anaesthesia (narcosis). Agents are also applied indiscriminately of species. Anaesthetic protocols of new species that are introduced to research or cultivation are generally based on drugs and dosages developed for the more established species. For example, in Norway, the anaesthetic protocols of Atlantic cod (*Gadus morhua*) and Atlantic halibut (*Hippoglossus hippoglossus*), which were introduced to fish farming in the 1980s, have thus been based on the protocols used for salmonids, which have been farmed since the 1960s and are the most important species in Norwegian fish farming.

Anaesthetic protocols of human and veterinary medicine comprise combinations of several drugs, each contributing to one or more of the effects needed in the anaesthesia. Different drugs are applied for induction and maintenance, and analgesic treatment is used both under and following surgical procedures. The drugs are selected on the basis of their properties, tailored both to the surgical intervention and to the physiological state of the patient or the animal. Pre-anaesthetic sedation used in order to avoid stress prior to anaesthesia is an integral part of veterinary protocols. By combining drugs of different properties, a more complete anaesthesia is obtained than what is possible with one single substance alone. Further, synergistic effects between different drugs may also permit a reduction in the dosage of each drug compared to individual administration. This may result in smoother induction and recovery and reduce the incidence of adverse effects (Rang et al. 2003).

Anaesthesia of fish

Anaesthetic agents are now used widely, ranging from light sedation that aims to reduce stress during handling and non-invasive procedures to full anaesthesia to avoid inflicting pain during surgery and larger interventions (Ackerman et al. 2005; Neiffer and Stamper 2009; Ross and Ross 2008; Summerfelt and Smith 1990). The most common route of administration is via inhalation. The anaesthetic agent is dispersed in the water and is absorbed across the gills. The effect is usually assessed by induction and recovery time, reflex reactions to external stimuli and responsiveness to handling.

The progress of induction and depth of anaesthesia are generally divided into distinct stages and 'planes' (Table 1). Common clinical indicators evaluated in order to determine stages and planes during anaesthesia of animals include behaviour, activity, corneal reflexes and pupil size, muscle tone, reflexes, respiratory rate, heart rate and blood pressure. As several of these indicators are difficult to assess in fish, it may be difficult to distinguish one stage or plane from the other, especially in situations where induction is rapid. In fish, the stages are therefore usually described by changes in swimming activity, balance, respiratory frequency and reactions to external stimuli (Table 1).

In order to obtain efficient surgical anaesthesia through inhalation using the agents described in this review, respiration will stop in spite of maintained heart rate and blood pressure (Kiessling et al. unpublished results). However, injecting a local anaesthetic at the site of incision will provide sufficient analgesic effect and allow surgical procedures at an anaesthetic depth where spontaneous respiration is maintained, corresponding to stage III, plane 2 (Table 1; Karlsson et al. 2011a; Kiessling et al. 2003, 2009).

Nociception in fish

Studies on rainbow trout and goldfish have demonstrated that fish possess the basic neural system necessary for nociception, i.e. perception of painful stimuli (Dunlop and Laming 2005; Sneddon 2002, 2003; Sneddon et al. 2003). Thus, to ensure the welfare of fish subjected to procedures that might inflict pain, anaesthetic agents with the ability to block nociceptive pathways are necessary. Of the agents assessed in this review, benzocaine and MS-222 possess this ability, as does isoeugenol. However, the route of administration may be of importance for the effect of this class of agents (local anaesthetics). Anaesthetic effect is produced locally at the site of administration, e.g. topically. Administration systemically has, on the other hand, been found inadequate for blocking peripheral nociceptive pathways in higher vertebrates (mice, rat, cat and man; Boas et al. 1982; Haegerstam 1979; Rigon and Takahashi 1996; Wiesenfeld-Hallin and Lindblom 1985; Woolf and Wiesenfeld-Hallin 1985). Haegerstam (1979) concluded that it was unlikely that peripheral pain pathways could be blocked by systemic i.v. injections

Table 1 Stages of anaesthesia in fish

Stage	Plane	Description stage/plane	Appearance	Swimming activity	Equilibrium	Responsiveness[a]	Muscle tone	Respiration	Heart rate
0		Normal	Normal	Normal	Normal	Normal	Normal	Normal	Normal
I		Light sedation	Disoriented	Reduced	Normal	Slightly reduced	Normal	Normal	Normal
II		Excitatory stage	Excited	Increased	Struggles to maintain balance	Normal or exaggerated	Normal	Irregular or increased	Irregular, may increase
III	1	Light anaesthesia	Anaesthetised	Stopped	Lost	Reacts to strong tactile stimuli	Decreased	Normal or decreased	Regular
	2	Surgical anaesthesia	Anaesthetised	Stopped	Lost	None	Relaxed	Shallow	Depressed
	3	Deep narcosis	Anaesthetised	Stopped	Lost	None	None	Nearly absent	Depressed
IV		Impending death	Moribund	Stopped	Lost	None	None	Stopped	Cardiac arrest

[a] Responsiveness refers to reaction to external stimuli. The stimulus may be visual or tactile

Adapted from Bell (1987), Burka et al. (1997), McFarland (1959), McFarland and Klontz (1969) and Summerfelt and Smith (1990)

of local anaesthetics. On the basis of these data, it is unclear whether a sufficient level of anaesthesia can be obtained peripherally when anaesthetic agents are administered to fish via inhalation. As the whole body surface of the fish is exposed to anaesthetics during bath immersions, i.e. topical administration in addition to inhalation, the agents will also have a peripheral effect. The absorption of small amounts of anaesthetic through the skin of the fish has been observed (Ferreira et al. 1984), but whether such amounts produce the peripheral analgesia needed for surgical procedures is not known. It is therefore important to determine whether a lack of response to nociceptive stimuli is due to a peripheral blockage of nociceptive afferent pathways or whether it is due to a general inhibition of the CNS. The latter may have severe welfare consequences as an inadequate blockage of nociceptive pathways may result in sensitisation and thus a reduction in threshold and an increased response to noxious stimuli. Work by Zahl et al. (2009, 2011) indicates that neither of the agents covered in this review administered via inhalation either individually or in combination blocks peripheral nociceptive pathways in fish. Kiessling et al. (1995, 2003) show that this is obtained when lidocaine is administered locally during combination anaesthesia with metomidate and MS-222 administered via inhalation. In the studies by Zahl et al. (2009, 2011), lower dosages in combination anaesthesia resulted in similar induction time as individually administrated agents. However, the fish displayed increased reactions to nociceptive stimulation although agents blocking nociception were used. This is possibly due to an enhanced central effect without a parallel increase in effect peripherally. Since all surgical procedures are associated with nociception, our conclusion must therefore be that any such procedure should never be conducted without local injections of anaesthetic agents with analgesic effect.

Anaesthetic agents

A wide range of anaesthetic agents is currently used for fish. Among the most common are MS-222, benzocaine, isoeugenol, metomidate, 2-phenoxyethanol and quinaldine (Ackerman et al. 2005; Neiffer and Stamper 2009; Ross and Ross 2008; Summerfelt and Smith 1990). Dosage recommendations are listed in Table 2. The sites of action in the nervous system, i.e. central or peripheral, for these drugs are not well known in fish, and most knowledge is based on that of higher vertebrates. There are many reasons for applying this knowledge in fish with caution. First, there are more than 30,000 different species of fish (Froese and Pauly 2011) that inhabit highly diverse environments, which vary in factors like salinity, temperature, chemistry and depth. Secondly, and possibly the most important and most often overlooked fact, the route of administration may differ both between different species and between different agents, *i.m.* for local anaesthetics vis-à-vis *i.v.* or inhalation for general anaesthetics. Finally, fish and higher vertebrates, particularly mammals, differ in the development of the central nervous system, while their peripheral nervous systems seem to be surprisingly similar (see e.g. Sneddon et al. 2003).

Metacaine and benzocaine

Metacaine (ethyl 3-aminobenzoate, tricaine methanesulphonate and MS-222) and benzocaine (ethyl 4-aminobenzoate) are the two most common anaesthetics for fish used in Norway (Anon 2007). They are approved for use in food fish production with a withdrawal period of 21 days. MS-222 and benzocaine are local anaesthetics that act by blocking voltage-sensitive sodium channels (Frazier and Narahashi 1975; Neumcke et al. 1981), thereby preventing the voltage-dependent increase in sodium conductance. This inhibits the initiation and propagation of action potentials in excitable cells; thus, local anaesthetics block most neurons and muscle cells and may cause paralysis in addition to blocking nociception. This class of drugs is used in human and veterinary medicine for local analgesia. When administrated to fish via bath immersion, they enter the circulation and produce general anaesthesia by inhibiting neural signal transmission ranging from the periphery to higher parts of the nervous system. The precise mechanism of action in the central nervous system is not fully understood (Hara and Sata 2007; Hedrick and Winmill 2003; Ueta et al. 2007). Many of the adverse effects caused by this class of drugs in man are related to effects on the CNS and the cardiovascular system (Rang et al. 2003).

A range of adverse effects of MS-222 and benzocaine have been reported in fish. The initial response to these agents is characterised by a period of raised heart

Table 2 Anaesthetic agents—dosage and exposure time recommendations by manufacturers/distributors

Anaesthetic agent	Sedation		Anaesthesia, light		Anaesthesia		Species
	Dosage (mg l^{-1})	Exposure (min)	Dosage (mg l^{-1})	Exposure (min)	Dosage (mg l^{-1})	Exposure (min)	
MS-222[a]	10–30	>480	30–80	>30	80–180	>10	Trout species
MS-222[a]	7–30	>240	30–80	>10	80–100	>5	Salmon species
MS-222[b]					60	5	Atlantic cod
Benzocaine[c]					30–40	2–5	Salmon and trout
Isoeugenol[d]	<7.5				7.5–10	12–15	Salmon species
Metomidate[e]	0.25–1	>480			5–10	>60	Various species
2-phenoxyethanol[f]					0.1–0.5[g]	3	Various species
Quinaldine[f]					15–40	2–4	Salmonid species

[a] Pharmaq AS, Oslo, Norway

[b] ScanVacc AS, Aarnes, Norway

[c] A.C.D Pharmaceuticals AS, Leknes, Norway

[d] Aqui-S, New Zealand Ltd

[e] Syndel International Inc., Vancouver, Canada

[f] Ackerman et al. (2005)

[g] The dosage of 2-phenoxyethanol is ml l^{-1}

rate and respiration accompanied by elevated levels of blood glucose, followed by a depression of cardiovascular and respiratory function that eventually may come to a complete stop (Hill et al. 2002; Houston et al. 1971a; Lochowitz et al. 1974; Randall 1962; Ryan 1992). This results in hypoxaemia, observed as changes in the partial pressure of arterial O_2 and CO_2, accompanied by lower blood glucose levels and increased levels of lactate, hematocrit and haemoglobin, and erythrocyte swelling (Bourne 1984; Hill and Forster 2004; Holloway et al. 2004; Houston et al. 1971b; Iwama et al. 1989; Ortuno et al. 2002; Ryan 1992; Soivio et al. 1977; Thomas and Robertson 1991; Velisek et al. 2009). Elevated plasma catecholamine levels have also been observed (Gingerich and Drottar 1989; Iwama et al. 1989; Wedemeyer 1970) Benzocaine has also been found to have immunodepressant effects (Cuesta et al. 2004; Ortuno et al. 2002).

Metomidate

Metomidate hydrochloride (methyl 3-(1-phenylethyl) imidazole-4-carboxylate hydrochloride) is a methyl analogue of the imidazole derivate etomidate, which activates and modulates inhibitory gamma-aminobutyric acid type A ($GABA_A$) receptors, thus affecting higher regions of the nervous system (Ashton and Wauquier 1985; Yang and Uchida 1996). Activation of the $GABA_A$ receptor triggers opening of a chloride ion-selective pore. The increased chloride conductance drives the membrane potential towards the reversal potential of the Cl^- ion, thus inhibiting the firing of new action potentials. $GABA_A$ receptor activation produces sedation and hypnosis, but limited analgesia and immobilisation (Grasshoff et al. 2006).

Etomidate is commonly used in human and veterinary medicine as a sedative and for inducing anaesthesia (Ching and Baum 2009; Darrouj et al. 2009; Falk and Zed 2004; Sams et al. 2008). One of the side effects of the drug in addition to respiratory and cardiovascular depression is suppression of adrenal steroidogenesis, thus leading to an inhibition of cortisol synthesis (Vanden Bossche et al. 1984; Wagner et al. 1984; man, cow and rat). Metomidate, in parity with etomidate in mammals, has also been found to have an inhibitory effect on cortisol in fish (Davis and Griffin 2004; Olsen et al. 1995; Small 2004; Thomas and Robertson 1991) but the efficiency and dose dependence of this inhibition are still unclear (Eliason et al. 2007; Olsen et al. 1995; Sandodden et al. 2001). However, a number of reports indicate cortisol release in spite of metomidate treatment (Kiessling, et al. 2009; Zahl et al. 2010). Other adverse effects of metomidate in fish are related to depressant effects on respiration and circulation that subsequently lead to hypoxaemia, observed as a decrease in partial pressure of O_2, elevated partial pressure of CO_2 and reduced pH of the blood (Hill and Forster 2004; Hill et al. 2002; Iwama et al. 1989).

2-Phenoxyethanol

2-Phenoxyethanol is a compound with antibacterial properties used as a preservative in vaccines, in dermatological products and as fixative for perfumes. To the best of our knowledge, the exact mechanism of its anaesthetic effect in fish has not been reported, but it has been suggested that it involves an expansion of neuronal cell membranes (Burka et al. 1997). 2-Phenoxyethanol has been found to inhibit the activity of excitatory *N*-methyl-D-aspartate (NMDA) receptors in *Xenopus* oocytes (Musshoff et al. 1999), and its anaesthetic effect may thus be related to the suppression of neural activity in higher regions of the nervous system. Activation of the NMDA receptors increases neuron excitability by lowering the threshold for firing of action potentials due to opening of ion channels that cause influx of positively charged ions, especially Ca^{2+}, Na^+ and K^+. NMDA receptor inhibition has been linked to an analgesic effect (Grasshoff et al. 2006).

Adverse effects of 2-phenoxyethanol include reduced ventilation, decreased heart rate and blood pressure, reduced blood partial pressure of O_2 accompanied by increased CO_2 and decreased blood pH and elevated plasma levels of adrenaline and glucose (Fredricks et al. 1993; Iwama et al. 1989; Lambooij et al. 2009; Ortuno et al. 2002). 2-Phenoxyethanol has also been found to have immunodepressant effects (Cuesta et al. 2004; Ortuno et al. 2002).

Isoeugenol

Isoeugenol (2-methoxy-4-prop-1-enyl-phenol) is an oily liquid found in ylang–ylang and other essential oils. It is structurally similar to eugenol, a widely used analgesic in dentistry that inhibits sodium, potassium and calcium channels, inhibits NMDA receptors and potentiates $GABA_A$ receptors (Aoshima and Hamamoto 1999; Lee et al. 2005; Li et al. 2007; Park et al. 2006; Wie et al. 1997). Like eugenol, isoeugenol is one of the constituents of clove oil, another widely used fish anaesthetic. Isoeugenol is the active ingredient in Aqui-S®, which has approval for use in food fish production in New Zealand, Australia and Chile, Costa Rica and Republic of Korea with no withdrawal period. In Norway, Aqui-S® is used for brood stock and in research.

Adverse effects of isoeugenol (Aqui-S®) in fish include reduced ventilation and depression of the cardiovascular system, which result in slower heart rate and decreased cardiac output, in addition to reduced blood pressure and vascular tone (Hill and Forster 2004; Hill et al. 2002; Rothwell and Forster 2005). Elevated plasma levels of catecholamines have also been observed, as has increased haematocrit (Hill and Forster 2004). Eugenol works synergistically with d-tubocurarine; it is therefore believed to block $nicotine_2$ receptors at the motor end plate in muscle (Brodin and Røed 1984). $Nicotine_2$ receptor blockade causes paralysis without analgesic or hypnotic effect. This effect has since been confirmed for isoeugenol by Ingvast-Larsson et al. (2003), who used a rat phrenic nerve–diaphragm muscle preparation.

Quinaldine

Quinaldine (2-methylquinoline) is a colourless oily liquid that has been used to anaesthetise fish since early in the last century. To the best of our knowledge, its precise mechanism of action has not been reported.

Quinoline family compounds possess antiseptic and antipyretic properties. They are widely used as parent compounds to make drugs, especially anti-malarial medicines, and are also used in fungicides, biocides, alkaloids, dyes, rubber chemicals and flavouring agents.

Fish anaesthetised with quinaldine respond by increasing their heart rate, followed by bradycardia and impaired respiratory function (Lochowitz et al. 1974). Elevated concentrations of plasma cortisol and glucose have also been observed, as have increased levels of serum IgM (Cuesta et al. 2004; Davis and Griffin 2004; Ortuno et al. 2002).

Combination anaesthesia

Current protocols of fish do not typically include administration of combinations of anaesthetics. Schoettger and Steucke (1970) examined the possible synergistic effects of combination anaesthesia consisting of a mixture of MS-222 and quinaldine in rainbow trout (*Oncorhynchus mykiss*) and northern pike (*Esox lucius*) and found that the mixture resulted in safer and more effective anaesthesia. The same mixture tested in 14 freshwater species showed that the dosages needed when administered in combination were considerably lower than when each drug was used alone (Gilderhus et al. 1973). However, it has also been found that not only efficacy but also toxicity increased when these drugs were combined (Dawson and Marking 1973). Synergistic effects have also been found in combination anaesthesia using a mixture of quinaldine sulphate and the benzodiazepine diazepam (Kumlu and Yanar 1999; Yanar and Kumlu 2001). Administered in combination with diazepam, quinaldine sulphate could be used in lower dosages in anaesthesia of both gilthead sea bream (*Sparus aurata*) and European sea bass (*Dicentrarchus labrax*). Hyperactivity and excitement, unwanted side effects of quinaldine sulphate, were reduced, and there were no mortalities (Kumlu and Yanar 1999; Yanar and Kumlu 2001). Combinations consisting of MS-222 anaesthesia and injections of local analgesics during surgery have been studied in koi carp (*Cyprinus carpio*; Harms et al. 2005). Injections of the opiate butorphanol during surgery reduced post-surgery behavioural changes, while the non-steroidal anti-inflammatory drug (NSAID) ketoprofen reduced muscle tissue damage. Protocols that include both pre-anaesthetic sedation and local analgesic treatment in addition to general anaesthesia have been successfully employed in anaesthesia of salmonids undergoing surgery for insertion of dorsal aorta and portal vein cannulae (Eliason et al. 2007; Karlsson et al. 2006; Kiessling et al. 1995, 2003). Pre-anaesthetic sedation with metomidate followed by full anaesthesia with MS-222 combined with injections of lidocaine at the site of incision improved recovery in comparison with MS-222 used individually, and fish quickly resumed their normal behaviour (Eliason et al. 2007; Karlsson et al. 2006; Kiessling et al. 1995, 2003). The response of Atlantic cod subjected to caudal artery cannulation using pre-anaesthetic sedation with metomidate has been found to be markedly different from that of salmonids, in that the former display a delay rather than a reduction in post-surgery cortisol release, and a prolonged recovery, measured as blood glucose, lasting from 3 to 5 days (Karlsson et al. 2011b). Zahl et al. (2009) found that pre-anaesthesia sedation of Atlantic cod with metomidate permitted a significant reduction in the dosage of anaesthetic to be made, with no change in induction or recovery time. However, the same authors report that Atlantic cod, subjected to pre-anaesthetic sedation with metomidate and the reduced dose of anaesthetic, displayed an increased response to external stimulation in the form of a caudal peduncle pinch. They also report that exposing Atlantic cod to pre-anaesthetic stress instead of pre-anaesthetic sedation resulted in the same induction time in spite of a lower dose of anaesthetic. These results clearly underline the fact that the effect of an anaesthetic agent varies with pre-anaesthetic stress and show that the use of an anaesthetic protocol that includes a combination of pre-sedation and a reduced anaesthetic dosage may produce an acceptable hypnotic effect without a sufficient analgesic effect.

Variations in response

There are substantial variations between fish species in response to anaesthetic agents, as well as large individual differences within each species. Variations may be the result of pharmacokinetic differences, usually described as what the body does to the drug, and by pharmacodynamic differences, usually described as what the drug does to the body. Both pharmacokinetic and pharmacodynamic differences in fish may be influenced by biological factors such as age, sex, life stage, body weight, growth rate, body composition, physiological condition and health status, as well as environmental factors such as water temperature, salinity, pH and oxygen level.

In the majority of assessments of anaesthesia in fish, the fish have been anaesthetised to a predetermined stage, often corresponding to loss of equilibrium, no response to external stimuli and reduced or even stopped respiration (Table 1, stage III, plane 3 to stage IV) and thereafter directly moved to clean water for recovery (Hoskonen and Pirhonen 2004; Mattson and Riple 1989; Mylonas et al. 2005; Stehly and Gingerich 1999; Sylvester and Holland 1982). As the induction time may differ between individual fish, exposure time has varied accordingly. Another approach is to impose an identical exposure time for all fish. When comparing the results of different studies, it is important to distinguish between these two approaches. The duration of the exposure should always be kept within an adequate margin of safety, as practical situations often involve anaesthesia of a large number of fish in one tank, with exposure time usually exceeding the induction time for individual fish.

Pharmacodynamics

The pharmacodynamic characteristics of a drug describe the consequences of the interaction between the drug and the receptor or other primary sites of action and include all effects and adverse effects. As described earlier, anaesthetic agents may act by binding to receptors or by interacting with ion channels. As several types of tissue may contain the target molecules of a drug, the primary responses may induce secondary responses or side effects that subsequently affect the primary response. Pharmacodynamic effects may also influence the pharmacokinetic processes. Thus, an effect on respiration and circulation will also be of importance for the absorption, distribution and elimination of a drug.

Pharmacodynamic properties studied in Atlantic cod (Zahl et al. 2009) and Atlantic halibut (Zahl et al. 2011) include wide differences within and between these two species as well as between the agents. Induction time, defined as time of loss of equilibrium, occurred within 3 min for all agents tested, thus easily meeting one criterion of suitability for fish anaesthetic agents (Bell 1987; Marking and Meyer 1985). The recovery time differed widely between the agents. Metomidate produced the longest recovery time in Atlantic cod and was together with isoeugenol and 2-phenoxyethanol one of the agents that gave rise to the longest recovery time in Atlantic halibut. This agrees with pharmacokinetic data that indicate that a slower elimination and a prolonged recovery might be the expected outcome in agents of higher lipid solubility (Guenette et al. 2007; Hansen et al. 2003; Kiessling et al. 2009; Kildea et al. 2004). The recovery time in Atlantic cod reported by Zahl et al. (2009, 2011) was within the range defined for suitable anaesthetics, i.e. 5 min, (Bell 1987; Marking and Meyer 1985) for all agents except for metomidate, but exceeded this in Atlantic halibut. However, as the parameter 'recovery time' was recorded differently in Atlantic halibut and in Atlantic cod, values for recovery time need to be compared with caution. In pelagic fish such as Atlantic cod, recovery time is normally defined as time of recovery of equilibrium, whereas in Atlantic halibut, a bottom dwelling flatfish, recovery time is more difficult to assess. In the work by Zahl et al. (2011), recovery time was set at when the halibut managed to right itself after having been placed upside down in a recovery tank.

The agents studied by Zahl et al. (2009, 2011) differed greatly in their capacity to depress responses to external stimulation, assessed during handling of the anaesthetised fish and by a caudal peduncle pinch. Reflex reactions to the pinch and responsiveness to handling were most effectively reduced by two local anaesthetics MS-222 and benzocaine. This is probably related to the mode of action, as these substances suppress signal transmission in both central and peripheral nervous system. The result also tallies well with findings in higher vertebrates, in which systemic injections of local anaesthetics did not block peripheral nociceptive pathways, but merely increased the response threshold. Suppression of nerve signals at central level may have adverse effects. After being inhaled by the fish, the agents enter the circulation, where they may be distributed throughout the body. By passing the blood–brain barrier, these agents affect the brain and the centres that control respiration and circulation. A depression of respiration and circulation would produce hypoxaemia, metabolic acidosis and changes in blood electrolytes and lead to extended recovery time, thus enhancing the effects of the anaesthetic.

Pharmacokinetics

In order to induce a biological response, a drug must be present in a sufficient concentration at the receptor site for a certain length of time. The concentration achieved at the receptor site depends on the pharmacokinetic properties of the drug, i.e. how it is absorbed, distributed, metabolised and eliminated. In fish, drug administration via bath immersion is equivalent to inhalation anaesthesia in human and veterinary medicine. The processes of absorption and elimination are determined by the diffusion rate through the gill epithelium, which mainly depends on the lipid solubility, ionisation and polarity of the drug. Following absorption, a fraction of the drug binds to plasma protein, mainly albumin, while the rest remains unbound and pharmacologically active. The unbound drug is then distributed to various parts of the body, commonly separated into compartments, where it exerts its effect. The main compartments are plasma, interstitial fluid, intracellular fluid and adipose tissue.

Lipid solubility and ionisation are important factors in determining the distribution of the drug. High lipid solubility results in rapid diffusion through the cell membranes and the possible accumulation of the drug within hydrophobic compartments such as the adipose tissue and the lipid bilayers of cell membranes, while ionised molecules and molecules with high polarity are less lipid-soluble and do not easily pass through biological membranes.

The ionisation of weak acids and bases is determined by the dissociation constant p*K*a, defined as the pH at which 50% of the acid or base is in ionic form (p*K*a, Table 3). When a compound is allowed to distribute itself between equal volumes of two immiscible liquids, the ratio of the concentration of the solute in two phases at equilibrium is called the partition coefficient (logP). For drugs, two phases used are octanol and an aquatic buffer with pH of 7.4 (Table 3). As the logP value describes the partition of neutral uncharged molecules, it is not an exact measure of the distribution of ionisable compounds such as many anaesthetic agents. Anaesthetic agents are usually associated with high lipid solubility, which may lead to accumulation in adipose tissue both during long exposures and following repeated exposures (Rang et al. 2003).

In fish, anaesthetics are mainly eliminated via the gills in unchanged form (Hayton et al. 1996; Hunn and Allen 1974; Hunn et al. 1968; Meinertz et al. 1991; Stenger and Maren 1974). Some may also undergo metabolism before elimination (Hayton et al. 1996; Meinertz et al. 1991; Ryan 1992; Stenger and Maren 1974).

Models are used to describe the pharmacokinetic processes. Plasma concentration–time curves are employed to estimate the pharmacokinetic parameters and perform compartmental analyses to describe drug distribution. In the simplest compartmental pharmacokinetic model, the body is regarded as a single compartment throughout which the drug is distributed evenly following absorption. This model, called the one-compartment open model with first-order elimination, is a simplification used to illustrate basic pharmacokinetic principles. A two-compartment open model includes a peripheral compartment representing the tissue besides the central compartment, which represents the plasma, thereby introducing a distribution process into the model. From the central compartment, the drug can be removed both by distribution to the peripheral compartment and by elimination. The process of distribution is described by the initial rapid decrease in plasma concentration of the drug and is called the α–phase. When the distribution is complete, the elimination of the drug is determined by the slower β-phase.

The distribution of a drug between the plasma and the rest of the body is described by the pharmacokinetic parameter volume of distribution (Vd). The Vd indicates how a drug relocates into body compartments relative to the blood, and is a non-physiological volume that corresponds to the volume that would be necessary to dilute the drug to produce the concentration present in the blood. Lipophilic drugs that have a tendency to accumulate in body fat thus have high Vd values, while drugs with a high degree of plasma protein binding have low Vd values. The rate at which the drug is removed from the blood is described by the parameter clearance (C), which is defined as the volume of blood or plasma that is cleared per unit time. Together with clearance, the volume of distribution determines the plasma half-life ($t_{1/2}$) of a drug. A study by Kiessling et al. (2009) revealed large differences in the pharmacokinetics of various anaesthetics. MS-222, the most hydrophilic of the agents tested (Table 3), was distributed rapidly, followed by rapid clearance and elimination. The plasma data were best described by a one-compartment open model with first-order elimination, which indicates that distribution was homogenous and elimination proportional to plasma concentration. The plasma data of benzocaine and isoeugenol were best described by a two-compartment open model, indicating a rapid transfer of drug from plasma to tissues followed by the slower process of elimination from the plasma. Both benzocaine and isoeugenol were found by Kiessling et al. (2009) to be rapidly distributed, but their rates of clearance and elimination were much slower than for MS-222. An initial phase of rapid distribution followed by a period of slow elimination has also been found in rainbow trout anaesthetised with benzocaine (Meinertz et al. 1996; Stehly et al. 1998). The pharmacokinetic properties of benzocaine in rainbow trout were best described by a two-compartment model when benzocaine was administered through bath immersion, while a three-compartment model best described the properties of the distribution and clearance process following intra-arterial injection (Meinertz et al. 1996; Stehly et al. 1998). Guenette

Table 3 log *P* (octanol:water), p*K*a and water solubility of anaesthetic agents

Anaesthetic	log *P*	p*K*a	Water solubility (mg l^{-1})
MS-222	1.8	3.8[b]	1.0×10^5
Benzocaine	1.9	2.5	1.3×10^3
Metomidate hydrochloride[a]	3.1	4.5[c]	6.3×10
2-phenoxyethanol	1.2	15.1	2.7×10^4
Isoeugenol	3.0	9.9	3.6×10^2
Quinaldine	2.6	5.7	5.0×10^2

Source: ChemIDplus, U.S. National Library of Medicine

[a] The values indicated for log *P*, p*K*a and water solubility are applicable to etomidate

[b] Source: Stenger and Maren (1974)

[c] Source: Levron and Assoune (1990)

et al. (2007) found that eugenol, which is similar to isolegenol, was well absorbed and eliminated by rainbow trout. The plasma concentration versus time curve was biphasic, which indicates two compartments. A biphasic plasma elimination curve was also found in the elasmobranch dogfish (*Squalus acanthias*) following arterial injection of MS-222 (Stenger and Maren 1974). MS-222 was excreted rapidly across the gills, mainly in unchanged form. Kiessling et al. (2009) found that artificial ventilation had no influence on the compartment analysis. However, elimination rates were faster in the groups of fish that received artificial ventilation, demonstrating the importance of the gills as a pathway for drug elimination. This is in agreement with the previous reports in other species (Hayton et al. 1996; Hunn and Allen 1974; Meinertz et al. 1991) and emphasises the importance of respiratory function, i.e. exchange of water over the gills, during recovery (Table 4).

Water temperature

Temperature is an important factor in determining the rate of physiological processes in ectothermic animals such as fish and thus plays an important role in the processes related to the uptake and elimination of drugs. A 10°C rise in water temperature (Q_{10}) typically doubles the basal metabolic rate of teleost fish (Clarke and Johnston 1999). This leads to a higher oxygen demand that is met by enhanced respiration, increased cardiac output and increased blood flow through the gills (Graham and Farrell 1989; Nilsson and Sundin 1998; Webber et al. 1998). Poorer oxygen solubility due to rising water temperature leads to an additional need to enhance respiration and blood flow.

Temperature increases have been reported to shorten induction and recovery time for a number of anaesthetic agents in several teleost species, including benzocaine in striped bass (*Morone saxatilis*; Gilderhus et al. 1991), 2-phenoxyethanol and clove oil in European sea bass and gilthead sea bream (Mylonas et al. 2005), isoeugenol in rainbow trout (Stehly and Gingerich 1999), clove oil in rainbow trout, brown trout (*Salmo trutta*), Atlantic salmon, whitefish (*Coregonus lavaretus*), perch (*Perca fluviatilis*) and roach (*Rutilus rutilus*) (Hoskonen and Pirhonen 2004; Woolsey et al. 2004), and benzocaine, MS-222, metomidate and 2-phenoxyethanol in Atlantic cod (Zahl et al. 2009). Reduced induction time at higher water temperatures has also been demonstrated in common carp (*Cyprinus carpio*), rainbow trout, fathead minnows (*Pimephales promelas*) and Atlantic halibut anaesthetised with MS-222 and in Atlantic halibut anaesthetised with benzocaine (Hikasa et al. 1986; Houston and Woods 1976; Sylvester and Holland 1982; Zahl et al. 2011).

The rapid induction time seen at higher water temperature may be related to higher basal metabolic rate at higher temperatures (Clarke and Johnston 1999; Saunders 1963; Schurmann and Steffensen 1997) and the corresponding increase in oxygen demand leading to increased respiration and circulation. This facilitates a rapid absorption and distribution of anaesthetic as well as faster clearance rates but may also lead to a greater amount being absorbed. While faster clearance rates of both benzocaine and isoeugenol at higher temperature were found in rainbow trout and silver perch (*Bidyanus bidyanus*) (Kildea et al. 2004; Stehly et al. 1998), significantly higher concentrations of isoeugenol were found in rainbow trout fillets after exposure at 17°C than at 7°C (Meinertz et al. 2006). Zahl et al. (2009) report that Atlantic cod made a quicker recovery at higher water temperature and had a wider therapeutic window, indicating a more rapid clearance and elimination. Atlantic halibut, on the other hand, seem to recover more slowly at higher temperatures, but in spite of longer recovery time the fish displayed stronger reactions to handling, indicating a lack of anaesthetic effect (Zahl et al. 2011). In large Atlantic cod (1,000 g) anaesthetised at a low water temperature (8°C), the dosage of both

Table 4 Effects of different anaesthetic agents in fish

Anaesthetic agent	Mechanism of action	Effects
Metacaine and Benzocaine	Blocks Na^+ channels	Blocks most neurons, glands and muscle cells (both striated, cardiac and smooth). May, in addition to nocicpetion, cause paralysis and respiratory depression. Adverse effects in fish include depression of cardiovascular and respiratory function, increased lactate and elevated catecholamine levels. Local analgesic effects is questionable when administered via inhalation
Metomidate	Activates $GABA_A$ receptors	Produces sedation and hypnosis. No nociception blockage. Adverse effects in fish include respiratory and cardiovascular depression and inhibition of cortisol synthesis
Isoeugenol	Inhibits Na^+, K^+ and Ca^{2+} channels Potentiates $GABA_A$ receptors Inhibits NMDA receptors	Produces analgesia and sedation. Adverse effects in fish include reduced ventilation, heart rate and cardiac output, blood pressure and vascular tone
2-phenoxyethanol	Not reported	Adverse effects in fish include reduced ventilation, heart rate and blood pressure. Efficiency is indicated as low in some species
Quinaldine	Not reported	Adverse effects in fish include reduced ventilation and heart rate. Efficiency is indicated as low in some species

metomidate and MS-222 had to be reduced, while for benzocaine no suitable dosage was found (Zahl et al. 2009). This was due to the fact that after five minutes of exposure to benzocaine the fish reacted both to handling and to the caudal peduncle pinch, but suffered respiratory arrest and died in the recovery bath. The findings in Atlantic cod agree with those in brown trout and goldfish, in which higher sensitivity was seen at lower water temperature during anaesthesia with benzocaine and 2-phenoxyethanol, respectively (Dawson and Gilderhus 1979; Weyl et al. 1996). However, the findings contradict with observations in striped bass, where higher dosages of benzocaine were used at lower water temperature for fish of large body weight (Gilderhus et al. 1991). In an evaluation of Aqui-S® in rainbow trout, reduced toxicity was found at higher temperature (Stehly and Gingerich 1999).

In the study on Atlantic halibut (Zahl et al. 2011), only MS-222 and benzocaine were tested at higher water temperature (15°C), but even so, the data indicate that in this species the central and peripheral effects of the anaesthetics are differently influenced by the temperature increase in comparison with Atlantic cod (Zahl et al. 2009). This might be explained in part by central effects in Atlantic halibut being more pronounced at higher water temperature, while the peripheral effects are less so, resulting in greater responsiveness and fish that regain equilibrium more slowly. Zahl et al. (2011) report that this effect is most prominent with benzocaine, which may reflect higher lipid solubility of the drug.

Anaesthetic agents are associated with high lipid solubility and a tendency to accumulate in tissues with a high content of fat (Rang et al. 2003). This might be a problem following repeated or long-duration exposures, for instance during larger surgical procedures.

Uptake of fat from the diet, storage in adipose tissue and the incorporation of lipids into membranes all vary with temperature. There is greater incorporation of unsaturated fatty acids and increased aggregation of hydrophobic substances in membranes as temperature falls (Berge et al. 1980; Cossins 1983; Ruyter et al. 2006; Sellner and Hazel 1982). Substances with higher lipid solubility such as benzocaine, isoeugenol and metomidate may therefore penetrate and possibly also accumulate in the lipid bilayer of cell membranes more readily at lower temperatures than more hydrophilic substances such as MS-222. Any accumulation would then alter the characteristics of the membranes and affect normal cell function by influencing transport mechanisms, stabilising the membranes and possibly inhibiting nervous signal propagation. If so, exposure time would influence this effect of membrane lipid changes and could explain some of the contradictory effects of temperature found in the literature. Anaesthetic depth has been associated with the concentration of anaesthetic agent in the brain. This has been reported for snapper (*Pagrus auratus*) and several freshwater species anaesthetised with MS-222, as well as sting ray (*Dasyatis sabina*) and lemon shark (*Negaprion brevirostris*) anaesthetised with quinaldine (Brown et al. 1972; Hunn 1970; Ryan 1992). The brain receives a large supply of oxygen-rich blood from the gills, and the blood–brain barrier is permeable to anaesthetic agents. Accumulation in the brain probably leads to an enhanced central effect, with depression of brain centres controlling respiratory and cardiovascular function as a consequence.

Body weight

The relationship between body weight and sensitivity to anaesthetics may be influenced by several characteristics related to body weight, such as age, body composition, growth rate and sexual maturity, as all of these characteristics influence the physiology of the fish. Since the major route of both uptake and elimination is through the gills, the rate of oxygen consumption, the ratio of body volume to gill surface area and the rate of gill perfusion are of great importance. These factors vary within and between species and with life stage and life style as well as with body size. The gill surface area decreases in relation to increased body weight (Muir 1969; Oikawa and Itazawa 1985; Oikawa et al. 1999). The basal metabolic rate falls relative to increasing body size, and larger animals thus have lower oxygen consumption relative to body size than smaller animals (Clarke and Johnston 1999; Schmidt-Nielsen 1984). The scaling relationship between metabolic rate and body weight is usually described by the equation $Vo_2 = a\ M_b^{0.75}$, where the metabolic rate (Vo_2) scales as the 0.75 power of body mass (M_b) and the constant a indicates the level of the resting metabolic rate (Schmidt-Nielsen 1984). The constant a may differ between species and varies in relation to lifestyle, and in ectotherm animals such as fish, it is highly dependent on ambient temperature. For teleost fish, a scaling exponent of 0.8 has been found (Clarke and Johnston 1999). Sedate and less active fish species have a lower resting metabolic rate than more active species (Morris and North 1984). Clarke and Johnston (1999) found that the resting metabolic rate of fish varied between taxonomic groups, gadoids having higher resting metabolic rates than pleuronectiformes and salmoniformes, although there was a large degree of overlap.

Reports regarding the importance of body weight for the effect of anaesthetic agents in fish are contradictory. Some studies show no relationship between body size and induction and recovery time (Houston and Woods 1972; Stehly and Gingerich 1999), while others indicate that such a relationship exists (Gilderhus and Marking 1987; Houston et al. 1976; Olsen et al. 1995). Hoskonen and Pirhonen (2004) observed a shortening of induction time with increasing body size in whitefish, but observed the opposite relationship in rainbow trout, and found no size-related differences in Atlantic salmon or brown trout. Weber et al. (2009) found that induction time increased with increasing body weight in Senegalese sole (*Solea senegalensis*) anaesthetised with 2-phenoxyethanol, metomidate and clove oil, while this was not found for MS-222. The dynamics of the recovery time were more complex and were only weight-related for MS-222 (Weber et al. 2009).

Zahl et al. (2009, 2011) found no uniform relationship between the effects of anaesthetic agents and the body weight of the fish in either Atlantic cod or Atlantic halibut. For Atlantic cod, the results suggest that the body weight of the fish, or factors related to body weight, is less important for the differences in effect than the properties of the anaesthetic agents themselves.

Stress

When fish are subjected to handling, such as crowding or dip-netting, they respond by attempting to escape, which increases oxygen consumption. Paralleling the alterations in behaviour, a cascade of physiological reactions makes up the physiological stress response, first termed the General Adaptation Syndrome (see e.g. Selye 1985). The physiological stress response in fish is similar to that described in other vertebrates and is characterised by three phases (see Wendelaar Bonga 1997, for details). Large species differences in the magnitude of the cortisol response to stress have been found (Barton 2002; Barton and Iwama 1991), and high and low responders within species have also been identified (Fevolden et al. 1999; Mommsen et al. 1999; Pottinger and Carrick 1999; Pottinger et al. 1992, 1994; Sumpter et al. 1986; Tanck et al. 2001; Trenzado et al. 2003).

Stressed animals display abnormal reactions to anaesthesia and may require larger doses for both induction and maintenance (Hall et al. 2001). In fish, increased physical activity while attempting to escape handling, followed by the cascade of physiological alterations of the stress response, contributes to facilitating anaesthetic uptake. The elevated levels of catecholamines and corticosteroids further stimulate ventilation and enhance cardiac output, in turn increasing gill blood flow (Farrell 1984; Graham and Farrell 1989; Nilsson and Sundin 1998). The perfusion of the gill lamellae increases and lamellae that in a normal resting state are barely or not perfused at all are recruited, expanding the respiratory surface area and facilitating diffusion, and thereby affecting the uptake of substances across the gill.

Several reports show that stress associated with handling of fish can be effectively reduced by anaesthetic agents. Metomidate has been found to reduce handling stress in Atlantic salmon, Chinook salmon (*Oncorhynchus tshawytscha*), hybrid striped bass (*Morone chrysops* x *Morone saxatilis*), channel catfish (*Ictalurus punctatus*) and red drum (*Sciaenops ocellatus*), but this may be related to the cortisol blocking side effects of this drug. Isoeugenol has been shown to be effective in reducing stress in Atlantic salmon and channel catfish and MS-222 inhibits the stress response to handling in rainbow trout and channel catfish (Davis and Griffin 2004; Iversen et al. 2003; King et al. 2005; Kreiberg and Powell 1991; Olsen et al. 1995; Small 2003, 2004; Small and Chatakondi 2005; Thomas and Robertson 1991). However, it has also been found that exposure to anaesthetics may itself induce a stress response. Of the agents discussed in this review, MS-222 has been found to induce stress in Atlantic salmon, rainbow trout, Atlantic cod, Atlantic halibut, channel catfish, striped bass, hybrid striped bass, gilthead sea bream and red drum, while isoeugenol and eugenol induce stress in Atlantic salmon, rainbow trout and hybrid striped bass, and benzocaine and metomidate have a similar effect in Atlantic salmon, Atlantic cod and Atlantic halibut (Barton and Peter 1982; Davidson et al. 2000; Davis and Griffin 2004; Davis et al. 1982; Iversen et al. 2003; Kiessling et al. 2009; Molinero and Gonzalez 1995; Olsen et al. 1995; Small 2003; Thomas and Robertson 1991, Zahl et al. 2010). However, confinement and relocation in connection with anaesthesia influence the responses to anaesthesia and may also mask the stress-inducing effects of the anaesthetic agents themselves (Hill and Forster 2004; Molinero and Gonzalez 1995; Thomas and Robertson 1991).

In a practical situation, stress, accompanied by the cascade of the physiological stress response, facilitates the uptake of anaesthetic by stimulating respiration and circulation and increasing the area of the blood/water interface. A doubling of oxygen consumption has been observed in coho salmon (*Oncorhynchus kisutch*), rainbow trout and Atlantic cod following handling, returning to basal levels within 1 h in coho salmon and 3 to 5 h in Atlantic cod (Barton and Schreck 1987; Davis and Schreck 1997; Saunders 1963). This probably results in an immediate and excessive uptake of anaesthetic, leading to faster induction and possibly also more deeply anaesthetised fish. In the study by Zahl et al. (2009), in which Atlantic cod were subjected to stress in the form of 30 s in air in a dip-net prior to anaesthesia, the fish displayed shorter induction time, became more deeply anaesthetised and recovered more slowly than unstressed fish. Moreover, the dosage of MS-222 had to be reduced in order to avoid mortality in the stressed group, but was too low to provide sufficient anaesthetic effect in the unstressed group. Both stress and sedation prior to anaesthesia may have marked practical consequences. This was well demonstrated by Kiessling and co-workers in a series of studies on feeding and growth after vaccination in salmon smolts anaesthetised with benzocaine with and without pre-anaesthetic metomidate sedation (Kiessling et al. 2001; Oppedal et al. 2000). They found that fish vaccinated using combination anaesthesia resumed normal feeding behaviour within a week, while fish subjected to the ordinary procedure with netting directly into the benzocaine bath without pre-sedation needed more than 3 weeks to resume normal feeding behaviour. At that time, the weights of the fish in the pre-sedated group were 22 per cent higher than the fish in the non-pre-sedated group.

References

Ackerman PA, Morgan JD, Iwama GK (2005) Anesthetics. CCAC guidelines on: the care and use of fish in research, teaching and testing. http://www.ccac.ca/en/CCAC_

Programs/Guidelines_Policies/GDLINES/Fish/Fish Anesthetics—ENG.pdf. Canadian Council on Animal Care, Ottawa CA (last accessed December 2010)

Anon (2007) Pharmaceutical use in Norwegian fish farming in 2001–2007, Wholesale-based drug statistics. Norwegian Institute of Public Health

Aoshima H, Hamamoto K (1999) Potentiation of $GABA_A$ receptors expressed in *Xenopus* oocytes by perfume and phytoncid. Biosci Biotechnol Biochem 63:743–748

Ashton D, Wauquier A (1985) Modulation of a GABA-ergic inhibitory circuit in the in vitro hippocampus by etomidate isomers. Anesth Analg 64:975–980

Barton BA (2002) Stress in fishes: a diversity of responses with particular reference to changes in circulating corticosteroids. Integrat Comparat Biol 42:517–525

Barton BA, Iwama GK (1991) Physiological changes in fish from stress in aquaculture with emphasis on the response and effects of corticosteroids. Ann Rev Fish Dis 1:3–26

Barton BA, Peter RE (1982) Plasma cortisol stress response in fingerling rainbow trout, *Salmo gairdneri* Richardson, to various transport conditions, anaesthesia, and cold shock. J Fish Biol 20:39–51

Barton BA, Schreck CB (1987) Metabolic cost of acute physical stress in juvenile steelhead. Trans Am Fish Soc 116: 257–263

Bell G (1987) An outline of anesthetics and anesthesia for salmonids: a guide for fish culturists in British Columbia. Canadian technical report of fisheries aquatic sciences no. 1534

Berge RK, Slinde E, Farstad M (1980) Discontinuities in Arrhenius plots due to formation of mixed micelles and change in enzyme substrate availability. FEBS Lett 109: 194–196

Boas RA, Covino BG, Shahnarian A (1982) Analgesic responses to i.v. lignocaine. Br J Anaesth 54:501–505

Bourne PK (1984) The use of MS-222 (tricaine methanesulphonate) as an anaesthetic for routine blood sampling in three species of marine teleosts. Aquaculture 36:313–321

Brodin P, Røed A (1984) Effects of eugenol on rat phrenic nerve and phrenic nerve-diaphragm preparations. Arc Oral Biol 29:611–615

Brown EAB, Franklin JE, Pratt E, Trams EG (1972) Contributions to the pharmacology of quinaldine (uptake and distribution in the shark and comparative studies). Comp Biochem Physiol 42A:223–231

Burka JF, Hammell KL, Horsberg TE, Johnson GR, Rainnie DJ, Speare DJ (1997) Drugs in salmonid aquaculture—a review. J Vet Pharmacol Ther 20:333–349

Ching KY, Baum CR (2009) Newer agents for rapid sequence intubation: etomidate and rocuronium. Pediatr Emerg Care 25:200–207

Clarke A, Johnston NM (1999) Scaling of metabolic rate with body mass and temperature in teleost fish. J Anim Ecol 68:893–905

Cossins AR (1983) Adaptive responses of fish membranes to altered environmental temperature. Biochem Soc Trans 11:332–333

Cuesta A, Meseguer J, Esteban MA (2004) Total serum immunoglobulin M levels are affected by immunomodulators in seabream (*Sparus aurata* L.). Vet Immunol Immunopathol 101:203–210

Darrouj J, Karma L, Arora R (2009) Cardiovascular manifestations of sedatives and analgesics in the critical care unit. Am J Therapeut 16:339–353

Davidson GW, Davie PS, Young G, Fowler RT (2000) Physiological responses of rainbow trout *Oncorhynchus mykiss* to crowding and anesthesia with AQUI-S (TM). J World Aqua Soc 31:105–114

Davis KB, Griffin BR (2004) Physiological responses of hybrid striped bass under sedation by several anesthetics. Aquaculture 233:531–548

Davis LE, Schreck CB (1997) The energetic response to handling stress in juvenile coho salmon. Trans Am Fish Soc 126:248–258

Davis KB, Parker NC, Suttle MA (1982) Plasma corticosteroids and chlorides in striped bass exposed to tricaine methanesulfonate, quinaldine, etomidate, and salt. Progress Fish Cultur 44:205–207

Dawson VK, Gilderhus PA (1979) Ethyl-*p*-aminobenzoate (Benzocaine): efficacy as an anesthetic for five species of freshwater fish, vol 87. Investigations in Fish Control, United States Department of the Interior

Dawson VK, Marking LL (1973) Toxicity of mixtures of quinaldine sulfate and MS-222 to fish, vol 53. Investigations in Fish Control, United States Department of the Interior

Dunlop R, Laming P (2005) Mechanoreceptive and nociceptive responses in the central nervous system of goldfish (*Carassius auratus*) and trout (*Oncorhynchus mykiss*). J Pain 6:561–568

Eliason EJ, Kiessling A, Karlsson A, Djordjevic B, Farrell AP (2007) Validation of the hepatic portal vein cannulation technique using Atlantic salmon *Salmo salar* L. J Fish Biol 71:290–297

Falk J, Zed PJ (2004) Etomidate for procedural sedation in the emergency department. Ann Pharmacother 38:1272–1277

Farrell AP (1984) A review of cardiac performance in the teleost heart: intrinsic and humoral regulation. Can J Zool 62: 523–536

Ferreira JT, Schoonbee HJ, Smit GL (1984) The uptake of the anesthetic benzocaine hydrochloride by the gills and the skin of three freshwater fish species. J Fish Biol 25:35–41

Fevolden SE, Røed KH, Fjalestad KT, Stien J (1999) Post-stress levels of lysozyme and cortisol in adult rainbow trout: heritabilities and genetic correlations. J Fish Biol 54: 900–910

Frazier DT, Narahashi T (1975) Tricaine (MS-222): Effects on ionic conductances of squid axon membranes. Eur J Pharmacol 33:313–317

Fredricks KT, Gingerich WH, Fater DC (1993) Comparative cardiovascular effects of four fishery anesthetics in spinally transected rainbow trout, *Oncorhynchus mykiss*. Compar Biochem Physiol C Pharmacol Toxicol Endocrinol 104: 477–483

Froese R, Pauly D (2011) FishBase, http://www.fishbase.org, version 10/2011

Gilderhus PA, Marking LL (1987) Comparative efficacy of 16 anesthetic chemicals on rainbow trout. North Am J Fish Manag 7:288–292

Gilderhus PA, Berger BL, Sills JB, Harman PD (1973) The efficacy of quinaldine sulfate: MS-222 mixtures for the anesthetization of freshwater fish, vol 54. Investigations in Fish Control, United States Department of the Interior

Gilderhus PA, Lemm CA, Woods LC (1991) Benzocaine as an anesthetic for striped bass. Progress Fish Cultur 53: 105–107

Gingerich WH, Drottar KR (1989) Plasma-catecholamine concentrations in rainbow trout (*Salmo gairdneri*) at rest and after anesthesia and surgery. Gen Comp Endocrinol 73: 390–397

Graham MS, Farrell AP (1989) The effect of temperature acclimation and adrenaline on the performance of a perfused trout heart. Physiol Zool 62:38–61

Grasshoff C, Drexler B, Rudolph U, Antkowiak B (2006) Anaesthetic drugs: linking molecular actions to clinical effects. Curr Pharm Des 12:3665–3679

Guenette SA, Uhland FC, Helie P, Beaudry F, Vachon P (2007) Pharmacokinetics of eugenol in rainbow trout (*Oncorhynchus mykiss*). Aquaculture 266:262–265

Haegerstam G (1979) Effect of i.v. administration of lignocaine and tetrodotoxin on sensory units in the tooth of the cat. Br J Anaesth 51:487–491

Hall LW, Clarke KW, Trim CM (2001) Veterinary anaesthesia. W. B. Saunders, London

Hansen MK, Nymoen U, Horsberg TE (2003) Pharmacokinetic and pharmacodynamic properties of metomidate in turbot (*Scophthalmus maximus*) and halibut (*Hippoglossus hippoglossus*). J Vet Pharmacol Ther 26:95–103

Hara K, Sata T (2007) The effects of the local anesthetics lidocaine and procaine on glycine and gamma-aminobutyric acid receptors expressed in xenopus oocytes. Anesth Analg 104:1434–1439

Harms CA, Lewbart GA, Swanson CR, Kishimori JM, Boylan SM (2005) Behavioral and clinical pathology changes in koi carp (*Cyprinus carpio*) subjected to anesthesia and surgery with and without intra-operative analgesics. Compar Med 55:221–226

Hayton WL, Szoke A, Kemmenoe BH, Vick AM (1996) Disposition of benzocaine in channel catfish. Aqua Toxicol 36:99–113

Hedrick MS, Winmill RE (2003) Excitatory and inhibitory effects of tricaine (MS-222) on fictive breathing in isolated bullfrog brain stem. Am J Physiol Regul Integrat Compar Physiol 284:R405–R412

Hikasa Y, Takase K, Ogasawara T, Ogasawara S (1986) Anesthesia and recovery with tricaine methanesulfonate, eugenol and thiopental sodium in the carp, *Cyprinus carpio*. Jpn J Veterin Sci 48:341–351

Hill JV, Forster ME (2004) Cardiovascular responses of Chinook salmon (*Oncorhynchus tshawytscha*) during rapid anaesthetic induction and recovery. Compar Biochem Physiol C Toxicol Pharmacol 137:167–177

Hill JV, Davison W, Forster ME (2002) The effects of fish anaesthetics (MS222, metomidate and AQUI-S) on heart ventricle, the cardiac vagus and branchial vessels from Chinook salmon (*Oncorhynchus tshawytscha*). Fish Physiol Biochem 27:19–28

Holloway AC, Keene JL, Noakes DG, Moccia RD (2004) Effects of clove oil and MS-222 on blood hormone profiles in rainbow trout *Oncorhynchus mykiss*, Walbaum. Aquacult Res 35:1025–1030

Hoskonen P, Pirhonen J (2004) Temperature effects on anaesthesia with clove oil in six temperate-zone fishes. J Fish Biol 64:1136–1142

Houston AH, Woods RJ (1972) Blood concentrations of tricaine methane sulphonate in brook trout, *Salvelinus fontinalis*, during anesthetization, branchial irrigation, and recovery. J Fish Res Board Can 29:1344–1346

Houston AH, Woods RJ (1976) Influence of temperature upon tricaine methane sulphonate uptake and induction of anesthesia in rainbow trout (*Salmo gairdneri*). Compar Biochem Physiol C Pharmacol Toxicol Endocrinol 54:1–6

Houston AH, Madden JA, Woods RJ, Miles HM (1971a) Some physiological effects of handling and tricaine methanesulphonate anesthetization upon brook trout, *Salvelinus fontinalis*. J Fish Res Board Can 28:625–633

Houston AH, Madden JA, Woods RJ, Miles HM (1971b) Variations in blood and tissue chemistry of brook trout, *Salvelinus fontinalis*, subsequent to handling, anesthesia, and surgery. J Fish Res Board Can 28:635–642

Houston AH, Corlett JT, Woods RJ (1976) Specimen weight and MS-222. J Fish Res Board Can 33:1403–1407

Hunn JB (1970) Dynamics of MS-222 in the blood and brain of freshwater fishes during anesthesia, vol 42. Investigations in Fish Control, United States Department of the Interior

Hunn JB, Allen JL (1974) Movement of drugs across gills of fishes. Annu Rev Pharmacol Toxicol 14:47–55

Hunn JB, Schoettger RA, Willford WA (1968) Turnover and urinary excretion of free and acetylated MS 222 by rainbow trout, *Salmo gairdneri*. J Fish Res Board Can 25:25–31

Ingvast-Larsson JC, Axén VC, Kiessling A (2003) Effects of isoeugenol on rat phrenic nerve-diaphragm preparations. Am J Veter Res 64:690–693

Iversen M, Finstad B, McKinley RS, Eliassen RA (2003) The efficacy of metomidate, clove oil, Aqui-S (™) and Benzoak (®) as anaesthetics in Atlantic salmon (*Salmo salar* L.) smolts, and their potential stress-reducing capacity. Aquaculture 221:549–566

Iwama GK, McGeer JC, Pawluk MP (1989) The effects of five fish anaesthetics on acid-base balance, hematocrit, blood gases, cortisol, and adrenaline in rainbow trout. Can J Zool Revue Canadienne de Zoologie 67:2065–2073

Karlsson A, Eliason EJ, Mydland LT, Farrell AP, Kiessling A (2006) Postprandial changes in plasma free amino acid levels obtained simultaneously from the hepatic portal vein and the dorsal aorta in rainbow trout (*Oncorhynchus mykiss*). J Exp Biol 209:4885–4894

Karlsson A, Rosseland B-O, Thorarensen H, Kiessling A (2011a) Changes in arterial oxygen tension and physiological status in resting and unrestrained Arctic charr *Salvelinus alpinus* (L.) exposed to mild hypoxia and hyperoxia. J Fish Biol 78:962–966

Karlsson A, Rosseland B-O, Massabuau JC, Kiessling A (2011b) Pre-anaesthetic metomidate sedation delays the stress response after caudal artery cannulation in Atlantic cod (*Gadus morhua*). Fish Physiol Biochem. doi: 10.1007/s10695-011-9516x

Kiessling A, Dosanjh B, Higgs D, Deacon G, Rowshandeli N (1995) Dorsal aorta cannulation: a method to monitor changes in blood levels of astaxanthin in voluntarily feeding Atlantic salmon, *Salmo salar* L. Aquac Nutr 1:43–50

Kiessling A, Johansson D, Axen C, Johansson B (2001) Anestesi och analgesi vid vaccinering av lax. In: Olsen RE, Hansen T (eds) Havbruksrapporten 2001, Fisken og havet, saernr.3-2001

Kiessling A, Olsen RE, Buttle L (2003) Given the same dietary carotenoid inclusion, Atlantic salmon, *Salmo salar* (L.) display higher blood levels of canthaxanthin than astaxanthin. Aquac Nutr 9:253–261

Kiessling A, Johansson D, Zahl IH, Samuelsen OB (2009) Pharmacokinetics, plasma cortisol and effectiveness of benzocaine, MS-222 and isoeugenol measured in individual dorsal aorta-cannulated Atlantic salmon (*Salmo salar*) following bath administration. Aquaculture 286:301–308

Kildea MA, Allan GL, Kearney RE (2004) Accumulation and clearance of the anaesthetics clove oil and AQUI-S (TM) from the edible tissue of silver perch (*Bidyanus bidyanus*). Aquaculture 232:265–277

King WV, Hooper B, Hillsgrove S, Benton C, Berlinsky DL (2005) The use of clove oil, metomidate, tricaine methanesulphonate and 2-phenoxyethanol for inducing anaesthesia and their effect on the cortisol stress response in black sea bass (*Centropristis striata* L.). Aquacult Res 36:1442–1449

Kreiberg H, Powell J (1991) Metomidate sedation reduces handling stress in chinook salmon. World Aqua 22:58–59

Kumlu M, Yanar M (1999) Effects of the anesthetic quinaldine sulphate and muscle relaxant diazepam on sea bream juveniles (*Sparus aurata*). Israeli J Aqua Bamidgeh 51: 143–147

Lambooij B, Pilarczyk M, Bialowas H, Reimert H, Andre G, van de Vis H (2009) Anaesthetic properties of Propiscin (Etomidaat) and 2-phenoxyethanol in the common carp (*Cyprinus carpio* L.), neural and behavioural measures. Aquacult Res 40:1328–1333

Lee MH, Yeon KY, Park CK, Li HY, Fang Z, Kim MS, Choi SY, Lee SJ, Lee S, Park K, Lee JH, Kim JS, Oh SB (2005) Eugenol inhibits calcium currents in dental afferent neurons. J Dent Res 84:848–851

Levron JC, Assoune P (1990) Pharmacokinetics of etomidate. Annales Françaises d'Anesthésie Et de Réanimation 9:123–126

Li HY, Park CK, Jung SJ, Choi SY, Lee SJ, Park K, Kim JS, Oh SB (2007) Eugenol inhibits K^+ currents in trigeminal ganglion neurons. J Dent Res 86:898–902

Lochowitz RT, Miles HM, Hafemann DR (1974) Anesthetic-induced variations in cardiac rate of the teleost, *Salmo gairdneri*. Gen Pharmacol 5:217–224

Marking LL, Meyer FP (1985) Are better anesthetics needed in fisheries? Fisheries 10:2–5

Mattson NS, Riple TH (1989) Metomidate, a better anesthetic for cod (*Gadus morhua*) in comparison with benzocaine, MS-222, chlorobutanol, and phenoxyethanol. Aquaculture 83:89–94

McFarland WN (1959) A study of the effects of anaesthetics on the behaviour and physiology of fishes. Publ Inst Mar Sci 6:22–55

McFarland WN, Klontz GW (1969) Anesthesia in fishes. Feder Proc 28:1535–1540

Meinertz JR, Gingerich WH, Allen JL (1991) Metabolism and elimination of benzocaine by rainbow trout (*Oncorhynchus mykiss*). Xenobiotica 21:525–533

Meinertz JR, Stehly GR, Gingerich WH (1996) Pharmacokinetics of benzocaine in rainbow trout (*Oncorhynchus mykiss*) after intraarterial dosing. Aquaculture 148:39–48

Meinertz JR, Greseth SL, Schreier TM, Bernardy JA, Gingerich WH (2006) Isoeugenol concentrations in rainbow trout (*Oncorhynchus mykiss*) skin-on fillet tissue after exposure to AQUI-S $^{(TM)}$ at different temperatures, durations, and concentrations. Aquaculture 254:347–354

Molinero A, Gonzalez J (1995) Comparative effects of MS 222 and 2-phenoxyethanol on gilthead sea bream (*Sparus aurata* L) during confinement. Compar Biochem Physiol A Physiol 111:405–414

Mommsen TP, Vijayan MM, Moon TW (1999) Cortisol in teleosts: dynamics, mechanisms of action, and metabolic regulation. Rev Fish Biol Fisheries 9:211–268

Morris DJ, North AW (1984) Oxygen consumption of five species of fish from South Georgia. J Exp Mar Biol Ecol 78:75–86

Muir BS (1969) Gill dimensions as a function of fish size. J Fish Res Board Can 26:165–170

Musshoff U, Madeja M, Binding N, Witting U, Speckmann EJ (1999) Effects of 2-phenoxyethanol on N-methyl-$_D$-aspartate (NMDA) receptor-mediated ion currents. Arch Toxicol 73:55–59

Mylonas CC, Cardinaletti G, Sigelaki I, Polzonetti-Magni A (2005) Comparative efficacy of clove oil and 2-phenoxyethanol as anesthetics in the aquaculture of European sea bass (*Dicentrarchus labrax*) and gilthead sea bream (*Sparus aurata*) at different temperatures. Aquaculture 246: 467–481

Neiffer DL, Stamper MA (2009) Fish sedation, anesthesia, analgesia, and euthanasia: considerations, methods, and types of drugs. ILAR J 50:343–360

Neumcke B, Schwarz W, Stampfli R (1981) Block of Na channels in the membrane of myelinated nerve by benzocaine. Pflügers Archiv Eur J Physiol 390:230–236

Nilsson S, Sundin L (1998) Gill blood flow control. Compar Biochemis Physiol Part A Mol Integrat Physiol 119: 137–147

Oikawa S, Itazawa Y (1985) Gill and body-surface areas of the carp in relation to body-mass, with special reference to the metabolism-size relationship. J Exp Biol 117:1–14

Oikawa S, Hirata M, Kita J, Itazawa Y (1999) Ontogeny of respiratory area of a marine teleost, porgy, *Pagrus major*. Ichthyol Res 46:233–244

Olsen YA, Einarsdottir IE, Nilssen KJ (1995) Metomidate anesthesia in Atlantic salmon, *Salmo salar*, prevents plasma cortisol increase during stress. Aquaculture 134: 155–168

Oppedal F, Johansson B, Kiesslin A (2000) Bedøvelse og vaksinering—økt appetitt ved bedre rutiner. 2000. Norsk Fiskeoppdrett no 9:24–26

Ortuno J, Esteban MA, Meseguer J (2002) Effects of four anaesthetics on the innate immune response of gilthead seabream (*Sparus aurata* L.). Fish Shellfish Immunol 12:49–59

Park CK, Li HY, Yeon KY, Jung SJ, Choi SY, Lee SJ, Lee S, Park K, Kim JS, Oh SB (2006) Eugenol inhibits sodium currents in dental afferent neurons. J Dent Res 85:900–904

Pottinger TG, Carrick TR (1999) A comparison of plasma glucose and plasma cortisol as selection markers for high and low stress-responsiveness in female rainbow trout. Aquaculture 175:351–363

Pottinger TG, Pickering AD, Hurley MA (1992) Consistency in the stress response of individuals of two strains of rainbow trout, *Oncorhynchus mykiss*. Aquaculture 103: 275–289

Pottinger TG, Moran TA, Morgan JAW (1994) Primary and secondary indices of stress in the progeny of rainbow trout (*Oncorhynchus mykiss*) selected for high and low responsiveness to stress. J Fish Biol 44:149–163

Randall DJ (1962) Effect of an anaesthetic on the heart and respiration of teleost fish. Nature 195:506

Rang HP, Dale MM, Ritter JM, Moore PK (2003) Pharmacology. Churchill Livingstone, London

Rigon AR, Takahashi RN (1996) The effects of systemic procaine, lidocaine and dimethocaine on nociception in mice. Gen Pharmacol 27:647–650

Ross LG, Ross B (2008) Anaesthetic and sedative techniques for aquatic animals. Blackwell, Oxford

Rothwell SE, Forster ME (2005) Anaesthetic effects on the hepatic portal vein and on the vascular resistance of the tail of the Chinook salmon (*Oncorhynchus tshawytscha*). Fish Physiol Biochem 31:11–21

Ruyter B, Moya-Falcon C, Rosenlund G, Vegusdal A (2006) Fat content and morphology of liver and intestine of Atlantic salmon (*Salmo salar*): Effects of temperature and dietary soybean oil. Aquaculture 252:441–452

Ryan S (1992) The dynamics of MS-222 anaesthesia in a marine teleost (*Pagrus auratus*: Sparidae). Compar Biochemis Physiol C Pharmacol Toxicol Endocrinol 101:593–600

Sams L, Braun C, Allman D, Hofmeister E (2008) A comparison of the effects of propofol and etomidate on the induction of anesthesia and on cardiopulmonary parameters in dogs. Veterinary Anaesth Analg 35:488–494

Sandodden R, Finstad B, Iversen M (2001) Transport stress in Atlantic salmon (*Salmo salar* L.): anaesthesia and recovery. Aquacult Res 32:87–90

Saunders RL (1963) Respiration of the Atlantic Cod. J Fish Res Board Can 20:373–386

Schmidt-Nielsen K (1984) Scaling: why is animal size so important?. Cambridge University Press, New York

Schoettger RA, Julin AM (1967) Efficacy of MS-222 as an anesthetic on four salmonids, vol 13. Investigations in Fish Control, United States Department of the Interior

Schoettger RA, Steucke EW (1970) Synergic mixtures of MS-222 and quinaldine as anesthetics for rainbow trout and northern pike. Progress Fish Cultur 32:202–205

Schurmann H, Steffensen JF (1997) Effects of temperature, hypoxia and activity on the metabolism of juvenile Atlantic cod. J Fish Biol 50:1166–1180

Sellner PA, Hazel JR (1982) Incorporation of polyunsaturated fatty acids into lipids of rainbow trout hepatocytes. Am J Physiol 243:R223–R228

Selye H (1985) The nature of stress. Basal Facts 7:3–11

Small BC (2003) Anesthetic efficacy of metomidate and comparison of plasma cortisol responses to tricaine methanesulfonate, quinaldine and clove oil anesthetized channel catfish *Ictalurus punctatus*. Aquaculture 218:177–185

Small BC (2004) Effect of isoeugenol sedation on plasma cortisol, glucose, and lactate dynamics in channel catfish *Ictalurus punctatus* exposed to three stressors. Aquaculture 238:469–481

Small BC, Chatakondi N (2005) Routine measures of stress are reduced in mature channel catfish during and after AQUI-S anesthesia and recovery. NA J Aqua 67:72–78

Sneddon LU (2002) Anatomical and electrophysiological analysis of the trigeminal nerve in a teleost fish, *Oncorhynchus mykiss*. Neurosci Lett 319:167–171

Sneddon LU (2003) Trigeminal somatosensory innervation of the head of a teleost fish with particular reference to nociception. Brain Res 972:44–52

Sneddon LU, Braithwaite VA, Gentle MJ (2003) Do fishes have nociceptors? Evidence for the evolution of a vertebrate sensory system. Proc R Soc Lond Ser B Biol Sci 270:1115–1121

Soivio A, Nyholm K, Huhti M (1977) Effects of anesthesia with MS 222, neutralized MS 222 and benzocaine on blood constituents of rainbow trout, *Salmo gairdneri*. J Fish Biol 10:91–101

Stehly GR, Gingerich WH (1999) Evaluation of AQUI-S (TM) (efficacy and minimum toxic concentration) as a fish anaesthetic sedative for public aquaculture in the United States. Aquacult Res 30:365–372

Stehly GR, Meinertz JR, Gingerich WH (1998) Effect of temperature on the pharmacokinetics of benzocaine in rainbow trout (*Oncorhynchus mykiss*) after bath exposures. J Vet Pharmacol Ther 21:121–127

Stenger VG, Maren TH (1974) Pharmacology of MS 222 (ethyl-*m*-aminobenzoate) in *Squalus acanthias*. Gen Pharmacol 5:23–35

Summerfelt RC, Smith LS (1990) Anaesthesia and surgery and related techniques. In: Schreck CB, Moyle PB (eds) Methods for fish biology. American Fisheries Society, Bethesda, pp 213–272

Sumpter JP, Dye HM, Benfey TJ (1986) The effects of stress on plasma ACTH, α-MSH, and cortisol levels in salmonid fishes. Gen Comp Endocrinol 62:377–385

Sylvester JR, Holland LE (1982) Influence of temperature, water hardness, and stocking density on MS-222 response in three species of fish. Progress Fish Cultur 44:138–141

Tanck MWT, Vermeulen KJ, Bovenhuis H, Komen H (2001) Heredity of stress-related cortisol response in androgenetic common carp (*Cyprinus carpio* L.). Aquaculture 199: 283–294

Thomas P, Robertson L (1991) Plasma-cortisol and glucose stress responses of red drum (*Sciaenops ocellatus*) to handling and shallow-water stressors and anesthesia with MS-222, quinaldine sulfate and metomidate. Aquaculture 96:69–86

Trenzado CE, Carrick TR, Pottinger TG (2003) Divergence of endocrine and metabolic responses to stress in two rainbow trout lines selected for differing cortisol responsiveness to stress. Gen Comp Endocrinol 133:332–340

Ueta K, Suzuki T, Sugimoto M, Uchida I, Mashimo T (2007) Local anesthetics have different mechanisms and sites of action at recombinant 5-HT_3 receptors. Reg Anesth Pain Med 32:462–470

Vanden Bossche H, Willemsens G, Cools W, Bellens D (1984) Effects of etomidate on steroid biosynthesis in subcellular fractions of bovine adrenals. Biochem Pharmacol 33: 3861–3868

Velisek J, Stejskal V, Kouril J, Svobodova Z (2009) Comparison of the effects of four anaesthetics on biochemical blood profiles of perch. Aquacult Res 40:354–361

Wagner RL, White PF, Kan PB, Rosenthal MH, Feldman D (1984) Inhibition of adrenal steroidogenesis by the anesthetic etomidate. N Engl J Med 310:1415–1421

Webber DM, Boutilier RG, Kerr SR (1998) Cardiac output as a predictor of metabolic rate in cod *Gadus morhua*. J Exp Biol 201:2779–2789

Weber RA, Peleteiro JB, Martin LOG, Aldegunde M (2009) The efficacy of 2-phenoxyethanol, metomidate, clove oil and MS-222 as anaesthetic agents in the Senegalese sole (*Solea senegalensis* Kaup 1858). Aquaculture 288:147–150

Wedemeyer GA (1970) Stress of anesthesia with MS-222 and benzocaine in rainbow trout (*Salmo gairdneri*). J Fish Res Board Can 27:909–914

Wendelaar Bonga SE (1997) The stress response in fish. Physiol Rev 77:591–625

Weyl O, Kaiser H, Hecht T (1996) On the efficacy and mode of action of 2-phenoxyethanol as an anaesthetic for goldfish, *Carassius auratus* (L.), at different temperatures and concentrations. Aquacult Res 27:757–764

Wie MB, Won MH, Lee KH, Shin JH, Lee JC, Suh HW, Song DK, Kim YH (1997) Eugenol protects neuronal cells from excitotoxic and oxidative injury in primary cortical cultures. Neurosci Lett 225:93–96

Wiesenfeld-Hallin Z, Lindblom U (1985) The effect of systemic tocainide, lidocaine and bupivacaine on nociception in the rat. Pain 23:357–360

Woolf CJ, Wiesenfeld-Hallin Z (1985) The systemic administration of local anaesthetics produces a selective depression of C-afferent fibre evoked activity in the spinal cord. Pain 23:361–374

Woolsey J, Holcomb M, Ingermann RL (2004) Effect of temperature on clove oil anesthesia in steelhead fry. NA J Aqua 66:35–41

Yanar M, Kumlu M (2001) The anaesthetics effects of quinaldine sulphate and/or diazepam on sea bass (*Dicentrarchus labrax*) juveniles. Turk J Veter Anim Sci 25:185–189

Yang J, Uchida I (1996) Mechanisms of etomidate potentiation of $GABA_A$ receptor-gated currents in cultured postnatal hippocampal neurons. Neuroscience 73:69–78

Zahl IH, Kiessling A, Samuelsen OB, Hansen MK (2009) Anaesthesia of Atlantic cod (*Gadus morhua*)—effect of pre-anaesthetic sedation and importance of body weight, temperature and acute stress. Aquaculture 295(1–2):52–59

Zahl IH, Kiessling A, Samuelsen OB, Olsen RE (2010) Anaesthesia induces stress in Atlantic salmon (*Salmo salar*), Atlantic cod (*Gadus morhua*) and Atlantic halibut (*Hippoglossus hippoglossus*). Fish Physiol. Biochemistry 36(3):719–730

Zahl IH, Kiessling A, Samuelsen OB, Hansen MK (2011) Anaesthesia of Atlantic Halibut (*Hippoglossus hippoglossus*)—effect of pre-anaesthetic sedation and importance of body weight and water temperature. Aquacult Res 42: 1235–1245

Achieving Consensus on Current and Future Priorities for Farmed Fish Welfare: A Case Study From the UK – Summary

Iain Berrill, Trevor Cooper, Craig MacIntyre, Tim Ellis, Toby Knowles, Emma Jones and James Turnbull

We developed an approach which helps different stakeholder groups to discuss farmed animal welfare and to then work together to agree how it might be safeguarded. The approach was designed to be used at stakeholder meetings and it focuses on improving communication between groups with different viewpoints. It also focuses on establishing useful and practical outcomes from these discussions. After the approach was developed, it was tested at a meeting in the UK, which brought together a wide range of stakeholders to discuss how farmed fish welfare might be safeguarded in the future. The approach was successful in improving communication between different stakeholders. Additionally, it allowed stakeholders to develop a prioritised list of ideas which described how we might further improve farmed fish welfare in the future, and also allowed stakeholders to propose possible practical ways of doing so.

Background

The welfare of farmed animals (including fish) has received considerable interest in recent years. Many different stakeholder groups (e.g. governmental groups, NGO's, farmers, veterinarians, ethicists, researchers) have a role to play in safeguarding the welfare of farmed animals. However, it can be very difficult to get these groups to agree (or reach "consensus") on how welfare can and should be safeguarded. This is often due to the various groups having strong and very different views on animal welfare. This can also result in a lack of communication between those groups, making it even more difficult for them to agree.

Scientific Significance

The developed approach can further the application of scientific results in practice and can help scientists to identify relevant research questions through more efficient communication with stakeholder groups.

Practical Application

The approach described in this paper has provided a useful mechanism to assist those with different views to come together and discuss contentious issues, and to help those individuals or groups to agree on how they might progress forward. Although we have applied it to the issue of farmed fish welfare, it can easily be applied to any subject where there are strongly opposing views. As such it provides a valuable tool to assist, and hopefully move forward, a wide range of both scientific and non-scientific debates.

Contact

I. K. Berrill
Institute of Aquaculture, University of Stirling,
Stirling FK9 4LA, UK
e-mail: iain.berrill@googlemail.com

Fish Physiol Biochem (2012) 38:219–229
DOI 10.1007/s10695-010-9399-2

Achieving consensus on current and future priorities for farmed fish welfare: a case study from the UK

I. K. Berrill · T. Cooper · C. M. MacIntyre · T. Ellis · T. G. Knowles · E. K. M. Jones · J. F. Turnbull

Received: 2 November 2009 / Accepted: 5 April 2010 / Published online: 21 April 2010

Abstract The welfare of farmed fish has attracted attention in recent years, which has resulted in notable changes within the aquaculture industry. However, a lack of communication between stakeholders and opposing ethical views are perceived as barriers to achieving consensus on how to improve farmed fish welfare. To address these issues, we developed an interactive approach that could be used during stakeholder meetings to (1) improve communication between different stakeholder groups, (2) build consensus on priorities for farmed fish welfare and (3) establish mechanisms to address welfare priorities. We then applied this approach during a meeting of stakeholders to identify current and future priorities for farmed fish welfare in the UK. During the meeting in the UK, stakeholders initially identified 32 areas that they felt were in need of development for future improvements in farmed fish welfare. These were further refined via peer review and discussion to the seven most important "priority" areas. Establishing a "*better understanding of what good fish welfare is*" emerged as the highest priority area for farmed fish welfare. The second highest priority area was "*the need for welfare monitoring and documentation systems*", with mortality recording proposed as an example. The other five priority areas were "[improved understanding of] *the role of genetic selection in producing fish suited to the farming environment*", "*a need for integration and application of behavioural and physiological measures*", "*the need for a more liberal regime in Europe for introducing new medicines*", "*a need to address the issues of training existing and new workers within the industry*", and "*ensuring best practise in aquaculture is followed by individual businesses*". Feedback from attendees, and the meeting outputs, indicated that the approach had been successful in improving communication between stakeholders and in achieving consensus on the priorities for farmed fish welfare. The approach therefore proved highly beneficial for future improvements in fish welfare in the UK.

I. K. Berrill (✉) · C. M. MacIntyre · J. F. Turnbull
Institute of Aquaculture, University of Stirling,
Stirling FK9 4LA, UK
e-mail: iain.berrill@googlemail.com

T. Cooper
Defra Innovation Centre, Northgate House,
21–23 Valpy Street, Reading RG1 1AR, UK

T. Ellis
Centre for Environment, Fisheries and Aquaculture
Science (Cefas), Barrack Road, The Nothe, Weymouth,
Dorset DT4 8UB, UK

T. G. Knowles
School of Veterinary Science, University of Bristol,
Langford House, Langford BS40 5DU, UK

E. K. M. Jones
Department for Environment, Food and Rural Affairs
(Defra), Area 4A, Nobel House, 17 Smith Square,
London SW1P 3JR, UK

Keywords Fish · Welfare · Stakeholder · Aquaculture · Consensus · UK

Introduction

The welfare of farmed fish has been the subject of increasing interest and debate in recent years (e.g. Arlinghaus et al. 2007; Huntingford et al. 2007; North et al. 2008; see Diseases of Aquatic Organisms Special Issue 2007: *Welfare of Aquatic Organisms*, http://www.int-res.com/abstracts/dao/v75/n2/). However, in order to better understand the welfare of farmed fish, one must have an appreciation of how changes in welfare can be measured and also how to weigh those changes against human interest (Huntingford et al. 2006). Despite this, there has, until recently, been a lack of communication between science and ethics disciplines with regard to animal welfare (Fraser 1999).

Scientific research provides a means to measure how fish respond to their environment, and consequently it allows us to study any resultant effects of that environment on their welfare. Although progress has been made in our ability to empirically measure indicators of fish welfare, primarily using functional responses (e.g. growth and cortisol; Turnbull et al. 2005; North et al. 2006), such measures require a degree of subjective interpretation, and conclusions based on those interpretations can be influenced by an individual's ethical stance (Appleby 1999).

Animal welfare raises many philosophical questions. For example, one might question whether animals have a moral standing (Sandøe et al. 1997) and if so what standard of animal welfare is morally acceptable (Sandøe et al. 2003). How people consider and evaluate such philosophical questions leads to different ethical viewpoints (for in depth reviews of ethics in relation to fish and animal welfare see Sandøe et al. 1997; Appleby 1999; Fraser 1999; Huntingford et al. 2006). Often, individuals do not follow wholly one specific ethical view, instead choosing to establish hybrid views which comprise of components of two or more approaches (Appleby 1999). Evidently, this leads to many disparate views regarding the ethics of animal welfare.

It is well recognised that many stakeholders (e.g. farmers, fish veterinarians, governmental and non-governmental organisations, research funding bodies, consumers, retailers, and ethicists) other than researchers have a role to play in safeguarding the welfare of farmed fish (North et al. 2008). However, achieving consensus between stakeholders on many aspects of farmed fish welfare and, more specifically, how it can be improved can be difficult: stakeholders have their own ways of defining and assessing welfare and have differing ethical stances. Appleby (1999) has highlighted this indicating that individuals use different approaches in coming to their conclusions regarding animal welfare and that they do not necessarily recognise that they are arguing from difference ethical stances. However, Appleby (1999) discussed possibilities for improving dialogue between those with different ethical views, suggesting that (1) there is already some overlap in opinions, (2) people should recognise their own and others' viewpoints, (3) people should identify principles they have in common and (4) they should take practical steps to improve communication. Previously, "focus groups" involving different farmed fish welfare stakeholder groups established that there is already some consensus between stakeholders on how improvements in welfare can be achieved for farmed fish (North et al. 2008). In light of this (and the suggestions made by Appleby 1999), we developed an approach which aimed to build consensus amongst stakeholders involved in farmed fish welfare, although it is envisaged that the approach might also be suitable for use in other industries. The approach was designed to (1) provide up-to-date information that could be understood and debated by all stakeholders, (2) encourage constructive debate between stakeholders, (3) identify important areas for development in farmed fish welfare, (4) establish consensus on the most important priorities for farmed fish welfare and (5) identify what actions might be taken to address those priorities. We then applied this approach during a meeting of UK stakeholders involved in farmed fish welfare. This paper describes the approach and the priorities for farmed fish welfare which were identified when it was tested at a meeting of stakeholders in the UK.

Methodology

The consensus-building approach

The approach used to develop consensus between stakeholders involved in farmed fish welfare was developed in collaboration with staff at the Defra Innovation Centre (Reading, UK). Staff at the Innovation Centre have considerable experience in developing approaches, and conducting workshops, aimed at achieving consensus and facilitating discussion between different stakeholders. For example, previous workshops have been used to develop an environmental strategy for pig husbandry and to ascertain what scientific evidence is required to underpin future governmental policies for sustainable food and farming. Approaches developed and tested at the Innovation Centre typically result in agreed actions, together with supporting analyses such as risk and benefit assessments, and requirements for further research. In addition, the Centre lays weight on tangible outputs, such as increased consensus, or the commitment of time and resources.

The Innovation Centre does not use any overarching theory in developing the consensus-building approaches that are used during stakeholder meetings. To our knowledge, no such theories exist. Further, it is doubtful whether any practical, general theory could be developed given the specific nature of some stakeholder engagement events and difficulties in formally defining success, particularly when stakeholders may be required to create new solutions and decide between them, trade off their interests, act on behalf of a perceived greater good, establish trust, commit resources, or assess their own ability to influence stakeholders that do not attend a particular meeting (Innovation Centre, *Personal Communication*).

Staff at the Innovation Centre develop practical consensus-building approaches based on their own experience. The high throughput of successful stakeholder meetings at the Innovation Centre (approximately 300 per year) allows new techniques to be explored and assessed, albeit in a qualitative manner rather than through controlled experimentation. The approaches are designed to mitigate the well-known difficulties which arise when groups seek to resolve complex issues, primarily behaviours such as group thinking, anchoring, dominance of certain individuals, and the failure to consider alternatives (Baron and Kerr 2003). Consensus-building approaches aim to encourage convergent thinking and avoid situations where stakeholders can take a particular stance without any appreciation for possible alternative views. They also focus on the dissemination of relevant knowledge amongst stakeholders in a manner that can be easily understood by all. Facilitators that oversee consensus-building workshops must be particularly sensitive to difficulties that can arise when stakeholders choose between options.

The Innovation Centre routinely monitors the satisfaction levels of both workshop sponsors (i.e. those who wish to use the results of the workshop) and attendees and uses these as a proxy for the success of a particular approach. Satisfaction levels amongst both sponsors and attendees are consistently high (Innovation Centre, *Personal Communication*). Finally, the Innovation Centre tracks the extent to which sponsors return to use its services, regarding this as another useful proxy illustrating the success of the consensus-building approaches that are developed (Innovation Centre, *Personal Communication*).

The consensus-building approach used for farmed fish welfare incorporates specific exercises that stimulate discussion and debate between stakeholders, allowing them to generate ideas and identify priorities for farmed fish welfare. It also incorporates regular peer review of ideas and voting. The approach is designed to have no influence on the direction of specific discussions. Further, when applied during stakeholder meetings, it is suggested that a meeting facilitator, who has no experience in farmed fish welfare and who will have no influence on the direction of discussions, oversees the completion of the approach.

The approach is divided into four sequential stages (Fig. 1) with each stage carried out by groups of regularly mixed stakeholders. The four stages are:

1. Information and improved communication between stakeholders.
2. Identifying important development areas for farmed fish welfare.
3. Better understanding the issues surrounding each development area.
4. Identifying practical actions to address each priority area.

Although each stage does not necessarily result in recordable outputs, all stages are considered important to the process of achieving consensus. Here, we describe in detail how the approach can be used with groups of approximately 30 stakeholders. However, if different numbers of stakeholders are present during the consensus-building process, group sizes in each stage can easily be adjusted accordingly.

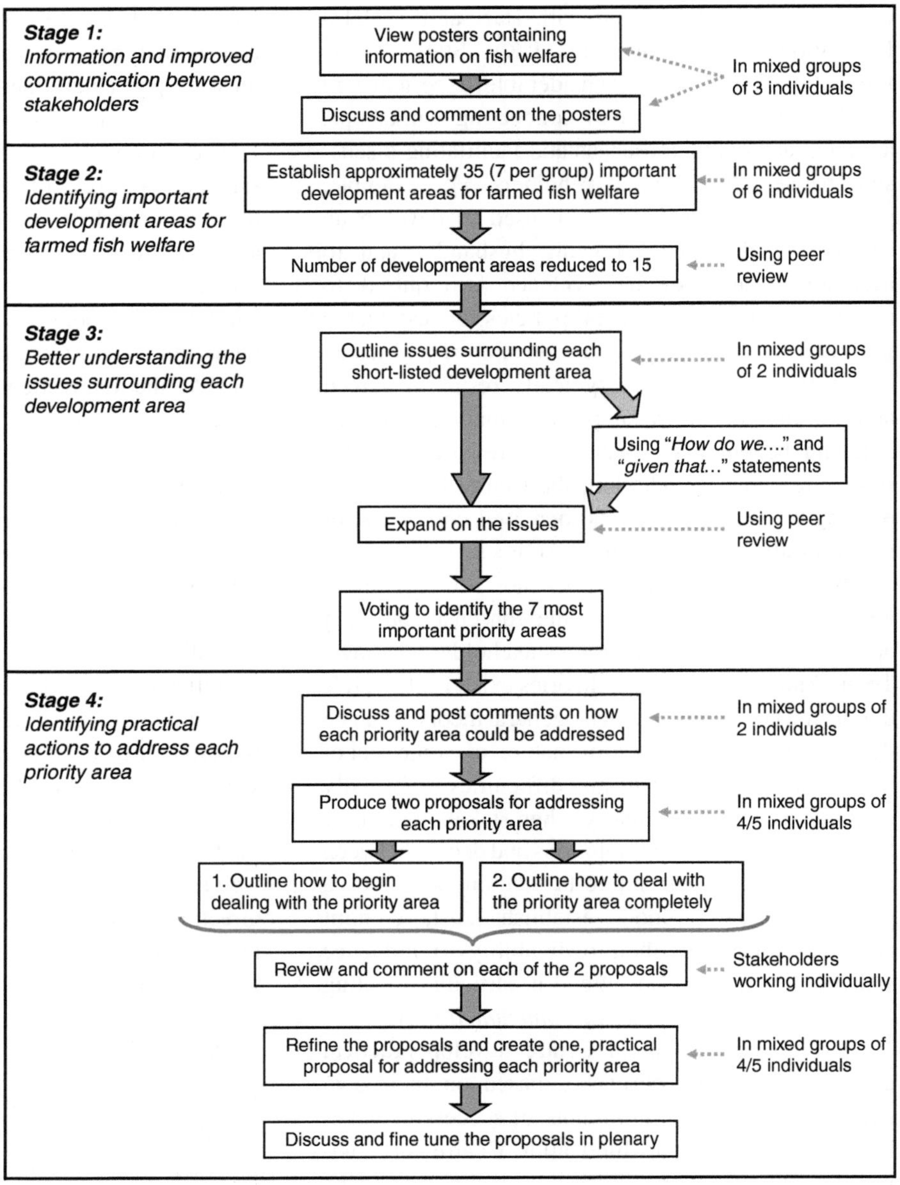

Fig. 1 The approach used to build consensus on current and future priorities for farmed fish welfare

Stage 1: Information and improved communication between stakeholders

To stimulate discussions between stakeholders, up-to-date information is provided on approximately ten pre-prepared posters. The posters should not attempt to provide all available information on farmed fish welfare nor should they present the views of only one stakeholder group. They should provide a range of appropriate and timely information on farmed fish welfare that will promote debate between stakeholders.

Working in mixed groups ($n = 10$) of three individuals, stakeholders are asked to view and discuss the posters and to post statements relating to farmed fish welfare. Anyone can post statements with the caveat (for clarity) that the other members of that group must understand the comments, even if they do not agree with them. Following the initial comment posting, all participants are then invited to respond to the posted comments. It is suggested that Stage 1 lasts for approximately 60 min. No outputs of this initial "warm-up" session are carried forward to subsequent stages, instead this process aims to improve communication between stakeholders and recognition of each others' views.

Stage 2: Identifying important development areas for farmed fish welfare

Groups ($n = 5$) of six mixed stakeholders are each asked to identify seven areas of farmed fish welfare which they feel need developing and improving (35 areas in total). Once this is completed, the ideas are reduced in number using a system of collective peer review: each group discusses the ideas of another group, removing the two development areas they consider the least important. This process is then repeated, with two further development areas discarded. If at any point, groups feel strongly that previously discarded areas should be re-instated, this is permitted. At the conclusion of the peer review process, any duplicated development areas are combined and the remaining areas (approximately 15) move forward into Stage 3. Stage 2 should last between 45 min and 1 h.

Stage 3: Better understanding the issues surrounding each important development area

Fifteen groups of two mixed stakeholders are each assigned one of the shortlisted development areas for farmed fish welfare and asked to outline the main problems associated with addressing that area. To facilitate this, groups create a statement which asks: "*how do we…*" address the development area, "*given that…*" the following issues are present (illustrated in Fig. 2a). These statements then undergo a process of peer review to further expand on the issues raised: each group review a statement created by another group and, where necessary, add to what the previous group has written (see example in Fig. 2b).

Whilst statements are created for the 15 shortlisted development areas, it is appropriate to focus discussions in Stage 4 on the areas stakeholders consider most important. For a meeting of 30 people, approximately seven priority areas is a manageable and appropriate number for discussion in Stage 4, although this number can be adjusted according to the specific needs of a meeting. Attendees are therefore invited to vote for the areas they feel are most important: each attendee is given seven votes (or more if more priority areas are to be considered in Stage 4) which can be distributed amongst the priority areas in any way they choose (i.e. multiple votes can be given to an area to emphasise importance). Using votes allows a measure of importance to be placed on each development area, irrespective of whether it progresses forward to Stage 4. Stage 3 should last between 45 min and 1 h.

Stage 4: Identifying practical actions to address each priority area

Meeting attendees, working in mixed pairs, are asked to discuss the implications of each of the seven priority areas identified during Stage 3, posting comments on how they could be addressed. Attendees are then divided into seven mixed groups (i.e. 4/5 individuals), with each group assigned a specific priority area and asked (using the posted comments as a guide) to produce two proposals for addressing that priority area: a first basic proposal outlining how stakeholders could begin addressing it, and a second more detailed one explaining how it could be addressed fully. Attendees, working individually, are then invited to review all of the proposals and post comments. Each group then return to their proposals and create one final, practical proposal which they feel is the most appropriate way of addressing the priority area at the present time. The final proposals have to be realistic, practical and representative of the comments that have been posted. To conclude this final session, the priority areas are presented to the audience by the facilitator and discussed, peer-reviewed and fine-tuned by the entire attendance. At this point, it may also be possible to identify candidates amongst the attendees who are willing to take practical steps to address each proposal after the meeting. Stage 4 should last between 1 h 45 min and 2 h.

Reporting

Using the approach described above, it is possible to establish and document priorities for farmed fish welfare and practical ways of addressing those priorities. Development areas of lower importance, identified and developed during Stages 2 and 3 but not carried forward into Stage 4, can also be reported. However, it is unlikely that the content of debates between stakeholders throughout the process can be documented. However, these discussions are important outputs from using this approach.

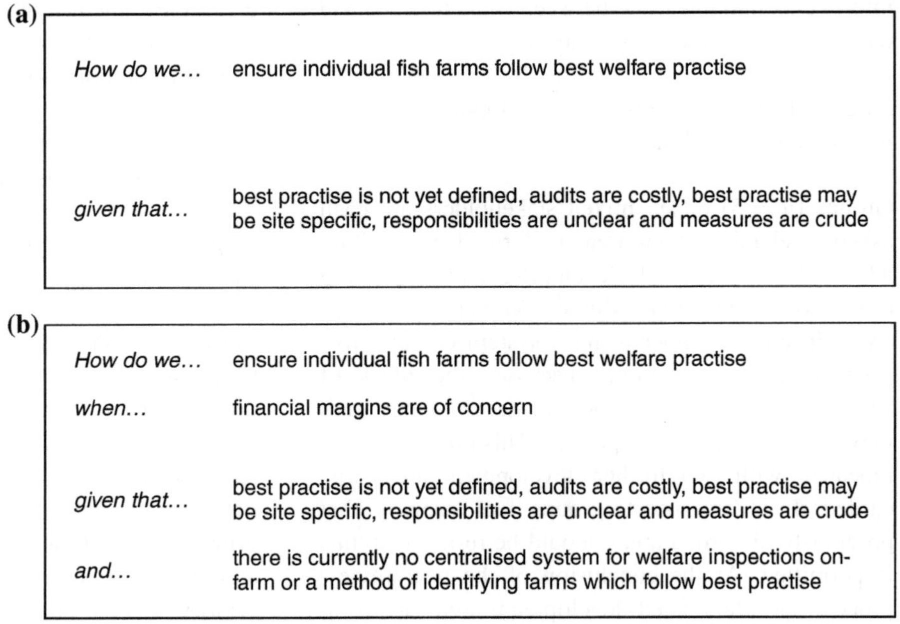

Fig. 2 An example of the statements created to better understand the issues surrounding each development area for farmed fish welfare. The example shown is a statement created during a meeting of farmed fish welfare stakeholders in the UK. **a** Shows a statement initially created by meeting attendees, **b** shows how the original statement was expanded by others

Testing the approach: establishing priorities for farmed fish welfare in the UK

The consensus-building approach described above was applied and tested at a meeting of UK stakeholders involved in farmed fish welfare. The meeting was held on 22nd November 2007 at the Defra Innovation Centre (Reading, UK). Representatives were invited from each of the following stakeholder groups: fish farmers (and trade association representatives), researchers, fish veterinarians, governmental and non-governmental organisations, research-funding bodies, consumers (i.e. consumer groups), retailers and ethicists. One hundred and forty-five individuals were originally invited to the meeting and 31 participants attended, with each stakeholder group represented by at least one individual (Table 1). Some participants represented more than one stakeholder group, for example a governmental representative who was also a veterinarian. On the day prior to the meeting, researchers had disseminated the results of various Defra-funded fish welfare research projects covering stress and welfare assessment, harvesting, humane slaughter, stocking density, water quality and fin damage. The participants had therefore been familiarised with some of the most recent research on farmed fish welfare prior to the meeting.

Prior to the meeting, nine posters were prepared for use during Stage 1 of the approach. The poster topics and content were developed by the authors of this paper in full consultation with one another. They were entitled:

Table 1 The number of representatives from each stakeholder group attending a meeting to establish priorities for farmed fish welfare in the UK

Stakeholder group	Number of representatives
Fish farmers/trade associations	4
Researchers	12
Veterinarians	3
Governmental organisations	9
Non-governmental organisations	4
Funders of research	2
Consumer groups	1
Retailers	1
Ethicists	1

Thirty-one people attended the meeting, with some participants representing more than one stakeholder group

1. Assessing farmed fish welfare is difficult (shown as an example in Fig. 3).
2. Fish compared with terrestrial animal welfare.
3. Stocking density is not a good way to control fish welfare.
4. Changes in fish welfare over the last 10 years.
5. Farmer training, competence and farm infrastructure affect welfare.

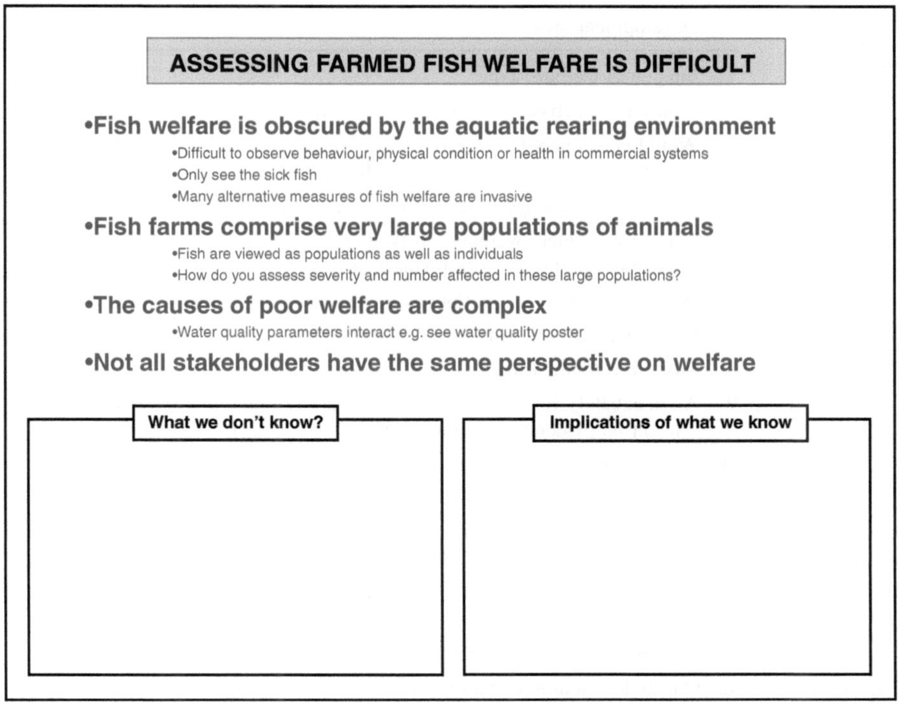

Fig. 3 An example of the posters used to stimulate discussions between meeting participants during Stage 1 of the consensus-building approach

6. Water quality effects on welfare.
7. Mortality and welfare in farmed fish.
8. Nearly all research has studied long-term effects on welfare.
9. Future fish welfare issues.

All stages of the approach were conducted as described above, except that group sizes were adjusted, where necessary, for each stage.

Results: farmed fish welfare in the UK

Identification and prioritisation of important areas for development

In total, 32 important areas for development for farmed fish welfare (including duplicates) were initially identified by stakeholders during Stage 2 of the approach. Following the process of peer review, 12 of those areas were discussed in more detail in during Stage 3. The 20 areas (with duplicates) that were deemed of lower importance and which were not taken forward for further discussion are provided in Table 2. The 12 areas taken forward into Session 3 are shown in Table 3. The 12 shortlisted areas were ranked in order of importance by voting (Table 3). The "*need for a better understanding of what good welfare is*" was seen as the most important area (16% of votes). The "*need for more data on mortalities*" also had a high number of votes (15%), and this area was deemed sufficiently similar to the "*need for welfare monitoring and documentation systems*" (8%) that the two were merged as one priority area for subsequent discussions. The "*role of genetic selection in producing fish suited to the farming environment*" and a "*need for integration and application of behavioural and physiological measures*" were also considered important (13 and 11% of votes, respectively). All other areas had <10% of the votes. The four development areas with the lowest number of votes were not considered in further discussions during Stage 4 (Table 3).

Identifying practical actions to address each priority area

During Stage 4, proposals were developed which explained how the seven priority areas for farmed fish

Table 2 Areas for development for improved farmed fish welfare in the UK which were considered of lower importance during a meeting of stakeholders

	Development area
1	Integrated welfare score
2	Produce a database of farm production, quality and survival data
3	Would consumers be prepared to pay more for "higher welfare" fish? A better price for fish might lead to on-farm welfare improvements
4	Improved knowledge transfer to consumers and stakeholders
5	Raising customer awareness of welfare within the farmed fish industry because the consumer is the ultimate financial driving force
6	Which are better (1) continuous good conditions, (2) mostly good conditions?
7	More methods and time for observing fish
8	More development relative to research
9	An adequately funded fish welfare research programme because we need a robust scientific evidence base on which to make decisions
10	Establish what do we all mean by "welfare" because opinions differ
11	Establish priorities for welfare research
12	How can we speed up the process of bringing medical products to the market because medicines have a direct benefit to fish welfare
13	Easy licensing is required for new and existing therapeutics because the lack of therapeutics is leading to welfare problems in farms, lakes etc
14	Our ability to measure the feelings/emotions of fish
15	Our understanding of variations in the requirements of different species
16	Research into on-farm systems for measuring production and welfare
17	Refinement of automatic percussive devices to minimise/eliminate incidence of poor/mis-stunning
18	Put in place official inspections for welfare at slaughter (as with mammals)
19	Investigate the impact of artificial lighting regimes on welfare
20	Address transport conditions

Duplicated development areas have not been removed from the table to avoid deleting areas which appear similar but are subtly different. The development areas are shown as written by the attendees, with only minor grammatical changes made to improve the clarity of the text

welfare could be addressed. These have been shown in Table 4. The final, practical proposals for addressing these most important priority areas at the present time are shown in Table 5.

Discussion

In fields where stakeholders have strongly opposing views, for example farmed fish welfare, a lack of informed discussion can act as a barrier to establishing consensus on important issues. Interactive approaches, such as the one described in the current paper, provide a mechanism by which we can improve communication between stakeholders with different stances, hopefully leading to unbiased consensus.

Established approaches have been used previously to gain unbiased stakeholder consensus in other areas of aquaculture (e.g. the Delphi approach; Bunting 2008). However, following guidance from experienced staff at the Innovation Centre, who have conducted nearly 1000 stakeholder engagement events, the approach we developed was considered most appropriate for building consensus on priorities in farmed fish welfare. The approach was designed so that:

- it had no influence in directing discussions between participants.
- the outputs of each session were peer-reviewed using groups of regularly mixed stakeholders, with group size varying according to the needs of the exercise.
- Continual debate between attendees was encouraged.

When the approach was tested at a meeting of UK stakeholders involved in farmed fish welfare, all

Table 3 The most important areas for development for farmed fish welfare in the UK as identified during a meeting of stakeholders

	Development area	Relative importance (%)
1	We need to know what good welfare is for various farmed species because you cannot improve welfare if you do not know what good welfare is	16.2
2	We need more data on mortalities	15.1
3	The role of genetic selection in producing fish more suited to the fish farming environment	13.0
4	Integration and application of behavioural and physiological welfare indices	10.8
5	The need for a more liberal regime in Europe for introducing (licensing) new medicines, which requires government commitment	9.7
6	Address the issues of staff training and new trained workers coming to the industry because well-trained, competent staff are important to welfare	8.1
7	The need for welfare monitoring and documentation systems because we need baseline data about the farming environment, mortalities, fish performance and husbandry practises to identify problems and research issues	8.1
8	How do we ensure that best practise for welfare in aquaculture is followed by individual businesses?	6.5
9	Risk assessment and contingency planning because welfare problems can be avoided (or corrected)	5.9
10	The development of welfare standards backed by actual scientific research and not just guesswork	3.2
11	We must understand the costs and benefits of fish welfare because improvements must be economically and environmentally sustainable	2.2
12	Disseminate more widely fish welfare research so it can be used and applied	1.1

The relative importance of each area was established using a system of voting. The development areas are shown as written by the attendees, with only minor grammatical changes made to improve the clarity of the text

Table 4 Proposals outlining how each of the most important priority areas for farmed fish welfare in the UK (as identified by stakeholders) could be addressed

	Priority area	Proposal 1	Proposal 2
1	We need to know what good welfare is for various farmed species	Establish a welfare checklist of practical indices of bad welfare	Define acceptable levels of performance. Investigate tests to identify emotional states
2	We need more data on mortalities The need for welfare monitoring and documentation systems	Benchmark mortalities by species, age, time of year, production system (using an anonymous one-off study)	Use production and welfare indicator benchmarking. Develop a centralised data-recording system. Ensure farm health plans take account of fish welfare and any benchmarking system
3	The role of genetic selection in producing fish more suited to the fish farming environment	An analysis of whether current selection programmes have welfare implications	Adopt selection criteria in addition to performance based measures and use more selection environments
4	Integration and application of behavioural and physiological welfare indices	Define behavioural and physiological welfare indices and inform farmers	Introduce an assured standard, with incentives
5	The need for a more liberal regime in Europe for introducing (licensing) new medicines, which requires government commitment	Re-evaluate 'old' products and review the cascade system with regard to withdrawal periods	Alignment between veterinary medical product regulation in the EU and USA leading to an eventual review of the legislative framework for veterinary medicines
6	Address the issues of staff training and new trained workers coming to the industry	Fish farming industry to publish a training syllabus with requirements. Appoint accreditation body. Publish Codes of Practice	Licence fish farms with conditions that require minimum competence standards to be demonstrated/testable/auditable. Create a common standard to be checked

Table 4 continued

	Priority area	Proposal 1	Proposal 2
7	How do we ensure that best practise for welfare in aquaculture is followed by individual businesses?	Create a best practice scheme for one species. Focus this at a very specific level (e.g. one developmental stage, one husbandry task etc.)	Expand for all life stages and all species. At this level, legislation may be required

Proposal 1 outlines what stakeholders could do to begin addressing each priority area. Proposal 2 outlines what stakeholders could do to completely address each priority area. The proposals are shown as written by the attendees, with only minor grammatical changes made to improve the clarity of the text

Table 5 Final practical proposals that UK stakeholders suggested as most appropriate for addressing the most important priority areas for farmed fish welfare at the present time

	Priority area	Final proposal
1	We need to know what good welfare is for various farmed species	Develop and validate a checklist of practical indices of bad welfare to be used on all fish farms
2	We need more data on mortalities The need for welfare monitoring and documentation systems	Seek agreement from the main trade associations and other stakeholders' representatives to use centrally shared data
3	The role of genetic selection in producing fish more suited to the fish farming environment	1. To hold a round table of trade associations on identifying and sharing information on genetic problems and opportunities 2. Conduct a broader stakeholder meeting to establish if there are problems and to increase awareness
4	Integration and application of behavioural and physiological welfare indices	Research and define behavioural welfare indices to add to current knowledge so that a minimum welfare standard can be agreed between stakeholders
5	The need for a more liberal regime in Europe for introducing (licensing) new medicines, which requires government commitment	Campaign to rationalise and streamline the international approval of veterinary medicines for use in the fish farming industry. The development of an appropriate briefing note will assist this process. This briefing note can then be used for parliamentary and other lobbying
6	Address the issues of staff training and new trained workers coming to the industry	Publish an agreed set of stockman competences which farmers have to demonstrate (by audit) that they actively use and is part of their written animal health and welfare plan. This would provide measurable baseline standards to inspect against
7	How do we ensure that best practise for welfare in aquaculture is followed by individual businesses?	Develop a pilot scheme to establish best practise from a representative sample of farms for a specific species during a specific phase and event. This will involve a working group of stakeholders to implement/construct this pilot scheme

The proposals are shown as written by the attendees, with only minor grammatical changes made to improve the clarity of the text

attendees actively voiced their views and opinions, irrespective of the fact that stakeholder groups were not evenly represented in terms of numbers. Consequently, the authors feel that the approach provided a valid representation of the views of all stakeholders and that it was a suitable mechanism for establishing consensus on the current and future priorities for farmed fish welfare.

Any approach which aims to achieve unbiased consensus amongst stakeholders must incorporate sufficient flexibility to allow unforeseen changes to be made when it is applied. When our approach was tested at the meeting of farmed fish welfare stakeholders in the UK, it was found to be sufficiently robust to allow changes to its structure without any detrimental effects on the quality of outputs. For example, changes were needed to account for different group sizes during each stage of the approach and to allow different numbers of ideas to be discussed and taken forward between the respective stages.

We believe (supported by feedback from attendees) that our approach has helped to advance communication and understanding between stakeholders involved in UK farmed fish welfare and that a similar outcome would be likely if it were applied in other countries or to other areas of aquaculture. The importance of improved communication between fish welfare stakeholders, achieved using this approach, cannot be overstated, and it is hoped that it will continue in the future. One aquaculture industry trade association representative did acknowledge (paraphrased) that lobby groups were correct in suggesting that the industry needed to address certain welfare issues, and welcomed dialogue with those lobby groups that worked with, rather than against, the industry.

Attendance at meetings aimed at improving fish welfare represents a significant time investment from a range of stakeholders. During the fish welfare meeting held in the UK, both the seven priority areas and the other important development areas are valuable outputs from the day. Each priority area was established through detailed, and sometimes lengthy, discussions between meeting attendees. It is not possible when reporting the outputs of such meetings to provide details of these discussions, and there is a danger that any brief summary would not accurately represent the views of all attendees present, which is vital for unbiased consensus building. The tangible output from using this approach, of priority areas developed and agreed by meeting attendees, will help stakeholders to focus their attention on specific aspects of farmed fish welfare in the future.

Acknowledgments The authors would like to thank all participants of the stakeholder meeting, to many to name individually, for their valuable involvement. Further thanks to Mrs Karen Carr for assistance during the meeting, to Dr Mark James for advice during the development of the consensus-building approach and to Prof. Anders Kiessling for valuable comments on an earlier version of this paper. This work was funded as part of the Department for Environment, Food and Rural Affairs (Defra) project AW1205.

References

Appleby MC (1999) Tower of Babel: variation in ethical approaches, concepts of welfare and attitudes to genetic manipulation. Anim Welf 8:381–390

Arlinghaus R, Cooke SJ, Schwab A, Cowx IG (2007) Fish welfare: a challenge to the feelings-based approach, with implications for recreational fishing. Fish Fish 8:57–71

Baron RS, Kerr NL (2003) Group process, group decision, group action. Open University Press, Buckingham

Bunting SW (2008) Horizontally integrated aquaculture development: exploring consensus on constraints and opportunities with a stakeholder Delphi. Aquac Int 16:153–169

Fraser D (1999) Animal ethics and animal welfare science: bridging the two cultures. Appl Anim Behav Sci 65:171–189

Huntingford FA, Adams C, Braithwaite VA, Kadri S, Pottinger TG, Sandøe P, Turnbull JF (2006) Current issues in fish welfare. J Fish Biol 68:332–372

Huntingford FA, Adams C, Braithwaite VA, Kadri S, Pottinger TG, Sandøe P, Turnbull JF (2007) The implications of a feelings-based approach to fish welfare: a reply to Arlinghaus et al. Fish Fish 8:277–280

North BP, Turnbull JF, Ellis T, Porter MJ, Migaud H, Bron J, Bromage NR (2006) The impact of stocking density on the welfare of rainbow trout (*Oncorhynchus mykiss*). Aquaculture 255:466–479

North BP, Ellis T, Bron J, Knowles TG, Turnbull JF (2008) The use of stakeholder focus groups to identify indicators for the on-farm assessment of trout welfare. In: Branson EJ (ed) Fish welfare. Blackwell, Oxford, pp 243–267

Sandøe P, Crisp R, Holtug N (1997) Ethics. In: Appleby MC, Hughes BO (eds) Animal welfare. CABI Publishing, Oxon, pp 3–17

Sandøe P, Christiansen SB, Appleby MC (2003) Farm animal welfare: the interaction of ethical questions and animal welfare science. Anim Welf 12:469–478

Turnbull JF, Bell A, Adams C, Bron J, Huntingford F (2005) Stocking density and welfare of cage farmed Atlantic salmon: application of a multivariate analysis. Aquaculture 243:121–132

Implementation of Welfare in Fish

Risk Assessment in Fish Welfare, Applications and Limitations – Summary

Christine Müller-Graf, Franck Berthe, Tomasz Grudnik, Ed Peeler and Ana Afonso

Animal welfare issues need to be assessed in a scientific and transparent way, which is not a simple task. One way is to use risk assessment techniques together with the available scientific knowledge, which includes scientific data and expert opinion, to present, summarise and analyse the welfare situation. A risk assessment describes the likelihood and consequences of an un-wanted event, the so called hazard. There are four steps to a risk assessment: (1) hazard identification – what kind of hazard are we talking about; (2) hazard characterization – what are the consequences of this hazard; (3) exposure assessment – how often is the hazard encountered and how many are affected and (4) the risk characterization – as a summary of all the information which went into the assessment and which will allow for instance the final ranking. These four steps are described using the example of farmed Atlantic salmon for aquaculture methods and farmed eel for stunning and killing. The results can be used for instance to rank the husbandry systems in terms of risk for fish welfare and to discuss which method may carry the highest risk for fish welfare. We also discuss the limits of this type of technique and review how it is applied.

Background

The European Union has since 1999 established new rules on animal welfare. Animals are recognized as sentient beings and EU institutions are asked to pay attention to the welfare requirements of animals. The application of risk assessment in animal welfare is a relatively new area. Here, we describe the technique of risk assessment in the context of fish welfare in general and then show how this can be used in aquaculture husbandry systems and for stunning/killing methods for fish.

Scientific Significance

Risk assessment methods have only recently been applied to fish welfare and this is the first publication to describe the methodology in more detail. The use of this methodology permits the transparent presentation of scientific data and expert knowledge which went into the assessment. The processing of the obtained information with this method makes it possible to use different types of information together and to carry out a structured assessment of fish welfare for a particular question, such as the ranking of aquaculture methods and stunning and killing methods with regard to the welfare risk.

Practical Application

We believe that risk assessment methodology provides a transparent approach to identify significant hazards and thus support recommendations for improved welfare.

Contact

A. Afonso
EFSA European Food Safety Authority, Lgo. N. Palli 5 A,
43121 Parma, Italy
e-mail: ana.afonso@efsa.europa.eu

Fish Physiol Biochem (2012) 38:231–241
DOI 10.1007/s10695-011-9520-1

Risk assessment in fish welfare, applications and limitations

Christine Müller-Graf · Franck Berthe · Tomasz Grudnik · Ed Peeler · Ana Afonso

Received: 5 October 2010 / Accepted: 28 May 2011 / Published online: 14 June 2011

Abstract The Treaty of Amsterdam, in force since 1 May 1999, has established new ground rules for the actions of the European Union (EU) on animal welfare. It recognizes that animals are sentient beings and obliges the European Institutions to pay full regard to the welfare requirements of animals when formulating and implementing Community legislation. In order to properly address welfare issues, these need to be assessed in a scientific and transparent way. The principles of risk assessment in terms of transparency and use of available scientific data are probably well suited for this area. The application of risk assessment for terrestrial and aquatic animal welfare is a relatively new area. This paper describes the work developed in the context of the European Food Safety Authority (EFSA) opinions on the application of a risk assessment methodology to fish welfare. Risk assessment is a scientifically based process that seeks to determine the likelihood and consequences of an adverse event, which is referred to as a hazard. It generally consists of the following steps: (i) hazard identification, (ii) hazard characterisation, (iii) exposure assessment and (iv) risk characterisation. Different approaches can be used for risk assessments, such as qualitative, semi-quantitative and quantitative approaches. These are discussed in the context of fish welfare, using examples from assessments done to aquaculture husbandry systems and stunning/killing methods for farmed fish. A critical review of the applications and limitations of the risk methodology in fish welfare is given. There is a need to develop appropriate indicators of fish welfare. Yet, risk assessment methodology provides a transparent approach to identify significant hazards and support recommendations for improved welfare.

Keywords Animal welfare · Farmed fish · Husbandry · Killing · Stunning · Risk assessment methodology

C. Müller-Graf
Bundesinstitut fürRisikobewertung, Thielallee 88-92, 14195 Berlin, Germany

F. Berthe · T. Grudnik · A. Afonso (✉)
EFSA European Food Safety Authority, Lgo. N. Palli 5 A, 43121 Parma, Italy
e-mail: ana.afonso@efsa.europa.eu

E. Peeler
CEFAS, Centre for Environment Fisheries and Aquaculture Science, Weymouth DT4 8UB, UK

Introduction

The Council Directive 98/58/EC[1] concerning the protection of animals kept for farming purposes lays down the minimum standards for the protection of animals bred or kept for farming purposes, including

[1] Council Directive 98/58/EC of 20 July 1998 concerning the protection of animals kept for farming purposes Official Journal L 221, 08/08/1998, pp. 0023–0027.

fish. The Treaty of Amsterdam,[2] in force since 1st May 1999, has established new ground rules for the actions of the European Union (EU) on animal welfare. It recognizes that animals are sentient beings and obliges the European Institutions to pay full regard to the welfare requirements of animals when formulating and implementing Community legislation. The Council of Europe has also in 2005 issued a recommendation on the welfare of farmed fish.[3]

Extensive research efforts on the welfare of farmed fish have been undertaken in the last decade particularly in Europe. Measures of physiological functioning, productivity, health and pathology and behaviour all form the basis of welfare assessment currently being used in field conditions. Due to the complex causal relationships among the various needs of farmed fish and their behavioural and physiological consequences, it is impossible to find one single measurement or welfare indicator that will cover all possible rearing systems, farmed species and potential hazards. When the welfare of fish or other animals is assessed, sets of measures, which might be physiological (Oliveira et al. 1999; Ellis et al. 2002), behavioural or pathological (Huntingford et al. 2006) may be used alone or in combinations. Whilst a single measure could indicate poor welfare, a range of measures will usually provide a more accurate assessment of welfare because of the variety of coping mechanisms used by the animals (Koolhaas et al. 1999; Huntingford and Adams 2005) and the various effects of the environment on individual species of fish. A compound welfare index along the lines of ecological diversity indices—using different measures of welfare (Morton and Griffiths 1985) or the use of a composite measurement scale methodology (EFSA 2008a) is yet be developed.

In order to properly address welfare issues, the risks of poor welfare need to be assessed in a scientific and transparent way. The principles of risk assessment in terms of transparency and use of available scientific data are probably well suited for this area.

[2] Treaty of Amsterdam amending the treaty on European Union, the treaties establishing the European communities and related acts, official journal, C 340, 10 November 1997.

[3] Recommendation concerning farmed fish adopted by the Standing Committee of the European Convention for the protection of animals kept for farming purposes on 5 December 2005.

This paper describes the work developed by EFSA on the application of a risk assessment methodology to fish welfare in farmed fish providing a critical review of its applications and limitations. Two examples, (i) welfare aspects of husbandry systems of Atlantic salmon (EFSA 2008b) and (ii) welfare aspects of the main systems of stunning and killing of farmed eels (EFSA 2009a), are used to illustrate the proposed methodology.

Methodology

The application of risk assessment for terrestrial animal welfare is a relatively new area. However, it is being developed for use in other species and has been employed to compare, for instance, welfare risks associated with different husbandry systems (EFSA 2006a; Müller-Graf et al. 2008; Paton et al. 2008; Paton and Martin 2006). In the context of fish health management, risk assessment has been used to determine the risk of disease introduction (OIE 2004a and spread with international trade in live animals or animal product (Baldock et al. 2008; Peeler et al. 2006; Kahn et al. 1999) and more broadly to support fish health policy (Peeler et al. 2007). To our knowledge, risk assessment methods have not been used to assess fish welfare with the exception of the work undertaken in the context of EFSA scientific opinions. In terms of fish welfare, the questions to address could, for instance, deal with the comparison of welfare risks in different husbandry systems, the comparison of different stunning and killing methods and their respective risk for the welfare of the fish or welfare risks associated with different transport methods. The description of the risk associated to a single hazard is also a possible application.

Risk assessment is a scientifically based process that seeks to determine the likelihood and consequences of an adverse event, which is referred to as a hazard. It generally consists of the following steps: (i) hazard identification, (ii) hazard characterisation, (iii) exposure assessment and (iv) risk characterisation. The final objective is to describe each step in a transparent way and provide a quantitative or qualitative statement of the associated risk.

Different models can be used for risk assessments, such as qualitative, semi-quantitative and quantitative approaches. Qualitative models describe the risk in a verbal manner. The expression of the overall risk probabilities poses a specific challenge, since it has to be ensured that all the parties concerned—risk assessor, risk managers—have the same understanding of the terms such as for example 'serious' or 'moderate' risk. Definitions of these qualitative terms may be useful and appropriate. A quantitative risk assessment uses quantitative information and can be either deterministic or stochastic. Quantitative models may be more transparent because of the numerical format and will allow simulations and expressions of distributions of the input variables—their ranges—and risk estimates, but may sometimes give a wrong impression of the precision without a helpful discussion of the model uncertainties (EFSA 2009b). Depending on the information available and the specific question, it may be also useful to express information using scores, i.e., on a semi-quantitative scale.

Definitions were also proposed by the Codex Alimentarius Commission (1995).

Quantitative risk assessment

A risk assessment that provides numerical expressions of risk and an indication of the attendant uncertainties.

Qualitative risk assessment

A risk assessment based on data which, whilst forming an inadequate basis for numerical risk estimations, nevertheless, when conditioned by prior expert knowledge and identification of attendant uncertainties, permits risk ranking or separation into descriptive categories of risk. Risk assessment methodologies have been developed to assess food safety risks in the context of the *Codex Alimentarius* (WHO 1999) or the risk of disease introduction (OIE 2004a, b). No standardized methodology exists in the field of animal welfare risk assessment (EFSA 2006b; Smulders 2009).[4]

EFSA assessment work on animal welfare follows the methodology proposed by the *Codex Alimentarius* (WHO 1999). A risk in animal welfare would be the result of the probability of a negative animal welfare effect (i.e. the adverse effect) and the severity of the hazard, consequential to the exposure to a hazard(s). The probability or likelihood of the hazard at a population level can also be taken into account (i.e. probability that the population is exposed to the hazard and the proportion of the population which is exposed). The degree of confidence in the final estimation of risk would depend on the variability, uncertainty, and assumptions identified and integrated in the different risk assessment steps.

Hazard identification

A hazard in animal welfare risk assessment is defined as a factor/condition with a potential to cause a negative animal welfare effect (or adverse effect). Hazards may be specific to (i) the species considered, (ii) their stage of development and iii) the husbandry system and specific production methods. In the case of the risk assessment on animal welfare aspects of husbandry systems of farmed Atlantic salmon (EFSA 2008b), the different production systems for each production life stages were identified. Table 1 represents the matrix used for animal welfare aspects of husbandry systems for farmed Atlantic salmon, each of the combinations constitutes a target population, for each of which hazards were identified and separately assessed. The duration of each life stage was also estimated as a percentage of the total production cycle.

Hazards associated with husbandry can have direct effects on animal or indirect effects by changing the animals' environment in ways that affect their abilities to fulfil basic needs consequently leading to adverse effects. The risk assessment concentrated on single factors without interactions and factors were grouped in different categories of hazards: abiotic, biotic, genetic, management and disease (Table 2). The hazard identification was done as specifically as possible, by going through all the different categories. Water temperature was, for example, considered as a hazard in the abiotic category. Three different hazards related to water temperature were considered: (i) rapid change in temperature, (ii) too high temperature and (iii) too low temperature. Other hazards in the abiotic section included the salinity, pH, flow rate and oxygen content of water. Hazards could occur to some or all life stage/production system combinations, e.g.,

[4] For further discussion of animal welfare methodology see also, http://www.daff.gov.au/__data/assets/pdf_file/0004/1046497/37-michael-paton.pdf.

Table 1 Production systems by life stages of farmed Atlantic salmon

Production system	Production life stages						
	Eggs	Alevins	Fry	Parr	Smolts	On growing	Brood stock
Trays	X	X					
Cylinders	X						
Tanks (re-circulated)			X	X	X		
Tanks (flow-through)			X	X	X		X
Freshwater cages				X	X		
Sea-water cages						X	X

grading fish by size or morphological characteristics was a management hazard applicable to only some life stages.

Hazard identification was also done for the assessment of methods in use for fish stunning and killing. The methods for stunning and killing eels (*Anguilla anguilla*) in Europe are as follows: (1) salt treatment, (2) ammonia treatment, (3) ice and salt and (4) electricity. For each stunning and/or killing method, hazards were systematically identified by breaking down the process into the sequence of activities (Fig. 1). Hazards were categorised as (i) pre-slaughter hazards (Table 2) associated with unloading and transport at the slaughter facility, (ii) pre-slaughter hazards associated with lairage (Table 3) and (iii) hazards associated with each of the stunning/killing method (Table 4) (EFSA 2009a).

Hazard characterisation

Hazard characterization is the qualitative and quantitative evaluation of the nature of the adverse effects associated with the hazard. The example described here belongs to a semi-quantitative risk assessment where the information used is based on expert opinion. The severity of the adverse effect describes the consequence of exposure of an individual to a hazard and was scored according to the scheme shown in Table 5.

Furthermore, the relationship between the severity of the hazard in terms of duration and the severity of the adverse effect affecting the individual needs to be described. The duration of the adverse effects, i.e., the consequences of the hazard, can be scored on a 0–100% scale considering the overall fish life span or the duration of the particular life stage. On the case of risk assessment for stunning and killing, the duration of the adverse effect was estimated on a categorical scale (Table 6).

A hazard is described not only by the magnitude of its adverse effect but also by the likelihood of the adverse effect occurring, once the hazard occurs not all individuals will experience the adverse effect. The likelihood of the adverse effect for an individual is represented by the proportion of the individuals in a population affected when the adverse effect occurs, this can be described either by percentage or as a score as shown in Table 7.

Each hazard will be characterized by a score which is the result of:

Hazard characterisation = (severity of adverse effect) $*$ (duration of the adverse effects) $*$ (likelihood of adverse effect)

The risk assessment for stunning/killing of farmed eels was based on two assumptions:

1. all fish exposed to the hazard experienced the same intensity and duration of the adverse effect.
2. in the absence of any evidence to the contrary, it is assumed that all fish exposed to the hazard experience the adverse effect.[5]

The uncertainty associated with the hazard characterization needs to be stated in order to evaluate the robustness of the assessment. Uncertainty may be associated with absence or limited scientific evidence or measurement error but also with lack of empirical evidence. Uncertainty was scored as an indication of the type of information available (see Table 8), whether there are different studies with differing

[5] If this assumption was not found to be sound for a particular hazard an additional parameter (probability that exposure resulted in the adverse effect) was used.

Table 2 Hazard identification for welfare risk assessment of husbandry systems of farmed Atlantic salmon

	Hazard identification	Hazard specification
Abiotic	Water flow	Too low/too high
	Light	Period/intensity
	Water depth	
	Water temperature	Rapid change/high/low
	Shape of tank/distortion of cage	
	Suspended solids	
	Storm impact/water vessel impact	
	Salinity	Too high/fluctuations
	pH	Too high or low in combination with Al
	Oxygen content	Too low
	Metals other than Al	Too high, pH dependent
	Environmental complexity	
	Carbon dioxide content	Too high
	Ammonia content	Too high, pH dependent
	Aluminium content	Too high, pH dependent
Biotic	Stocking density	High/low
	Intra-specific interaction	
	Predators	
	Other invasive species (e.g. algae)	
	Mixing fish from different origins	
	Inter-specific interactions	
Feeding	Nutrients	Surplus/deficiency
	Vegetable protein	
	Lack of food	Short time/long time
	Dietary toxins	
	Feed additives for fish	
Management	Sorting of fish	Frequency/methods
	Lack of staff training	
	Lack of bio security	
	Impact of lack of monitoring	Health/biomass
	Handling	
	Genetic selection	Growth/disease
Diseases	Furunculosis	
	Winter ulcers	
	Saprolegnia infection	
	Infectious pancreatic necrosis	
	Infectious salmon anaemia	
	Sea lice infestation	
	Pathology as a result of jellyfish	
	Eye lesions	

conclusions, but also whether the scientific information has been published or not. This is done for all individual fields in hazard characterisation and exposure assessment; at the end of the risk assessment, a sum of all the uncertainty scores is presented. Uncertainty is different from variability, which can be biological inherent and may occur between two different populations, for example, due

Fig. 1 Farmed eels stunning and killing process (adapted from EFSA 2009a)

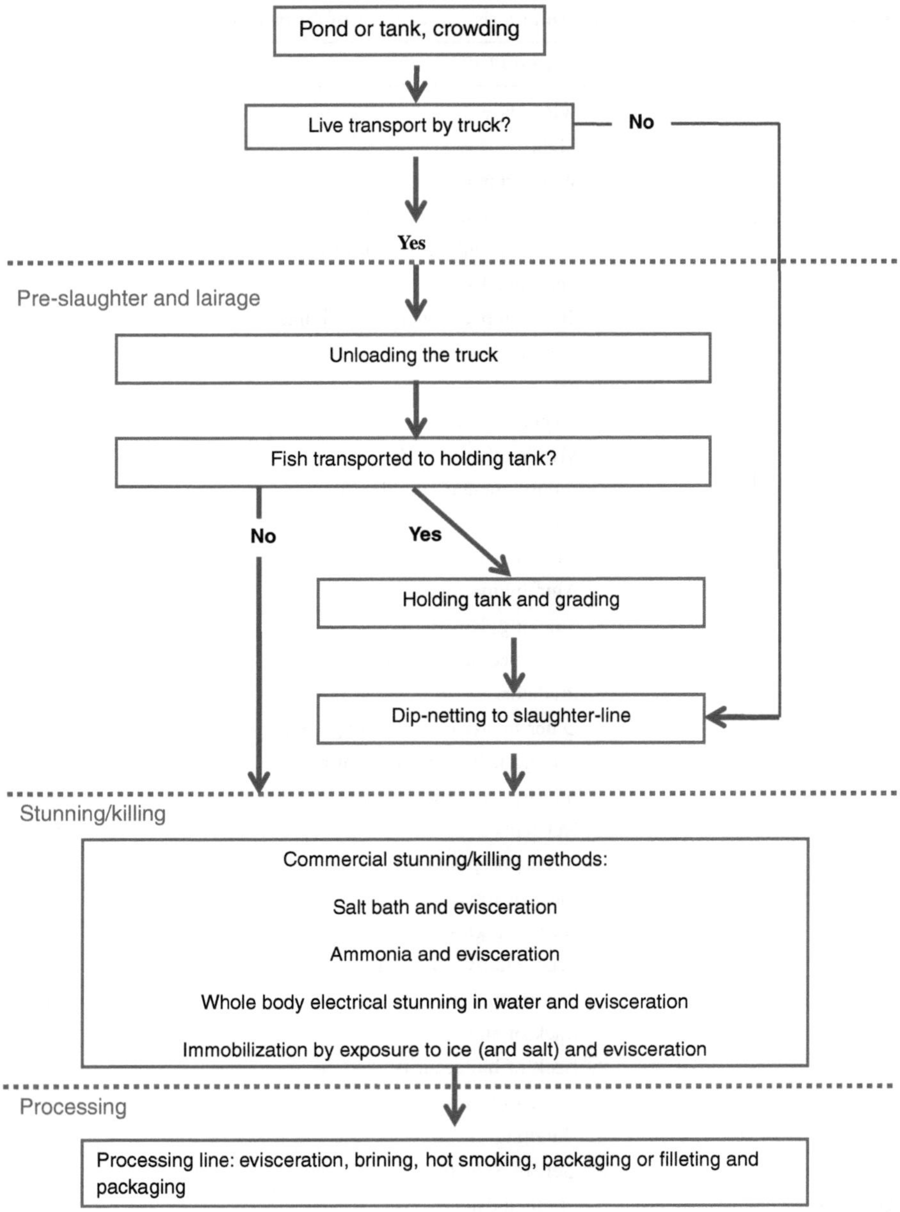

to a slightly different genotype. However, on a practical level, it is quite difficult to separate these two and they are often mixed together.

Exposure assessment

Exposure assessment is the qualitative, semi-quantitative or quantitative evaluation of the probability of a specific scenario of exposure. It takes into account the frequency and duration of exposure to one or several hazards during the particular life stage of the fish. The frequency of exposure was considered as a score (Table 9), that is, how often a particular hazard would be encountered. The duration of the hazard for a given life stage was described as a percentage, to describe for how long the hazard would occur within that particular life stage of the fish.

$$\text{Exposure assessment} = (\text{frequency of exposure to the hazard}) * (\text{duration of hazard})$$

This information was also characterized by an uncertainty score.

Table 3 Hazard identification in welfare risk assessment: pre-slaughter hazards associated with stunning and killing of eels

	Hazard
Pre-slaughter	Unloading with a drop
	Unloading without a drop
	Jumping out of the container
	Piling up in transport containers at slaughter facility
	Trapped in shutter of the transport tank
	Sudden change in temperature
	Sudden exposure to daylight
Lairage	Unloading from the transport container
	Sudden change in temperature
	Poor water quality
	High density
	Grading before stunning
	Jumping out of holding tank
	Dip-netting (before slaughter)

Table 4 Hazard identification in welfare risk assessment: slaughter hazards associated with stunning and killing of eels

Stunning/killing method	Hazard
Salt treatment	Salt treatment
	Evisceration
Ammonia treatment	Ammonia treatment
Ice and salt	Ice and salt (2% salt and 25% ice water) for 16 h
	Evisceration
Electricity	Insufficient current/voltage
	Evisceration

Risk characterisation

The risk characterisation or risk estimate aims to give information to the risk manager to evaluate a specific situation regarding risk for poor welfare. It is calculated for each hazard and expresses the animal welfare burden in the considered population.

$$\text{Risk characterisation} = \text{hazard characterisation} * \text{exposure assessment}$$

The various scores were standardized across the calculations. That is, if for instance, a severity of 3 is scored in a system which uses 4 categories, the standardized score would be 3/4.

Take the example of a biotic factor, i.e., high stocking density as a welfare hazard for on-growing Atlantic salmon in sea cages:

Hazard characterization	Categories	Exposure assessment	Categories
Severity = 3	4		
Likelihood = 2	5	Frequency = 2	5
Duration of adverse effect = 32	100	Duration of exposure = 40	100

Risk characterisation = (severity of adverse effect*duration of the adverse effects*likelihood of adverse effect*frequency of exposure to the hazard*duration of hazard)*100 = (3/4*2/5*32/100*2/5*40/100)*100 = 1.65.

If more quantitative information than risk scores is available, the risk could also be presented as a distribution with minimum, mean/median and maximum value.

Table 5 Severity of the adverse effect on welfare associated with the hazard

Evaluation	Score	Explanation
Negligible	0	No pain, malaise, frustration, fear or anxiety as evidenced by measures of the normal range of behavioural observations, physiological measures and clinical signs
Mild	1	Minor changes from normality and indicative of pain, malaise, fear or anxiety
Moderate	2	Moderate changes from normality and indicative of pain, malaise, fear or anxiety
Substantial	3	Substantial changes from normality and indicative of pain, malaise, fear or anxiety
Severe	4	Extreme changes from normality and indicative of pain, malaise, fear or anxiety, that if persist would be incompatible with life

Table 6 Duration categories for adverse effects arising from hazards associated with pre-slaughter and stunning/killing of farmed fish

Duration (min)	Score
<5[a]	1
5–15	2
>15–60	3
>60	4

[a] Adverse effects with a duration of less than one second are not scored

Table 7 Likelihood of adverse effect occurring (i.e. proportion of population affected)

Evaluation	Score	Explanation
Negligible	0	The adverse effect would almost certainly not occur
Extremely low	1	The adverse effect would be extremely unlikely to occur
Very low	2	The adverse effect would be very unlikely to occur
Low	3	The adverse effect would be unlikely to occur
Moderate	4	The adverse effect would occur with an even probability
High	5	The adverse effect would be very likely to occur

The scores of the semi-quantitative risk assessment provide a ranking of risk factors. The identification of the more relevant hazards is an important outcome of the risk assessment. The scores may help to compare between different production systems and thus evaluating whether certain production systems may be more welfare friendly than others. For each of the different production stages, the scores within a

Table 9 Frequency of exposure to the hazard

Evaluation	Score	Explanation
Negligible	0	The exposure would almost certainly not occur
Extremely low	1	The exposure would be extremely unlikely to occur
Very low	2	The exposure would be very unlikely to occur
Low	3	The exposure would be unlikely to occur
Moderate	4	The exposure would occur with an even probability
High	5	The exposure would be very likely to occur

life stage may also be added for and their sums compared.

The risk assessment on the welfare aspects of different husbandry systems in farmed Atlantic salmon (EFSA 2008b) made possible a ranking of risks for each husbandry system. It was concluded that the highest welfare risk was associated to disease in all life stages. Inadequate management was an important poor welfare risk, but with considerably lower scores. Biotic factors were more a risk for brood stock and abiotic hazards, i.e., mainly water quality, and were of concern for all life stages. A second objective of this work was to compare different husbandry systems. This was achieved by summing the risk scores for all the hazards arising for each system/life stage. The overall scores did not demonstrate large differences between the welfare risk of the different husbandry systems assessed.

The information included in the risk assessment is made more transparent to other experts and stakeholders and inform a debate about the scientific information on which the assessment is based.

Table 8 Uncertainty scores used in describing evidence on the welfare risks associated with stunning/killing of farmed fish

Evaluation	Score	Explanation
Low	1	Solid and complete data available: strong evidence in multiple references with most authors coming to the same conclusions
Medium	2	Some or only incomplete data available: evidence provided in small number of references; authors' conclusions vary
		Solid and complete data available from other species which can be extrapolated to the species considered
High	3	Scarce or no data available: evidence provided in unpublished reports, or based on observation or personal communications; authors' conclusions vary considerably

Hazards which have a high uncertainty point to research needs.

Limitations of risk assessment methodology in fish welfare

The described method is an initial development to deal with risk assessment of fish welfare on a more structured and transparent way. There are limitations to the method which need to be considered in the discussion of the results. Some are common to the risk assessment of animal welfare in general and some more specific to fish.

Measuring of welfare is impaired by the absence of validated welfare indicators. Some attempts have been made by trying to develop indices for terrestrial animal welfare (Barutssek 1999; Bracke et al. 1999a, b), and various publications have used their own index for assessing the welfare of a given species. Extensive research has been conducted for the development of such indicators in terrestrial animals (Welfare quality)[6] and some progress is being done on the study of possible indicators for the assessment welfare in farmed fish (Hoyle et al. 2007). Unfortunately, measuring one single parameter—such as stress hormone levels in fish (Ellis et al. 2004)—does not necessarily capture the welfare status of the fish. A range of tools are needed to measure the welfare of animals which can be adapted to the appropriate groups or taxonomic ranks of animals.

Quantitative estimations of the impact on welfare from different hazards are missing. Applying the risk assessment methodology to fish welfare showed that very limited information is available in the peer reviewed literature regarding fish welfare risk estimates (EFSA 2009c). There are often huge data gaps depending on species, life stage and condition. Furthermore, some of the life cycles of fish are complex and not very well understood. Available studies are frequently done for a different purpose and it is difficult to extract and apply the relevant information to fish welfare risk assessment. Since not all scientists are familiar with the type of information needed for fish welfare risk assessment, there is not always enough consideration to publish the information in a suitable format. Clearly, the scope of the risk assessment has to be defined and it has to be decided beforehand whether to deal with the whole life span of a fish species or a specific life stage.

Welfare risk assessment so far depends often on the opinion of experts who are familiar with abnormal behaviour indicating welfare problems. Expert opinions can of course be subjective. Several methods of dealing with this type of bias, such as the Delphi panel and other methods can be used and, therefore, in the absence of solid quantitative data, the use of expert opinion is an acceptable option (Algers 2009, Bracke et al. 2008).

The interactions between different hazards have so far not been addressed adequately by the risk assessment welfare methodology. Whilst each specific hazard is described separately, there are very few occasions where only a single factor is involved in any welfare issue relating to environmental conditions. For instance, water quality is the result of several interacting factors. One solution is to describe the same factor under different scenarios, but the results are cumbersome to analyse. Furthermore, continuous variables and correspondingly changing hazards are difficult to integrate in a semi-quantitative risk assessment. For example, threshold for high and low water temperature for a fish species may be defined, but they cannot take into account the whole continuous scale of water temperatures and their effects.

The risk scores are linked to populations, but observations of individual fish are needed to assess severity. Observation of individuals in aquaculture conditions is problematic compared with individual terrestrial animals, and studies available are often the result from laboratory observations of selected fish.

Genetic or phenotypic differences cannot be easily taken into account. They are included in the variability described in the risk assessment. Variability (the biological range of different expressions) and uncertainty (due to lack of knowledge) need to be separated, which is not always easy. Uncertainty and variability are not easy to integrate in a qualitative or semi-quantitative risk assessment; more generally, they are flagged and indicated. In a fully quantitative risk assessment, ranges around input parameters provide estimates of variability and uncertainty.

[6] Welfare Quality®: Science and society improving animal welfare in the food quality chain, EU funded project FOOD-CT-2004-506508.

Risk estimate scores have no inherent value and can only be used in a meaningful way to rank or compare hazard or the sums of hazards for between different species, life stages or production systems.

The problem of death as endpoint is another open question. Some animal welfare scientists consider death not a welfare problem, but the suffering of the animal, independent of whether it will lead up to death. However, when calculating the effect of a hazard over a certain period—life time or life stage—it makes a difference whether to calculate the hazard over a 'potential' or 'actual' life time. A quick death will lead to a lower risk score even though the hazard may be a serious problem. No clear solution has been found. As a compromise, both possibilities can be computed or hazards which lead to quick death are flagged, since they may point to problems connected with welfare. In any case, it is useful to indicate which hazards lead to mortality and which hazards result in morbidity since both may be indicators of poor welfare. The current risk assessment approach only considers negative impact. Welfare is not only matter of 'absence of' but also of presence of positive factor and in the long run; a risk–benefit analysis would provide a more complete analysis of welfare issues on which to compare welfare systems. However, for the time being the methods to include negative and positive impacts are lacking.

Conclusions

Overall, the specific welfare question will determine the approach. The more clearly defined question is the better the experts can make their judgment or the collection of data can be defined. However, for fish welfare, some broad questions on for instance transport of fish or husbandry systems may be asked to inform risk managers and legislators. Whilst fish welfare is an important area of welfare research, the scope and application still need better definition. Risk assessment approach is of value to identify data gaps and can be used for prioritization of research needs. Fish welfare must be assessed through an explicit process based on both scientifically derived data and value-based assumptions. Risk assessment methodology provides a good approach to identify significant hazards and support recommendations for improved welfare. However, there is a need to develop appropriate and inclusive performance/welfare indicators. Welfare assessment for fish has to overcome the problem of the diversity of fish species—if they are evaluated together or individually—and the overall lack of suitable scientific data.

Acknowledgments The authors wish to express their appreciation to all working group experts and the members of the Animal Health and Welfare panel of EFSA involved in the drafting of EFSA opinions on fish welfare. Their expertise made possible the development of the methodology here discussed.

References

Algers B (2009) A risk assessment approach to animal welfare. In: Smulders FJM, Algers B (eds) Welfare of production animals: assessment and management of risks. Food safety assurance and veterinary public health, vol 5. Wageningen Academic Publishers, The Netherlands, pp 223–237

Baldock FC, More SJ, Peeler EJ (2008) An introduction to import risk analysis for aquatic. Fish Vet J 10:1–25

Barutssek H (1999) A review of the animal needs index (ANI) for assessment of animal's well-being in the housing systems for Austrian proprietary products and legislation. Levest Prod Sci 61:179–192

Bracke MBM, Metz JHM, Spruijt BM (1999a) Overall animal welfare reviewed. Part 2: assessment tables and schemes. Neth J Agric Sci 47:230–293

Bracke MBM, Spruijt BM, Metz JHM (1999b) Overall animal welfare assessment reviewed. Part 1: is it possible? Neth J Agric Sci 47:279–291

Bracke MBM, Edwards SA, Engel B, Buist WG, Algers B (2008) Expert opinion as 'validation' of risk assessment applied to calf welfare. Acta Vet Scand 50:29

Codex Alimentarius (Codex Alimentarius Commission) (1995) Application of risk analysis to food standards issues. Report of the Joint FAO/WHO expert consultations. Geneva, Switzerland. 13–17 March 1995

EFSA (2006a) Opinion of the scientific panel on animal health and welfare (AHAW) on a request from the commission related with the risks of poor welfare in intensive calf farming systems. http://www.efsa.europa.eu/EFSA/Scientific_Opinion/ahaw_op_ej366_calveswelfare_en1,0.pdf

EFSA (2006b) summary report, scientific colloquium of food producing animals, 1–2 December 2005, p 156. http://www.efsa.europa.eu/en/science/colloquium_series/no4_animal_diseases.html

EFSA (2008a) Basic information for the development of the animal welfare risk assessment guidelines. Available on line http://www.efsa.europa.eu/en/scdocs/doc/130ar.pdf. Accessed 29 Sept 2010

EFSA (2008b) Scientific opinion of the panel on animal health and welfare on a request from the European commission on animal welfare aspects of husbandry systems for farmed Atlantic salmon. The EFSA Journal 736:1–31. Available on line http://www.efsa.europa.eu/en/scdocs/doc/736.pdf. Accessed 28 Sept 2010

EFSA (2009a) Scientific Opinion of the Panel on Animal Health and Welfare on a request from the European commission on welfare aspect of the main systems of stunning and killing of farmed eel (*Anguilla anguilla*). EFSA J 1014: 1–42. Available on line http://www.efsa.europa.eu/en/scdocs/doc/1014.pdf. Accessed 28 Sept 2010

EFSA (2009b) Guidance on good practice in conducting scientific assessments in animal health using modelling. EFSA J 7(12):1419. doi:10.2093/j.efsa.2009.1419

EFSA (2009c) Statement of EFSA prepared by the AHAW panel on: knowledge gaps and research needs for the welfare of farmed fish. EFSA J 1145:1–7. http://www.efsa.europa.eu/en/scdocs/scdoc/1145.htm

Ellis T, North B, Scott AP, Bromage NR, Porter M, Gadd D (2002) The relationships between stocking density and welfare in farmed rainbow trout. J Fish Biol 61:493–531

Ellis T, James JD, Stewart C (2004) A non-invasive stress assay based upon measurement of free cortisol released into the water by rainbow trout. J Fish Biol 65(5):1233–1252

Hoyle I, Oidtmann B, Ellis T, Turnbull J, North B, Nikolaidis J, Knowles TG (2007) A validated macroscopic key to assess fin damage in farmed rainbow trout (*Oncorhynchus mykiss*). Aquaculture 270:142–148

Huntingford F, Adams C (2005) Behavioural syndromes in farmed fish: implications for production and welfare. Behaviour 142:1207–1221

Huntingford FA, Adams C, Braithwaite VA, Kadri S, Pottinger TG, Sandoe P, Turnbull JF (2006) Current issues in fish welfare. J Fish Biol 68:332–372

Kahn SA, Wilson DW, Perera RP, Hayder H, Gerrity SE (1999) Import risk analysis on live ornamental finfish. Australian Quarantine and Inspection Service, Canberra, p 172

Koolhaas JM, Korte SM, de Boer SF, Van der Vegt BJ, Van Reenen CG, Hopster H, de Jong IC, Ruis MAW, Blokhuis HJ (1999) Coping styles in animals: current status in behaviour and stress-physiology. Neurosci Bio Behav Rev 23:925–935

Morton DB, Griffiths PHM (1985) Guidelines on the recognition of pain, distress and discomfort in experimental animals and an hypothesis for assessment. Vet Rec 116:431–436

Müller-Graf C et al. (2008) Risk assessment in animal welfare—EFSA approach. 6th world congress on alternatives and animal use in the life sciences, 2007 Tokyo, Japan, Japanese Society for Alternatives to Animal Experiments (JSAAE), pp V789–V794

OIE (2004a) Handbook on import risk analysis for animals and animal products, vol 1. Introduction and qualitative risk analysis, p 57

OIE (2004b) Handbook on import risk analysis for animals and animal products, vol 2. Quantitative risk assessment. p 126

Oliveira RF, Canario AVM, Bsharry R (1999) Hormones, behaviour and conservation of littoral fishes: current status and prospects for future research. In: Almada VC, Oliveira ARF, Gonçalves EJ (eds) Behaviour and conservation of littoral fishes. Instituto Superior de Psicologia Apilcada, Lisboa

Paton M, Martin PAJ (2006) Can composite indices and risk assessment be used to evaluate animal welfare? 11th Symposium of the International society for veterinary epidemiology and economics ISVEE 2006.V 375

Paton M et al. (2008) Practical uses of risk assessment method in animal welfare. AAWS international animal welfare conference, 2008

Peeler E, Thrush M, Paisley L, Rodgers C (2006) An assessment of the risk of introducing the fish parasite *Gyrodactylus salaris* to infected territories in the European Union with the movement of live atlantic Salmon (*Salmo salar)* from coastal waters. Aquaculture 258(1–4):187–197

Peeler EJ, Murray AG, Thebault A, Brun E, Giovaninni A, Thrush MA (2007) The application of risk analysis in aquatic animal health management. Pre Vet Med 81:3–20

Smulders FJM (2009) A practicable approach to assessing risks for animal welfare—methodological considerations. In: Smulders FJM, Algers B (eds) Welfare of production animals: assessment and management of risks. Food safety assurance and veterinary public health, Vol 5. Wageningen Academic Publishers, The Netherlands, pp 239–274

WHO (1999) Codex alimentarius commission. Principles and guidelines for the conduct of microbiological risk assessment, CAC/GL-30

Fish Welfare Assurance System: Initial Steps to Set Up an Effective Tool to Safeguard and Monitor Farmed Fish Welfare at a Company Level – Summary

Hans van de Vis, Marnix Poelman, Bert Lambooij, Marie-Laure Bégout and Maciej Pilarczyk

We used the existing HACCP system to set up a quality assurance (QA) system for fish welfare in aquaculture. HACCP is an internationally agreed approach, used in industry as a management tool for food safety, based on scientific principles. We adapted the HACCP concept to a Fish Welfare Assurance System (FWAS) to safeguard and monitor welfare of farmed fish. In our study we focused on the first steps towards development of the FWAS for on growing of sea bass, carp and European eel.

We developed the FWAS for four major hazards, which differ with respect to species and technology used for on growing. These hazards have the potential to cause impaired welfare of the selected species. To prevent, eliminate or reduce the four major hazards to acceptable levels, we established at which steps control is essential and which control measures can be applied. We present criteria/target levels that serve as standards by which each critical step is measured. To perform measurements to assess the process of on growing, we proposed methods for monitoring it. In addition, we described a plan for implementation of the FWAS at a company level.

Being based on HACCP, which is a system applicable to various food processing and distributing companies and specialty shops, FWAS has a large level of flexibility and, therefore, its implementation is practicable.

Background

Current ethical concerns about aquaculture, which are broader than welfare of fish alone, drive the preparation of certification schemes that may include consideration of fish welfare. A certificate is a declaration that states that certain standards are met by the audited company or entity. Experiences, however, show that a process-oriented system focused on prevention, rather than inspection only, is a better approach for control of a process. Quality assurance (QA) provides such a system focused on the control of processes.

Scientific Significance

This is the first reported study on the development of a Fish Welfare Assurance System to monitor and safeguard welfare of fish at a farm level.

Practical Application

Our study shows that the structured approach of the FWAS will provide a more timely response to fish welfare problems that may occur, as the basic premise is prevention rather than inspection. An example: as sea bass is a more "nervous" type of fish than carp, handling can provoke a stronger stress response than in carp. This hazard can be minimized by using best practices, which requires providing personnel at the farm with detailed handling instructions in a standard operation procedure (SOP). The application of this SOP is identified as a critical control point and the target level is met when sea bass is handled as specified in the SOP.

Record-keeping will provide evidence that the desired level of control was obtained. The FWAS places the responsibility for safeguarding and monitoring welfare of fish on the company itself. Hence, the FWAS tool can contribute to a pro-active response to societal concerns regarding welfare of fish.

Contact

J. W. van de Vis
IMARES, Wageningen UR, PO Box 77,
4400 AB Yerseke, The Netherlands
e-mail: hans.vandevis@wur.nl

Fish Physiol Biochem (2012) 38:243–257
DOI 10.1007/s10695-011-9596-7

Fish welfare assurance system: initial steps to set up an effective tool to safeguard and monitor farmed fish welfare at a company level

J. W. van de Vis · M. Poelman · E. Lambooij · M.-L. Bégout · M. Pilarczyk

Received: 3 January 2011 / Accepted: 22 December 2011 / Published online: 26 January 2012

Abstract The objective was to take a first step in the development of a process-oriented quality assurance (QA) system for monitoring and safeguarding of fish welfare at a company level. A process-oriented approach is focused on preventing hazards and involves establishment of critical steps in a process that requires careful control. The seven principles of the Hazard Analysis Critical Control Points (HACCP) concept were used as a framework to establish the QA system. HACCP is an internationally agreed approach for management of food safety, which was adapted for the purpose of safeguarding and monitoring the welfare of farmed fish. As the main focus of this QA system is farmed fish welfare assurance at a company level, it was named Fish Welfare Assurance System (FWAS). In this paper we present the initial steps of setting up FWAS for on growing of sea bass (*Dicentrarchus labrax*), carp (*Cyprinus carpio*) and European eel (*Anguilla anguilla*). Four major hazards were selected, which were fish species dependent. Critical Control Points (CCPs) that need to be controlled to minimize or avoid the four hazards are presented. For FWAS, monitoring of CCPs at a farm level is essential. For monitoring purposes, Operational Welfare Indicators (OWIs) are needed to establish whether critical biotic, abiotic, managerial and environmental factors are controlled. For the OWIs we present critical limits/target values. A critical limit is the maximum or minimum value to which a factor must be controlled at a critical control point to prevent, eliminate or reduce a hazard to an acceptable level. For managerial factors target levels are more appropriate than critical limits. Regarding the international trade of farmed fish products, we propose that FWAS needs to be standardized in aquaculture chains. For this standardization a consensus on the concept of fish welfare, methods to assess welfare objectively and knowledge on the needs of farmed fish are required.

Keywords Fish · Welfare assurance · Aquaculture

J. W. van de Vis, M. Poelman and M. Pilarczyk contributed equally to this study.

Wageningen Aquaculture is a consortium of IMARES (Institute for Marine Resources and Ecosystem Studies) and AFI (Aquaculture and Fisheries Group, Wageningen University), both part of Wageningen University and Research Centre (WUR).

J. W. van de Vis (✉) · M. Poelman
IMARES, Wageningen UR, PO Box 77,
4400 AB Yerseke, The Netherlands
e-mail: hans.vandevis@wur.nl

E. Lambooij
Livestock Research, Wageningen UR, P.O. Box 65,
8200 AB Lelystad, The Netherlands

M.-L. Bégout
Ifremer, Place Gaby Coll, BP 7, 17137 L'Houmeau,
France

M. Pilarczyk
PAS, Institute of Ichthyobiology and Aquaculture, ul.
Kalinowa 2, 43-520 Zaborze, Poland

Introduction

In Europe there is a trend towards increased concern for the welfare of animals at all levels where they come into contact with humans; recently this concern has expanded to include the welfare of fish, as reviewed by Braithwaite and Huntingford (2004) and Lund et al. (2007).

Animal welfare is a multisided concept. The two major issues here are the definition of animal welfare and how best to measure it objectively (Broom 1991). The concept of fish welfare will not be discussed further in this chapter, as this is the subject of the first paper in this issue (Cottee 2012).

Due to the continued growth of the aquaculture industry and an increased focus on fish welfare (Ashley 2007), it is vital to monitor and safeguard the welfare of farmed fish. At present most fish farmers monitor water quality parameters, fish behaviour, growth and health status (tissue damage, parasites etc.) on a daily or at least routinely basis. However, while through this approach, the symptoms of impaired welfare are observed, the origin or cause of the welfare problems may not be detected, for example, observation that the appetite of fish is lower than expected suggests a welfare problem.

Regarding commercial drivers of animal welfare management, current ethical concerns in Europe are broader than the animal welfare alone. These concerns are reflected by certification schemes that aim for sustainable production schemes for farmed animals (Vessier et al. 2008). In organic certification schemes, for example, animal welfare is included as part of the basic production philosophy, alongside environmental and human health, food safety and quality (FAO 2007). Recently, the EU has developed standards for organic aquaculture for a coherent and consistent aquaculture policy. Commission Regulation 710/2009 (EC 710/2009) sets conditions for the aquatic production environment and for impacts on other species. It deals with the separation of organic and non-organic units and specifies animal welfare conditions, including maximum stocking densities. Certification under the EU standards for organic aquaculture boils down to conformity assessment; that is, a process of testing and auditing is conducted to assess the degree of compliance of the entity to be certified to specific standards (FAO 2007). Testing and auditing procedures are not focused on preventive measures, nor control of the entire process. Generally, however, certification schemes that include consideration of animal welfare originate from non-governmental organizations (NGOs), retailers and the aquaculture industry (FAO 2007; Vessier et al. 2008). The Freedom Food certification scheme in the UK is exceptional in this regard, as it focuses on animal welfare alone, and uses the Five Freedoms (Brambell 1965) as a framework (RSPCA 2010).

In contrast to the use of certification schemes, a strategy based on a thorough analysis of farming conditions, considering fish species and their life stages, and approaching this by reviewing the entire process at a fish farm level is proposed. This is much more likely to provide improved control of aquaculture practices in order to safeguard the welfare of farmed fish. Quality assurance (QA) provides such a system focused on control of processes; given the increased concern for farmed fish welfare in Europe, the development of QA system to safeguard welfare is therefore timely.

Within the scope of this study we present the initial steps of establishing a QA system for on growing of sea bass (*Dicentrarchus labrax*), carp (*Cyprinus carpio*) and European eel (*Anguilla anguilla*).

Quality assurance

It is known from the food industry and other manufacturing industries that QA is focused on identification of hazards and establishment of critical steps that need to be controlled in a given process to prevent, eliminate or minimize a hazard to an acceptable level.

Prior to the introduction of the concept of quality assurance in the food industry, quality control (QC) systems were used. Food products were manufactured for an entire shift or day and placed in a storage area to be analysed for defects. Production lots that exceeded the maximum allowable number of defects in the sample were rejected (Sperber 2005). Despite these QC systems, serious food safety accidents occurred in the food industry that had developed rapidly since World War II (Sperber 2005). Due to the inability of QC systems in the food industry to detect hazards that occur at a low incidence, they were replaced by a quality assurance system, known as Hazard Analysis Critical Control Points (HACCP). HACCP was originally developed as a food safety management system for use by NASA in manned space exploration in order to avoid the risk of illness of an astronaut (Quinn and Marriott 2002).

The HACCP system is the internationally agreed approach for food safety control (Codex Committee on Food Hygiene 1993: Codex Committee on Food Hygiene 1997). As of 1 January 2006, the EU mandated all food manufacturing facilities that produce for the European market to incorporate HACCP into their food safety systems (EC Directive No 952/2004).

Compared with QA in the food industry generally, establishment of standardized systems for safeguarding the welfare of farmed fish at the fish farm has not yet been carried out. Contrary to fish, for pigs, a QA system for housing of these animals was established (Von Borell et al. 2001). For the pig housing QA system, which is based on the HACCP concept, these authors established criteria to be used primarily by the farmer as part of an internal on farm assessment scheme.

For QA in aquaculture, the hazards, preventive measures and critical steps that require control in aquaculture are clearly different to those used in the food industry and the production of pigs. The present paper covers these issues in respect to fish welfare in aquaculture.

In this chapter we show how the HACCP approach, which consists of seven principles (Snyder 1991), can be used as a framework for fish welfare assurance at farms. These seven principles are the following:

1. Conduct a Hazard Analysis and Risk Assessment (RA)
2. Determine Critical Control Points (CCPs)
3. Establish Target Levels (TLs) and Critical Limits (CLs) for each CCP
4. Establish monitoring procedures at each CCP
5. Establish Corrective Actions
6. Establish Verification Procedures
7. Establish a Record-Keeping System.

Farmed fish welfare assurance

A farmed Fish Welfare Assurance System (FWAS) must be supported by a foundation of prerequisite programs. Well-functioning prerequisite programs, such as a sound cleaning and sanitation plan and other standard operation procedures, simplify the implementation of a FWAS and strengthen it. In the following sections we focus primarily on adaptations regarding the first 4 principles of HACCP, that is, hazard and risk analysis, preventive measures, determination of CCPs, establishing TLs and CLs and monitoring procedures for CCPs. The remaining 3 principles will be presented briefly, as procedures for corrective actions, verification procedures and record-keeping systems are common practice in manufacturing industry and in the production of livestock.

Conduct a hazard analysis and risk assessment

A hazard in animal welfare risk assessment is a factor with potential to cause a negative animal welfare effect (EFSA 2008a). In this first stage, hazards that may be detrimental to fish welfare have to be determined and preventive measures identified, which can be applied to control these hazards. Contrary to the factors that affect welfare, a husbandry system or a factor can be beneficial for the animals. These benefits have to be taken into account in assessing the welfare. However, a model for risk–benefit has not been developed for fish by EFSA, and therefore, the existing model for risk assessment is used in this paper (Müller-Graf et al. 2008).

Hazards can be categorized as biotic, abiotic, managerial and environmental factors that may impair fish welfare. Interactions between these factors can also influence fish welfare. The factors depend on the fish species, life history stage, sex, physiological stage and the system (e.g. ponds, cages, flow-through or recirculation) used for production.

A non-fulfilled need has potential to cause a welfare effect. Studies and understanding of the needs of individual farmed fish regarding their preferences for housing (including water quality) and which behavioural expressions are important to them are presently very limited (EFSA 2009). A need is a requirement, which is part of the basic biology of an animal, to obtain a particular resource or respond to a particular environmental or bodily stimulus (Broom and Johnson 1993; Broom 2008). In general, knowledge about the needs of farmed fish is limited, and therefore, these data were not included in this study.

The hazard analysis must also consider factors which may be beyond the immediate control of the farmer, such as the process of stunning and killing at the slaughterhouse. Information on the methods used at various slaughterhouses may, however, influence, for example, the decision to which slaughterhouse the live fish will be transported.

In general, farming systems inevitably introduce a number of stressors to the organism, which may impair fish welfare if not properly managed. In this paper we use the definition of Wendelaar Bonga (2007) for stress and stressors. This author defined stress as a condition in which the dynamic equilibrium of an animal is threatened or disturbed as a result of the action of intrinsic or extrinsic stimuli, commonly defined as stressors. Potential stressors may include inappropriate water chemistry (e.g. NH_3, NO_2^-, NO_3^-, pH, dissolved oxygen, CO_2) water temperature, fish handling, physical damage to fish, diseases or disease treatments and inappropriate nutrition. It is impossible to completely avoid many of the procedures known to induce stress responses in fish farming; netting, crowding, grading and transport are integral components of farming routines, and at best, the farmer can minimize the effects of these stressors (Pickering 1993).

During the hazard analysis, the potential significance of each hazard should be assessed by considering its risk and severity. Risk is an estimate of the likely occurrence of a hazard.

Guidelines are available for risk assessments (RAs) to describe and quantify the hazard of introduction of infections in food. In the case of animal welfare, however, RAs cannot be performed in the same manner as is done in food safety issues (Müller-Graf et al. 2008). This is due to the lack of knowledge on the severity of impaired welfare for a fish and on the probability of exposure to a hazard during fish production (Müller-Graf et al. 2008). Nevertheless, there is a need for a method to rank hazards for various purposes, for example, to prioritize management measures at a fish farm. EFSA developed a semi-quantitative RA for ranking of hazards that may occur in aquaculture, as described below.

Experts scored four parameters to assess the importance of a hazard (EFSA 2008a):

1. the probability of a given target population to be exposed to a particular hazard has been scored as frequency of occurrence of the hazard [extremely low (1) to high (5)];
2. the proportion of the population affected [scored (1–20%) to 5 (81–100%)]
3. the consequences of exposure have been scored by severity of the effect in the individual [scored limited (1) to very severe (4)]
4. the duration of the effect (calculated as duration of effect of hazard in days/hazard causing an effect with the longest duration in days).

In addition, mortality was scored to provide more information on the consequences of a hazard. Score low (1) is given as mark when 1–20% of the population died and very high (5) in case of 81–100% mortality.

A linearity of the severity scores (e.g. 2 days of impaired welfare caused by a hazard with severity of 2 is equivalent to 1 day of impaired welfare due to a hazard with severity 4) was assumed, and no interaction between hazards that are mutually exclusive was considered. Each parameter is standardized by dividing it by the maximum score, with the overall score for each hazard being the product of the standardized scores obtained for the four parameters. In the EFSA reports on the major fish species farmed in Europe the risk scores for each life stage were ranked and tabulated by the uncertainty mortality scores. The uncertainty score is an indication of the quality of the information available. A single uncertainty score (low (1) for solid and complete data to max (3) for scarce or no data available) gives the overall uncertainty of all parameters.

Determine critical control points (CCPs)

A CCP is a point, step or procedure during on growing at which control can be applied, and as a result, deterioration of fish welfare can be prevented, eliminated or reduced to an acceptable level.

The identification of a CCP is facilitated by the application of a decision tree. The decision tree is a series of questions that the Fish Welfare Assurance team asks about each step in the process of on growing and associated hazards, and the control measures identified (see Fig. 1).

Application of the decision tree should be flexible, according to whether the operation is for production at farm, transport or slaughter.

If a hazard can be prevented, then the process at the farm needs to be adapted. In practice, various

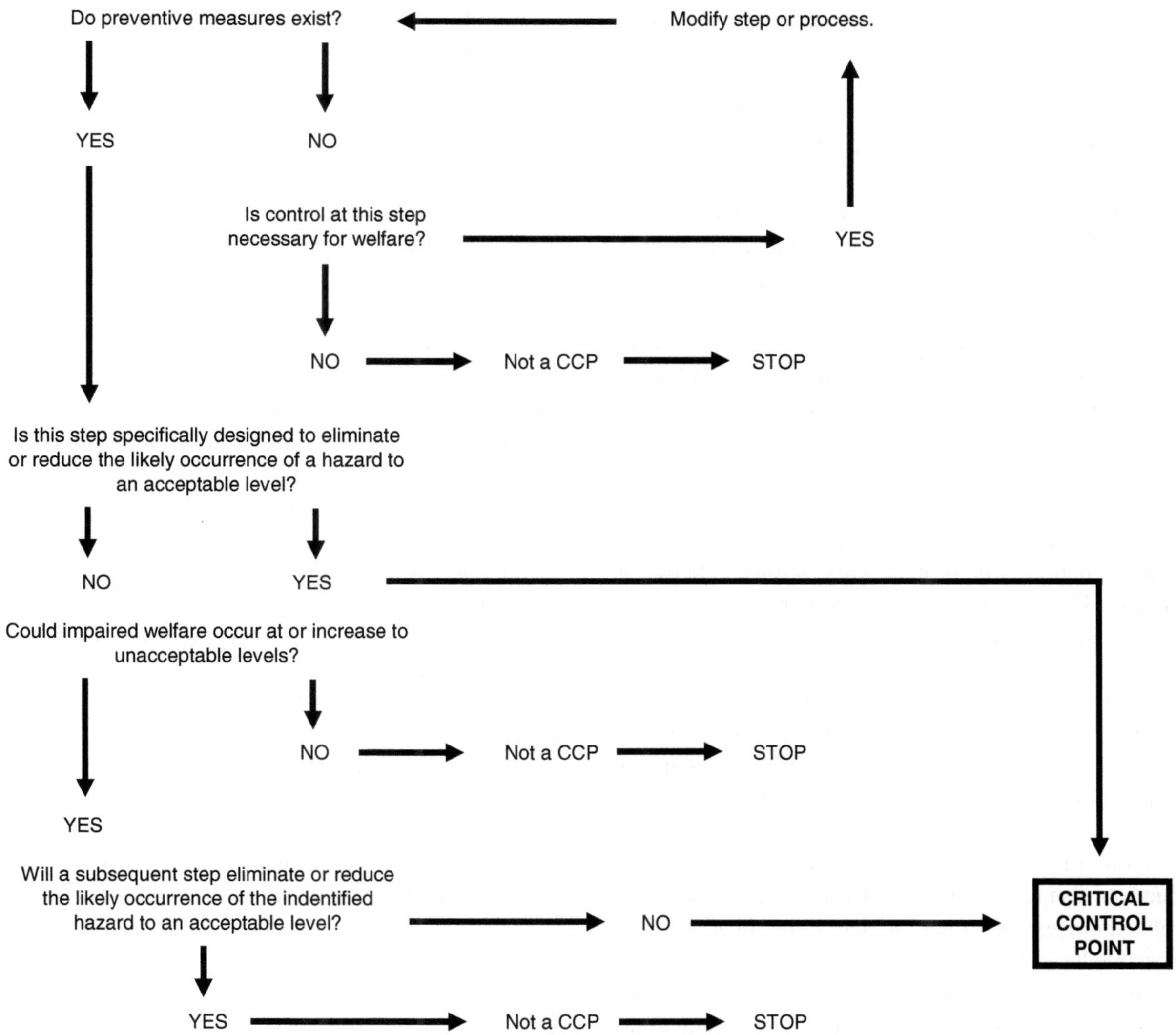

Fig. 1 Decision tree used in the determination of CCPs

preventive measures are identified by farmers and implemented. The hazard that is caused by a lack of oxygen, for example, can be prevented by aeration or oxygenation. Exposure of fish to predators can be avoided by having adequate antipredator control measures in place at a farm.

Selection of the site where a farm is to be established is a preventive measure. By careful selection, farmers avoid environmental conditions that result in low water quality due to, for example, poor water flow for fish that are grown in cages (EFSA 2008a).

Occurrence of diseases is a major hazard for fish welfare. However, absence of diseases does not necessarily imply good welfare, as is well known for terrestrial livestock. For health it is essential to focus on preventive measures. More detailed information about fish diseases is given by Segner et al. (2012) in this issue.

Establish target levels and critical limits for each CCP

A critical limit is the maximum or minimum value to which biotic, abiotic, managerial and environmental factors must be controlled at a critical control point to prevent, eliminate or reduce a hazard to an acceptable level.

For managerial factors, target levels are more appropriate than critical limits. A managerial factor is, for instance, that a fish like Atlantic salmon is removed from the water. Here, the target level is that salmon from the water is removed when this is really necessary, for instance to vaccinate them. For a fish species like common carp, which can tolerate low oxygen levels, thus for carp this target level can be less strict (Wedermeyer 1997).

Water quality is of paramount importance for welfare of farmed fish, as the animals are in intimate contact with the water through their gills and skin. Suboptimal water quality is a stressor and can affect physiology, behaviour, growth rate, feed intake and conversion and lead to pathological changes, damage to organs and mortality. (MacIntyre et al. 2008).

Fish have certain tolerance limits with regard to water quality, which may vary strongly between species. For welfare it is essential to have a profound understanding about water quality factors and which ranges will allow fish to change/adapt behaviour and physiology. The adaptive process for actively maintaining stability through change is known as the concept of allostasis (Korte et al. 2007). It is assumed that for welfare these ranges may be narrower than those reported in various studies on farming of A. salmon (EFSA 2008a). As little is known about the needs of fish under farming conditions, these welfare-related ranges are difficult to determine. In addition, a possible hazard to fish welfare originating from water quality relates not only to the absolute levels, but also to the rate of change and the number of interactions (EFSA 2008b).

Establish monitoring procedures at each CCP

A scheduled programme of measurements and assessments at the farm is essential for monitoring purposes in order to ensure that the process is under control at each critical control point.

For Fish Welfare Assurance it is crucial to monitor CCPs at a farm level. For this so-called Operational Welfare Indicators (OIWs) are required. In the Wellfish project (Cost Action 867) OWIs are defined as *measurements or assessments of welfare that can be made practically at commercial aquaculture facilities, as opposed to the biological measures of welfare that are often not practical or suitable for use at a commercial aquaculture facility.*

A preferred set of OWIs for a fish farm will depend on the variables such as farmed species, life stage and husbandry system and for open systems environmental conditions, such as temperature and photoperiod. OWIs that are associated with severe physical injury, severe infectious disease, or high mortality, can be direct indicators of failure in a major critical step/procedure (or several of these). For FWAS, OWIs that are predictive of the potential for subsequent severe deterioration of welfare are required.

With regard to the methods of prediction, ongoing data analysis can provide information that predicts the occurrence of a hazard at an early stage; a cumulative sum (CUSUM) control chart is such an example of a tool for the establishment of an early warning system. This method was originally developed by Page (1954) and plots the cumulative sum of the deviations from a target value using samples from all prior observations. CUSUM control charts are often used in quality management, to monitor ongoing processes. Baer et al. (2008) showed that this tool could predict increasing mortality rates at a turbot farm long before these rates exceeded a threshold level.

At a fish farm the following measurements/assessments are among those which may be used as OWIs: behavioural observations (including feed intake), physical injury, occurrence of visible infections, mortality and water quality data.

Establish corrective actions

It is necessary to establish the actions to be undertaken when monitoring indicates a deviation from an established critical limit. For corrective actions contingency planning needs to be taken into account.

Establish verification procedures

Set up a process to ensure that the FWAS works effectively and continues to work effectively through any changes made.

Establish a record-keeping system

The proposed approach requires that a fish farmer maintains certain documents, including the hazard analysis, a written quality assurance plan, and records documenting the monitoring of critical control points, critical limits, verification activities and the handling of deviations.

Implementation of FWAS

For the implementation of FWAS in aquaculture, we propose to follow the approach of the Codex Alimentarius for implementation of food safety assurance in the food industry. The Codex Alimentarius identifies 5 application steps that precede the 7 principles or stages of the HACCP concept; thus, a total of 12 stages are identified (Codex Alimentarius 1997).

The 5 preliminary stages are considered relevant for implementation of FWAS at a company level and have therefore been adapted and are given below.

Assemble a fish welfare assurance team

The company must put a team together which has the relevant knowledge and expertise for the fish species and technology used for on growing.

Describe the fish species concerned

Within the scope of quality assurance, this boils down to: a description of the fish species and what is known about current criteria regarding husbandry and the needs of the animal with respect to water quality, feeding regime and handling, detailed information regarding the technology used for on growing.

Identify intended use of the fish species produced

The intended use of the fish is relevant. For food fish, other criteria are used for water quality than in case of a zoo. In a zoo, fish have to be very visible for visitors. Therefore, the water in the fish tanks needs to be very clear, and the light intensity should be sufficiently high. For food fish, such strict criteria for husbandry are not used. Hence, the intended use of a fish species may be relevant with respect to their welfare.

Construct a flow diagram

The flow diagram, which should be constructed by the FWAS team, makes it easier to understand how all steps and conditions used in the farming process influence the welfare of the stock during grow out.

Conduct on-site verification of the flow diagram

Once a flow diagram has been produced, it needs to be checked for accuracy. In order to achieve this, the farms need to be visited during operating hours for verification and that all recorded details show what actually happens. It is essential to verify the latter as out-of-date operation manuals may be used or variation in work practices occur when another manager is in control.

FWAS to protect welfare of on growing of sea bass, carp and eel

As stated previously, welfare issues are species specific and depend on gender and life history and physiological status of the animals, environment and technologies used in aquaculture systems. In this section the initial stages of the set-up of a Fish Welfare Assurance System for on growing of fish are described. We chose sea bass, common carp and European eel, as these species differ with respect to biology and technology used for on growing.

Sea bass

Fish

Sea bass occur naturally in the Mediterranean Sea, the Black Sea and along the North Eastern Atlantic coasts, from Norway to Senegal. The fish species is found in coastal waters down to 100 m depth, but they commonly inhabit estuarine areas and coastal lagoons which they enter during summer and then return to the sea during the winter period (Tortonese 1984). Both juveniles and adults often overwinter in Mediterranean coastal lagoons. The species is a eurythermal and euryhaline, tolerating temperatures between 2 and 32°C and surviving in freshwater and high salinity waters. Young sea bass inhabit coastal waters, are gregarious and also form schools with other fish. Juveniles ingest mainly small crustaceans and small fish. Consumption of crabs and shrimps is predominant in adult fish, which are less gregarious than juveniles. Juveniles and adult fish change their feeding rhythms (diurnal and nocturnal) on a seasonal basis (EFSA 2008c; Sanchez-Vazquez et al. 1995, 1998).

It is known that sea bass is a more 'nervous' type of fish than instance carp (Flik et al. 2006; Varsamos et al. 2006). Hence, inadequate feed formulation and disturbance can elicit a stress response in sea bass, which can be stronger than, for instance, in carp.

On growing in sea cages

EFSA (2008c) estimated that the most common intensive growing system for sea bass in Europe is floating net cages. Market weight for sea bass is on average 350 g, but considerable size range can be observed.

Most cages consist of plastic-framed pens with a diameter of 10–30 m, which are circular or rectangular and can withstand strong wave action. For feeding and harvesting, mechanized systems are used.

The cages may be located in lagoons, sheltered bays or semi-exposed and offshore conditions (Basurco 2004). Originally, cages tended to be placed at well-protected, largely enclosed sites. Nowadays cage systems are located in somewhat more exposed sites where stronger and more resistant equipment is required to withstand heavy weather conditions, allowing for the exploitation of sea areas more distant from the coasts (FEAP 2010).

Applying FWAS for on growing of sea bass in sea cages

Handling issues, inadequate feed formulation and storage, disturbances and poor adaptation to feed distribution are the four most important hazards for on growing of sea bass in sea cages (EFSA 2008c). These hazards and their sources were identified in an elaborate study on welfare aspects of husbandry systems for sea bass (EFSA 2008c). In Table 1, an overview is presented on the application of FWAS for safeguarding and monitoring welfare of on growing sea bass, regarding the four hazards identified.

Handling (not according to best practice), which is the highest ranked hazard, was judged by EFSA (2008c) to occur with a high probability (scored 5), affecting the entire population (scored 5) for approximately 30 days with moderate severity (scored 2) and with a low uncertainty of the effect of the hazard (scored 1). This hazard can be prevented or minimized by instruction of personnel on handling of fish at the farm (see Table 1). The use of standard operation procedures (SOPs) for handling is a CCP. The target level is met when handling of the live fish is carried out as specified in the SOPs. Here, an OWI is monitoring that workers, indeed, carry out their task as specified in the SOPs. When necessary, remedial teaching of staff is carried out as a corrective action. Verification is carried out by inspection of operations or registration thereof. Documenting procedures used and data obtained is necessary for keeping records.

For monitoring and safeguarding the welfare of on growing sea bass regarding the three other hazards, the approach used in QA is similar to the one specified for the highly ranked hazard. Inadequate feed formulation and storage was ranked as the second hazard, because it affects a high proportion of the population, has a relatively high probability (3), severity score (3), but above all because of a high estimated duration (150 days). For the latter hazard the uncertainty of effect was high (scored 3). Disturbances of fish were judged inevitable (probability 5), affecting a high proportion of the population (4) with a limited severity (1) for 30 days and a low uncertainty (scored 1). Poor adaptation to feed distribution was ranked as the fourth hazard (probability 2), proportion of population 2, severity 3 and duration of 30 days with a medium uncertainty score (2). Other hazards that were identified, such as predation, were ranked lower.

Common carp

Fish

It is known that common carp naturally occurs in summer-warm lakes and slow flowing rivers. Carps are omnivorous fish that prefer a habitat where they can search for food organisms (EFSA 2008b; Hepher and Pruginin 1981). In spring they emerge from a state of virtual hibernation, during which they do not feed, to commence feeding at a rapidly increasing level. These animals grow best in waters where the temperature exceeds 20°C for at least 2 months of the year. Carps are eurythermal, as they can tolerate temperatures below 2°C during winter as well as short periods above 30°C at the height of summer. Water temperatures of approx. 30°C can lead to very rapid growth rates (EFSA 2008b). The fish are tolerant to low oxygen levels, except in the first few months of the life cycle of the species (EFSA 2008b).

Table 1 A first step to apply FWAS for on growing of sea bass in cages

Step	Hazard	Source	Preventive measures	CCPs	Critical limit/ target level	OWI			Corrective actions	Verification	Records
						How	When	Who			
On growing phase	Handling issues not in accordance with best practice	Personnel were not instructed. Standard operation procedures are not available	Instruct personnel on handling of fish	Use a standard operation procedure for handling of fish	Handling is carried out as described in standard operation procedure	Check that handling is performed as specified	At handling	Manager	Train staff	Inspection of operations, or validations of critical limits, or review recorded data	Document procedures used and data measured
On growing phase	Inadequate feed formulation and storage conditions	Commercial feeds are variable and may be deficient	Produce feeds of uniform quality with known formulation. Seasonal variation in fish needs to be taken into account. Use standard operation procedures for storage of feed	Purchase feed from a manufacturer that specifies the formulation of feed. Manufacturer provides instructions on storage conditions of feed	Purchased feed is accompanied with a certificate on the formulation Storage is in accordance with standard operation procedures	Certificate is available Procedure for storage is used	Upon arrival of feed. At storage	Manager	Select another feed producer Train staff	Inspection of operations, or validations of critical limits, or review recorded data	Document procedures used and data measured
On growing phase	Disturbances	Interaction human-fish, during ordinary farming procedures	Sound management to avoid unnecessary interaction human-fish	A detailed plan on handling of fish	Handle fish only when necessary	Check that tasks are carried out as specified	During interaction human-fish	Manager	Train staff	Inspection of operations, or validations of critical limits, or review recorded data	Document procedures used and data measured
		Presence of predators and boats	Place protective nets	Protective nets are present	Protective nets are adequate and maintained	Inspect presence and state of the nets	State is inspected regularly	Workers	Install new protective nets		
On growing phase	No adaptation to feed distribution mode	Fish adapts poorly to conditions in sea cages	Minimize stress caused by repeated handling of juveniles. At pre-on growing	Set up a standard operation procedure for handling of fish. At pre-on growing	Handling is carried out as described in standard operation procedure	Check that handling is performed as specified	At transfer to sea cages	Manager	Train staff	Inspection of operations, or validations of critical limits, or review recorded data	Document procedures used and data measured

For feeding and reproduction, carps do not usually migrate long distances. The species seem to prefer to move in small groups, searching for food or suitable spawning areas. (EFSA 2008b).

On growing in ponds

Common carp are cultured primarily in earthen ponds. The culture intensity ranges from extensive (low stocking densities, with no supplemental feeding or fertilization) to relatively intensive (high stocking density, feeding complete diets, mechanical aeration, management of water quality, including phytoplankton blooms) (EFSA 2008b).

Across the EU, carp ponds cover an estimated area over 200,000 ha. Due to climate, production of table-size carp (1.5–2.5 kg) in pond culture generally takes 3 years (EFSA 2008b). The fish are harvested at the end of each production season (autumn) and restocked after overwintering.

For purposes of the present paper, only a fraction of the production cycle is considered. In this paper we cover the on growing phase in extensive systems.

Applying FWAS for extensive on growing of carp in ponds

EFSA (2008b) identified hazards and their sources for the life stages of carp, as well as criteria for, for example, water quality. A risk assessment was performed to rank hazards.

Using the concept of FWAS for the four major hazards identified by EFSA, results for on growing of carp ponds are provided in Table 2. For on-grown fish kept in extensive conditions, the semi-quantitative risk assessment by EFSA (2008b) resulted in the following ranking of the first four major hazards: rapid change in water temperature [probability 3, proportion population affected 4, severity 3, estimated duration 393 days and low level of uncertainty of effect (1)]; too low water oxygen content [probability 2, proportion population affected 4, severity 4, estimated duration 393 days and low level of uncertainty of effect (1)]; a water flow that is too low [probability 1, proportion population affected 4, severity 4, estimated duration 393 days and medium level of uncertainty of effect (2)]; water pH too low or high [probability 3, proportion population affected 4, severity 3, estimated duration 393 days and low level of uncertainty of effect (1)].

Predation was the eighth highest ranked hazard (EFSA 2008b); however, in some farms, this is likely to be one of the most important welfare issues.

Eel

Fish

The European eel (*Anguilla anguilla*) occurs naturally all over Europe from the elver stage on. Eels are able to live in water with a temperature in the range of 3–30°C (Quéro 1997). European eel is not only caught in the wild but also farmed. The species is a catadromous, long lived, and the life cycle involves several metamorphic changes before the final adult spawning migration to the Sargasso Sea. Reproduction is a singular event (Tesch 2003).

Contrary to other fish species, the gills of the fish are completely covered with a skin sack with one small opening near to each pectoral fin. Once filled with water, these sacks function as breathing organs when an eel is outside the water. Due to this feature, an eel is able to remain out of water for a considerable time. The blood of eels is oxygenated not only by the gills, but also by the skin and the swim bladder.

On growing of eel in recirculating aquaculture systems

Recirculating aquaculture systems (RAS) are aquaculture production systems that involve reuse of culture water after internal purification. All designs of RAS remove waste solids, oxidize ammonia and nitrite/nitrogen, remove carbon dioxide, and aerate or oxygenate the water before returning it to the fish tank (Losordo et al. 1998). More intensively stocked systems or systems culturing sensitive species may require additional treatment processes such as fine solids removal, dissolved organics removal and water disinfection. Irrespective of their exact design, all RAS are mechanically sophisticated and biologically complex compared to "flow-through" tank or cage production systems. On growing of eels in RAS is performed at approx. 25°C and lasts for 18–24 months.

Table 2 A first step to apply FWAS for extensive on growing of common carp in ponds

Step	Hazard	Source	Preventive measures	CCPs	Critical limit/ Target level	OWI			Corrective actions	Verification	Records
						How	When	Who			
On growing phase	Rapid water temperature change	Volume and depth of ponds are too small and this results in water T fluctuations due to changes in ambient conditions (day vs. night)	Keep fish in ponds with greater volume and depth	Specify the volume and depth of a pond in relation to fluctuations in ambient temperatures	Rapid temperature changes should be lower than 3°C/h	Monitor that fish are kept in ponds with depth and volume as specified	Prior to start of sorting and grading	Workers	Transfer fish to ponds with depth and volume as specified	Inspection of operations, or validations of critical limits, or review recorded data	Document procedures used and data measured
On growing phase	Water oxygen content too low	Can be associated with collapse algal bloom. Freezing of water surface. High temperatures in August and September	Aeration and increase inflow from a clean source of water	Surface of the water and oxygen level in the water and its temperature	20–30% oxygen level	Inspect for gasping fish	Regularly during on growing	Workers	Aeration and replace the water or a part thereof by water from a clean natural flow or another pond	Inspection of operations, or validations of critical limits, or review recorded data	Document procedures used and data measured
		Handling disturbs bottom deposits causing anoxia in the water	Instruct personnel on handling	Use a standard operation procedure	Handling is carried out as described in standard operation procedures	Check that handling is performed as specified	Regularly during on growing	Workers	Train staff	Inspection of operations, or validations of critical limits, or review recorded data	Document procedures used and data measured
On growing phase	Water flow too low	Inadequate design of inflow or blocked inflow	Install an adequately designed inflow	Inflow meets specifications	Specifications related to volume and depth of pond	Inspect specifications installed inflow	Inspection inflow regularly for presence of blockage	Manager	Install newly designed inflow or clean blocked inflow	Inspection of operations, or validations of critical limits, or review recorded data	Document procedures used and data measured

Table 2 continued

Step	Hazard	Source	Preventive measures	CCPs	Critical limit/ Target level	OWI			Corrective actions	Verification	Records
						How	When	Who			
On growing phase	Water pH too low or too high. Too high pH in relation to ammonia	High discharge of snow smelt water lead to a low pH. Too high pH due to algal bloom	Increase inflow of water with pH within critical limits	Visual inspection of environment and pH analysis of water in ponds	Below 5.2 mortality pH Above 10.3 lethal. Optimum pH between 7 and 8	Measure pH and T	On regular basis when the discharge may occur	Workers	Replace the water or a part thereof by water from a clean natural flow or another pond	Inspection of operations, or validations of critical limits, or review recorded data	Document procedures used and data measured
		Excessive amount of soft sediment may lead to high ammonia levels	When pond is drained sediment needs to be removed	Use a standard operation procedure for cleaning of a pond	After draining pond is cleaned as specified	Monitor cleaning procedure	Prior and during cleaning	Management	Train staff	Review recorded data	Document procedures used and data measured

Applying FWAS for on growing of eels in recirculating aquaculture systems

The parasite *Pseudodactylogyrus* sp. forms a problem during on growing, but is less severe compared to its infection effects at the juvenile stage. Nevertheless, EFSA (2008d) ranked it as the most important hazard for on growers (probability 5, affecting the entire population (5) for maximal 400 days with a moderate severity (2) and a low uncertainty of effect (1). For this hazard and the three other ones presented here, an overview of the application of FWAS is given in Table 3. In this table we describe OWIs for monitoring CCPs to minimize or avoid the four major hazards identified by EFSA (2008d).

The second major hazard is the herpes virus disease (probability 3, affecting the entire population (5) for maximal 42 days with a severe effect (3) and a high uncertainty of effect (3), as the glass eels that survived vaccination with the virus are more or less protected from further infection (EFSA 2008d). The herpes virus is present on all farms; however, disease occurs on only about 50% of farms (where exposure of juveniles did not result in a sufficiently high level of immunity). When outbreaks occur, most eels are affected, and mortalities of 21–40% can occur. The disease is worsened by poor environmental conditions (EFSA 2008d).

Handling and vibriosis are the third and fourth hazard, respectively. Handling may affect the entire population (score 5) with a high probability (5) throughout the entire grow out phase (400 days) with a severe effect (3). As little is known about the effects of handling on welfare of on growing eel, the uncertainty is high (3) and mortality is in the range of 1–20%. Contrary to handling, for vibriosis, the probability of occurrence of the hazard is low (1), the severity is low (1), the uncertainty of effect is low (1), and mortalities of 41–60% can occur.

Conclusions

The FWAS presented in this paper has strengths in its flexibility. QA Systems can be easily adapted, when deemed necessary. Regarding the implementation of FWAS, we expect that this is practicable, as in various manufacturing industries where QA systems have been incorporated.

Table 3 A first step to apply FWAS for on growing of European eel in recirculating aquaculture systems

Step	Hazard	Source	Preventive measures	CCPs	Critical limit/ Target level	OWI			Corrective actions	Verification	Records
						How	When	Who			
On growing	External parasites-Dactylogyrus	Water in system, other infected fish	Adequate hygiene to prevent spreading and treatment with formaldehyde	Use a standard operation procedure for maintaining hygiene and treatment with formaldehyde	Absence of dactolygysus in fish or its presence reduced to a level that average growth is not affected	Visual Inspect for gill and skin lesions. Observe feed intake and behaviour	Regularly Observe fish during feeding and during sorting	Workers	Treatment with mebendazole and disinfection of tanks, pipes, pumps and filter system. Consult a veterinarian	Inspection of operations, and inspection of fish during feeding and sorting	Document procedures used and data measured
On growing	Herpes virus disease	Water in system, other infected fish	Vaccination with infected fingerlings from preceding year Minimize stress during on growing	Use a standard operation procedure for vaccination and hygiene at the farm	Morbidity should not exceed 10%	Visual Inspect fish for haemorrhages around head, gill cover and pectoral fin	Regularly Observe fish during feeding and during sorting	Workers	In case of an outbreak lower water T to 20°C to slow down spreading of the virus and allow eel to improve immune system. Reduce stress, as this may lead to a morbidity of 50%	Inspection of operations, and inspection of fish during feeding and sorting	Document procedures used and data measured
On growing	Handling	Interaction fish with humans	Supply oxygen to water during crowding Pump fish instead of netting them or use knotless nets	Use a standard operation procedure for grading	Absence of or minimal stress, skin damage and demucinization	Inspect fish for lesions and deviations in behaviour	Inspection after handling	Workers	Train staff	Inspection of operations, or validations of critical limits, or review recorded data	Document procedures used and data measured
On growing	Vibrio	Outbreak is associated with poor husbandry conditions or grading	Standard operation procedures for adequate husbandry and conditions for grading	Use a standard operation procedure for husbandry and grading	Personnel are given instructions on the use of the standard operation procedures	Inspect fish during and after grading for lesions	Regularly	Workers	Treat fish with antibiotics Instruct personnel on use of standard operation procedures	Inspection of operations, validations of critical or limits, or review recorded data	Document procedures used and data measured

Regarding the international trade of farmed fish products, we foresee that standardization of the FWAS concept may be required in the near future. In addition, issues such as management responsibility and compatibility between FWAS and other quality systems commonly used have to be taken into account.

At present, standardization of the FWAS concept may not be achieved easily, as fish welfare is a multisided concept, there is a lack of standardized methods to assess welfare objectively, detailed studies and profound understandings of the needs of farmed fish are absent and criteria/target levels corresponding to needs are not available (EFSA 2009). For farmed warm-blooded animals the level of understanding is similar to that regarding fish. Nevertheless, within the EU Integrated Project Welfare Quality (Anonymous 2010), a consensus on protocols to assess welfare of warm-blooded animals at the farm level was achieved. The established protocols show that, despite gaps in knowledge on animal welfare of warm-blooded animals, further standardization is also possible for aquaculture.

When harmonization regarding fish welfare assurance in aquaculture is achieved, a new International Standard may be designed, similar to ISO 22000 which was published on 1 September 2005, as an Internal Standard for food safety management to ensure safe food supply chains worldwide (Frost 2005).

Acknowledgments This study was partly financed by the Dutch Ministry of Agriculture, Nature and Food Quality. The authors would like to thank Dr. Sunil Kadri of Aquaculture Innovation, Glasgow, Scotland fur useful comments on the manuscript.

References

Anonymous (2010) Website www.welfarequality.net, visited in April 2010

Ashley PJ (2007) Fish welfare: current issues in aquaculture. Appl Anim Behav Sci 104:199–235

Baer A, Schulz C, Krieter J (2008) Long time survey of mortality rates and growth performance of tubot (Psetta maxima) depending on environmental conditions in recirculation systems. In: Proceedings aquaculture Europe 2008, Krakow Poland, September 15–18, 2008, pp 50–51

Basurco B (2004) Introduction to Mediterranean aquaculture. Options Méditerr Ser B 49:9–13

Braithwaite VA, Huntingford FA (2004) Fish and welfare: do fish have the capacity for pain perception and suffering? Anim Welf 13:S87–S92

Brambell FWR (1965) Report of the technical committee to enquire into the welfare of animals kept under intensive husbandry systems. Command report 2836, Her Majesty's Stationary Office, London, 85 pp

Broom DM (1991) Animal welfare- concepts and measurement. J Anim Sci 69:4167–4175

Broom DM (2008) Welfare assessment and relevant ethical decisions: key concepts. Annu Rev Biomed Sci 10:T79–T90

Broom DM, Johnson KG (1993) Stress and animal welfare. Kluwer/Springer, Dordrecht (reprinted with corrections 2000)

Codex Committee on Food Hygiene (1993) Codex guidelines for the application of the hazard analysis critical control point (HACCP) system: adopted by the 20th session of the joint FAO/WHO Codex Alimentarius Commission. Food Control 5:210–213

Codex Committee on Food Hygiene (1997) HACCP System and guidelines for its application. Food Hygiene Basic Texts, Food and Agriculture Organisation of the United Nations, World Health Organization, Rome, pp 33–45

Cottee SY (2012) Are fish the victims of 'speciesism'? A discussion about fear, pain and animal consciousness. Fish Physiol Biochem. doi:10.1007/s10695-010-9449-9

EC 710/2009: Commission Regulation EC, 710/2009 of 5 August 2009 amending Regulation (EC) No 889/2008 laying down detailed rules for the implementation of Council Regulation (EC) No 834/2007, as regards laying down detailed rules on organic aquaculture animal and seaweed production. Official Journal of the European Union L 204/15

EC Directive No 952/2004: Regulation (EC) No 852/2004 of the European parliament and of the council of 29 April 2004 on the hygiene of foodstuff. Official Journal of the European Union L 139/1 and L 226/3

EFSA (2008a) Scientific report on animal welfare aspects of husbandry systems for farmed Atlantic salmon (Question No EFSA-Q-2006–033). Annex I EFSA J 736:1–122

EFSA (2008b) Scientific report of EFSA on animal welfare aspects of husbandry systems for farmed common carp (Question No EFSA-Q-2006–148). Annex I EFSA J 843: 1–81

EFSA (2008c) Scientific report of EFSA on Animal welfare aspects of husbandry systems for farmed European sea bass and Gilthead sea bream (Question No EFSA-Q-2006–149). Annex I EFSA J 844:1–89

EFSA (2008d) Scientific report of EFSA on Animal welfare aspects of husbandry systems for farmed European eel (Question No EFSA-Q-2006–149). Annex I EFSA J 809: 1–48

EFSA (2009) Scientific opinion of the panel on animal health and welfare on request from the European Commission on General Approach to fish welfare and to the concept of sentience in fish. Annex I EFSA J 954:1–26

FAO (2007) A qualitative assessment of standards and certification schemes applicable to aquaculture in the Asia-Pacific region. Asia-Pacific Fishery Commission, Bangkok

FEAP (2010) Website www.feap.info. Last visited March 10, 2010

Flik G, Klaren PHM, Van den Burg EH, Metz JR, Huising MO (2006) CRF and stress in fish. Gen Comp Endocrinol 146:36–44

Frost R (2005) ISO 22000 is the first in family of food safety management system standards. ISO Management Systems, November–December, 16–19. http://www.iso.org/iso/iso-22000_food_safety.pdf

Hepher B, Pruginin Y (1981) Commercial fish farming. Wiley, New York, USA

Korte SM, Olivier B, Koolhaas JM (2007) A new animal welfare concept based on allostasis. Physiol Behav 92:422–428

Losordo TM, Masser MP, RakocyJ (1998) Recirculating aquaculture tank production systems: an overview of critical considerations. SRAC Publication No. 451

Lund V, Mejdell CM, Röcklinsberg H, Anthony R, Håstein T (2007) Expanding the moral circle: farmed fish as objects of moral concern. Dis Aquat Org 75:109–118

MacIntyre C, Ellis T, North BP, Turnbull JF (2008) The influences of water quality on the welfare of farmed trout: a review. In: Branson E (ed) Fish welfare. Blackwells, London, pp 150–178

Müller-Graf C, Candiani D, Barbieri S, Ribó O, Afonso A, Aiassa E, Have P, Correia S, De Massis F, Grudnik T, Serratosa J (2008) Risk assessment in animal welfare: EFSA's approach. AATEX 14(Special issue):789–794

Page ES (1954) Continuous inspection schemes. Biometrika 41:100–114

Pickering AD (1993) Husbandry and stress. In: Muir JF, Roberts RJ (eds) Recent advances in aquaculture IV. Blackwell Scientific Publications, Oxford, pp 155–169

Quéro J-C (1997) Les poissons de mer des peches françaises. Delachaux et Niestlé SA, Paris, pp 245–246

Quinn BP, Marriott NG (2002) HACCP plan development and assessment: a review. J Muscle Foods 13:313–330

RSPCA (2010) RSPCA welfare standards for farmed salmon. RSPCA, Horsham, UK. http://content.www.rspca.org.uk/cmsprd/Satellite?blobcol=urldata&blobheader=application%2Fpdf&blobkey=id&blobnocache=false&blobtable=MungoBlobs&blobwhere=1232991903901&ssbinary=true

Sanchez-Vazquez FJ, Madrid JA, Zamora S (1995) Circadian rhythms of feeding activity in sea bass, *Dicentrarchus labrax* L.: dual phasing capacity of diel demand-feeding pattern. J Biol Rhythm 10:256–266

Sánchez-vázquez FJ, Azzaydi M, Martínez FJ, Zamora S, Madrid JA (1998) Annual rhythms of demand-feeding activity in sea bass: evidence of a seasonal phase inversion of the diel feeding pattern. Chronobiol Int 15:607–622

Segner H, Sundh H, Buchmann K, Douxfils J, Sundell KS, Mathieu C, Ruane N, Jutfelt F, Toften H, Vaughan L (2012) Health of farmed fish: its relation to fish welfare and its utility as welfare indicator. Fish Physiol Biochem. doi: 10.1007/s10695-011-9517-9

Snyder OP (1991) HACCP in the retail food industry. Dairy Food Environ Sanit 11:73–81

Sperber WH (2005) HACCP does not work from farm to table. Food Control 16:511–514

Tesch FW (2003) The eel. Blackwell Science, Oxford

Tortonese E (1984) Serranidae. In: Whitehead PJP, Bauchot ML, Hureau JC, Nielsen J, Tortonese E (eds) Fishes of the North-Eastern Atlantic and the mediterranean. UNESCO, Paris, pp 793–795

Varsamos S, Xuereb B, Connes T, Flik G, Spanings-Pierrot C (2006) Pituitary hormone mRNA expression in European sea bass *Dicentrarchus labrax* in seawater and following acclimation to fresh water. J Endocrinol 191:473–480

Vessier I, Butterworth A, Bock B, Roe E (2008) European approaches to ensure good animal welfare. Appl Anim Behav Sci 113:279–297

Von Borell E, Bockisch F-J, Büscher W, Hoy S, Krieter J, Müller C, Parvizi N, Richter T, Rudovsky A, Sundrum A, Van den Weghe H (2001) Critical control points for on-farm assessment of pigs housing. Livest Prod Sci 72:177–184

Wedermeyer GA (1997) Effects of rearing conditions on the health and physiological quality of fish in intensive culture. In: Iwama GK, Pickering AD, Sumpter JP, Schreck CB (eds) Fish stress and health in aquaculture. Society for experiment biology, seminar series 62. Cambridge University Press, Cambridge, pp 35–71

Wendelaar Bonga SE (2007) The stress response in fish. Physiol Rev 77:592–625

Effect of Nutritional Status and Sampling Intensity on Recovery after Dorsal Aorta Cannulation in Free-Swimming Atlantic Salmon (*Salmo salar* L.) – Summary

Brankica Djordjevic, Trond Kristensen, Øyind Øverli, Bjørn Rosseland and Anders Kiessling

We describe how repeated blood sampling from non-stressed and normally eating fish can be performed. We show that by using a permanent cannula, which is a standard technique in humans, this is feasible. In fish, living in water, this procedure includes additional challenges compared to man, such as creating an aseptic environment in spite of the aquatic environment.The holding environment is also important with regard to normal behaviour and feed intake. This paper also includes this aspect.
Analysis of blood plasma revealed marked differences between baseline level of cannulated fish and uncannulated control fish, indicating that cannulation can be a preferred method to obtain representative resting values in fish.

Background

To study cause and effect between a specific factor and a specific effect, controlled experiments are a prerequisite. In practice, for instance at a fish farm, husbandry conditions are complex and may include , too many unknown variables, to allow the distinction "what causes what". For an experimental setup in research control of conditions is essential. This is especially true when studies demand invasive procedures as repeated blood sampling, as the procedure itself may cause stress. In humans these problems are known and can, at least in part, be counteracted by the experimenter explaining the procedure to the person exposed to sampling. This is naturally no option for fish. Researchers need other ways to ensure working with fish in a rested state and thereby more representative animals. It is well known that not only the capture (netting) of the fish is stressful, but the anaesthetic procedure used in fish will in itself cause stress, possibly affecting the outcome of the analysis. Furthermore, non-aseptic conditions when performing repeated sampling may cause secondary inflammations with pain and stress as result, reducing the precision of the experiment. Moreover, these inflammations lead to poor welfare.

Scientific Significance

It is important to note that in a well-designed experiment baseline levels of substances in blood plasma of fish of a control group should indicate a rested state. A sound experimental setup is crucial to meet high scientific standards for experimental research and to comply with ethical standards. Such a setup is a challenge when fish are subjected to repeated blood sampling. Hence, we addressed this challenge in our paper and showed that for repeated blood sampling cannulation of fish could be a preferred approach.

Practical Application

Whenever live fish are used for experimental purposes the three R's (Replacement, Reduction and Refinement) rule applies. Refinement refers to methods that minimize potential discomfort and thereby enhance animal welfare. Our paper contributes to Refinement by providing evidence that cannulation may be the preferred method to obtain representative resting values in fish that are subjected to repeated blood sampling.

Contact

A Kiessling
Department of Wildlife, Fish and Environmental Studies,
Swedish University of Agricultural Sciences,
901 83 Umeå, Sweden
e-mail: Anders.kiessling@slu.se

Fish Physiol Biochem (2012) 38:259–272
DOI 10.1007/s10695-009-9362-2

Effect of nutritional status and sampling intensity on recovery after dorsal aorta cannulation in free-swimming Atlantic salmon (*Salmo salar* L.)

B. Djordjevic · T. Kristensen · Ø. Øverli · B. O. Rosseland · A. Kiessling

Received: 30 April 2009 / Accepted: 30 September 2009 / Published online: 25 October 2009

Abstract Recovery from implantation of a cannula in the dorsal aorta (DA) of Atlantic salmon (*Salmo salar*) was studied in relation to nutritional status and sampling intensity. The incentive for the study was the inconsistency between published reports and our own experience of recovery and longevity of fish exposed to this protocol. In two studies using starved fish, blood (0.3 ml) was sampled 0, 1 and 24 h after DA-cannulation, and thereafter at 48 and 72 h and thereafter once weekly for four weeks. In a third study using fed fish, four consecutive samples (0, 3, 6 and 12 h after a meal) were obtained twice a week over a four-week period. All fish displayed a sharp increase in pCO_2 and haematocrit (Hct) during surgery, followed by a marked raise in cortisol, glucose, sodium and potassium (1 h). pCO_2, pH and Hct approached baseline levels as early as the 1 h post-surgery sample, while this was not the case for cortisol and electrolytes before the 24 h post-surgery sample. Glucose did not display any significant changes post surgery. From then on, all variables displayed minor but non-significant ($P > 0.05$) changes indicating a steady state close to baseline values for unstressed fish. This pattern was independent of sampling procedure, i.e. repeated single or multiple samples and thus volume of blood removed. Nutritional status (fed vs. starved) did not affect post-surgical recovery pattern. Only K^+ and Hct displayed consistent and significant post-prandial patterns. We found marked differences between baseline level of cannulated fish and uncannulated control fish, in pH, K^+ and Hct indicating that cannulation may be the preferred method to obtain representative resting values in fish.

Keywords Cortisol · Anaesthesia · Haematocrit · Glucose · pCO_2 · Tank design

B. Djordjevic · Ø. Øverli · A. Kiessling (✉)
Department of Animal and Aquacultural Sciences, Norwegian University of Life Sciences, P.O. Box 5003, 1432 Aas, Norway
e-mail: anders.kiessling@umb.no

T. Kristensen
The Norwegian Institute for Water Research, 0411 Oslo, Norway

T. Kristensen · B. O. Rosseland
Department of Ecology and Natural Resource Management, Norwegian University of Life Sciences, P.O. Box 5003, 1432 Aas, Norway

B. Djordjevic
Aquaculture Protein Centre (APC), P.O. Box 5003, 1432 Aas, Norway

A. Kiessling
Department of Wildlife, Fish and Environmental Studies, Swedish University of Agricultural Sciences, 901 83 Umeå, Sweden

Introduction

Dorsal aorta (DA) cannulation has frequently been utilized as a tool in whole-animal models for physiological observations in teleost fish. The technique allows repeated blood sampling from the same fish compared to the alternative of serial slaughter and was first described by Smith and Bell (1964). Most studies that have utilized this method have focused on short-term physiological measurements such as blood volume (Conte et al. 1963), evaluation of haematological parameters (Soivio et al. 1975), endocrinology (Zohar 1980), post-prandial levels of plasma amino acids (Ok et al. 2001), degree and time course of changes in catecholamine and cortisol concentrations in cannulated black box-confined rainbow trout (Gamperl et al. 1994a) and immunological measures (Lo et al. 2003). In a series of experiment, Kiessling et al. introduced long-term studies of free swimming and voluntarily feeding cannulated fish (Eliason et al. 2007, 2009; Hamre et al. 2001; Karlsson et al. 2006; Kiessling et al. 1995, 2003, 2006, 2009; Olsen et al. 2005a; Sunde et al. 2003)

The DA cannulation method involves several procedures that are known to stress the fish, such as handling, sedation and surgery, in addition to isolating the fish in individual tanks. The presence of the catheter and the sampling protocol may also induce stress directly or due to loss of blood. There are inconsistencies in the literature regarding recovery time from the stress of cannulation surgery and regarding pre- and post-surgery cortisol levels. Gamperl et al. (1994a) showed that cortisol levels in cannulated rainbow trout took 6 days to return to control values, while Zohar (1980) reported that only 3 days after cannulation are needed for the rainbow trout to regain its homeostatic balance. In species other than salmonids, recovery time after DA cannulation has been shown to be even longer. Lo et al. (2003) reported that the grouper (*Epinephelus malabaricus*) required 9 days for plasma cortisol to return to its normal range, while Mazik et al. (1994) showed that channel catfish (*Ictalurus punctatus*) needed 7 days. Much of the difference may be a result of inter-species differences, seasonal variation, stage of development or environmental conditions, but also to the dissimilarities in the experimental set-up, including sampling, anaesthetic-, aseptic- and surgical protocols. Information is also scarce regarding the effects of nutritional status and sampling intensity on the recovery and longevity of homeostasis in the animal.

Therefore, even if the DA cannulation technique has hitherto been used with great success compared to serial slaughter, it needs to be more thoroughly evaluated, particularly if it is used for complex long-term studies involving nutritional, pharmacological, toxicological, physiological or other metabolic variables. This is in order to maximize the welfare of the animal, to be able to distinguish between treatment and experimentally induced effects, and to minimize the latter.

DA-cannulated fish are often confined in a square box or pipe of almost the same size as the fish itself. The rationale for this is to allow the researcher easy access to the cannula during sampling (reviewed in Axelsson and Fritsche 1994). Such tank/box/pipe systems make nutritional studies ambiguous, as the fish do not display normal feeding behaviour and growth under such confined conditions.

The experimental set-up, developed for long-term studies of voluntary feeding and free-swimming salmonids, as a result of the series of experiments by Kiessling et al. described previously, includes tank design and specific protocols for aseptic surgery, anaesthesia, analgesics and blood sampling (Kiessling et al. 2003). The fish kept according to this protocol display normal free swimming and voluntary feeding behaviour fairly similar to that of fish kept in production-like settings (Kiessling et al. 1995).

The main aim of this study was to present systematically obtained blood variables indicative of physiological balance and stress during short- and long-time recovery in free-swimming Atlantic salmon subjected to the previous protocol and to compare these values with those already discussed in the literature. The longevity of the method was also evaluated against the nutritional status of the fish (fed or starved) and sampling procedure (repeated single or post-prandial multiple samplings). Differences in physiological responses were measured and presented as changes in circulating plasma cortisol, blood pH, haematocrit (Hct), partial pressures of carbon dioxide (pCO_2), electrolytes (Na^+, K^+, Cl^-) and glucose in whole blood.

Materials and methods

Fish and experimental facilities

This study was carried out at the Norwegian Institute for Water Research (NIVA) research station at

Solbergstrand (Akershus County, Norway). Three experiments were performed, two in early and late spring (March and May) and one during the winter (January). The experiments were denoted NF1 (Not Fed 1), NF2 (Not Fed 2) and F1 (Fed 1), respectively. The time difference between experiments occured due to the limited number of tanks, resulting in experiments being carried out with 2 months intervals. Photoperiod and water temperature variations were kept to a minimum, in order to minimize any confounding seasonal effects (see below).

Atlantic salmon (*Salmo salar* L.) ranging in weight from 800 to 1,000 g (NF1, March experiment, $n = 8$; NF2, May experiment, $n = 8$; F1, January experiment, $n = 8$) were kept prior to cannulation in a round 4-m^3 holding tank supplied with seawater from a depth of 60 m (average for NF1: Temperature 6.5°C, salinity 34.5 g l^{-1}; NF2: 6.7°C, 34.7 g l^{-1}, F1: 9.8°C, 34.7 g l^{-1}). A 12 h light/12 h dark regime (lights on at 08.00 am) was used both before and during the experiments. Fish were fed a 1% body mass ration of a standard commercial diet 5 days a week (Optiline V 2000—50 A 9.0 mm pellets, Skretting, Norway). One week before surgery, the fish were moved from the holding tank to experimental isolation tanks containing 0.3 m^3 water of the same quality as that in the holding tank (water exchange rate 2 l min^{-1}; see Fig. 1 for tank dimensions). Tanks were physically separated from each other and placed on vibration-absorbent material, so that the experimenter could move between tanks and blood samples be taken without alerting the fish. Each tank was equipped with its own light (halogen), water supply and an aquarium pump (50 Hz, 375 l h^{-1}), creating a water current of approximately 10 cm s^{-1} along the perimeter of the tank (Fig. 1), independent of incoming water flow. Inside the tank, a PVC cover (49 cm× 36 cm) was placed 4 cm above the surface, above the water outlet. The cover served the purpose of providing shade, and the combination of shade and current kept the fish voluntarily in a position that permitted undisturbed sampling (Fig. 1). The lights were attached to one corner of the tank in order to create a reflection that blocked the view of the rest of the room and thereby the experimenter. All other possible disturbances in the tank room were minimized.

The NF1 and NF2 fish were starved throughout the experiment. The F1 fish were fed for at least 1 week to resume normal feeding behaviour after transfer to rearing in isolation and before DA cannulation. Fish not resuming normal feeding within 7 days of acclimatization were replaced.

A control group (C1) was also sampled for plasma cortisol and blood variables. These fish ($n = 8$) were transferred and kept in the experimental tanks for 7 days. They were sampled by the "grab and stab" method (i.e. netted as fast as possible, stunned by

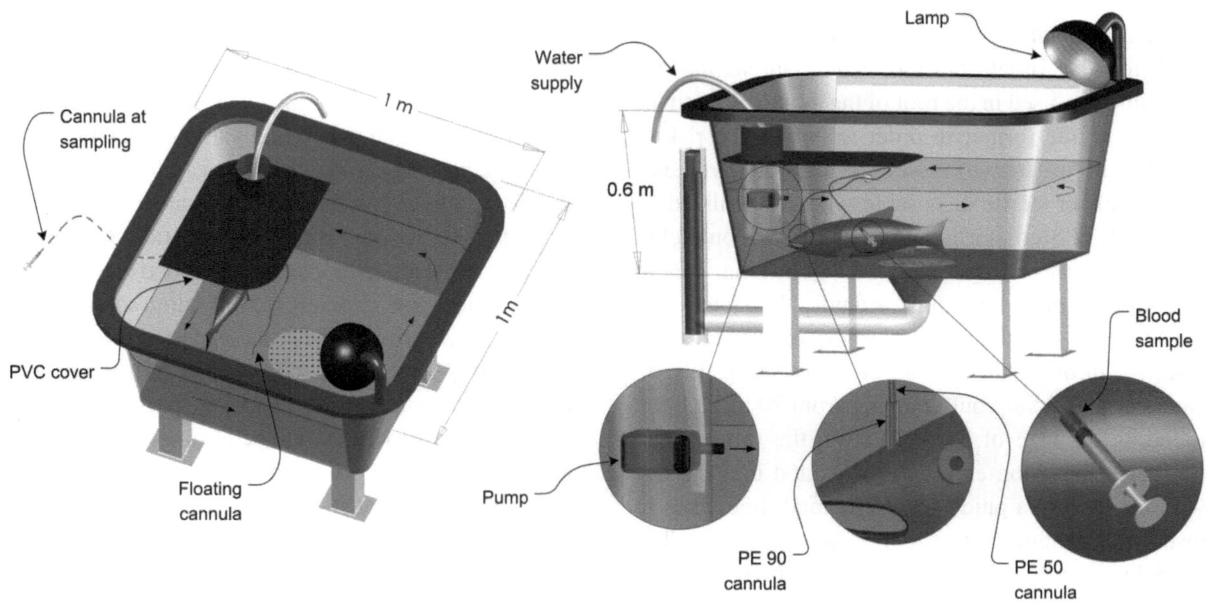

Fig. 1 Tank set-up for the cannulation studies

percussion and sampled by aspiration from the caudal vein/artery vessels).

Cannulation procedure

All fish were sedated, anaesthetized and cannulated following the procedure of Kreiberg and Powell (1991) and Soivio et al. (1975), respectively, with modifications described by Kiessling et al. (1995, 2003) using 0.5 mg kg^{-1} metomidate (metomidate hydrochloride; Syndell Ltd, Victoria, B.C. Canada) as pre-anaesthetic sedation administered directly in the experimental tank, followed by 60 mg l^{-1} Metacanium (MS 222, tricaine methane sulphonate; Norwegian Medical Depot, Bergen, Norway) anaesthetic in a separate bath. This bath was equipped with an aquarium pump connected to a silicone hose that was kept in the mouth of the fish in order to ventilate the gills during cessation of respiratory movements and also ensure as similar exposure to the anaesthetic as possible between individual fish. Cessation of the coughing reflex was used as a criterion for surgical anaesthesia. The fish was then transferred to a V-shaped surgical table protected with a soft wet cloth. During surgery, the fish were also covered by wet cloths, and aerated seawater with 40 mg l^{-1} MS 222 was constantly pumped over the gills except for 20 s during the actual incisions.

The sites of surgical incisions (roof of the mouth and in the snout) were identified and injected with 0.1 ml lidocaine (Xylocain®, 20 mg ml^{-1}, Astra, Södertälje, Sweden). Lidocaine with adrenaline (5 μg ml^{-1}) was used for the snout incision, while adrenaline-free lidocaine was used in the roof of the mouth close to the DA-vessel, the former in order to reduce the risk of bleeding in case a small vessel was hit. All instruments were of either sterile disposable type, or sterilized by heat using a Steri sterilizer (Steri 250, Simon Keller AG, Burgdorf, Switzerland). A 4-cm-long PE 90 cannula (Intramedic, Clay Adams, Parsipan, NJ, USA) with a circular stopper on the ventral side was inserted through the snout. A 70-cm-long PE 50 cannula, with a stop bulb ranging from 70 to 100 mm (pending the size of the fish) from the thinned and sharpened insertion end, was then fitted into the DA with the help of a guide wire. The total time taken to insert the cannula into the DA was 4–5 min. The cannula was then filled with 150 IU heparin (heparin aqua solution, injection quality; Nycomed Pharma Ltd, Oslo, Norway) in an isotonic sodium chloride solution (injection quality; Fresenius Kabi, Uppsala, Sweden), before being sealed by heat. Mucus from an untreated area was gently spread over all affected skin area before the fish was returned to the tank. Recovery was monitored until normal breathing and swimming behaviour had resumed, normally within 0.5–2 min and 10–15 min, respectively.

Sampling procedure and physiological analysis

Blood (0.3 ml) from the eight individual fish per repeat was sampled from DA immediately post-surgery (0 h) and then after 1, 24, 48, 72, 150, 300, 420 and 528 h (h) in NF1. In NF2, the sample intervals were the same, but ended at 420 h post-surgery. In the F1 experiment, the sampling schedule was adjusted to the feeding schedule and to the post-prandial repeated multiple samplings (described in the next section). The F1 fish were thus sampled at 0, 1, 24, 96, 147, 330, 450 and 498 h post-surgery, leaving out the 48 and 72 h samples in order to allow fish to resume their normal feeding behaviour post-surgery. In order to simplify interpretation for the reader, the sampling points in F1 fish, at 96, 147, 330, 450 and 498 h in the Table 1 and Fig. 2 are presented as 72, 150, 300, 420 and 528 h.

At sampling, the 600–700 mm cannula, which normally floated freely in the water, was gently caught with a thin pin and taken through the hole in the side of tank (Fig. 1). Sampling was initiated by cutting off the tip of the cannula and emptying the saline by gravity before attaching a syringe. After discarding the first 50 μl of blood, a new syringe was attached in order to obtain a representative blood sample. Hct was analysed, using heparinized microcapillary tubes in duplicates, in a portable microhematocrit centrifuge (Compur Microspin, Bayer Diagnostics, Germany). Before heat-sealing the cannula, all blood was replaced by injecting approximately 0.3 ml saline solution with 150 IU of heparin. Plasma was separated immediately by centrifugation at 2°C at 750*g* for 3 min and kept on dry ice until storage at −80°C. A portable i-STAT clinical analyser was used to determine Na^+, K^+, Cl^-, glucose, pCO2 and pH in whole blood (Bingham et al. 1999; Harrenstien et al. 2005; Jacobs et al. 1993). The pH (Tables 1, 2; Fig. 2) and pCO2 (Tables 1, 2; Fig. 2) values were corrected for the temperature difference between ambient water temperature and the temperature-adjusted (37°C) values displayed by the

Table 1 Changes (Ls-mean ± SE) in blood variables after DA cannulation

Variable	Exp.	Time after DA cannulation								
		0	1	24	48	72	150	300	420	528
Cortisol (ng ml^{-1})	F1	39 ± 9.9 a	92 ± 9.2 b	42 ± 10.4 aA	–	28 ± 13.3 a	20 ± 12.5 a	23 ± 16 a	20 ± 13.5 a	18 ± 16.4 a
	NF1	35 ± 5.8 a	95 ± 5.8 b	15 ± 5.8 acB	25 ± 6.3 ac	16 ± 6.3 ac	15 ± 6.3 ac	–	11 ± 6.9 ac	9 ± 6.9 c
	NF2	49 ± 6.3 a	115 ± 6.3 b	40 ± 6.3 aA	25 ± 6.3 a	24 ± 6.8 a	23 ± 6.8 a	22 ± 7.3 a	18 ± 6.8 a	–
pH	F1	7.15 ± 0.03 aA	7.85 ± 0.03 b	7.89 ± 0.03 bA	–	7.86 ± 0.03 b	7.89 ± 0.03 b	7.87 ± 0.03 b	7.86 ± 0.04 b	7.87 ± 0.03 b
	NF1	7.59 ± 0.06 aB	7.99 ± 0.06 b	7.75 ± 0.06 aB	7.82 ± 0.06 ab	7.79 ± 0.06 ab	7.72 ± 0.06 ab	7.78 ± 0.06 ab	7.87 ± 0.06 ab	7.80 ± 0.08 ab
	NF2	7.45 ± 0.03 aB	7.89 ± 0.03 b	7.91 ± 0.03 bA	7.89 ± 0.03 b	7.91 ± 0.03 b	7.87 ± 0.03 b	7.84 ± 0.03 b	7.88 ± 0.03 b	–
Hct	F1	38 ± 1.1 aA	26 ± 1.0 b	26 ± 0.9 b	–	24 ± 0.9 b	24 ± 1.0 b	20 ± 1.1 b	21 ± 1.1 b	21 ± 1.2 b
	NF1	40 ± 0.8 aAB	24 ± 0.8 b	26 ± 0.8 b	25 ± 0.8 b	22 ± 0.8 bc	21 ± 0.8 bc	20 ± 0.8 bc	19 ± 0.9 c	20 ± 0.9 c
	NF2	43 ± 1.1 aB	27 ± 1.1 b	26 ± 1.1 b	22 ± 1.1 b	24 ± 1.1 b	22 ± 1.1 b	23 ± 1.2 b	23 ± 1.2 b	–
Na^+ (mmol l^{-1})	F1	154 ± 1.3 abA	157 ± 1.3 bA	151 ± 1.2 aA	–	149 ± 1.1 a	151 ± 1.2 a	151 ± 1.4 a	151 ± 1.5 a	151 ± 1.5 a
	NF1	157 ± 1.9 abAB	163 ± 1.6 bAB	159 ± 1.6 abB	155 ± 1.7 ab	152 ± 1.7 a	151 ± 1.7 a	151 ± 1.7 a	152 ± 1.9 a	151 ± 1.9 a
	NF2	168 ± 1.8 acB	169 ± 1.8 acB	153 ± 1.8 bA	154 ± 1.8 b	156 ± 1.8 bc	159 ± 2.0 ac	163 ± 2.0 cA	159 ± 2.0 bcA	–
K^+ (mmol l^{-1})	F1	2.5 ± 0.13 a	3.1 ± 0.13 b	2.5 ± 0.11 a	–	2.4 ± 0.11 a	2.5 ± 0.12 a	2.5 ± 0.14 ab	2.4 ± 0.14 a	2.6 ± 0.14 ab
	NF1	1.9 ± 0.24 a	3.3 ± 0.21 b	2.4 ± 0.21 ab	1.8 ± 0.22 a	2.2 ± 0.22 a	2.5 ± 0.22 ab	2.8 ± 0.22 ab	2.4 ± 0.24 ab	2.3 ± 0.24 ab
	NF2	2.9 ± 0.17 ab	3.5 ± 0.17 b	2.3 ± 0.17 a	2.4 ± 0.17 a	2.2 ± 0.18 a	2.3 ± 0.19 a	2.9 ± 0.19 ab	2.7 ± 0.18 a	–
Cl^- (mmol l^{-1})	F1	140 ± 0.4 A	140 ± 0.4 A	140 ± 0.3 A	–	139 ± 0.3 A	139 ± 0.4 A	140 ± 0.5	138 ± 0.5	140 ± 0.4
	NF1	150 ± 1.4 acB	150 ± 1.2 acB	148 ± 1.2 acB	150 ± 1.2 aB	145 ± 1.2 cB	148 ± 1.4acB	140 ± 1.4 bc	139 ± 1.4 bc	140 ± 1.4 bc
	NF2	149 ± 1.0 acB	150 ± 1.0 aB	144 ± 1.0 cAB	139 ± 1.1 bC	137 ± 1.1 bA	137 ± 1.3 bA	139 ± 1.3 b	137 ± 1.1 b	–
Glucose (mmol l^{-1})	F1	4.25 ± 0.36	5.11 ± 0.34	4.21 ± 0.32	–	4.78 ± 0.30	5.06 ± 0.34	5.08 ± 0.39	4.10 ± 0.39	5.17 ± 0.39
	NF1	3.63 ± 0.4	4.95 ± 0.40	4.49 ± 0.40	4.01 ± 0.43	3.87 ± 0.44	3.01 ± 0.44	3.45 ± 0.44	4.17 ± 0.47	3.47 ± 0.48
	NF2	4.04 ± 0.37	4.24 ± 0.37	4.10 ± 0.37	3.51 ± 0.37	3.57 ± 0.37	3.27 ± 0.39	3.14 ± 0.37	3.30 ± 0.37	–
P_{CO2} (mmHg)	F1	4.94 ± 0.25 a	4.06 ± 0.23 bA	3.56 ± 0.22 b	–	3.91 ± 0.21 bA	3.50 ± 0.23 bA	3.35 ± 0.27 b	3.45	3.49 ± 0.27 b
	NF1	3.69 ± 0.18	3.08 ± 0.15 B	3.01 ± 0.15	3.27 ± 0.16	3.37 ± 0.18 A	3.31 ± 0.16 A	3.29 ± 0.18	3.03 ± 0.18	2.93 ± 0.18
	NF2	4.27 ± 0.11 a	2.98 ± 0.11 bB	2.86 ± 0.12 b	2.86 ± 0.11 b	2.09 ± 0.11 cB	2.98 ± 0.13 bB	3.15 ± 0.12 b	3.18 ± 0.12 b	–

Note: Values of the same variable and experiment followed by none or the same lower case letter do not differ significantly ($P > 0.05$). Values followed by none or the same upper case letter do not differ significantly ($P > 0.05$) between experiments (F1, NF1, NF2) for a given variable at a given time

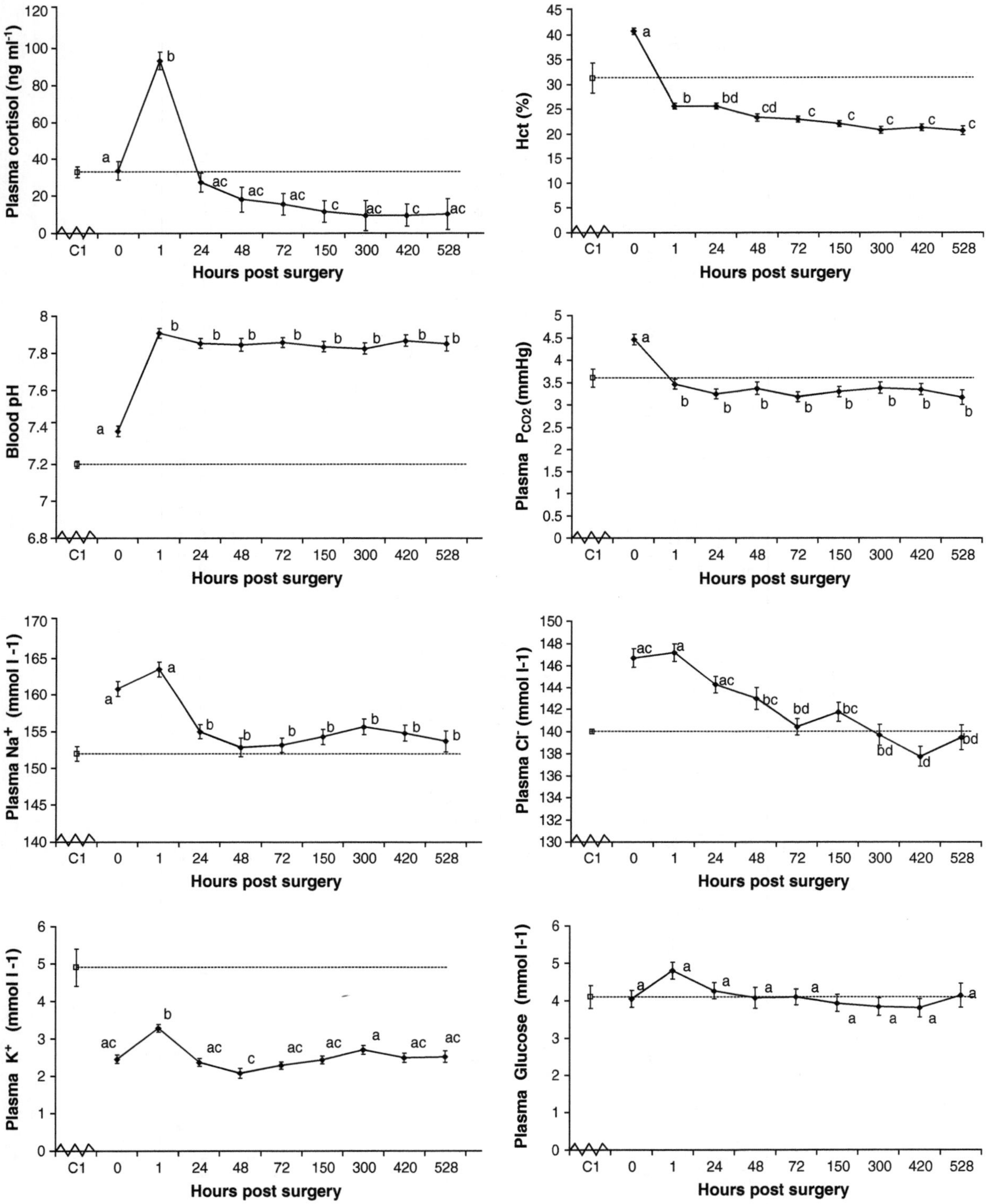

Fig. 2 Acute and long-term changes in blood variables in free swimming and fed/starved DA-cannulated Atlantic salmon. *Values* are presented as Ls-mean ± SE of the three experiments. *C1* represent levels taken by aspiration in uncannulated fish. *Same letters* denote insignificant differences ($P < 0.05$) between sample times

Table 2 Post-prandial levels of blood/plasma variables (Ls-mean of 8 meals ± SE) in DA-cannulated Atlantic salmon

Variables	Time after feeding (hours)				*P*-value
	0	3	6	12	
Cortisol (ng ml^{-1})	28 ± 2.4	23 ± 2.4	24 ± 2.5	30 ± 2.5	0.09
pH	7.83 ± 0.01	7.87 ± 0.01	7.85 ± 0.01	7.86 ± 0.01	NS
Hct (%)	23 ± 0.2 a	22 ± 0.3 b	21 ± 0.3 b	20 ± 0.3 c	<0.0001
pCO_2 (mmHg)	3.7 ± 0.08	3.8 ± 0.08	3.7 ± 0.09	3.9 ± 0.09	NS
Na^+ (mmol l^{-1})	151.3 ± 0.3	151.0 ± 0.3	150.7 ± 0.3	150.0 ± 0.3	NS
K^+ (mmol l^{-1})	2.67 ± 0.04 a	2.55 ± 0.04 b	2.44 ± 0.04 b	2.51 ± 0.04 b	<0.0001
Cl^- (mmol l^{-1})	139.7 ± 0.14	139.5 ± 0.14	139.3 ± 0.16	139.6 ± 0.16	NS
Glucose (mmol l^{-1})	5.2 ± 0.16	5.1 ± 0.16	5.1 ± 0.16	5.5 ± 0.18	NS

Values of the same variable followed by none or the same lower case letter do not differ significantly ($P > 0.05$). A P value > 0.1 is denoted NS

instrument in accordance with the i-STAT procedure (i-STAT Corporation, Windsor Center Drive,East Windsor, NJ, USA). Plasma cortisol was analysed using a radioimmunoassay kit (Orion Diagnostica, Espoo, Finland, Spectria Cortisol Ria test).

Diet component of the F1 experiment

This study was performed as a part of a larger experiment that included a comparison of plasma fatty acid and amino acid profiles in response to partial (25% of wet weight) replacement of fish meal (FM) by soybean meal (SBM) and pea protein concentrate (PPC). Each diet was fed for 1 week in the following order: FM, PPC, SBM and FM. The results were analysed independently of diet and meal size. However, all three diets had almost identical gross energy (23.1–23.7 MJ kg^{-1}), dry matter (943–946 g kg^{-1}), crude protein (450–455 g kg^{-1}) and crude fat content (246–276 g kg^{-1}). No significant differences ($P > 0.05$) in the blood variables measured were found as an effect of diet or meal size. The data were therefore treated as a single population in the statistical evaluation. Fish were fed twice per day and sampled for blood on day 5 and 7 every week. Blood samples (0.3 ml) were taken at 0, 3, 6 and 12 h after feeding following the sampling procedure described in the previous paragraph. Fish started to be fed 24 h after DA cannulation, with the first post-prandial sampling point 96 h after DA cannulation (0 h after feeding). On the days of blood sampling, the fish were fed only in the morning. The number of pellets eaten was recorded for each fish and converted to grams.

Statistical analysis

Blood data were either analyzed by the main factorial model for fixed effects (general linear model) or for repeated measures using the mixed procedures for linear models in the Statistical Analysis System (SAS) for PC (ver. 8. 2). The main factors included in the models were sample time after surgery (categorical), experiment (categorical), fish as block (categorical) and body weight as covariate (continuous). Repeated analysis was only performed per experiment as each experiment was composed of different individual fish. Diet (categorical), body weight (continues) and meal size (continues) did not contribute significantly ($P > 0.05$) to the model and were therefore omitted from the statistical model. Groups were compared by the ad hoc variance test (F-test) and corrected for multiple comparisions by Tukey using the least-squares means procedure when significant effects were found in the main model. The level of statistical significance was set at $P < 0.05$. All data were tested for normality by a normal probability plot (univariate procedure).

Results

Changes in all measured variables post-DA surgery are presented for each experiment and sample time in Table 1. The combined ls-mean value of all three experiments is presented in Fig. 2, including the average value of uncannulated, free-swimming Atlantic salmon (C1). Cortisol displayed a significant peak in the 1 h post-surgery sample in all three experiments

(Table 1). The level of cortisol, expressed by individual (Table 1) or by combined (Fig. 2) experiments, did not vary significantly from 24 h and onwards. The blood pH and pCO_2 displayed an inverse response to DA surgery in all three experiments, with a low and high pH and pCO_2 value, at 0 h, respectively. These variables were stable from then onwards (Fig. 2). pH equalled (F1 fish, Table 1) or was close to (NF1 and NF2, Table 1) values of C1 fish (Fig. 2) at 0 h sample and from then on deviated from the C1 fish. The opposite was observed for the pCO_2 values (Fig. 2). Hct values recorded at 0 h were significantly higher than those seen from 1 h post-surgery and onwards, when analysing each experiment separately (Table 1). Combined data from all three experiments showed a continuing slow and significant reduction in the Hct until the 72 h sample (Fig. 2). Blood Na^+ and Cl^- concentrations paralleled each other with a higher level than C1 fish at 0 h sample, displaying a peak at 1 h and from then onwards approaching the C1 fish level at 24 and 72 h, respectively (Fig. 2). Blood K^+ showed a significant peak at 1 h post-surgery, stabilizing from 24 h sample and onwards (Table 1; Fig. 2) at a level well below C1 fish. Blood glucose concentrations were elevated 1 h after surgery in all individual experiments, however, not significantly ($P > 0.05$) so. F1 fish very often displayed a higher value of blood glucose than either of the unfed groups of fish (NF1 and NF2). However, this was never manifested as significant at any sampling point (Table 1).

A statistical analysis of post-prandial changes in all variables combined by linear regression (GLM procedure) is given in Table 2. Hct and K^+ showed small but significant post-prandial patterns. Both, Hct and K^+ decreased from 0 to 12 h post-prandial (Table 2). Cortisol indicated ($P = 0.09$) a possible u-shaped post-prandial pattern (Table 2). All other variables measured post-prandially demonstrated a P value > 0.1 and were therefore deemed completely random (Table 2).

Discussion

Cortisol

Resting (pre-stress) levels of cortisol can vary considerably (2–42 ng ml^{-1}) (Gamperl et al. 1994b). Variations in resting cortisol titres may be due to a number of "non-stress" factors, such as nutrition status, temperature, physical activity or time of day (Barton and Iwama 1991). The C1 plasma cortisol levels in the present experiments were in the upper range of those reported in the literature, which may have been due to such "non-stress" factors. The C1 fish (uncannulated fish) were kept in isolation for 7 days before sampling, in order to follow the same procedure as for the experimental cannulated fish and to allow them to become acclimatized to the isolation. In the recent experiment (Doyon et al. 2005), rainbow trout were individually isolated in 120 l tanks or confined to 1.5 l boxes for 4, 24 or 72 h. The findings, that confined rainbow trout displayed higher cortisol values than isolated one, suggest that confinement could be more stressful than isolation. One would therefore expect fish in our experimental set-up to display resting values in the lower range. However, in our experiment, the fish were kept in a current (Fig. 1) and had a choice of different "niches" within the tank environment (either remaining below the PVC cover or not; Fig. 1). We therefore suggest as an explanation that the apparent high resting level of cortisol in the fish in this experiment is not indicative of stress but could relate to e.g. physical activity or endogenous diurnal cycles of plasma cortisol as natural variation causes (Bonga 1997). In fact, there are certain numbers of experiments reporting resting levels even higher than 50 ng ml^{-1} (Barton and Iwama 1991; Gamperl et al. 1994b). The higher resting levels could arise due to differences in assay specificity, husbandry conditions, capture procedure, and other external and internal factors (Bonga 1997; Pottinger et al. 2000). Under certain circumstances, low cortisol levels may even indicate a stress condition (Vijayan and Leatherland 1990), illustrating the difficulty of using a cortisol as a sole variable to distinguish between a stressed and a non-stressed state.

In all three experimental series, plasma cortisol at 1 h after surgery reached levels comparable to that reported by Gamperl et al. (1994a), i.e. around 100 ng ml^{-1} plasma (Fig. 2). The time frame is also in accordance with earlier studies, showing the highest elevation in plasma cortisol around 0.5–1 h after a stressful disturbance (Barton and Iwama 1991; Kiessling et al. 2009). Gamperl et al. (1994a) found that plasma cortisol returned to baseline levels 7 days after surgery, using fish confined in black PVC boxes.

In our experiment, circulating cortisol reached low levels within 24–48 h of surgery independent of nutritional status, and then declining further to the baseline resting plasma levels reported by Barton and Iwama (1991), Einarsdottir and Nilssen (1996), and Olsen et al. (1995). A return of plasma cortisol levels close to pre-stress levels (C1 fish) only 24 h after surgery is much faster than previously reported (see above and Sect. "Introduction"). However, other work by us using DA-cannulated salmon (Kiessling et al. 2009; fish exposed to anaesthesia stress) and rainbow trout (*Oncorhynchus mykiss;* unpublished results; fish exposed to surgery stress) has indicated that the plasma level of cortisol will start to level out already as early as 3–6 h following exposure to a stressor. Due to the absence of samples between 1 and 24 h, we were unable to confirm these observations in the present study.

The plasma levels of cortisol in this study were not influenced by the nutritional status of the fish. Starved (NF1, March and NF2, May) and fed (F1, January) fish all had very similar levels post-surgery and onwards. This is in accordance with some previous studies (Barton et al. 1988; Farbridge and Leatherland 1992; Milne et al. 1979) that reported no effect of starvation on post-stress plasma cortisol concentrations. Pottinger and Brierley (1997) suggested that *Salmonidae* are adapted to long periods of food deprivation and as long as sufficient reserves have been laid down, these fish can survive prolonged periods of starvation with no adverse effects. On the other hand, Reddy and Leatherland (2003) and Olsen et al. (2005b) reported that the post-stress plasma cortisol level in rainbow trout was elevated in starved compared to fed fish. The contradictory data concerning the starvation effect on plasma cortisol in fish are probably a result of several factors, such as duration of starvation, season, stage of fish development as well as individual variations in plasma cortisol dynamics and metabolic clearance rate. For example, Kiessling et al. (1990) showed that long-term starvation induced completely different metabolic responses in immature and mature rainbow trout. Another possible factor is habituation, as fed fish may be more used to activity around the tank and might therefore react less to a stressor than a fish less used to activity in the vicinity of the tank.

The slight differences in the cortisol responses in the March (NF1), May (NF2) and January (F1) experiments might also be due to the minor differences in water temperature, with the highest temperature in the January experiment. Seasonal and diurnal patterns are also known to influence osmoregulatory and metabolic functions in teleosts (Bradford and Taylor 1987; Pankhurst and Sharples 1992). However, in the present study, the light regime was kept constant, and the temperature pattern was the opposite of that intuitively expected, which would speak against seasonal effects as an explanatory factor. The water for the present experiments was supplied from 60 m depth, with temperature being more influenced by water currents than seasonal changes.

The post-prandial cortisol data in the F1 experiment (Table 2) suggest that the repeated multiple samplings did not produce any severe stress as manifested in plasma cortisol titres, but only transitory short-term responses. On the basis of Table 2, a post-prandial u-shape plasma cortisol pattern could be suggested, although not significant ($P = 0.09$). There have been studies reporting elevated cortisol levels due to the anticipation of food (Reddy and Leatherland 2003), stress during feeding (Small 2005) and/or specific dynamic action (SDA) or behavioural and thereby hormonal changes associated with feeding activity (Boujard and Leatherland 1992; Bry 1982; Holloway et al. 1994;). Also, circadian-like patterns in physiological and behavioural activities of teleost fish, as a response to the light–dark cycle, have been reported (de Pedro et al. 2005; Holloway et al. 1994; Pavlidis et al. 1999; Small 2005). It is therefore tempting to speculate that slight cortisol increase at 0 and 12 h post-prandial sample (taken at 09 a.m. and 21 p.m., respectively) in this study could be interpreted as food anticipation and scotophasic response.

Haematocrit

The initial Hct increase observed at 0 h after DA cannulation is normally considered to be a combined effect of erythrocyte swelling (Nikinmaa 1986), initiated by a raised concentration of plasma catecholamines (Olsen et al. 1995) and release of erythrocytes from the spleen (Pearson and Stevens 1991). The 1 h values are comparable to those reported in unstressed salmonids by Gallaugher et al. (1995). The values observed by us from then on correspond well to the Hct values reported by Kiessling et al. (1995, 2003) in cannulated salmon. Increases in Hct during and after MS-222 anaesthetics have been indicated by other studies (Houston et al. 1971; Lowe-Jinde and Niimi 1983; Reinitz and Rix 1977; Soivio et al. 1977). On the basis of our data and those described previously, we suggest that in salmonids, an Hct value above 30% could be indicative of a catecholamine release induced by stress at sampling. In fact, elevated Hct values could be used as a qualitative, and possibly, with further research, also quantitative, indicator of catecholamine release in salmonids and possibly in other fish species too.

Detailed study of Fig. 2 gives us the impression of a continued decline in Hct till the end of the experiment. However, this apparent decline is not significant beyond the 72 h sample, neither in a combined (Fig. 2) or in a individually analysed experiments (Table 1). This apparent decline might be due to several factors. A noticeable effect of repeated blood sampling on Hct has been reported (Deng et al. 2000; Soivio et al. 1975; Soivio et al. 1972) and explained by haemodilution. Against this speaks the study of Schultz and Hayton (1991), who reported that 10% of the total blood volume might be withdrawn without adverse effects on the fish, a volume much larger than that sampled in the present experiment. On the other hand, the slow decrease in Hct over time may be due to red blood cells accumulating in the spleen (Pearson and Stevens 1991), i.e. an adaptation to a low stress state. Our own experience, as well as the data presented here, supports this latter explanation. First, we have tried to replace sampled blood, volume for volume, as suggested by several authors (Axelsson and Fritsche 1994; Gallaugher et al. 1995; McLean and Ash 1989), with saline, plasma or whole blood without any effect on Hct (Kiessling et al. 1995). Secondly, a slow accumulation in the spleen is consistent with the slow reduction in plasma cortisol observed to parallel the decrease in Hct (Fig. 2). It has also been suggested that the slow decrease in Hct could be caused by long-term bleeding at the site of incision of the cannula in the DA. We argue against such an interpretation in the present study, partly on the grounds of the continued decrease in cortisol values, partly due to stable electrolyte and glucose values. Furthermore, no major bleeding was seen around the incision at the end of the experiment. DA cannulation carried out in accordance with Soivio et al. (1975) involves only minor tissue trauma.

In contrast to the weekly sampling scheme of NF1 and NF2, multiple sampling within a short time, as tested in the F1 experiment (Table 2), indicated that erythrocyte depletion caused by sampling may occur if the fish is not given ample time to recover. The post-prandial drop in Hct could also be a natural effect due to redistribution of blood from external tissue (muscle, skin, etc.) to the digestive tract (Altimiras et al. 2008; Thorarensen et al. 1993). The relatively stable level of electrolytes and cortisol speaks in favour of this later explanation. Further studies are therefore needed in order to clarify the regulation mechanisms of Hct in salmon during digestion (Table 2) and also the underlying causes of the long- and short-term decrease observed during the weekly and multiple samplings (Table 1; Fig. 2).

Blood glucose, pH, electrolytes and pCO_2

A plasma glucose level have been shown to be elevated 1–8 h after the onset of stress and is a common indicator of the secondary stress responses in fish (Farbridge and Leatherland 1992; Pickering et al. 1982; Vijayan and Moon 1992; Waring et al. 1992). In our study, we did not detect any significant plasma glucose changes neither in the starved nor in the fed fish (Tables 1, 2). The explanation could be that we have simply missed the peak due to the absence of the samples between 1 and 24 h. The numerical increase at 1 h post-surgical sample seen in all three experiments (Table 1; Fig. 2) could be speculated as the beginning of a peak occurring between the 1 and 24 h sample. Neither did the post-prandial glucose levels exhibited any significant changes (Table 2). del Sol Novoa et al. (2004) found that post-prandial plasma glucose in rainbow trout peaked at 9 h in the study conducted at water temperatures of 14.1–19.5°C. The absence of a marked post-prandial blood glucose peak in our experiment at 12 h may therefore be due to slower uptake rate caused by lower water temperature (9.8°C) than in the experiment of del Sol Novoa et al. (2004).

The elevated pCO_2 level in blood at 0 h may reflect inadequate secondary lamella ventilation relative to the metabolic rate during sedation. This underlines the importance of maintaining adequate water flow over the gills during surgery, in order to avoid excessive accumulation of anaerobic metabolites. A concomitant decrease in blood pH was also seen at that point in time. The hypercapnia and pH drop were compensated for within 1 h after surgery and were maintained stable by the fish throughout the experiments (Table 1; Fig. 2, respectively). The fact that C1 fish displayed a similar pH level as fish sampled 0 h post-surgery indicates an immediate reduced CO_2 transfer, as the fish is removed out of the water, likely caused by a collapse of the secondary lamella.

Ionic imbalance in fish caused by stress has been demonstrated previously (Soivio et al. 1977; Soivio and Oikari 1976; Willford 1970). Our results agree that the variations in the blood Na^+ and Cl^- concentration during recovery after MS 222 anaesthesia may be due to the water moving from the plasma into the tissues during anaesthesia, causing osmoregulatory disorders (Soivio et al. 1974). An elevated plasma K^+ levels in stressed situations were observed in studies with Coho salmon (Redding and Schreck 1983) and in rainbow trout (Turner et al. 1983) and thought to be the result of the blood acidification affecting the shift in K^+–H^+ exchange mechanisms between intra- and extracellular fluid. pH compensation is achieved by regulating plasma $HCO3^-$ levels directing H^+ and $HCO3^-$ effluxes, which are believed to be tied to the influx of Na^+ and Cl^- (Perry and Gilmour 2006) and thus coupled to the activity of the Na^+, K^+-ATPase, which maintains intracellular Na^+ at low levels and intracellular K^+ at high levels. The elevated K^+ level at 1 h sample was observed in parallel with the cortisol peak, and it is has been reported that cortisol exerts osmoregulatory effects in fish (McCormick 2001) probably by binding to the intracellular receptors. The consistently lower plasma K^+ in cannulated compared to C1 fish is intriguing, but could be the result of a methodological error. Caudal sampling by expiration will involve a mix of venous and arterial blood. The needle has to enter the caudal aorta through the caudal vein. In the fish size used in this study, the needle opening will be located in both vessels. Fish that is quickly netted will struggle and thereby perform vigorous muscle contractions. Thus likely leading to the venous blood containing significantly elevated levels of K^+, liberated from the contracting muscle.

The levels of pCO_2, Na^+, Cl^- and pH (Table 2) were not significantly affected by post-prandial repeated multiple samplings during the four-week sampling period. However, we found that post-prandial blood levels of K^+ fluctuated in a consistent manner, with time in the course of digestion (Table 2). This finding contradicts reports that plasma K^+ concentration is very stable in fish reared in seawater (Halver and Hardy 2002), as the fish can easily absorb this element from the surrounding water. It is therefore tempting to speculate that the decrease in K^+ is the result of active absorption from the blood at a higher rate than can be replenished by exogenous uptake.

Comparing recovery levels of variables such as pH, K^+ and Hct, with the levels detected in the C1 fish (Fig. 2) obtained by "grab and stab", indicated that the values of C1 fish did not represented a resting value. This, in combination with the report by Kiessling et al. (2009), concluding that also anaesthesia in itself has marked physiological effects suggest to us that cannulation may be the preferred method to obtain true resting values of these variables in fish.

Conclusions

Our study employed DA cannulation, including pre-sedation, local analgesics and aseptic procedures, in combination with adequate holding conditions. The tank set-up, with partial cover and oriented current, served to position the fish permanently in its preferred location. This in turn allowed repeated collection of blood samples (0.3 ml) from fish that were unaware of the presence of the experimenter. A transient stress response was reversed within 24–48 h of surgery. From 24 or 48 h all variables showed some fluctuations, but were all within expected ranges for unstressed fishes (Eliason et al. 2007; Powell and Nowak 2003). The implantation of the cannula itself did not add additional stress beyond that caused by sedation and handling. The nutritional status of the fish was also without effect on recovery, as we saw the same pattern in both fed and starved fish. Similarly, we concluded that neither single nor multiple sampling caused either an acute or an accumulated allostatic load even during multiple repeated samplings within a four-week period. We also concluded that the use of implanted cannula is the preferred method when the objective is to obtain resting values of variables such as pH, K^+ and Hct.

Thus, this protocol allows for long-term physiological monitoring, with particular reference to individual variation in, for instance, nutritional requirements, endocrine dynamics, metabolism and clearance of exogenous substances as xenobiotics, pharmaca and time course of physiological responses to varying environments and external stimuli.

Acknowledgments The authors would like to thank the staff of the NIVA Research Station at Solbergstrand for their support and help. We also thank Carlos Salas for his help with the illustrations. Brita Dalsgard for laboratory assistance. Support in kind from the Department of Animal and Aquacultural Sciences, Norwegian University of Life Sciences, the Norwegian Institute for Water Research (NIVA), and financial support the Research Council of Norway (grant no 153202/120) to NIVA and the Aquaculture Protein Centre, respectively, are gratefully acknowledged.

References

Altimiras J, Claireaux G, Sandblom E, Farrell AP, McKenzie DJ, Axelsson M (2008) Gastrointestinal blood flow and postprandial metabolism in swimming sea bass *Dicentrarchus labrax*. Physiol Biochem Zool 81(5):663–672

Axelsson M, Fritsche R (1994) Cannulation techniques. In: Hochachka PW, Mommsen TP (eds) Biochemistry and molecular biology of fishes. Elsevier, Amsterdam, pp 17–36

Barton BA, Iwama GK (1991) Physiological changes in fish from stress in aquaculture with emphasis on the response and effects of corticosteroids. Annu Rev Fish Dis 1:3–26

Barton BA, Schreck CB, Fowler LG (1988) Fasting and diet content affect stress-induced changes in plasma glucose and cortisol in juvenile chinook salmon. Prog Fish-Cult 50(1):16–22

Bingham D, Kendall J, Clancy M (1999) The portable laboratory: an evaluation of the accuracy and reproducibility of i-STAT (c). Ann Clin Biochem 36:66–71

Bonga SEW (1997) The stress response in fish. Physiol Rev 77(3):591–625

Boujard T, Leatherland JF (1992) Circadian rhythms and feeding time in fishes. Environ Biol Fishes 35(2):109–131

Bradford CS, Taylor MH (1987) Semilunar changes in estradiol and cortisol coincident with gonadal maturation and spawning in the killifish *Fundulus heteroclitus*. Gen Comp Endocrinol 66(1):71–78

Bry C (1982) Daily variations in plasma cortisol levels of individual female rainbow trout˙ *Salmo gairdneri*: evidence for a post-feeding peak in well-adapted fish. Gen Comp Endocrinol 48(4):462–468

Conte FP, Wagner HH, Harris TO (1963) Measurement of blood volume in the fish (*Salmo gairdneri gairdneri*). Am J Physiol 205(3):533–540

de Pedro N, Guijarro AI, Lopez-Patino MA, Martinez-Alvarez R, Delgado MJ (2005) Daily and seasonal variations in haematological and blood biochemical parameters in the tench, *Tinca tinca* Linnaeus, 1758. Aquac Res 36(12): 1185–1196

del Sol Novoa M, Capilla E, Rojas P, Baró J, Gutiérrez J, Navarro I (2004) Glucagon and insulin response to dietary carbohydrate in rainbow trout (*Oncorhynchus mykiss*). Gen Comp Endocrinol 139(1):48–54

Deng DF, Refstie S, Hemre GI, Crocker CE, Chen HY, Cech JJ, Hung SSO (2000) A new technique of feeding, repeated sampling of blood and continuous collection of urine in white sturgeon. Fish Physiol Biochem 22(3):191–197

Doyon C, Trudeau VL, Moon TW (2005) Stress elevates corticotropin-releasing factor (CRF) and CRF-binding protein mRNA levels in rainbow trout (*Oncorhynchus mykiss*). J Endocrinol 186(1):123–130

Einarsdottir IE, Nilssen KJ (1996) Stress responses of Atlantic salmon (*Salmo salar* L.) elicited by water level reduction in rearing tanks. Fish Physiol Biochem 15(5):395–400

Eliason EJ, Kiessling A, Karlsson A, Djordjevic B, Farrell AP (2007) Validation of the hepatic portal vein cannulation technique using Atlantic salmon *Salmo salar* L. J Fish Biol 71(1):290–297

Eliason EJ, Djordjevic B, Trattner S, Pickova J, Karlsson A, Farrell AP, Kiessling A (2009) The effect of hepatic passage on postprandial plasma lipid profile of rainbow trout (*Oncorhynchus mykiss*) after a single meal. Aquac Nutr. doi:10.1111/j.1365-2095.2009.00693.x

Farbridge KJ, Leatherland JF (1992) Plasma growth hormone levels in fed and fasted rainbow trout (*Oncorhynchus mykiss*) are decreased following handling stress. Fish Physiol Biochem 10(1):67–73

Gallaugher P, Thorarensen H, Farrell AP (1995) Hematocrit in oxygen transport and swimming in rainbow trout (*Oncorhynchus mykiss*). Respir Physiol 102(2–3):279–292

Gamperl AK, Vijayan MM, Boutilier RG (1994a) Epinephrine, norepinephrine, and cortisol concentrations in cannulated seawater-acclimated rainbow trout (*Oncorhynchus mykiss*) following black-box confinement and epinephrine injection. J Fish Biol 45(2):313–324

Gamperl AK, Vijayan MM, Boutilier RG (1994b) Experimental control of stress hormone levels in fishes—techniques and applications. Rev Fish Biol Fish 4(2):215–255

Halver JE, Hardy RW (2002) Fish nutrition. Academic Press, London

Hamre K, Kolas K, Sandnes K, Julshamn K, Kiessling A (2001) Feed intake and absorption of lipid oxidation products in Atlantic salmon (*Salmo salar*) fed diets coated with oxidised fish oil. Fish Physiol Biochem 25(3):209–219

Harrenstien LA, Tornquist SJ, Miller-Morgan TJ, Fodness BG, Clifford KE (2005) Evaluation of a point-of-care blood analyzer and determination of reference ranges for blood parameters in rockfish. JAVMA-J Am Vet Med Assoc 226(2):255–265

Holloway AC, Reddy PK, Sheridan MA, Leatherland JF (1994) Diurnal rhythms of plasma growth hormone, somatostatin, thyroid hormones, cortisol and glucose concentrations in rainbow trout, *Oncorhynchus mykiss*, during progressive food deprivation. Biol Rhythm Res 25(4):415–432

Houston AH, Madden JA, Woods RJ, Miles HM (1971) Some physiological effects of handling and tricaine methanesulphonate anesthetization upon the brook trout, *Salvelinus fontinalis*. J Fish Res Board Can 28(5):625–633

Jacobs E, Vadasdi E, Sarkozi L, Colman N (1993) Analytical evaluation of i-STAT portable clinical analyzer and use by nonlaboratory health-care professionals. Clin Chem 39(6):1069–1074

Karlsson A, Eliason EJ, Mydland LT, Farrell AP, Kiessling A (2006) Postprandial changes in plasma free amino acid levels obtained simultaneously from the hepatic portal vein and the dorsal aorta in rainbow trout (*Oncorhynchus mykiss*). J Exp Biol 209(24):4885–4894

Kiessling A, Johansson L, Kiessling K-H (1990) Effects of starvation on rainbow trout muscle. I: histochemistry, metabolism and composition of white and red muscle in mature and immature fish. Acta Agric Scand 40:909–924

Kiessling A, Dosanjh B, Higgs D, Deacon G, Rowshandeli N (1995) Dorsal aorta cannulation: a method to monitor changes in blood levels of astaxanthin in voluntarily feeding Atlantic salmon, *Salmo salar,* L. Aquac Nutr 1(1):43–50

Kiessling A, Olsen RE, Buttle L (2003) Given the same dietary carotenoid inclusion, Atlantic salmon, *Salmo salar* (L.) display higher blood levels of canthaxanthin than astaxanthin. Aquac Nutr 9(4):253–261

Kiessling A, Dosanjh B, Koppe W, Higgs D (2006) Relationship between blood and muscle levels of astaxanthin in dorsal aorta cannulated Atlantic salmon. Aquaculture 254(1–4):653–657

Kiessling A, Johansson D, Zahl IH, Samuelsen OB (2009) Pharmacokinetics, plasma cortisol and effectiveness of benzocaine, MS-222 and isoeugenol measured in individual dorsal aorta-cannulated Atlantic salmon (*Salmo salar*) following bath administration. Aquaculture 286(3–4):301–308

Kreiberg H, Powell J (1991) Metomidate sedation reduces holding stress in chinook salmon. World Aquaculture 22:58–59

Lo WY, Chang CF, Song YL (2003) Evaluation of dorsal aorta cannulation for immunological studies of grouper (*Epinephelus malabaricus*). Fish Shellfish Immunol 14(4):289–303

Lowe-Jinde L, Niimi AJ (1983) Influence of sampling on the interpretation of haematological measurements of rainbow trout, *Salmo gairdneri.* Can J Zool 61:396–402

Mazik PM, Plakas SM, Stehly GR (1994) Effects of dorsal aorta cannulation on the stress response of channel catfish (*Ictalurus punctatus*). Fish Physiol Biochem 12(5):439–444

McCormick SD (2001) Endocrine control of osmoregulation in teleost fish. Soc Integr Comp Biol 41(4):781–794

McLean E, Ash R (1989) Chronic cannulation of the hepatic portal vein in rainbow trout, *Salmo gairdneri*: a prerequisite to net absorption studies. Aquaculture 78(2):195–205

Milne R, Leatherland J, Holub B (1979) Changes in plasma thyroxine, triiodothyronine and cortisol associated with starvation in rainbow trout (*Salmo gairdneri*). Environ Biol Fishes 4(2):185–190

Nikinmaa M (1986) Control of red cell pH in teleost fishes. Ann Zool Fenn 23(2):223–235

Ok IH, Bai SC, Park GJ, Choi SM, Kim KW (2001) The patterns of plasma free amino acids after force-feeding in rainbow trout *Oncorhynchus mykiss* (Walbaum) with and without dorsal aorta cannulation. Aquac Res 32:70–75

Olsen YA, Einarsdottir IE, Nilssen KJ (1995) Metomidate anaesthesia in Atlantic salmon, *Salmo salar*, prevents plasma cortisol increase during stress. Aquaculture 134(1–2):155–168

Olsen RE, Kiessling A, Milley JE, Ross NW, Lall SP (2005a) Effect of lipid source and bile salts in diet of Atlantic salmon, *Salmo salar* L., on astaxanthin blood levels. Aquaculture 250(3–4):804–812

Olsen RE, Sundell K, Mayhew TM, Myklebust R, Ringo E (2005b) Acute stress alters intestinal function of rainbow trout, *Oncorhynchus mykiss* (Walbaum). Aquaculture 250(1–2):480–495

Pankhurst NW, Sharples DF (1992) Effects of capture and confinement on plasma cortisol concentrations in the snapper, *Pagrus auratus*. Aust J Mar Freshw Res 43(2):345–355

Pavlidis M, Greenwood L, Paalavuo M, Molsa H, Laitinen JT (1999) The effect of photoperiod on diel rhythms in serum melatonin, cortisol, glucose, and electrolytes in the common dentex, *Dentex dentex*. Gen Comp Endocrinol 113(2):240–250

Pearson MP, Stevens ED (1991) Size and hematological impact of the splenic erythrocyte reservoir in rainbow-trout, *Oncorhynchus mykiss*. Fish Physiol Biochem 9(1): 39–50

Perry SF, Gilmour KM (2006) Acid–base balance and CO2 excretion in fish: unanswered questions and emerging models. Respir Physiol Neurobiol 154(1–2):199–215

Pickering AD, Pottinger TG, Christie P (1982) Recovery of the brown trout, *Salmo trutta* L., from acute handling stress: a time-course study. J Fish Biol 20(2):229–244

Pottinger T, Brierley I (1997) A putative cortisol receptor in the rainbow trout erythrocyte: stress prevents starvation-induced increases in specific binding of cortisol. J Exp Biol 200(14):2035–2043

Pottinger TG, Carrick TR, Appleby A, Yeomans WE (2000) High blood cortisol levels and low cortisol receptor affinity: is the Chub, *Leuciscus cephalus*, a Cortisol-Resistant Teleost? Gen Comp Endocrinol 120(1):108–117

Powell MD, Nowak BF (2003) Acid–base and respiratory effects of confinement in Atlantic salmon affected with amoebic gill disease. J Fish Biol 62(1):51–63

Redding JM, Schreck CB (1983) Influence of ambient salinity on osmoregulation and cortisol concentration in yearling coho salmon during stress. Trans Am Fish Soc 112(6):800–807

Reddy PK, Leatherland JF (2003) Influences of photoperiod and alternate days of feeding on plasma growth hormone and thyroid hormone levels in juvenile rainbow trout. J Fish Biol 63(1):197–212

Reinitz GL, Rix J (1977) Effect of tricaine methanesulfonate (MS-222) on hematocrit values in rainbow trout (*Salmo gairdneri*). Comp Biochem Physiol C 56(2):115–116

Schultz IR, Hayton WL (1991) Experimental design for pharmacokinetic data analysis using compartmental pharmacokinetic models. In: Mayes MA, Barron MG (eds) Aquatic toxicology and risk assessment. American Society for Testing and Materials, Philadelphia, pp 139–148

Small BC (2005) Effect of fasting on nychthemeral concentrations of plasma growth hormone (GH), insulin-like growth factor I (IGF-I), and cortisol in channel catfish (*Ictalurus punctatus*). Comp Biochem Physiol B Biochem Mol Biol 142(2):217–223

Smith LS, Bell GR (1964) A technique for prolonged blood sampling in free-swimming salmon. J Fish Res Board Can 21(4):711–717

Soivio A, Oikari A (1976) Haematological effects of experimental stress on a teleost, *Esox lucius* L. J Fish Biol 8:397–411

Soivio A, Westman K, Nyholm K (1972) Improved method of dorsal aortal catheterization: haematological effects followed for three weeks in rainbow trout (*Salmo gairdneri*). Finn Fish Res 1:11–21

Soivio A, Malkonen M, Tuurala O (1974) Effects of asphyxia and MS 222 anaesthesia on the circulation of the kidney in *Salmo gairdneri* Richardson. A microscopical study. Ann Zool Fenn 11:271–275

Soivio A, Nynolm K, Westman K (1975) A technique for repeated sampling of the blood of individual resting fish. J Exp Biol 63(1):207–217

Soivio A, Nyholm K, Huhti M (1977) Effects of anaesthesia with MS 222, neutralized MS 222 and benzocaine on the blood constituents of rainbow trout, *Salmo gairdneri*. J Fish Biol 10(1):91–101

Sunde J, Kiessling A, Higgs D, Opstvedt J, Venturini G, Rungruangsak-Torrissen K (2003) Evaluation of feed protein quality by measuring plasma free amino acids in Atlantic salmon (*Salmo salar* L.) after dorsal aorta cannulation. Aquac Nutr 9(6):351–360

Thorarensen H, Gallaugher PE, Kiessling AK, Farrell AP (1993) Intestinal blood flow in swimming chinook salmon *Oncorhynchus tshawytscha* and the effects of hematocrit on blood flow distribution. J Exp Biol 179(1):115–129

Turner JD, Wood CM, Clark D (1983) Lactate and proton dynamics in the rainbow-trout (*Salmo gairdneri*). J Exp Biol 104(MAY):247–268

Vijayan MM, Leatherland JF (1990) High stocking density affects cortisol secretion and tissue distribution in brook charr, *Salvelinus fontinalis*. J Endocrinol 124(2):311–318

Vijayan MM, Moon TW (1992) Acute handling stress alters hepatic glycogen metabolism in food-deprived rainbow-trout (*Oncorhynchus mykiss*). Can J Fish Aquat Sci 49(11):2260–2266

Waring CP, Stagg RM, Poxton MG (1992) The effects of handling on flounder (*Platichthys flesus* L.) and Atlantic salmon (*Salmo salar* L.). J Fish Biol 41(1):131–144

Willford WA (1970) Effect of MS-222 on electrolyte and water content in the brain of rainbow trout. Bur Sport Fish Wildl Invest Fish Cont 43:3–7

Zohar Y (1980) Dorsal aorta catheterization in rainbow trout (*Salmo gairdneri*) I. Its validity in the study of blood gonadotropin patterns. Reprod Nutr Dev 20(6):1811–1823

Effects of Production Intensity and Production Strategies in Commercial Atlantic Salmon Smolt (*Salmo salar* L.) Production on Subsequent Performance in the Early Sea Stage – Summary

Torstein Kristensen, Trond Haugen, Trond Rosten, Anders Fjellheim, Åse Åtland and Bjørn Rosseland

Juvenile Atlantic salmon is produced in land-based freshwater facilities and from there transferred to cages in the sea. Variability in mortality rates and growth during the early sea stage was large, with up to 50-60% mortality in some extreme cases, demonstrating an improvement potential in terms of fish welfare. The analysis showed some effects of production intensity on subsequent fish growth and mortality at sea, but the general picture was that high-intensity production did not result in poorer results after sea-transfer. In fact, our results suggest that intermediate production intensity with poor environmental control (too much or too little available oxygen) resulted in reduced performance of the fish.

Background

During the juvenile stages up until 70-120 g size, Atlantic salmon is produced in land-based fresh water facilities. During the last decade a dramatic increase in the numbers of fish produced at each site implies an intensification of the production. Less available space and fresh water for each individual fish may carry welfare consequences. This issue was addressed using data on water quality, production intensity, fish growth and mortality from a large number of commercial production sites.

Scientific Significance

This work presents industry data that provide a frame of reference and help define further experimental work in the field. It also demonstrates the improvement potential the industry has in terms of fish welfare.

Practical Application

Within the general public, the assumption of poorer animal welfare in high intensity production systems vs. more "natural" systems is often made. This work indicates that this is not generally the case in salmon smolt production, and may therefore contribute to a more balanced debate on the topic.

Contact

T. Kristensen
Norwegian Institute for Water Research (NIVA),
P.O. Box 1266, Pirsenteret, 7462 Trondheim, Norway
e-mail: torstein.kristensen@niva.no

Fish Physiol Biochem (2012) 38:273–282
DOI 10.1007/s10695-011-9566-0

Effects of production intensity and production strategies in commercial Atlantic salmon smolt (*Salmo salar* L.) production on subsequent performance in the early sea stage

T. Kristensen · T. O. Haugen · T. Rosten · A. Fjellheim · Å. Åtland · B. O. Rosseland

Received: 30 March 2010 / Accepted: 14 October 2011 / Published online: 26 October 2011

Abstract A data set from commercial Atlantic salmon (*Salmo salar* L.) producers on production intensity and production strategies in smolt tanks ($N = 63$–94) was obtained during 1999–2006. The effects of production intensity on subsequent fish mortality and growth during the early sea phase (90 days) were examined by principal component analysis and subsequent generalized linear model analysis. Levels of accumulated metabolites (CO_2, total ammonia nitrogen and NH_3), and information provided by producers (production density (kg fish m^{3-1}), specific water use (l kg fish^{-1} min^{-1}) and oxygen drop (mg l^{-1}) from tank inlet to tank outlet), were used as predictor variables. In addition, several other welfare relevant variables such as disease history, temperature during freshwater and sea stage; season (S1) or off-season (S0) smolt production; and the use of seawater addition during the freshwater stage were analyzed. No strong intensity effects on mortality or growth were found. CO_2 levels alone ($P < 0.001$, $R^2 = 0.16$), and in combination with specific water use ($R^2 = 0.20$), had the strongest effect on mortality. In both cases, mortality decreased with increasing density. For growth, the intensity model with the most support ($R^2 = 0.17$) was O_2 drop, density and their interaction effects, resulting in the best growth at low and high intensity, and poorer growth at intermediate levels. Documented viral disease outbreaks (infectious pancreatic necrosis and two cases of pancreas disease) in the sea phase resulted in significantly higher mortalities at 90 days compared with undiagnosed smolt groups, although mortalities were highly variable in both categories. The temperature difference between the freshwater stage and seawater had a small, but significant, effect on growth with the best growth in groups stocked to warmer seawater ($P = 0.04$, $R^2 = 0.06$). S0 and S1 smolt groups did not differ significantly in growth, but the mortality was significantly ($P = 0.02$) higher in S1 groups. Seawater addition as a categorical variable had no significant effects, but when analyzed within the seawater addition group, intermediate salinities (15–25 ppt) gave the best results on growth ($p = 0.04$, $R^2 = 0.15$). Production intensity had small explanatory power on subsequent seawater performance in the analyzed smolt groups. If anything, the analysis shows a beneficial effect of intensive production strategies on subsequent performance.

T. Kristensen · T. O. Haugen · B. O. Rosseland
Department of Ecology and Natural Resource Management, Norwegian University of Life Sciences (UMB), P.O. Box 5003, 1432 Aas, Norway

T. Kristensen (✉) · T. O. Haugen · T. Rosten · A. Fjellheim · Å. Åtland
Norwegian Institute for Water Research (NIVA), P.O. Box 1266, Pirsenteret, 7462 Trondheim, Norway
e-mail: torstein.kristensen@niva.no

Present Address:
T. Rosten
SINTEF Fisheries and Aquaculture, P.O. Box 4762, Sluppen, 7465 Trondheim, Norway

Analysis of the various production strategies indicates better survival of S0 compared with S1 smolt groups, improved growth when stocked in seawater warmer than freshwater, and a negative effect of viral disease outbreaks on survival. The results clearly demonstrate the difficulty of extrapolating results from experimental work on fish welfare and production intensity variables to commercial production. On the other hand, the presented results may simply demonstrate that the traditional fish welfare criteria growth and mortality may not suffice to evaluate welfare consequences of suboptimal water quality or production strategies in the aquaculture industry.

Keywords Atlantic salmon · Aquaculture · Fish welfare · Water quality · Stocking density · Production intensity

Introduction

Commercial production of Atlantic salmon (*Salmo salar*) smolts has undergone dramatic increases in numbers of fish produced, scale of production units and advancements in production technology during recent decades. In Norway, the annual production has increased from about 130 million in 2000 to an estimated 250 in 2010 (R Nystøyl, pers comm.), while the number of production sites has remained stable or declined (Fig. 1). Most Norwegian production sites were constructed in the 1980s, a period where both government regulations and production technology limited production to <1 million smolts annually per facility. The facilities were dimensioned and their freshwater sources chosen accordingly. Later, production has been increased largely by a combination of increased production capacity pr site and increased production intensity. Recirculating aquaculture systems (RAS) are also increasingly applied, although their overall production capacity is still modest compared with traditional flow-through production (Bergheim et al. 2009).

Production intensification can be visualized as a stepwise process where a limiting water quality factor is being counteracted by implementing new technology. From the initial simple flow-through systems, production capacity was increased dramatically by adding pure oxygen to the water. Reducing specific water use to below 0.3 l kg fish^{-1} min^{-1} causes CO_2 accumulation to levels that have been shown to cause effects (Fivelstad et al. 2003b). CO_2 degassing by aeration is utilized, to increase production capacity further, with a resulting potential for ammonia, and especially ammonium to cause welfare problems (Thurston et al. 1984; Fivelstad et al. 1993; Wicks et al. 2002).

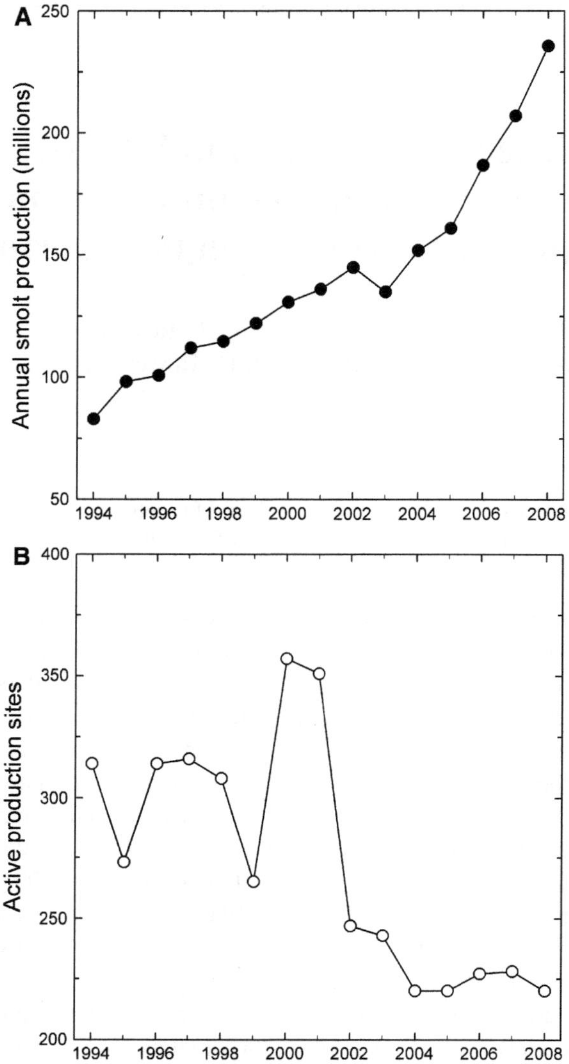

Fig. 1 Official Norwegian statistics on annual smolt production (**a**) and number of active smolt production sites (**b**) in Norway from 1994 to 2008. *Source*: Directorate of Fisheries

The production intensity concept can be viewed either as density, i.e., as a function the number/biomass of individuals confined to a given water volume (kg m^{3-1}), or as specific water use, i.e., the rate of water exchange relative to fish biomass (L kg $fish^{-1}$ min^{-1}). In practical terms, the two are mostly interconnected, but there are ongoing discussions about their relative importance in terms of fish welfare (Person-Le Ruyet et al. 2008). Reduced specific water use causes degradation of water quality by metabolite accumulation, thereby potentially affecting fish welfare, while high densities may compromise welfare through both social interactions and physical effects (Person-Le Ruyet et al. 2008).

The sensitivity to a range of stressors during smoltification of salmonids often results in delayed responses in the sea stage to conditions experienced during the freshwater stage. This has been demonstrated for metal exposure (Monette and McCormick, 2008; Kroglund et al., 2007), pesticides (Moore et al. 2003, 2007, Waring and Moore 2004) and production intensity. Banks (1992) found that increased rearing density in raceway production of coho salmon (Oncorhynchus kisutch) produced small but significant reductions in percent survival of released smolts and reduced $Na^{+}-K^{+}$ ATPase activity in the highest density. Banks (1994) described negative effects of increased density in combination with reduced water flow during the freshwater stage of Chinook salmon (Oncorhynchus tshawytscha), and significant effects were seen on subsequent performance in sea-water challenge tests and post-release performance.

Reduced growth during exposure to elevated CO_2 concentrations is frequently reported, but growth during the subsequent sea phase is most often not significantly affected (Martens et al. 2006; Hosfeldt et al. 2008). Hosfeldt et al. (2009) reported no negative effects of density up to 86 kg m^{3-1} t on growth or physiology of smolting Atlantic salmon when specific water use was kept constant (0.3 l kg $fish^{-1}$ min^{-1}). Similar findings have been demonstrated repeatedly in salmonids (e.g., North et al. 2006). However, the social interactions between individuals (e.g., fin erosion) constitute an additional welfare concern (Person-Le Ruyet et al. 2008), and disease dynamics may be affected by the number of available hosts (Banks 1994).

In addition to the direct effects of metabolite accumulation at low specific water use, indirect effects may arise as a consequence of water chemistry alterations such as pH decrease and aluminium mobilization (Fivelstad et al. 2003a) or hyperoxic conditions in the tank water due to insufficient oxygenation control (Hosfeldt et al. 2008; Espmark and Baeverfjord 2009). The effects of hyperoxia may range from increased growth rates at low levels (Hosfeldt et al. 2008) to behavioural changes and reduced growth (Espmark and Baeverfjord 2009), alterations in antioxidant systems (Olsvik et al. 2005), compromised immune-competence (Toften et al. unpublished data) and mortalities (Lygren et al. 2000) at higher levels. Delayed effects of hyperoxia exposure have also been shown (Brauner et al. 2000).

A number of fish welfare issues have been raised with regard to applied technology such as water heating during early life stages, light manipulation for off-season smolt production (S0), water oxygenation and CO_2 aeration, seawater addition and recirculation production.

Scientific efforts to determine the effect-limits of a number of variables have resulted in both adjustments in production strategies and legislation. In Norway, the legislative authorities (Norwegian Food Safety Authority) have developed guidelines for maximum and minimum concentrations of important water quality variables having adverse effects on fish health (Aquaculture Act, FOR 2008-06-17 No. 822, annotations to § 21). Kristensen et al. (2009) and Rosten et al. (2010) present data on current status for various intensity variables in Norwegian commercial salmon aquaculture. In general, high variability between production sites and a substantial proportion of sites with production intensity above recommended limits is reported. This indicates potential welfare challenges in the industry.

Aside from the more fine-tuned fish welfare evaluation tools available in scientific experiments, data on production intensity and production strategies, mortality, growth rates and disease outbreaks are available from the industry. In this paper, we present data collected from Norwegian smolt producers participating in annual surveys from 1999 to 2006 on water quality, production intensity and production strategies during the last freshwater phase, linked to mortality rates, disease history and growth in the early sea phase.

Materials and methods

Data

The presented results are based on water quality monitoring data obtained by the Norwegian Institute for Water Research (NIVA). Norwegian smolt producers participating in annual water quality monitoring projects from 1999 to 2006 (WQ 1999–2006) sampled water from the outlet of smolt tanks approximately 1 week prior to sea transfer. Analysis was conducted at NIVA using accredited methods described in Kristensen et al. (2009). A questionnaire containing information on the same smolt groups with regard to mortality, disease history and growth performance in the early sea phase (2, 30 and 90 days) was submitted by the participating smolt producers. Levels of accumulated metabolites (CO_2, total ammonia nitrogen (TAN) and NH_3), and information provided by producers (production density (kg fish m^{3-1}), specific water use (l kg fish^{-1} min^{-1}) and oxygen drop (mg l^{-1}) from tank inlet to tank outlet), was used as predictor variables. In addition, several other welfare relevant variables such as disease history (Infectious pancreatic necrosis (IPN) outbreaks); temperature during freshwater and sea stage; season (S1) or off-season (S0) smolt production; and the use of seawater addition during the freshwater stage were analyzed. Of a total number of more than 140 participating producers, sufficient information on fish performance and other data was provided by 63–94 smolt groups, depending on the variables in question.

Statistical treatment

Statistical analysis was performed in R v2.10.1 (R Development Core Team 2010). Owing to the colinearity issues among many of the focal predictor variables involved in this study, a principal component analysis (PCA) was undertaken (Jolliffe 2002). The aim of this analysis was to reduce the dimensionality of the predictor variables while retaining as much as possible of the variation present in the data set. This could be achieved by either transforming to a new set of variables, the principal components (PCs), which are uncorrelated, or to use some of the initial variables with low or no colinearity issues as revealed from the PCA.

Two sets of response variables were analyzed: mortality (at 48 h, 30 days and 90 days) and mean smolt group growth during the 90-day period as specific growth rate (SGR). The quantification was conducted using generalized linear models (GLM, McCullagh and Nelder 1989). This modelling approach allows for both continuous and categorical predictors. We also explored combined effects of the various predictors (using combinations of variables as suggested from the PCA), and the most supported model was selected using Akaike's Information Criterion (AIC) (Burnham and Anderson 1998). Categorical predictor variables were tested using one-way anova on log-transformed data.

Results

Production intensity

Based on the PCA analysis (Table 1), the variables CO_2, NH_3 and density could be used together without inferring the problems of colinearity. Many of the variables loaded relatively similar on PC1 (Table 1), and further use of the principal components was therefore considered as inappropriate.

Table 1 Variance table from PCA analysis of production intensity factors

Principal components	PC1	PC2	PC3	PC4	PC5	PC6
Standard deviation	1.554	1.087	0.996	0.785	0.758	0.469
Proportion of variance	0.402	0.197	0.165	0.103	0.095	0.036
Cumulative proportion	0.402	0.599	0.765	0.867	0.963	1.000
Factor loading						
NH_4^+	−0.50	0.40	−0.05	0.34	−0.23	−0.64
NH_3	−0.21	0.74	0.35	−0.28	0.31	0.34
CO_2	−0.55	−0.12	−0.20	0.40	−0.22	0.66
O_2-drop	−0.36	−0.09	−0.67	−0.38	0.51	−0.11
density	−0.32	−0.44	0.55	0.25	0.57	−0.14
Specific water use	0.42	0.29	−0.29	0.66	0.46	0.06

Mortality at 90 days post-sea transfer (PST) was the most sensitive response variable, and further analysis was continued on this mortality response variable only. Mortality decreased significantly with increasing CO_2 concentration (Fig. 2) with a simple regression model with Z as mortality as ratio: $\arcsin(\text{sqrt}\ (Z)) = 0.53–0.13*\ln(CO_2)$ ($P < 0.001$, $R^2 = 0.16$, F:16.65 on 1 and 88 DF). The most AIC-supported model for mortality 90 days PST was a function of CO_2 and specific water use without interactions between the two (Fig. 3), with $P < 0.001$ and $R^2 = 0.20$ (F:10.84 on 2 and 87 DF).

Growth rate at 90 days PST was insensitive to all single intensity variables. The most AIC-supported model was a function of O_2 drop and density, with interaction effects between the two (Fig. 4). This model explained about 17% of the SGR variation ($P < 0.001$, $R^2 = 0.17$, F: 6.257 on 3 and 90 DF) with the best SGR on low O_2 drop and density and high O_2 drop and density, respectively.

Temperature

Neither temperature at 90 days PST or in smolt tanks at the time of release were significantly correlated with 90 days PST growth or mortality. However, the temperature difference between the two time points gave weak but significant a negative correlation with SGR (Fig. 5), indicating a beneficial, although weak, effect of sea transfer at high sea temperatures relative to freshwater temperatures ($P = 0.045$, $R^2 = 0.06$, F: 4.194 on 1 and 62 DF).

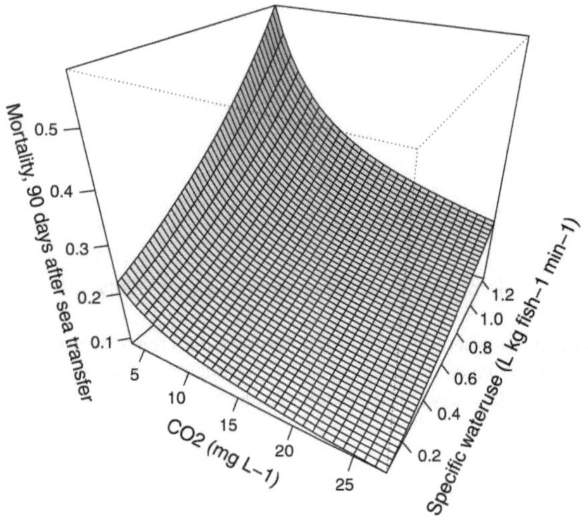

Fig. 3 Estimated combined effects of CO_2 and specific water use in the final freshwater stage on mortality rates at day 90 post-sea transfer in commercial smolt groups

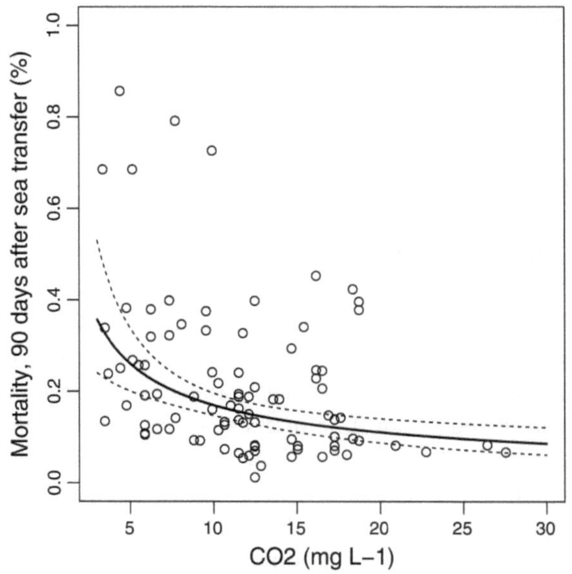

Fig. 2 Estimated effect of CO_2 in the final freshwater stage on mortality rates at day 90 post-sea transfer in commercial smolt groups ($N = 94$). Broken lines are 95% confidence interval (CI)

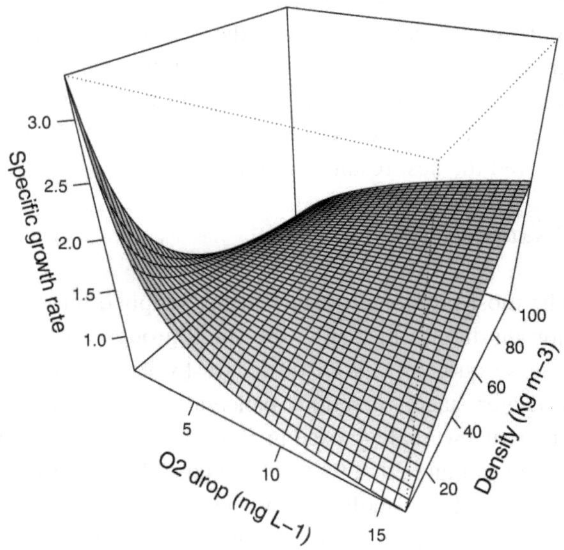

Fig. 4 Estimated combined effects of O_2 drop and density in the final freshwater stage on specific growth rate (% bw day^{-1}) at day 90 post-sea transfer in commercial smolt groups

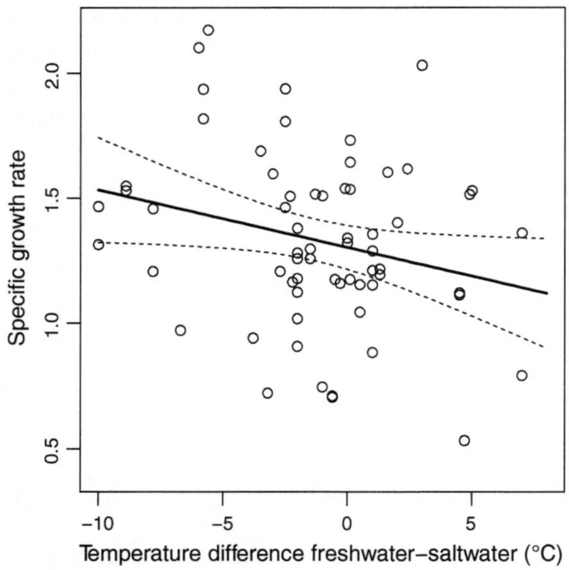

Fig. 5 Estimated effect of temperature difference between the final freshwater stage and 90 post-sea transfer on specific growth rate (% bw day^{-1}) in commercial smolt groups ($N = 63$). Broken lines are 95% CI

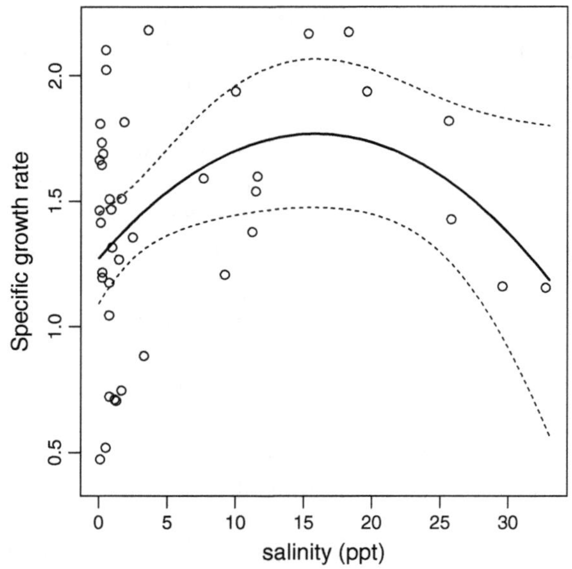

Fig. 6 Estimated effect of salinity in the final freshwater stage in smolt groups produced using seawater addition on specific growth rate (% bw day^{-1}) at day 90 post-sea transfer in commercial smolt groups ($N = 42$). Broken lines are 95% CI

Salinity during smolt production

Of a total of 94 smolt groups with data for this variable, 42 groups came from tanks with sea-water addition while the remaining 52 were produced using freshwater only. When used as a categorical variable, salinity had no significant effects on the 90 day PST growth or mortality. When the effects of salinity on SGR were analyzed within the sea-water addition groups, a significant polynomial relationship was found (Fig. 6) ($P = 0.048$, $R^2 = 0.14$, F:3.28 on 2 and 38 DF). Intermediate salinities of 15–25 ppt showed the best results on SGR.

Disease history

The most prominent feature of the supplied information on disease history in the smolt groups was viral disease outbreaks, both in the freshwater and in the seawater stage. Infectious pancreatic necrosis (IPN) outbreaks were diagnosed in 50 of 94 smolt groups in freshwater and 20 of 94 smolt groups 90 days PST. In addition, two smolt groups were diagnosed with pancreas disease (PD) within 90 days PST. No interaction effects were found in this variable between the freshwater and seawater stage. Viral disease outbreak 90 days PST, when used as a categorical variable, resulted in significantly higher mortality rates, but had no significant effect on SGR (Fig. 7).

Timing of sea transfer

In the data set, 29 groups were off-season (S0) smolts transferred to sea in the autumn, while 65 were smoltified and transferred in spring (S1). S0 smolts had significantly lower mortality rates, while no significant effects were found on SGR (Fig. 7).

Discussion

Data quality

The accuracy of questionnaire-based information is inherently difficult to validate. In the presented case, the combination of measured variables and questionnaire-based information offers some insurance of the density and production intensity data. Generally, significant correlations between measured water quality variables (CO_2 and TAN) and the density, O_2 drop and specific water use information were found. In the case of mortalities and growth at sea, accuracy may vary. However, due to the economical importance of these variables, they are assumed to be closely

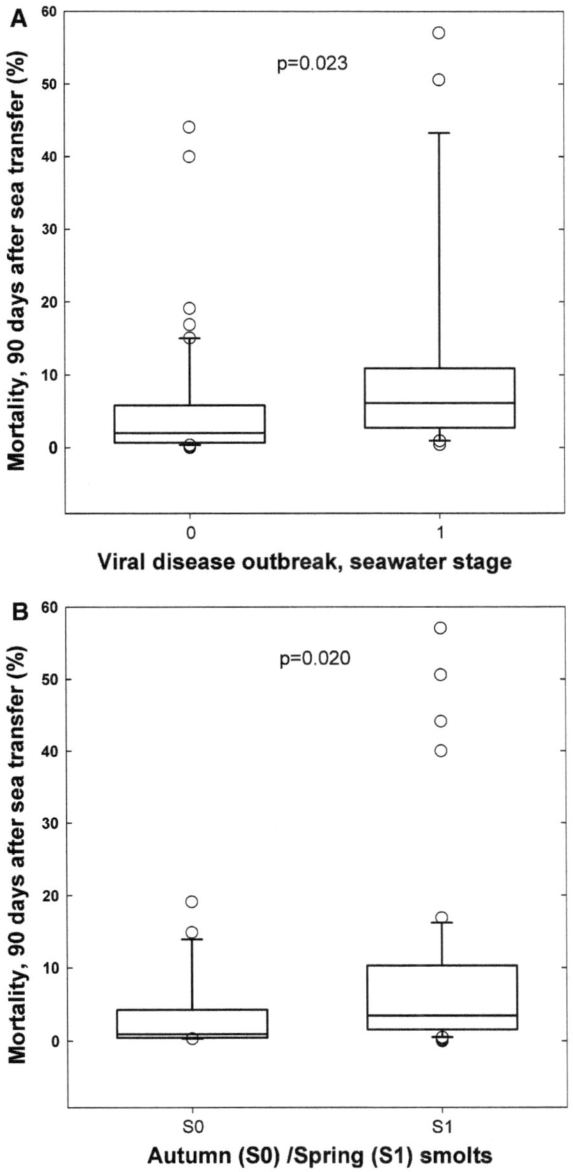

Fig. 7 Median, 25–75 percentile (*box*) and 5–95 percentile (*bars*) of mortality (%) during the first 90 days after sea transfer. The categorical variables used for comparison were (**a**) viral disease outbreak during seawater stage and (**b**) smolt groups stocked as either S0 (autumn) or S1 (spring) smolts. Significant differences are indicated by p values in the graphs (one-way Anova)

monitored. The available data set may not be sufficiently large to identify all relevant relationships with sufficient power, and there are also some counterintuitive results that may in part be explained by limitations in the data set. The lack of temperature effects on SGR is an unlikely result, but this may be caused by other variables overwhelming the physiological effects of temperature on growth.

Production intensity

The effects of density variables explained only a small proportion of the observed variance in growth and mortality 90 days PST, with the best supported model of CO_2 and specific water use explaining only 20% of the variation in mortality. The two variables are naturally partly connected, but the practice of CO_2 degassing and the use of water qualities with highly differing pH and buffer capacities (Kristensen et al. 2009, Fig. 6) makes the correlation between them rather weak ($P < 0.001$, $R^2 = 0.17$, F:19.19 on 1 and 92 DF). The direction of the response (Fig. 3) is also counterintuitive, given the vast amount of documented effects on of both CO_2 level and specific water use (Banks 1992, 1994; Fivelstad et al. 1999, 2003b and 2004). However, the effects on mortality and growth extended into the early sea stage in an aquaculture setting are rarely seen (Fivelstad et al. 2003b, 2004, Hosfeldt et al. 2008, 2009). Most reports on delayed sea-stage responses are on supplementary fish releases (Banks 1992, 1994). However, the reports on reduced performance in sea-water challenge tests (Banks 1992), viral challenge tests (Toften et al. *unpublished*) and reduced NKA activity (Banks 1994) still suggests negative effects.

In view of current oxygenation practices in salmonid aquaculture, an interaction effect between O_2 drop and density on SGR seems reasonable, although the model only explains about 17% of the variation in 90 days PST SGR (Fig. 4). The best SGR is obtained in cases with low O_2 drop and low density, i.e., a well O_2 regulated, low-density environment. The poorest SGR in the model occurs in a low-density, high O_2 drop situation, i.e., a hyperoxygenated environment where much of the O_2 is lost to air. The other situation that gives good SGR is a high-density, high O_2 drop situation. In this case, the fish receives sufficient O_2 and avoid periodic or chronic sublethal hypoxia. High levels of hyperoxia have been shown to depress growth during exposure (Lygren et al. 2000) and affect disease resistance (Toften et al. unpublished) and seawater tolerance post-sea transfer (Brauner et al. 2000). On the other hand, more moderate levels of hyperoxia resulted in increased FW growth and no significant differences in subsequent SW growth Hosfeldt et al. (2008).

Temperature

The lack of a significant relationship between 90 days PST SGR and actual temperature in the sea stage is quite surprising and suggests strong influences by other variables on growth. The variability in this parameter (median: 10, range 3.0–17.8°C) is also high. Sea transfer at times with higher sea temperatures than freshwater temperatures may be beneficial to growth (Fig. 5), but no significant effects on mortality were found. Overall, the effects of temperature are weak in the material.

Salinity during smolt production

Sea-water addition during smolt production is a common practice in Norway, and salinity levels range from 0.1 to 32.7 ppt in the data (Fig. 6). Beneficial effects on SGR at intermediate salinity may result from a number of effects, including enhanced smoltification by gradual salinity transition, beneficial temperatures and improved water quality by increased available water volume at the smolt stage. The cost of osmoregulation is reduced at intermediate salinities, and this should theoretically improve the scope for growth (Boef and Payan 2001). For Atlantic salmon, reports on salinity effects on growth are ambiguous. Fivelstad et al. (2004) reports improved growth at 4.5 ppt compared with lower salinities, and the best growth at high salinity combined with high specific flow rates. Handeland and Stefansson (2002) reported no beneficial effects of salinity acclimation, and a reduced growth after sea transfer. Imsland et al. (2010) found better growth in freshwater than seawater in smolts.

Disease history

It is not surprising to find a significant effect on mortality 90 days PST in diagnosed viral outbreak smolt groups (Fig. 7). The number of affected fish groups, in both the freshwater phase and early sea phase, shows the severe challenges viral infections pose in salmonid aquaculture. Several studies on interactions between intensive smolt production and subsequent IPNV susceptibility after sea transfer have documented negative effects of intensive rearing (Toften et al., unpublished). No such interactions were found in this study. Smolt groups without diagnosed disease also showed very variable mortality rates (median: 2.1%, range: 0.02–44.1%). The reasons quoted by the producers were diverse and unquantifiable, but the numbers point to a general problem with high mortalities at the early sea phase in the industry.

Timing of sea transfer

S0 smolt groups had similar SGR and significantly lower 90 days PST mortality rates compared with S1 smolt groups (Fig. 7). The number of S0 groups was lower (29 vs. 65), so this pattern may not fully represent the actual situation in the industry. However, the artificial situation of stimulating smoltification through photo-manipulation at decreasing temperatures (Handeland and Stefansson 2001) does not seem to affect subsequent performance. This finding supports Lysfjord et al. (2004) who found S0 smolts to outperform S1 smolts in the early sea stage. It should also be noted that the S0 groups consist of rapid growing under-yearlings, i.e., the upper modal group in a parr population, and that this growth pattern may be maintained throughout the production cycle.

Conclusions

Production intensity had small explanatory power on subsequent seawater performance in the analyzed smolt groups. If anything, the analysis shows a beneficial effect of intensive production conditions on subsequent performance. Analysis of the various production strategies indicates better survival of S0 compared with S1 smolt groups, improved growth when stocked in seawater warmer than freshwater, and a negative effect of viral disease outbreaks in the sea stage on survival. The results clearly demonstrate the difficulty of extrapolating results from experimental work on fish welfare and production intensity variables to commercial production. On the other hand, the presented results may simply demonstrate that the traditional fish welfare criteria growth and mortality may not suffice to evaluate welfare consequences of suboptimal water quality or production strategies in the aquaculture industry.

Acknowledgments The authors would like to thank all participating industry partners for supplying the data. Data compilation and analysis were funded by NIVA and the Research Council of Norway through contract no. 165239/S40.

References

Banks J (1992) Effects of density and loading on coho salmon during hatchery rearing and after release. Prog Fish-Cult 54:137–147

Banks J (1994) Raceway density and water flow as factors affecting spring Chinook salmon (Oncorhynchus tshawytscha) during rearing and after release. Aquaculture 119:201–217

Bergheim A, Drengstig A, Ulgenes Y, Fivelstad S (2009) Production of Atlantic salmon smolts in Europe-Current characteristics and future trends. Aquacultural Eng 41:46–52

Boef G, Payan P (2001) How should salinity influence fish growth? Comp Biocem Physiol C 130:411–423

Brauner CJ, Seidelin M, Madsen SS, Jensen FB (2000) Effects of fresh water hyperoxia and hypercapnia exposures and their influences on subsequent seawater transfer in Atlantic salmon (*Salmo salar*) smolts. Can J Fish Aquat Sci 57:2054–2064

Burnham KP, Anderson DR (1998) Model selection and inferences. Springer Verlag, New York

Espmark Å, Baeverfjord G (2009) Effects of hyperoxia on behavioural and physiological variables in farmed Atlantic salmon (*Salmo salar*) parr. Aquac Int 17:341–353

Fivelstad S, Kallevik H, Iversen HM, Møretrø T, Våge K, Binde M (1993) Sublethal effects of ammonia in soft water on Atlantic salmon smolts at a low temperature. Aquaculture Int 1:157–169

Fivelstad S, Olsen AB, Kløften H, Ski H, Stefansson S (1999) Effects of carbon dioxide on Atlantic salmon (*Salmo salar* L.) smolts at constant pH in bicarbonate rich freshwater. Aquaculture 178:171–187

Fivelstad S, Waagbø R, Zeitz SF, Hosfeld ACD, Olsen AB, Steffansson S (2003a) A major water quality problem in smolt farms: combined effects of carbon dioxide, reduced pH and aluminium on Atlantic salmon (*Salmo salar* L.) smolts: physiology and growth. Aquaculture 215:339–357

Fivelstad S, Olsen AB, Åsgård T, Baeverfjord G, Rasmussen T, Vindheim T, Stefansson S (2003b) Long-term sublethal effects of carbon dioxide on Atlantic salmon smolts (*Salmo salar* L.): ion regulation, haematology, element composition, nephrocalcinosis and growth parameters. Aquaculture 215:301–319

Fivelstad S, Bergheim A, Hølland PM, Fjermedal AB (2004) Water flow requirements in the intensive production of Atlantic salmon (*Salmo salar* L.) parr–smolt at two salinity levels. Aquaculture 231:263–277

Handeland SO, Stefansson SO (2001) Photoperiod control and influence of body size on off-season parr-smolt transformation and post-smolt growth. Aquaculture 192:291–307

Handeland SO, Stefansson SO (2002) Effects of salinity acclimation on pre-smolt growth, smolting and post-smolt performance in off-season Atlantic salmon smolts (*Salmo salar* L.). Aquaculture 209:125–137

Hosfeld CD, Engevik A, Mollan T, Lunde TM, Waagbø R, Olsen AB, Breck O, Stefansson S, Fivelstad S (2008) Long-term separate and combined effects of environmental hypercapnia and hyperoxia in Atlantic salmon (*Salmo salar L.*) smolts. Aquac 280:146–153

Hosfeld C, Hammer J, Handeland S, Fivelstad S, Stefansson S (2009) Effects of fish density on growth and smoltification in intensive production of Atlantic salmon (*Salmo salar L.*). Aquac 294:236–241

Imsland AK, Våge KA, Handeland S, Stefansson S (2010) Growth and osmoregulation in Atlantic salmon (*Salmo salar* L.) smolts in response to different feeding frequencies and salinities. Aquaculture Res 42:469–479

Jolliffe IT (2002) Principal component analysis, 2nd edn. Springer-Verlag, New York

Kristensen T, Åtland Å, Rosten T, Urke HA, Rosseland BO (2009) Important influent-water quality parameters at freshwater production sites in two salmon producing countries. Aquacultural Eng 41:53–59

Kroglund F, Finstad B, Steffansson S, Nilsen TO, Kristensen T, Rosseland BO, Teien HC, Salbu B (2007) Exposure to moderate acid water and aluminum reduces Atlantic salmon post-smolt survival. Aquaculture 273:360–373

Lygren B, Hamre K, Waagbø R (2000) Effect of induced hyperoxia on the antioxidant status of Atlantic salmon *Salmo salar* L. fed three different levels of dietary vitamin E. Aquac Res 31:401–407

Lysfjord G, Jobling M, Solberg C (2004) Atlantic salmon smolt startegy affects body composition and early seawater growth. Aquaculture 237:191–205

Martens LG, Witten PE, Fivelstad S, Huysseune A, Sævareid B, Vikeså V, Obach A (2006) Impact of high water carbon dioxide levels on Atlantic salmon smolts (*Salmo salar* L.): Effects on fish performance, vertebrae composition and structure. Aquaculture 261:80–88

McCullagh P, Nelder JA (1989) Generalized linear models, 2nd edn. Chapman and Hall, London

Monette MY, Bjornsson BT, McCormick SD (2008) Effects of short-term acid and aluminum exposure on the parr-smolt transformation in Atlantic salmon (*Salmo salar*): Disruption of seawater tolerance and endocrine status. Gen Compar Endocrinol 158:122–130

Moore A, Scott AP, Lower N, Katsiadaki I, Greenwood N (2003) The effects of 4-nonylphenol and Atrazine on Atlantic Salmon (*Salmo salar* L) Smolts. Aquaculture 222:253–263

Moore A, Lower N, Mayer I, Greenwood L (2007) The impact of a pesticide on migratory activity and olfactory function in Atlantic salmon (*Salmo salar* L.) smolts. Aquaculture 273:350–359

North BP, Turnbull JF, Ellis T, Porter MJ, Migaud H, Bron J, Bromage NR (2006) The impact of stocking density on the welfare of rainbow trout (Oncorhynchus mykiss). Aquaculture 255:466–479

Olsvik PA, Kristensen T, Waagbø R, Rosseland BO, Tollefsen KE, Bæverfjord G, Berntssen MHG (2005) SOD, CAT and GSH-Px mRNA expression and lipid peroxidative stress in liver of Atlantic salmon *Salmo sal*ar exposed to hyperoxic conditions during smoltification. Comp Biochem Physiol C 141:314–323

Person-Le Ruyet J, Labbe L, Le Bayon N, Severe A, Le Roux A, Delliou H, Quemener L (2008) Combined effects of water quality and stocking density on welfare and growth of rainbow trout (*Oncorhynchus mykiss*). Aquatic Living Resour 21:185–195

R Development Core Team (2010) R: a language and environment for statistical computing, version 2.10.1.R foundation for statistical computing. Vienna, Austria

Rosten T, Kristensen T, Knight CM, Suciu R (2010) Landbased intensive aquaculture of sturgeons: can we learn something from the salmon experiences? J Appl Ichthyoly (submitted)

Thurston RV, Russo RC, Luedtke RE, Smith CE, Meyn EL, Chakoumakos C, Wang K (1984) Chronic toxicity of Ammonia to rainbow trout. Trans Am Fish Soc 113:56–73

Waring CP, Moore A (2004) The effect of atrazine on Atlantic Salmon (*Salmo salar*) smolts in fresh water and after sea water transfer. Aquat Toxicol 66:93–104

Wicks BJ, Joensen R, Tang Q, Randall DJ (2002) Swimming and ammonia toxicity in salmonids: the effect of sub lethal ammonia exposure on the swimming performance of coho salmon and the acute toxicity of ammonia in swimming and resting rainbow trout. Aquatic Toxicol 59:55–69

Background Information on the Network

The project "Welfare of fish in European aquaculture" a European network (COST Action 867)

Cost, an intergovernmental framework for European Cooperation in Scientific and Technical research, sponsored the European network "Welfare of fish in European aquaculture" (Cost action 867) than ran from March 2006 to March 2011. This book is a major deliverable of the European network (acronym Wellfish).

The principal aim of the network was to bring all relevant stakeholders together 1) to harmonise the fish welfare concept throughout Europe and 2) increase awareness of relevant stakeholders with regard to fish welfare and 3) to provide a synthesis to contribute to a secure foundation for the development of operational welfare indicators (OWIs)–measurements or assessments of welfare that can be made practically at aquaculture facilities –, in order to monitor and safeguard the welfare of farmed fish.

The Wellfish project was chaired by prof. Anders Kiessling with prof. Marco Saroglia as vice chair.

The activities were divided into 3 Working Groups in order to achieve the overall objectives stated above as follows:

- WG1 Aspects of fish biology relating to welfare (chair dr.BørgeDamsgård, and vice chair dr. Ana Roque)
- WG2 Operational welfare indicators in farmed fish (chairprof. dr. James Turnbull and vice chair dr. Ian Duncan)
- WG3 Welfare and management (chair dr. Sunil Kadri and vice chairdr. Hans van de Vis).

The first WG on biological processes and fish welfare collated fundamental and comparative knowledge on the importance of different factors, e.g. brain physiology, stress physiology, health, behaviour, growth and environmental conditions, which are a prerequisite for the development, understanding and validation of fish welfare indicators in situations relevant to experimental and production environments. The second WG formulated welfare indicators and produced OWI–matrices including production experience, environmental factors and key management procedures. The fish welfare and management (WG 3)synthesised the interests of the main stakeholder groups in fish welfare, namely research, regulatory authorities and industry, to formulate recommendations, control and support tools for decision makers in government and industry.

At the start of the Wellfish project national representatives of the countries that participated in this network formed the management committee (MC). During the first meeting the MC appointed the chair and vice chair of the project, the chairs and vice chairs of the WGs. During the lifetime of the network the MC was responsible for overall co-ordination and critical appraisal of on-going and planned activities to meet the general objectives of the Wellfish project.The MC consisted of the following participants from Cost and non-Cost countries:

Chair

Prof. Anders Kiessling
Department of Animal and Aquacultural Sciences, Norwegian University of Life Sciences,
PO Box 5003 1432 AAs Norway

Present address

Department of Wildlife, Fish and Environmental Studies, Swedish University of Agricultural Sciences, 901 83 Umeå, Sweden
anders.kiessling@slu.se

Vice Chair

Prof. Marco Saroglia
University of Insubria Department of Biotechnology and Molecular Sciences, via JH Dunant n 3, I-21100 Varese, Italy
marco.saroglia@uninsubria.it

Domain Committee Rapporteur

Dr. Ragni Ofstad
Nofima, Osloveien 1, 1430 Aas, Norway
ragni.ofstad@nofima.no

Belgium

Prof. Patrick Kestemont
Research Unit in Environmental and Evolutionary Biology (URBE), The University of Namur (FUNDP), Rue de Bruxelles 61, 5000 Namur, Belgium
patrick.kestemont@fundp.ac.be

Bulgaria

Mr.Krasimir Hadzhiyski
Pontika Fish farming and aquaculture research, 17 Liubotran Str., Office 2, 1407 Sofia, Bulgaria
kzh@abv.bg

Cyprus

Dr George Anastasiades
Department of Fisheries and Marine Research 101 Vithleem Street, 1416 Nicosia, Cyprus
ganastasiades@dfmr.moa.gov.cy

Dr. Ana Afonso Polyviou
Public Health Department Veterinary Services, 1417 Nicosia, Cyprus
aafonso@vs.moa.gov.cy

Present address

EFSA European Food Safety Authority, Lgo. N. Palli 5 A, 43121 Parma, Italy
ana.afonso@efsa.europa.eu

Czech Republic

Dr. Frantisek Vacha
The University of South Bohemia, Research Institute of Fish Culture and Hydrobiology, Studentska 13, 37005 Ceské Budejovice, Czech republic
vacha@zf.jcu.cz

Denmark

Prof. Kurt Buchmann
Department of Veterinary Disease Biology, Section of Biomedicine University of Copenhagen, Stigbojlen 7, 1870 Frederiksberg, Denmark
kub@life.ku.dk

Prof. David John Mckenzie
Danish Institute for Fisheries Research, PO Box 101, 9850 Hirtshals, Denmark
dmc@dfu.min.dk

Dr. Erik Höglund, MC Substitute Member
DTU Aqua, National Institute of Aquatic Resources, Section for Aquaculture
Technical University of Denmark, Nordsøen Forskerpartk, PO Box 101, 9850 Hirtshals, Denmark
erh@aqua.dtu.dk

Finland

Dr. Susanna Airaksinen
Fisheries Research Institute, Itänen Pitkäkatu 3, FIN-20540 Turku, Finland
susanna.airaksinen@rktl.fi

Dr. Esa-Matti Lilius
University of Turku, Vatselankatu 2, FI-20014 Turku, Finland
esalil@utu.fi

Mr. Markus Kankainen, MC Substitute Member
Fisheries Research Institute, Itänen Pitkäkatu 3, FIN-20540 Turku, Finland
markus.kankainen@rktl.fi

Former Yugoslav Republic of Macedonia

Prof.Misho Hristovski
University SS. Cyril and Methodius, Faculty of Veterinary Medicine, Lazar Pop-Trajkov 5-7, 1000 Skopje Former Yugoslav Republic of Macedonia
hristovskim@fvm.ukim.edu.mk

France

Dr. Jean-Charles Massabuau
UMR 5805 EPOC-OASU, Station Marine d'Arcachon, Université de Bordeaux Place du Docteur Peyneau, 33120 Arcachon Cedex, France
jc.massabuau@epoc.u-bordeaux1.fr

Dr. Marie-Laure Bégout
Ifremer-Station de La Rochelle, Place Gaby Coll, BP 7, 17137 L'Houmeau, France
mlbegout@ifremer.fr

Dr. Guy Claireaux
UMR 5554, Institut des Sciences de l'Evolution de Montpellier, Université de Montpellier 2, Station Méditerranéenne de l'Environnement Littoral, 2 Rue des chantiers, F-34200 Séte, France
guy.claireaux@univ-montp2.fr

Greece

Dr. Michael Pavlidis
University of Crete, PO Box 2208,Vasilika Vouton, 71409 Heraklion, Greece
pavlidis@biology.uoc.gr

Dr. Constantinos Mylonas
Hellenic Center for Marine Research, Institute of Aquaculture, Ex US Military Base of Gournes, PO Box 2214, 71003 Iraklion, Crete, Greece
mylonas@her.hcmr.gr

Hungary

Dr. Zsigmond Jeney
Research Institute for Fisheries, Aquaculture and Irrigation, Anna liget 8, H-5540 Szarvas, Hungary
jeneyz@haki.hu

Mr. Szergey Blohin
Double Delta Company Limited, 12 Hullam Street, 4100 Bererttyoujfalu, Hungary
sergey.blokhin@t-online.hu

Dr. Olga Bukovszakaja MC Substitute Member
Ejszak Alfoldi Innovation Company Limited Derek u. 130, 8/24. 4031 Debrecen Hungary
olga.bukovskaya@t-online.hu

Iceland

Dr. Thorleifur Agustsson
Icelandic Fisheries Laboratories, Skulagata 4, 101 Reykjavik, Iceland
tolli@rf.is

Prof. Helgi Thorarensen
Hólar University College, Hólar, Hjaltadalur, IS-551 Sauðárkrókur, Iceland
helgi@holar.is

Dr. Sigurdur Bogason
Icelandic Fisheries Laboratories, Skulagata 4, 101 Reykjavik, Iceland
sigboga@rf.is

Ireland

Dr. Neil Ruane
Marine Institute Marine Environment & Food Safety Services, Rinville, Oranmore, Co. Galway, Ireland
neil.Ruane@marine.ie

Mr. Dave Garforth
IFQC Limited, Rivercourt Business Centre, River Lane, Dundalk, Co. Louth Ireland
davegarforth@ifqc.ie

Israel

Dr. Iris Meiri Ashkenazi, MC Substitute Member
Israel Oceanographic and Limnological Research, National Center for Mariculture, Northern Beach, 88000 Eilat, Israel
imeiri@ocean.org.il

Dr. Amos Tandler
Israel Oceanographic and Limnological Research, National Center for Mariculture, Northern Beach, 88000 Eilat, Israel
Tandler@ocean.org.il

Dr. Ariel Diamant
Israel Oceanographic and Limnological Research, National Center for Mariculture, Northern Beach, 88000 Eilat, Israel
diamant@ocean.org.il

Italy

Prof. Marco Saroglia
University of Insubria Department of Biotechnology and Molecular Sciences, via J.H. Dunant 3, I-21100 Varese, Italy
marco.saroglia@uninsubria.it

Dr. Giuseppe Lembo
COISPA Tecnologia & Ricerca, Stazione Sperimentale per lo Studio delle Risorse del Mare, Via Dei Trulli 18-20, 70045 Bari, Italy
lembo@coispa.it

The Netherlands

Dr. Hans van de Vis
IMARES Wageningen UR, PO Box 68, 1970 AB IJmuiden, The Netherlands
hans.vandevis@wur.nl

Dr. Johan Schrama
Aquaculture and Fisheries, Wageningen University, PO Box 338, 6700 AH Wageningen, The Netherlands
johan.schrama@wur.nl

Prof. Johan Verreth
Aquaculture and Fisheries, Wageningen University, PO Box 338, 6700 AH Wageningen, The Netherlands
johan.verreth@wur.nl

Norway

Prof. Anders Kiessling
Department of Animal and Aquacultural Sciences, Norwegian University of Life Sciences,
PO Box. 5003 1432 Aas, Norway

Present address

Department of Wildlife, Fish and Environmental Studies, Swedish University of Agricultural Sciences, 901 83 Umeå, Sweden
anders.kiessling@slu.se

Dr. Hilde Toften
Norwegian Institute of Fisheries and Aquaculture Research, Muninbakken 9-13, PO box 6122,NO- 9291 Tromsø, Norway
hilde.toften@fiskeriforskning.no

Poland

Dr. Maciej Pilarczyk
Polish Academy of Sciences, Institute of Ichthyobiology and Aquaculture ul, Kalinowa 2, 43-520 Zaborze, Poland
maciej.pilarczyk@fish.edu.pl

Portugal

Dr. Luis Conceiçaõ
Centre of Marine Sciences of Algarve (CCMAR) Aquaculture Research Group, University of Algarve, Campus de Gambelas, 8005-139 Faro,
Portugal
lconcei@ualg.pt

Prof. Rui Oliveira
ISPA, Unidade de Investigação em Eco-Etologia, Integrative Behavioural Biology Group, Rua Jardim do Tabaco 34, 1149-041 Lisbon, Portugal
ruiol@ispa.pt

Romania

Dr.Violeta-Elena Simion
The Veterinary Medicine Faculty, Spiru Haret University, Masina de Paine Street, No 47, sect.2 774275 Bucharest, Romania
ushmv_simion.violeta@spiruharet.ro

Dr. Elena Mitranescu
University of Agronomical Sciences and Veterinary Medicine, Faculty of Veterinary Medicine, 105 Spl. Independentei Street, Bucharest, Romania.
mitranescuelena@gmail.com

Dr. Florin Furnaris, MC Substitute Member
University Of Agronomical Sciences and Veterinary Medicine, Faculty of Veterinary Medicine 105 Spl. Independentei Street, Bucharest, Romenia.
ffurnaris@gmail.com

Spain

Dr. F. Javier Sanchez
Department of Physiology and Pharmacology, Faculty of Biology, University of Murcia Campus Espinardo,30100 Murcia, Spain
javisan@um.es

Dr. Ana Roque
Institut de Recerca i Tecnologia Agroalimentàries (IRTA), Pathology group Aquaculture, Ctra. Poble Nou Km5,5. 43540 Sant Carles de la Ràpità, Spain
ana.roque@irta.es

Sweden

Prof.Eva Brännäs
Swedish University of Agriculture Sciences, Dept. of Wildlife, Fish and Environmental Studies,Inst för vilt, fisk och miljö, Skogsmarksgränd, 901 83 Umeå Sweden
eva.brannas@vabr.slu.se

Prof. Jörgen Johnsson
Göteborg University, Institutionen för biologi och miljövetenskap, Medicinaregatan 18, 413 90 Göteborg, Sweden
jorgen.johnsson@zool.gu.se

Switzerland

Prof. Helmut Segner
Center for Fish and Wildlife Health,Vetsuisse Faculty, University of Bern, Laenggass-Strasse 122, PO Box 8466, CH-3001 Bern, Switzerland
helmut.segner@itpa.unibe.ch

Dr. Lloyd Vaughan
Institute of Veterinary Pathology, Vetsuiss Faculty, University of Zurich, Winterthurerstrasse 268, CH-8057 Zurich, Switzerland
vaughanl@vetpath.uzh.ch

Turkey

Prof. Hijran Yavuzcan Yildiz
Ankara University, Faculty of Agriculture, Department of Fisheries and Aquacuture, 06110 Ankara, Turkey
yavuzcan@ankara.edu.tr

United Kingdom

Prof. James F. Turnbull
Institute of Aquaculture University of Stirling, Stirling FK9 4LA, Scotland, United Kingdom
jft1@stir.ac.uk

Dr.Sunil Kadri
University of Glasgow, School of Life Science, Room 220, Bower Building, University of Glasgow, Glasgow, G12 8QQ, Scotland, United Kingdom
s.kadri@bio.gla.ac.uk

Non-COST Participants

Australia

Dr. John Purser
National Centre for Marine Conservation and Resource Sustainability, Australian Maritime College, University of Tasmania, Launceston Campus, Maritime Way, Newham, Tasmania 7248, Australia
j.purser@amc.edu.au

New Zealand

Dr. Neill Herbert
Leigh Marine Laboratory, University of Auckland, Box 349, Warkworth, Northland 0941,Newzealand
n.herbert@auckland.ac.nz

Acknowledgments

The editors thankfully acknowledge the reviewers' useful comments, which significantly helped us improve the papers in this book. The authors modified and improved their manuscripts according to comments and questions of the following the reviewers.

A. Alanara, B. Algers, H. Andersen, D. Baron, M. Bracke, A.Canario, A.P. Farrell, A. Fernö, A. Figueras, H.J.T. Goos, A. Hellström, F. Huntingford, M. Jobling, S. Kadri, R. Katersky, A. Kiessling, T. Kitano, P. Klaren, T. Kristiansen, E. Lambooij, A. Lindberg, T. Lundh, L. Madsen, C. Martyniuk, T.P. Mommsen, C. Mylonas, K. Okuzawa, Ø. Øverli, J.V. Planas, T. Poppe, D.M. Power, P. Prunet, D. Raubenheimer, B. Roth, T. Sabo-Attwood, K. Sundell, S.J. Wendelaar Bonga and S. Winberg.

We would like to thank Mrs. Alexandrine Cheronet and Mrs. Judith Terpos of Springer, and Mrs. Gerda Sulmann of Fijnzinnig for their contribution to turn volume 38, Issue 1, 2012 of the journal *Fish Physiology and Biochemistry* into this book.

MIX
Papier aus verantwortungsvollen Quellen
Paper from responsible sources
FSC® C105338

If you have any concerns about our products,
you can contact us on
ProductSafety@springernature.com

In case Publisher is established outside the EU,
the EU authorized representative is:
Springer Nature Customer Service Center GmbH
Europaplatz 3, 69115 Heidelberg, Germany

Printed by Libri Plureos GmbH
in Hamburg, Germany